Animal Nutrition

McGraw-Hill Publications in the Agricultural Sciences

John R. Campbell, *University of Illinois, Consulting Editor in Animal Science*
Carl Hall, *College of Engineering, Washington State University, Consulting Editor in Agricultural Engineering*

Brown: *Farm Electrification*
Campbell and Lasley: *The Science of Animals That Serve Mankind*
Campbell and Marshall: *The Science of Providing Milk for Man*
Christopher: *Introductory Horticulture*
Crafts and Robbins: *Weed Control*
Cruess: *Commercial Fruit and Vegetable Products*
Dickson: *Diseases of Field Crops*
Eckles, Combs, and Macy: *Milk and Milk Products*
Edmonds, Senn, Andrews, and Halfacre: *Fundamentals of Horticulture*
Jones: *Farm Gas Engines and Tractors*
Kipps: *The Production of Field Crops*
Kohnke: *Soil Physics*
Kohnke and Bertrand: *Soil Conservation*
Krider and Carroll: *Swine Production*
Lassey: *Planning in Rural Environments*
Laurie and Ries: *Floriculture: Fundamentals and Practices*
Laurie, Kiplinger, and Nelson: *Commercial Flower Forcing*
Maynard, Loosli, Hintz, and Warner: *Animal Nutrition*
Metcalf, Flint, and Metcalf: *Destructive and Useful Insects*
Muzik: *Weed Biology and Control*
Smith and Wilkes: *Farm Machinery and Equipment*
Sorensen: *Animal Reproduction: Principles and Practices*
Thompson and Troeh: *Soils and Soil Fertility*
Thompson and Kelly: *Vegetable Crops*
Thorne: *Principles of Nematology*
Treshow: *Environment and Plant Response*
Walker: *Plant Pathology*
Warwick and Legates: *Breeding and Improvement of Farm Animals*

Animal Nutrition

Seventh Edition

Leonard A. Maynard, A.B., Ph.D., Sc.D. (Hon.)
Late Emeritus Professor of Nutrition and Biochemistry,
Cornell University and a
Member, National Academy of Sciences

John K. Loosli, B.S., M.S., Ph.D.
Visiting Professor, University of Florida
Emeritus Professor of Animal Nutrition
Cornell University

Harold F. Hintz, B.S., M.S., Ph.D.
Associate Professor of Animal Nutrition
Cornell University

Richard G. Warner, B.S., M.S., Ph.D.
Professor of Animal Nutrition, Cornell University

McGraw-Hill, Inc.
New York St. Louis San Francisco Auckland Bogotá
Caracas Lisbon London Madrid Mexico City Milan
Montreal New Delhi San Juan Singapore
Sydney Tokyo Toronto

Printed and bound by Book-mart Press, Inc.

13 14 15 16 17 18 BKM BKM 9 9 8 7 6

Library of Congress Cataloging in Publication Data

Main entry under title:

Animal nutrition

   (McGraw-Hill publications in the agricultural sciences)
   Sixth ed. published in 1969 by L. A. Maynard and J. K. Loosli.
   Includes indexes.
   1. Animal nutrition.   I. Maynard, Leonard Amby, date
SF95.M35   1979   636.08'52   78-18172
ISBN 0-07-041049-6

This book was set in Times Roman by Cobb/Dunlop Publisher Services Incorporated.
The editor was C. Robert Zappa; the production supervisor was Donna Piligra.
Printed and bound by Impresora Donneco Internacional S. A. de C. V.

A Tribute to
Leonard Amby Maynard
(1887–1972)

L. A. Maynard was born November 8, 1887 on a farm in Hartford, New York. In 1911 he graduated, cum laude, from Wesleyan University. After working two years as a chemist in the Agricultural Experiment Stations in Iowa and Rhode Island, he enrolled in the Graduate School at Cornell University, completing the Ph.D. degree in chemistry in 1915. He immediately accepted a position as Assistant Professor in the Department of Animal Husbandry, there he equipped the Laboratory of Animal Nutrition for chemical and biological studies. His research included studies on the requirements of minerals, amino acids and vitamins; the development of purified diets; utilization and metabolism of protein, minerals and lipids in feed supplies especially for lactating animals. Rats, guinea pigs, rabbits, sheep and goats were often used in pilot experiments as a guide to large animal studies. Much of the research was carried out as part of a training program for graduate students who were attracted to Cornell University because of his ability in teaching and inspiring students in nutrition research.

In 1939 Dr. Maynard was appointed the first Director of the U.S. Plant, Soil and Nutrition Laboratory at Ithaca on a half time basis so he could continue in research and teaching. In 1941 he helped organize and became the first Director of the School of Nutrition at Cornell and in 1944 was appointed head of the newly organized Department of Biochemistry.

v

During World War I he served with the AEF in France (1917–1919) as a Major in Chemical Warfare Service. In World War II he was a nutrition expert for the Inter-allied Food Missions with the Federal War Food Administration and with the New York State Emergency Food Commission. As Chairman of the National Research Council Committee on Animal Nutrition in 1942 he organized subcommittees to prepare Recommended Nutrient Allowances for various farm animals similar to the report for humans of the Food and Nutrition Board, of which he was a member and later chairman. This series first published for poultry in 1944 now includes reports for many other animals as shown in the appendix tables. These reports have been joint efforts of scientists in the U.S. and Canada. In addition to many scientific journal articles and bulletins, Dr. Maynard authored a book on *Better Dairy Farming* with S. E. Savage and *Animal Nutrition,* first published in 1937.

Dr. Maynard received many honors. He was elected to membership in the National Academy of Sciences in 1944; was awarded honorary Sc.D. degrees by Wesleyan University (1945), and Rhode Island State University in 1958; and the Republic of Guatemala elected him to the *Order of Rodolfo Robbes* in 1959; Honorary member of both Omicron Nu and the American Dietary Association. In 1942–1943 he served as president of both the American Institute of Nutrition and the American Society of Animal Production. He received the Borden Award in nutrition in 1945 and the Osborne and Mendel Award by American Institute of Nutrition in 1954 and also the Award of Distinction by the American Grocery Association in 1952.

Of all his accomplishments, perhaps his greatest influence has come through his training of students and his book on *Animal Nutrition.*

Contents

Preface

In the preface of the first edition published in 1937 Maynard stated the purpose of the book to be: "To present the principles of nutrition and their applications to feeding practice." This objective was also followed in later revisions called for as new knowledge accumulated and remains the purpose of the present one.

In the ten years since the sixth edition was published, nutritional science has been moving forward at an ever-increasing pace. Additional essential trace elements have been discovered and, of more importance perhaps, our knowledge of how all the various nutrients function specifically in life processes has been greatly enhanced. This enhancement reflects the epoch-making discoveries in cellular metabolism made by the biochemist, which have produced more precise information regarding nutrient requirements, not only to prevent physical signs of deficiency, but also to ensure the normal functioning of the various body processes concerned. In the present edition an attempt has been made to portray these developments and point out their significance in feeding practice.

The general order of presentation has been preserved, but the chapters on digestion, the role, metabolism and requirements of water and energy have been reorganized and presented more in line with the other chapters. All chapters have been updated. Material which has become less important has been deleted to make room for a consideration of the wealth of new information, without markedly increasing the size of the new edition. The authors have endeavored to take account of the tremendous number of significant research reports which have been published since the sixth edition was prepared, but they have found that in attempting to keep up with the literature in such a broad field they may have run a losing race. They can only hope they have performed creditably.

Many new literature citations have been included and many older ones deleted. These citations (1) show the historical development of the field and acquaint the student with the classic papers, (2) provide authority for statements made when new facts are involved or if the facts remain in dispute, (3) furnish sources from which the student may obtain more detailed information on a specific topic, and (4) serve the teacher by listing references from which he may select assigned readings. For this last purpose it is believed that many of those listed under Supplementary Literature at the end of each chapter should be particularly useful. Apart from those given throughout the book to show the historical development and to document new findings, the literature citations have been made on the basis of their usefulness in illustrating the points under discussion and of their availability to those students who are the principal users of the text. This latter consideration has meant giving preferences to papers in the English language and in journals having a wide circulation. The student should appreciate that such a selection has resulted in an unbalanced representation of research publications on a worldwide basis.

The metric system of weights and measures is used throughout this edition. When tabular data from the literature, expressed in the English system, have been reproduced, they have been converted to the metric system, with a few exceptions. A table of metric and English equivalents is presented as Appendix Table X.

It may be of interest to record here the authorship of the various editions of this work: The first (1937), second (1947), and third (1951) editions were authored by Leonard A. Maynard; the fourth (1956), fifth (1962), and sixth (1969) editions, by Leonard A. Maynard and John K. Loosli.

In preparing this latest revision the authors have been greatly aided by suggestions received from teachers using the book, including many former students. Their assistance is gratefully acknowledged. The authors are indebted to their Cornell colleagues who have aided in the revision of some of the topics, particularly William Hansel, H. F. Schryver, L. P. Krook, D. B. McCormick and A. Bensadoun, and to H. D. Wallace and R. L. Shirley, University of Florida and D. R. Mertens, University of Georgia. Appreciation is also expressed to Mrs. Reha J. Loosli for assistance with revising and typing several chapters and preparing the glossary and indexes, to Caroline Bloomquist, Rebecca Freund, Yoshi Nakayama, Barbara Warner, and Sandy Hintz for secretarial assistance.

John K. Loosli
Harold F. Hintz
Richard G. Warner

Chapter 1

The Expanding
Field of Nutrition

A hungry people listens not to reason nor are its demands turned aside by prayers.

Seneca (4 B.C.–65 A.D.)

Nutrition involves various chemical reactions and physiological processes which transform foods into body tissues and activities. It involves the ingestion, digestion, and absorption of the various nutrients, their transport to all body cells, and the removal of unusable elements and waste products of metabolism.

The great French chemist Lavoisier[1] is frequently referred to as the founder of the science of nutrition. He established the chemical basis of nutrition in his famous respiration experiments carried out before the French Revolution. His studies led him to state, "La vie est une fonction chimique," (life is a chemical process). Thereafter chemistry became an important tool in nutrition studies. Through its application in physiological studies the old idea that the nutritive value of food resided in a single "aliment" was proved wrong in the first quarter of the nineteenth century. The need for protein, fat, and carbohydrates became recognized. For the remainder of the century, nutritional science and practice were concerned primarily with these nutrients and a few mineral elements. Calcium, chloride, fluoride, iron, magnesium, potassium, sodium, and sulfur were known and considered to be important in the body although critical research proving their essentiality was meager or lacking.[2] The large expansion in the nutrition field has occurred during the last sixty years with the discovery of the vitamins, of the role of amino acids, and of several more essential mineral elements. Today we know that the body needs over forty different nutrients, in contrast to the three recognized a century ago. The objective of nutrition is to provide all essential nutrients in adequate amounts and in optimum proportions.

While much of our current knowledge has resulted from direct attacks on evident nutritional and health problems of animals and humans, equally important discoveries have come from more basic studies of the functioning of the animal organism, of the physiological and biochemical changes involved, and of the effects of various dietary variables. In many instances, under inspired leadership, research has shown the way to better practice before the need for it was recognized. An outstanding example of such a contribution is the feeding experiment with single plants planned by Babcock[3] and carried out at the Wisconsin Experiment Station over sixty years ago.

During growth and reproduction five-month-old heifer calves were fed rations formulated from a single plant source. Grain and forage from either corn or wheat plant were combined to be equal in the organic nutrients known at the time and salt was provided. Weight gain was similar the first year but the corn-fed heifers were sleeker and more vigorous. Large differences became evident during reproduction. Each of the corn-fed heifers produced a normal calf, but from the wheat group all calves either were dead at birth or died soon after and the cows produced only one third as much milk as those fed corn. Exhaustive studies of the feeds and animal tissues failed to explain the results and it was concluded that the wheat plant contained something toxic or that it lacked something supplied by corn.[4] This was before the vitamins had been identified and very little was known about mineral requirements.

This experiment made it clear that there were marked differences in nutritive values which could not be detected by any chemical means available at the time and that the current scientific bases for formulating rations were seriously inadequate. More important, the experiment led to the conviction that simplified diets must be used for the solution of nutrition problems. It stimulated the use of the purified-diet method (Sec. 13.3), which resulted in the discovery of the first vitamin (Sec. 11.1) in 1913 and which has been so largely responsible for the newer knowledge of nutrition.

The modern discoveries in nutrition have resulted from studies with a wide variety of species. The contributions of the laboratory rat to our knowledge of vitamins, amino acids, and minerals have been enormous. The discovery of insulin and of the role of nicotinic acid in the prevention and cure of pellagra exemplifies the debt that we owe to the dog. Guinea pig experiments showed us the specific cause of scurvy and how to prevent it. The pioneer work that led to the discovery of thiamin was carried out with the chick, and this species has continued to help solve many puzzles in the field of vitamins. Monkeys, mice, and hamsters all have contributions to their credit. Even the lower forms, particularly bacteria, have played a large role in the discovery of growth factors, in the assay of our foods for various nutrients, and in explaining how these nutrients function in metabolism. Today the nutritional scientist realizes that basic, or pilot, experiments with one of these various species, selected in accordance with the objective of the study, provide the best approach for the solution of many of the problems in the nutrition of man and farm animals, although the final answers must be obtained with the animal species concerned.

The expanding developments in the field of nutrition have resulted from the application of the knowledge and techniques of many different sciences. Physiologists and biochemists have long worked as a team in studying the body's need for food and how this food is metabolized. But the advancement in many areas was slow until the aid of scientists in other fields was obtained. The identification of the various vitamins lagged until organic chemists became interested in their isolation and synthesis. Thanks to their efforts, commercial sources of many of them have become available both for further experimental work and also for use in feeding practice. Physicists have given us radiographs, the spectrograph, isotopes, chromatography, and other tools and have shown us how they can be used for the advancement of nutrition. X-ray crystallography has joined with computers to enable molecular biochemists to unravel the structure of certain proteins and thus help explain their functions. Geneticists have discovered breed differences in nutritive requirements and in the efficiency of food utilization. They have even developed new strains of certain lower forms that will detect specific vitamin and amino acid deficiencies in our foods. Microbiologists have assisted greatly in the discoveries of the nutritional roles that bacteria play in the rumen of the cow and sheep and in the intestine of other species. Microbiological and chemical methods have greatly speeded up the development of our knowledge regarding the vitamin and amino acid content of foods. Food technology has made large contributions in developing special feed ingredients and additives and in processing animal products to improve their usefulness in human nutrition.

The significance of these contributions from the various fields of science will become apparent in the discussions in the succeeding chapters. Modern students of nutrition must have an appreciation of the important relations of these basic sciences in the development of present knowledge and to the solution of the remaining problems. They must have a real working knowledge of chemistry and physiology, particularly, if they are to understand current developments and be able to apply them in practice. Such knowledge will also help them to evaluate the significance of new facts as they are discovered. Likewise, an appreciation of the history of earlier discoveries can stimulate future progress.

The essentiality of several nutrients has been a comparatively recent discovery because they are needed in trace amounts only. Here we think first of the vitamins, but 0.1 mg of cobalt or selenium per day makes the difference between life and death in a sheep. A lack of that minute amount was responsible for tremendous livestock losses in certain parts of the world before their specific cause was discovered. We express protein requirements in kilos or grams, but a few milligrams or micrograms of certain other nutrients are just as important for health and production.

Studies of some of the "trace" mineral elements have shown that the character of the soil on which we grow our food crops plays an important role in determining their nutritive value.[5] We have also learned that varieties of the same crop differ in nutritional quality and that various cultural factors have an influence here also. Thus, animal and human nutrition ties back into agriculture and to the soil, stressing the importance of yields of nutrients as distinguished from yields of a crop per unit of land.

We have learned that the soil may contribute toxic elements to our food supply as well as essential ones. Some of the vitamins, as well as the minerals that are essential in small amounts, may prove harmful at higher intakes. Poor nutrition can result from too much as well as too little. Further, a suitable balance between certain nutrients is important.

The recent developments have served to stress the interrelationships between human and animal nutrition. The foods of both humans and animals are products of the soil from captured solar energy and contain the same essential nutrients. The metabolic processes which absorbed nutrients undergo for the support of various body functions are largely identical, whatever the species. While animals concentrate the nutrients of food crops into more nutritious and palatable forms for the human diet, they waste basic food resources in the process if they are fed cereals and other foods which humans can eat.

1.1 Animals as Producers of Human Food

Table 1.1 shows the percentage of gross energy and protein in the feed of domestic animals that is recovered in the edible products animals produce. The efficiency of animals as food producers varies widely depending on their genetic ability, the adequacy of nutrition and management, and other factors. Animals contribute greatly to human food supplies by transforming products of little or no value into nutritious human food. They produce meat and milk on land that is too dry or poor to produce food crops. The world's land area of about 13.4 billion hectares is only 10 percent arable, 22 percent permanent pastures and meadows, 30 percent forests, and the remainder nonproductive. Domestic ruminants utilize much of the pastures and meadows and some of the forests to

TABLE 1.1. Efficiency of Converting Animal Feeds into Human Foods

Animal product	Percentage recovered	
	Energy	Protein
Cattle (milk)	15–20	15–36
Poultry (eggs)	10–18	10–30
Pigs (pork)	14–20	14–20
Poultry (broilers)	6–11	17–23
Cattle (beef)	3–8	4–15

produce meat and milk. In many countries animals provide traction power to produce and transport crops.

The most successful human diet in terms of the optimum nutrition of people is one that contains animal products. There is a real justification for an animal industry, though its extent may be governed by economic considerations. Thus producers need to have some understanding of the factors which govern the usefulness of animal products in human nutrition. Milk, meat, and eggs are foods of superior nutritive value.

There is much concern internationally that even in the next decade population growth will outrun the world's capacity to produce food, because of limited land, water, and energy. Some people, however, believe that the rising demand can be satisfied by improved technology, through a coordinated increase in both crop and animal production and fuller utilization of marginal lands and waste products. Wittwer[6] reviewed the potentials for increasing food production and pointed out that animals can either add to the total food supply or be directly competitive. Dogs, cats, horses, birds, and other pets compete for human food supply. In the United States the food these pets eat each year would feed some 40 million people. In contrast, most of the milk and meat produced in the world by cattle, buffaloes, sheep, and goats results from grazing nonarable land and from the use of waste by-products and crop residues. Thus they add to human food supply. The animals themselves represent about a year's supply of reserve food which could be called upon if needed. There is a need for further research to improve the use of present waste materials and to conserve energy in producing food.

Over the past seventy-five years as the result of research, the yields of food crops have doubled or tripled and the productivity of animals has increased likewise. Before the vitamins and essential minerals were discovered and made available, chickens confined in buildings without access to green feed ceased laying eggs during the long cold winters of the Northern United States and Northern Europe. High losses and poor performance of pigs was common during winter months, and the growth rate and milk yields of cattle was greatly reduced. With modern adequate housing, good management, and balanced diets, highest levels of production are now achieved at all seasons in confinement feeding systems.

Fifty years ago better dairy cows averaged 1,200 to 1,500 kg of milk per year. Now many herds produce up to 9,000 kg, and two cows have produced over 23,000 kg of milk in a year, indicating the future potential. Formerly, steers were 5 to 6 years of age before they reached 500 kg in body weight and were fat enough for slaughter, and they needed some 15 kg of feed for each kilogram of gain. Now on high-grain diets they reach the same weight at 12 to 15 months of age and use only 5 to 6 kg of feed per kilogram of gain. Even since 1940 the efficiency of animals has been markedly increased through selective breeding and improved nutrition. For example, now a 1.5-kg broiler can be produced in 8 weeks with 3.0 kg of feed compared with 12 weeks and 5.0 kg of feed previously.

Despite the large advances that typify the expanding field of nutrition, there are many "unknowns" for further study. It is doubtful whether all the essentials of an adequate diet for any species have been discovered. Certainly, much more needs to be learned regarding the quantitative requirements of our farm animals for some of the more recently discovered nutrients. This fact is evident in the recently published reports, referred to later, setting forth nutrient requirements for farm animals. Much more information is also needed about the quantitative occurrence of the recently discovered nutrients in our foodstuffs and of the availability of these nutrients.

There are about 1,140 million cattle, 940 million sheep, 500 million pigs, 340 million goats, and 70 million horses in the world.[7] The developing countries, where most of the malnutrition occurs, have 70 percent of all cattle and buffaloes, 63 percent of all sheep and goats, and 60 percent of all pigs, yet they produce only 34 percent of the beef, 21 percent of the milk, 50 percent of the sheep and goat meat, and 37 percent of the pork. It is clear that one way to increase protein supplies would be to improve the feeding and management of the domestic animals in these countries.

The experience in China[8] might serve as an example of a method to achieve this. Large numbers of animals, primarily sheep and goats, are produced in the areas of China unsuited to intensive agriculture. In the more densely populated areas, animal production is limited to pigs, poultry, a limited number of dairy cattle, and the horses, donkeys, and buffaloes needed for power and transportation. Some 250 to 260 million swine are produced in regions of intensive agriculture, four times more than in the United States. The animals are fed waste materials not suitable for human food. They grow slowly, take longer to reach slaughter weight, but they contribute important amounts of protein rather than competing for human food supplies. Such a system in which the animals are controlled and fed residues with limited supplements would not use edible foods and would improve animal performance in comparison with the very low productivity in many underdeveloped areas.

Nutrition workers who are true scientists recognize the limitations of present knowledge. They realize that findings which have later proved to be inadequate have been responsible for practical recommendations which did not prove effective. They are becoming more conservative accordingly. But overenthusiasm or error has characterized many of the popular articles in the field of nutrition, and some of those who have food products for sale seem to have no inhibitions at all. Claims of superiority for so-called *natural* foods or those grown with *organic fertilizers* and without use of insecticides are largely without scientific evidence. Yet consumers are duped into paying higher prices for these foods. The excessive levels of vitamins, such as massive doses of vitamins C and D and daily supplements added to already adequate diets will not improve performance or health of animals or human beings, and in some cases harmful effects have occurred. It is not surprising that people without special knowledge of nutrition are puzzled by what they read and that reasoning or actual experience may convince them that the field is overexpanded. There has been an overexploitation of present knowledge, but the need for research to fill in the many gaps in this knowledge cannot be denied, and the accomplishments of the recent decades make it evident that further intensive studies will prove highly beneficial to human and animal welfare. There should be a curtailment of premature conclusions and recommendations for practice and an expansion of critical research by competent workers who have the patience to carry through the long-time experiments demanded for the complete solution of current problems.

From the discussions that follow the student will become conscious of the many gaps, and even of some contradictions, in present knowledge. Such a situation exists in every scientific field as one approaches its borders, but these borders are being constantly extended by research, and the new ground gained is being consolidated. Continued activities in these directions are particularly needed in the field of nutrition. In the words of the late Professor Hart,[9] of the University of Wisconsin, we should "chart all the factors in nutrition, organic and inorganic, and study their distribution, physiology, pathology, and interplay. Put the need for these factors on a quantitative basis with optimum allowance for the complete cycle of the animal's life." Thus will the field of nutrition continue to expand for the benefit of the livestock industry and human welfare.

The greatest usefulness of a book on nutrition is, as Armsby[10] stated in 1880, not simply to give recipes for better feeding of human beings or animals but in so "elucidating our knowledge of the unchanging natural laws, chemical and physiological, of nutrition that the attentive student shall be able to adapt to the varying conditions in which he may be placed to appropriate intelligently the results of new investigations and follow or take part in advances of the science."

Animals are such agreeable friends—they ask no questions, they pass no criticism.

George Eliot. (1819–1880)

NOTES

1. Antoine Lavoisier (1743–1794) introduced the balance and thermometer into nutrition studies. He discovered that combustion was an oxidation, and he showed that respiration in the body involved the combination of carbon and hydrogen with oxygen from inspired air and that the quantities of oxygen absorbed and carbon dioxide given off depended on the food intake and the work done. With Laplace, he designed a calorimeter by means of which it was demonstrated that respiration is the essential source of body heat. The science of nutrition was undoubtedly set back many years when Lavoisier's career was ended by the guillotine.

2. C. M. McCay, Notes on the History of Nutrition Research, Hans Huber, Bern, 1973. Clive M. McCay (1898–1967), besides research activities in many other fields, devoted over thirty years to studies of the influence of diet and growth rate upon life span and the physiology and pathology of aging, which won him national and international renown. He was born in Indiana and received his bachelor's degree at the University of Illinois and his doctor's degree at the University of

California. After two years of study at Yale University as a National Research Council Fellow, he received an appointment as assistant professor and later as professor at Cornell, where he served until ill health forced his retirement in 1962.

3. Stephen M. Babcock (1843–1931), who is most widely known for his invention, the Babcock test, made many pioneer contributions in the fields of dairy chemistry and animal nutrition. Following six years at the New York Experiment Station, he served for twenty-five years as chemist and assistant director at the Wisconsin Experiment Station.

4. E. B. Hart et al., Physiological effect on growth and reproduction of rations balanced from restricted sources, *Wisconsin Agr. Expt. Sta. Research Bull.* 17, 1911.

5. W. H. Allaway, The effect of soils and fertilizers on human and animal nutrition, *U.S. Dept. Agr. Information Bull.* 378, 1975.

6. S. H. Wittwer, Food production technology and the resource base, *Science,* **188**:579–584, 1975.

7. H. A. Jasiorowski, The developing world as a source of beef for world markets, in A. J. Smith (ed.), Beef Cattle Production in Developing Countries, Univ. Edinburgh Press, Edinburgh, 1976, pp. 2–28.

8. G. F. Sprague, Agriculture in China, *Science,* **188**:549–555, 1975.

9. Edwin B. Hart (1874–1953) was born in Ohio, received his bachelor's degree at the University of Michigan, and then spent seven years as chemist at the New York State Experiment Station, with the exception of two years of study in Germany. In 1906 he became head of the department of agricultural chemistry (later renamed biochemistry) at the university, where he had a long and distinguished career as teacher and investigator. Among his many contributions to the field of nutrition was the discovery of the essentiality of copper.

10. Henry Prentiss Armsby (1853–1921), following periods of service at the New Jersey, Connecticut, and Wisconsin Experiment Stations, became director of the newly established Pennsylvania Experiment Station at State College in 1887. In 1907 the Institute of Animal Nutrition was established at this institution with Armsby as director, and here he served until his death, winning lasting fame for himself and his institute. During his postgraduate study at Leipzig, he became interested in the respiration experiments being carried out by Kühn and others at Möckern, and this resulted in his construction of a respiration calorimeter for farm animals and in the inauguration of his studies of heat production in cattle. These epoch-making studies, which led to the development of the net-energy system of evaluating feeds, are frequently referred to in this book.

SUPPLEMENTARY LITERATURE

Blaxter, K. L.: Conventional and unconventional farmed animals, *Proc. Nutrition Soc.,* **34**:51–56, 1975.

Hodgson, R. E.: Place of animals in world agriculture, *J. Dairy Sci.,* **54**:442–447, 1971.

Janick, J., C. H. Noller, and C. L. Rhykerd: The cycles of plant and animal nutrition, *Scientific American,* **235**:74–87, 1976.

Chapter 2

The Animal Body and Its Food

The human is an omnivorous animal, eating food from both vegetable and animal sources. The pig and rat are other examples of omnivores. Herbivorous animals, such as cattle and other ruminants, rabbits, guinea pigs, and horses, and omnivores like the pig, poultry, etc., that live largely on plants are the species that contribute most to human food supply. Carnivorous animals, such as dogs, cats, lions, and birds of prey are generally not acceptable as human food, but there are some exceptions. A brief consideration of the chemical composition of the animal body in relation to the composition of its food is useful to give a general picture of the nutrition process, the detailed aspects of which are presented in later chapters.

COMPOSITION OF THE ANIMAL BODY

Over a century ago the famous English scientists Lawes and Gilbert[1] performed the pioneer and laborious task of analyzing the entire bodies of farm animals. Since that time, many similar studies have been made by other workers, with the result that we have a large body of data regarding the composition of various species at different ages and in varying states of nutrition. From these data the figures given in Table 2.1 have been assembled to provide a general

9

TABLE 2.1. Percentage Composition of the Animal Body[a]

Species	Water	Protein	Fat	Ash	Fat-free			Dry, fat-free	
					Water	Protein	Ash	Protein	Ash
Calf, newborn	74	19	3	4.1	76.2	19.6	4.2	82.2	17.8
Calf, fat	68	18	10	4.0	75.6	20.0	4.4	81.6	18.4
Steer, thin	64	19	12	5.1	72.6	21.6	5.8	79.1	20.9
Steer, fat	43	13	41	3.3	72.5	21.9	5.6	79.5	20.5
Sheep, thin	74	16	5	4.4	78.4	17.0	4.6	78.2	21.8
Sheep, fat	40	11	46	2.8	74.3	20.5	5.2	79.3	20.7
Pig, 8 kg	73	17	6	3.4	78.2	18.2	3.6	83.3	16.7
Pig, 30 kg	60	13	24	2.5	79.5	17.2	3.3	84.3	15.7
Pig, 100 kg	49	12	36	2.6	77.0	18.9	4.1	82.4	17.6
Hen	57	21	19	3.2	70.2	25.9	3.9	86.8	13.2
Rabbit	69	18	8	4.8	75.2	19.6	5.2	79.1	20.9
Horse	61	17	17	4.5	73.9	20.6	5.5	79.2	20.8
Human	60	18	18	4.3	72.9	21.9	5.2	80.7	19.3
Mouse	66	17	13	4.5	75.4	19.4	5.2	79.1	20.9
Rat	65	22	9	3.6	71.7	24.3	4.0	86.0	14.0
Guinea Pig	64	19	12	5.0	72.7	21.6	5.7	79.3	20.7

[a] Less contents of digestive tract.

picture of the gross composition of mature animals in a good state of nutrition and to show that different species are very similar, expressed on a dry, fat-free basis.

2.1 Water and Organic Substances

The data in Table 2.1, given in round numbers, show large variations according to age and nutritional state. On a percentage basis, the water content shows a large decrease with age in early life. In the case of cattle, for example, the water content is approximately 95 percent for the embryo shortly after conception, 75 to 80 percent at birth, 66 to 72 percent at five months, and 40 to 65 percent in the mature animal. The variations for a given age are due primarily to the nutritional state as reflected in the store of fat. Very fat animals may have as little as 40 percent of water. The percentage of fat normally increases with age, but it is highly variable, depending upon the level of food intake. Its variation affects the percentages of the other constituents, and this is particularly true for water. Missouri workers, for example, found a thin steer to contain 18 percent of fat and 57 percent of water, in contrast to 41 percent of fat and 42 percent of water for a very fat animal.

The chemical composition of the physically separated muscle and fat (adipose) tissue is variable. For example, separated visible fat may contain from 30 to 90 percent ether-extractable lipid, and separated muscle may contain as little as 2.0 percent fat for chicken breast or up to 15 to 20 percent in well-marbled beef loin.

In view of the wide differences in fat content which may occur, much less variable figures for the other constituents are obtained by expressing them on a fat-free (protoplasmic) basis. On this basis Reid and associates[2] found that the gross composition of the animal body, less the contents of the digestive tract, averaged 72.9 percent water, 21.6 percent protein, and 5.3 percent ash. These figures are subject to little variation after the animal is nearly full grown, though there is a slight decrease in water throughout life. In contrast to the extreme variability noted for fat (range 3 to 46 percent), the protein and ash were practically constant when expressed on a water-free and fat-free basis, averaging 80 and 20 percent, respectively. The only species difference shown in the table which may be considered significant are the higher protein values on the dry, fat-free basis for the pig, rat, and hen, reflecting smaller relative size of skeleton.

The data of Table 2.1 do not reveal the very small amount of carbohydrate which is present in the body. Though occurring as much less than 1 percent at any given moment, it is constantly being formed and broken down in metabolism and thus performs a multitude of vital functions.

The chemical groups which make up the gross composition of the body are not evenly distributed throughout the various organs and tissues but are more or less localized according to their functions. Water is an essential constituent of every part of the body, but its quantitative distribution varies greatly in different parts. Blood plasma contains 90 to 92 percent, muscle 72 to 78 percent, bone approximately 45 percent, and the enamel of the teeth 5 percent.

Proteins are present in every cell and, as such, are the principal constituent, other than water, of the organs and soft structures of the body such as the muscles, tendons, and connective tissues. Most of the fat is localized in the adipose tissue, or fat depots, which occur under the skin, around the intestines, around the kidneys and other organs; but it is also present in the muscles, bones, and elsewhere. In fact every cell contains substances classed with the fats. The small amount of carbohydrate, e.g., glucose and glycogen, present in the body is found principally in the liver, muscles, and blood.

Since most body composition values are expressed on an empty body weight basis, it is interesting to note that the contents of the digestive tracts vary considerably among animal species. After animals were fasted for 20 to 24 hours, but allowed access to water, the values, expressed as percentage of body weight were: dogs, 2.3 ± 1.2; rats, 3.2 ± 1.1; pigs, 3.2 ± 0.7; horses, 9.6 ± 3.4; rabbits, 11.3 ± 1.2; guinea pigs, 12.5 ± 3.4; sheep, 15.9 ± 5.0; and steers, 15.9 ± 3.4.

2.2 Mineral Composition

The mineral matter of the body comprises a large number of elements present in varying amounts in different parts, according to the functions they perform. The percentages of the principal mineral constituents of the body are indicated by the following data:

Element	Percent	Element	Percent
Calcium	1.33	Chlorine	0.11
Phosphorus	0.74	Magnesium	0.041
Sodium	0.16	Sulfur	0.15
Potassium	0.19		

Source: A. G. Hogan and J. L. Nierman, Studies in animal nutrition. VI. The distribution of the mineral elements in the animal body as influenced by age and condition, *Missouri Agr. Expt. Research Bull.* 107, 1927.

These data are averages of analyses of 18 steers of varying ages. They are expressed as a percentage of the entire body less the contents of the digestive tract. It is noted that, aside from calcium, the elements occur as fractions of a percent only. Despite their small amounts, they are absolutely essential to life. These average data for the steer are subject to variation according to age and state of fattening. The data for other species show a similar pattern, though differing quantitatively, as is to be expected.

Calcium, the mineral occurring in largest amounts in the body, is present almost entirely in the bones and teeth as phosphate and hydroxide. The phosphorus which is combined with calcium to form the skeleton accounts for approximately 80 percent of the body supply. The remainder is widely distributed in combination with certain proteins and fats and as inorganic salts. Sulfur occurs throughout the body as a part of the protein molecule. Sodium, potassium, and chlorine are present almost entirely as inorganic salts in the various fluids. Most of the magnesium is present in the bones, but it is also found widely distributed elsewhere in the body.

In addition to the elements listed in the table, there are many others which are present in smaller amounts, some of which are known to be necessary for life. Iron is an essential constituent of the hemoglobin of the blood and occurs in lesser amounts throughout the organs and in the various tissues. Iodine, copper, zinc, manganese, cobalt, selenium, fluorine, chromium, and molybdenum are essential for either structural or metabolic purposes. Recent studies with laboratory animals over several generations suggest that tin, vanadium, silicon, nickel, and bromine may have essential physiological roles. The practical importance of these has not been demonstrated and it appears doubtful that a deficiency might limit animal production. Boron, aluminum, and arsenic are among the additional elements which have been reported as normally occurring in the body, though they have no known function.

2.3 The Blood

From the standpoint of nutrition, the composition of the blood is of special importance in that it is the medium by which oxygen and the nutrients are carried to the various parts of the body and by which the waste products of metabolism are removed. The blood comprises from 5 to 10 percent of the body weight, depending upon the species and nutritive state. Values for farm animals at different ages, determined by the use of isotopic phosphorus, have been reported by Hansard and coworkers.[3] The figure for birds is higher than for mammals. The blood volume is related primarily to the active tissues of the body. Thus the larger the amount of adipose tissue, the lower is the percentage of blood for the body as a whole.

The corpuscles make up from 30 to 45 percent of the blood, depending upon the species. This percentage value is the *hematocrit value*. The solid matter of the red corpuscles consists almost entirely of the iron-containing protein hemoglobin. In certain lower forms, however, the protein of the corpuscles contains an element other than iron as its respiratory pigment. The *Pinna squamosa*, a shellfish, has a protein called pinnaglobulin which contains manganese. Lobsters, crabs, and snails have the copper-containing hemocyanin.

The plasma contains 10 percent of solids, more than half of which are proteins. The remainder consists principally of various fatty substances, sugar, nonprotein nitrogen compounds, and inorganic salts. The principal inorganic elements are sodium and chlorine, with potassium, calcium, magnesium, phosphorus, and others occurring in much smaller amounts. Most of the sodium and chlorine are combined together, but various other combinations of these and other elements occur, such as sodium bicarbonate, disodium phosphate, and potassium chloride.

2.4 Muscle and Other Tissues

All movements of the body and of the organs and tissues which take part in life processes depend upon muscle action. Thus muscle tissue is distributed throughout the body. Skeletal muscle, which comprises about one-half the total body, contains about 75 percent of water. Protein makes up 75 to 80 percent of the dry matter. The remainder consists principally of fat, with small amounts of carbohydrate (glycogen) and mineral matter. The *epithelial tissues,* which comprise the skin, hair, feathers, the linings of the alimentary tract, respiratory tract, and genitourinary tract and occur elsewhere in the body, consist primarily of keratin. This is a highly insoluble protein which provides the protective and resistant qualities needed. *Connective tissue* is found in cartilage, tendons, ligaments, and the matrix of bone and provides an intercellular binding substance throughout the body. It consists of insoluble protein fibers, usually collagen, imbedded in a matrix or ground substance. The brain and nerves consist of *nervous tissue,* which is comprised mainly of various lipids and of complexes of lipid, protein, and carbohydrate. In later chapters, mention is made of certain specific compounds in these various tissues.

2.5 Estimation of Gross Body Composition

Data on the gross composition of the body provide specific information on its stage of development and state of nutrition (nutriture) not obtainable by merely weighing the animal or its products. Thus slaughter and chemical analysis are techniques frequently employed in feeding experiments (Sec. 13.8). The obvious limitation of these procedures, apart from being laborious, is that data on a given animal can be obtained only once, although what is most desired is information on its changing body composition as a result of the ration fed. Much recent study has therefore been given to the estimation of the gross body composition of the living animal, and very useful procedures have been developed accordingly.

The methods are based on the recognition that a highly predictable inverse relationship exists between the concentrations of water and of fat in the body and that in the fat-free, water-free body the proportions of protein and ash remain constant. Methods and equipment have been developed to determine the body water, fat, or protein (lean body mass) of animals. Body fat can be determined by densitometric methods.[4] Density is expressed as weight per unit volume or specific gravity. The volume of the animal is measured by displacement when it is submerged in water. In comparison with water at 20°C having a value of 1.000, fat cattle would be 0.900 or slightly higher and lean beef would be 1.300. The method is more applicable to animal carcasses than to live animals, but it has been used in studies with people. Equipment has been developed to use ultrasonics to measure the fat thickness in various parts of the animal and these measurements can be correlated with total body fat.[5] Lean muscle mass can be estimated from whole body counting of ^{40}K to calculate protein.[6] All of these methods need to be calibrated by comparisons with more direct measures. Body water can be determined by any of several methods based on dilution techniques. These involve the injection of compounds known to go into solution in body water and the quantitative determination of the dilution of a marker used, after equilibrium has been reached. The compounds most commonly employed are antipyrine and its analogs and isotopes deuterium and tritium. After the water content has been measured, the contents of fat, protein, and ash can be calculated. The method is illustrated for cattle by the report of Reid and coworkers.[7] The following equation is given for calculating fat content from the measured water:

$$Y = 355.88 + 0.355X - 202.91 \log X$$

where Y = fat content (percent) and X = water content (percent). The protein and ash contents are calculated from the finding that the fat-free dry body contains 80.3 percent of protein and 19.7 percent of ash.

Similar studies for sheep and swine have been reported.

COMPOSITION OF PLANTS AND THEIR PRODUCTS

Food must supply nutrients which can be used to build and renew the components of the body and to form its products such as milk, eggs, and wool and it must furnish energy for the processes involved. After weaning, most farm animals obtain all of their food supply from plants. While there are certain animal species which are entirely carnivorous, the plant kingdom is the original and essential source of all animal life, because plants are able to utilize the energy of the sun to build substances which will nourish the animal. Plants make use of carbon dioxide, water, nitrates, and other mineral salts to form carbohydrates, fats, and proteins which the animals must have to build their bodies and which are broken down in life processes. Thus plants store and animals dissipate energy.

2.6 Plants and Their Parts

Plants contain the same substances that are found in the animal body, but the relative amounts present are very different. Plants also show much larger differences in composition among species than do animals. The composition of certain typical plants and plant products is given in Table 2.2. These data are presented for the purpose of comparing them with data previously given for the animal body (Table 2.1) and of illustrating certain useful generalizations regarding differences in composition among plants and their various products.

The analyses of three green plants are given to show the general composition of the living plants at the stage when vegetative growth is practically completed but before the seed has formed. These data reveal the fact that the principal constituent of living plants is water, as is true of the animal body. This water content decreases as the seed is matured. The striking difference in the composition of plants and animals is the fact that the dry matter of plants consists principally of carbohydrate. This constituent serves as both structural and reserve material, while in animals protein comprises the structure of the soft tissues and fat is the reserve. Thus, although the animal body contains only a trace of carbohydrate, this nutrient is the principal constituent of the food of most species. It serves as a source of energy, either currently or as a reserve in the form of fat, into which it is readily transformed.

TABLE 2.2. Percentage Composition of Selected Feeds

	Water	Protein	Fat	Carbohydrates	Ash	Calcium	Phosphorus
Green plants							
Corn (*Zea mays*)	66.4	2.6	0.9	28.7	1.4	0.09	0.08
Alfalfa	74.1	5.7	1.1	16.8	2.4	0.44	0.07
Timothy	72.4	3.5	1.2	20.7	2.2	0.16	0.10
Dried plant products							
Alfalfa leaves	10.6	22.5	2.4	55.6	8.9	2.22	0.24
Alfalfa stems	10.9	9.7	1.1	74.6	3.7	0.82	0.17
Corn grain	14.6	8.9	3.9	71.3	1.3	0.02	0.27
Corn stover	15.6	5.7	1.1	71.4	6.2	0.50	0.08
Soybean seeds	9.1	37.9	17.4	30.7	4.9	0.24	0.58
Timothy hay	11.4	6.3	2.3	75.6	4.5	0.36	0.15

Source: Atlas of Nutritional Data on United States and Canadian Feeds, National Academy of Sciences, Washington, D.C., 1971.

The data for dried plant products, representing the moisture basis to which they are reduced after curing for storage, are given to bring out certain generalizations regarding differences in composition among the various parts of the plant. The figures for corn stover and corn grain provide a comparison between the vegetative portion of the plant and its seed. The data for the soybean illustrate some characteristics of legume seeds, and those for the alfalfa products bring out certain differences between leaves and stems.

Protein is primarily a constituent of active tissues, and thus leaves are much richer in this nutrient than are stems, as the data for alfalfa show. Leafy, legume hays such as alfalfa and the clovers always contain more protein than the grass hays such as timothy. As the plant matures, there is a movement of protein from the vegetative parts to the seed to provide for the requirements of growth during germination. Thus, at maturity, the seed contains a higher percentage of protein than the rest of the plant, as is indicated by the figures for corn grain and corn stover.

Fat is also higher in the leaves than in the stems and generally is highest in the seeds, where it serves as a condensed reserve of energy for later germination. In most seeds, of which corn and other cereals are examples, the principal store of energy is in the form of carbohydrates, but oil-bearing seeds, such as the soybean, cottonseed, and flax, contain their reserve primarily as fat, as their name implies. These seeds are used as commercial sources of oil, leaving the extracted meals as by-products for animal feeding. Oil-bearing seeds are also much higher in protein than are the cereal seeds.

In all plant products, with the exception of the oil-bearing seeds, carbohydrate is the principal constituent, as it is in the plant as a whole. The nature of this carbohydrate differs markedly according to whether it is serving as a reserve or structural element. In seeds, it occurs principally as starch, which is the reserve carbohydrate, while in stems and to a much lesser extent in leaves, a considerable proportion of it is present as cellulose, the principal structural carbohydrate. The outer coats of seeds also contain cellulose as a structural and protective element. Since cellulose and related compounds, classed by the nutrition chemist as *crude fiber, acid-detergent fiber,* or *cell walls* are much less digestible than starch, or cell contents, the various parts of plants differ markedly in nutritive value according to their digestibility. Feeds which are high in cellulose and related compounds and thus of low digestibility, such as hay, straw, and silage, are classed as *roughages.* The term *concentrates* is used to denote those low in crude fiber and highly digestible. Here are included the seeds and most of their by-products.

2.7 Mineral Matter

The amount of mineral matter in plants is highly variable in different species as well as in the different plant parts. From the standpoint of animal nutrition, we are particularly interested in the fact that the percentage distribution of the mineral elements of plants differs markedly from that in animals. To illustrate this fact, data for calcium and phosphorus in plants, the elements which make up over 70 percent of body ash, are given in Table 2.2. With the exception of the legumes, which are always rich in calcium, these elements make up a rather small part of the ash of plants. Both are generally exceeded by potassium. Calcium is primarily associated with the vegetative portion of the plant, and the leaf is richer than the stem. Without exception, seeds are low in calcium compared with the other parts of the plant, though oil-bearing seeds are higher than others. In contrast to calcium, phosphorus is richer in the seeds than in the rest of the plant. Leaves are richer than stems. The calcium and phosphorus content of the vegetative part of the plant is influenced by soil and other cultural factors.

2.8   By-product Feeds

The feeds of animals obtained from plants consist not only of forage crops, seeds, and roots but also of by-products arising from the processing of various plant materials, notably seeds, in the manufacture of products used for human food and for industrial purposes. The bran and middlings which arise from flour milling, gluten feed, which is a by-product of cornstarch manufacture, and the meals which are the residues of the extraction of oil from oil-bearing seeds are all familiar examples of the very large number of by-product feeds. Their composition is usually very different from that of the seed or other material from which they arise. This is illustrated by the figures in Table 2.3, which are taken in a condensed form from data presented by Osborne and Mendel.[8] While these data must be considered as approximate only in view of the rather large percentage of undetermined material, they serve to show the large differences in composition among the different parts. The endosperm consists very largely of starch, the reserve material, and contains very little of the less digestible carbohydrates. In contrast, the seed coats are characterized by a high content of cellulose and related compounds which provide the needed protective qualities. They are also richer in protein, the B-vitamins, fat, and mineral matter than the endosperm or the seed as a whole. The embryo is especially rich, compared with the other parts, in protein and fat and is lowest of all in cellulose. Most of the vitamin content of the entire kernel is found in the seed coats and embryo.

TABLE 2.3.   Percentage Distribution of Nutrients in the Parts of the Wheat Kernel

Part of wheat kernel	Protein	Fat	Starch, sugar, etc.	Cellulose, pentosans, etc.	Ash	Undetermined
Whole kernel	11.3	2.2	66.4	8.0	2.0	10.1
Endosperm	11.2	1.2	81.4	2.1	0.4	3.7
Seed coats	17.6	8.3	7.0	43.9	8.6	14.6
Embryo	40.3	13.5	24.3	1.7	4.8	15.4

Source: Thomas B. Osborne and Lafayette B. Mendel, The nutritive value of the wheat kernel and its milling products, *J. Biol. Chem.*, **37**:557–601, 1919.

Thus the milling of wheat leaves a feed for animals which is richer in protein, fat, mineral matter, and vitamins than the entire kernel but which is somewhat less digestible because of the larger amount of the higher carbohydrates. It is the endosperm which provides the white flour for human food. A yield of approximately 70 percent is obtained, which means that a portion of the endosperm is left behind with the seed coats and embryo which constitute the by-products of the milling process.

While wheat by-products have a high feeding value, this is by no means true for all by-product feeds. Oat-mill feed, for example, which is the residue from oatmeal production, contains less than half as much protein and over twice as much fiber as the seed itself, because it consists mostly of the hull. It is therefore of low digestibility and nutritive value. On the other hand, the extraction of soybeans, cottonseed, and flaxseed to obtain their oils for human food or industrial use provides products that are highly digestible and of special value for their protein content.

Half or more of the total dry matter of most food crops consists of inedible residues such as stalks and straw from cereal grains, vines and tops from legumes, and root crops. These can be important sources of nutrients for animals, but often they are burned or allowed to rot. Research has also shown that waste material from chicken laying and broiler houses and even from cattle feed lots can serve as a potential replacement for usual animal feeds.[9]

Similarly, many important feeds, used mostly in swine and poultry rations, result from the processing of animal products. Here are included tankage, meat scraps, fish meal, milk by-products, hydrolized feathers, hair, leather scraps, and many others. A knowledge of the processes from which by-product feeds arise and thus of their makeup in terms of the different parts of the original material is a very helpful guide to their composition and feeding value.

NOTES

1. Agricultural science owes a tremendous debt to John B. Lawes (1814–1900) and Joseph H. Gilbert (1817–1901) for their pioneer work in the fields of agronomy and animal nutrition, begun in 1843 and continued for over half a century. The experiment station which they founded at Rothamsted, England, rapidly gained fame throughout the world, and it remains today an outstanding center of research in the plant sciences. Their studies of the composition of the animal body are published under the title: Experimental enquiry into the composition of the animals fed and slaughtered as human food, *Trans. Roy. Soc. (London)*, **149**:493–680, 1859.

2. J. T. Reid, G. H. Wellington, and H. O. Dunn, Some Relationships among the major chemical components of the bovine body and their application to nutritional investigations, *J. Dairy Sci.*, **38**:1344–1359, 1955.

3. Sam L. Hansard and coworkers, Blood volume of farm animals, *J. Animal Sci.*, **12**:402–413, 1953.

4. A. M. Pearson, R. W. Purchas, and E. P. Reineke, Theory and potential usefulness of body density as a predictor of body composition, in Body Composition in Ani-

mals and Man, Publ. 1958, National Academy of Sciences, Washington, D.C., 1968, pp. 153–169.

5. R. G. Westervelt, J. R. Stouffer et al., Estimating fatness in horses and ponies, *J. Animal Sci.,* **43**:781–785, 1976.

6. B. C. Breidenstein, T. G. Lohman, and H. W. Norton, Comparison of the potassium-40 method with other methods of determining carcass lean muscle mass in steers, in Body Composition in Animals and Man, Publ. 1958, National Academy of Sciences, Washington, D.C., 1968, pp. 393–412.

7. Reid and coworkers, *loc. cit.*

8. Thomas B. Osborne (1859–1929), chemist of Connecticut Agricultural Experiment Station at New Haven, and Lafayette B. Mendel (1872–1935), professor of physiological chemistry at Yale University, collaborated in nutrition research for over twenty years. Their outstanding discoveries, particularly in the fields of proteins and vitamins, which are frequently referred to in this book, assure them lasting recognition as pioneers in developing the newer knowledge of nutrition. In addition to their joint work, Osborne became the leading authority of the world on the vegetable proteins, while Mendel made many important contributions on various aspects of nutritional physiology and was an inspiring teacher to a host of students who are now carrying on his work in many laboratories.

9. A. N. Bhattacharya and J. C. Taylor, Recycling animal waste as a feedstuff: A review, *J. Animal Sci.,* **41**:1438–1457, 1975.

SUPPLEMENTARY LITERATURE

Babatunde, G. M., W. G. Pond, and associates: Effects of plane of nutrition, sex and body weight on the chemical composition of yorkshire pigs, *J. Animal Sci.,* **26**:718–726, 1967.

Davey, R. J., and B. Bereskin: Genetic and nutritional effects on carcass chemical composition and organ weights of market swine, *J. Animal Sci.,* **46**:992–1000, 1978.

Ellenberger, H. B., J. A. Newlander, and C. H. Jones: Composition of the bodies of dairy cattle, *Vermont Agr. Expt. Sta. Bull.* 558, July 1950.

Fat Content and Composition of Animal Products, National Academy of Sciences, Washington, D.C., 1976, p. 243.

Harlan, J. R.: The plants and animals that nourish man, *Scientific American,* **235**:89–97, 1976.

McCelland, T. H., B. Bonaiti, and St. C. S. Taylor: Breed differences in body composition of equally mature sheep, *Animal Prod.,* **23**:281–293, 1976.

Mitchell, H. H., and associates: The chemical composition of the adult human body and its bearing on the biochemistry of growth, *J. Biol. Chem.,* **158**:625–637, 1945.

Oliphant, J. M.: Feeding dried poultry waste for intensive beef production, *Animal Prod.,* **18**: 211–217, 1974.

Oltjen, R. R., and D. A. Dinius: Production practices that alter the composition of foods of animal origin, *J. Animal Sci.,* **41**:703–722, 1975.

Reid, J. T.: Body composition of animals: Interspecific, sex and age peculiarities, and the influence of nutrition, *Festskrift til Knut Breirem,* Oslo, Norway, 213–238, 1972.

Stouffer, J. R.: Techniques for the estimation of the composition of meat animals, in Techniques and Procedures in Animal Science Research, American Society of Animal Science, 1969, pp. 207–219.

Wedin, W. F., H. J. Hodgson, and N. L. Jacobson: Utilizing plant and animal resources in producing human food, *J. Animal Sci.,* **41:**667–686, 1975.

Young, R. J.: Nutritional potential for recycling waste, *Federation Proc.,* **33:**1933–1946, 1974.

Chapter 3

Digestive Processes in Different Species

Digestion involves a series of processes in the alimentary tract by which feeds are broken down in particle size and finally rendered soluble so that absorption is possible. This is accomplished by a combination of mechanical and enzymatic processes. Microorganisms provide important enzymes not secreted by mammalian tissues.

3.1 Organs of Digestion

Mouth. The relative importance of the mouth and its components (teeth, tongue, cheeks, and salivary glands) varies with the species. In most species the functions of the mouth are to bring in feed, mechanically break it up, and mix it with saliva which acts as a lubricant to facilitate swallowing. For horses the principal organs of prehension are the lips whereas the lips of ruminants are less mobile and their main prehensile organ is the tongue. Pigs effectively use the pointed lower lip. Dogs and cats hold food with their forelimbs but the movement of the head and jaws brings the food into the mouth.

There are large differences among animal species in the extent to which they masticate their food, determined for the most part by the kind of food they eat and the physical structures of the mouth and teeth. For example, ruminants grind thoroughly the grass or other forage they consume, although much of this occurs during rumination when the boluses are regurgitated and remasticated rather than at the time the forage is consumed. In contrast, some ruminants such as cattle usually swallow small grain seeds with very little effective chewing. Thus, their grain should be ground or cracked (Sec. 3.9). Goats and sheep are generally more efficient in the chewing of grain than cattle.[1] Horses can chew large kernels such as oats or corn but not small kernels such as wheat and milo. Carnivores often swallow large chunks of meat with little mastication. Poultry have no teeth and they swallow their food whole. Any grinding is done by the action of the grit in the bird's gizzard.

In most species the principal salivary glands are the parotid, mandibular (submaxillary), and sublingual. Smaller glands are found in the cheeks (buccal glands) and lateral areas of the lips (labial glands).

The parotid secretes saliva continuously in ruminants but not in the horse where the stimulus of mastication is necessary. But the salivary reflex cannot be conditioned by the sight of food in the horse as it can be in dog and human. The amount of secretion varies from 2 to 3 liters per day for sheep to 10 to 12 liters for ponies to 130 to 180 liters for cattle. Saliva has functions other than lubrication. It dissolves the water-soluble components which enables the components to reach the taste buds. Saliva in species such as man and rat contains ptyalin (α-amylase) which splits $\alpha 1 : 4$ glucosidic linkages. Ptyalin is probably not of great significance in domestic animals.

In ruminants saliva can be a source of bicarbonate phosphate buffer for the rumen and provides a mechanism of recycling urea.

Esophagus.   Food passes from the mouth to the stomach or forestomach via the esophagus. It contains four layers: an outer layer of connective tissue, a layer of muscle, the submucosa, and mucosa. In dogs, cattle, and sheep the muscular layer consists of only striated muscle fibers, but in pigs, horses, and humans the portion adjacent to the stomach is composed of smooth muscle. In contrast to most species, the horse seldom vomits.

Forestomach.   Ruminants differ from other mammals in having a greatly enlarged forestomach consisting of three additional stomach compartments (Fig. 3.1). Of these the rumen and reticulum have more than 50 percent of the total capacity of the digestive tract. This large capacity is essential to allow feed retention so that microorganisms can break down cellulose and other complex carbohydrates which mammalian enzymes cannot hydrolyze. The omasum and abomasum (true stomach) each represent about 6 to 8 percent of the capacity of the digestive tract. Thus, the four compartments hold 60 to 65 percent of the total volume compared to 25 percent for the small intestine, 10 percent in the large intestine, and less than 5 percent in the cecum.

Several nonruminant animals such as the kangaroo, hamster, hippopotamus, peccary, sloth, and certain primates, for example, languar, colobus, and proboscis monkeys, have a voluminous compartmentalized or sacculated stomach that serves as the primary site of microbial activity.

Chickens and turkeys also have some structural peculiarities of the diges-

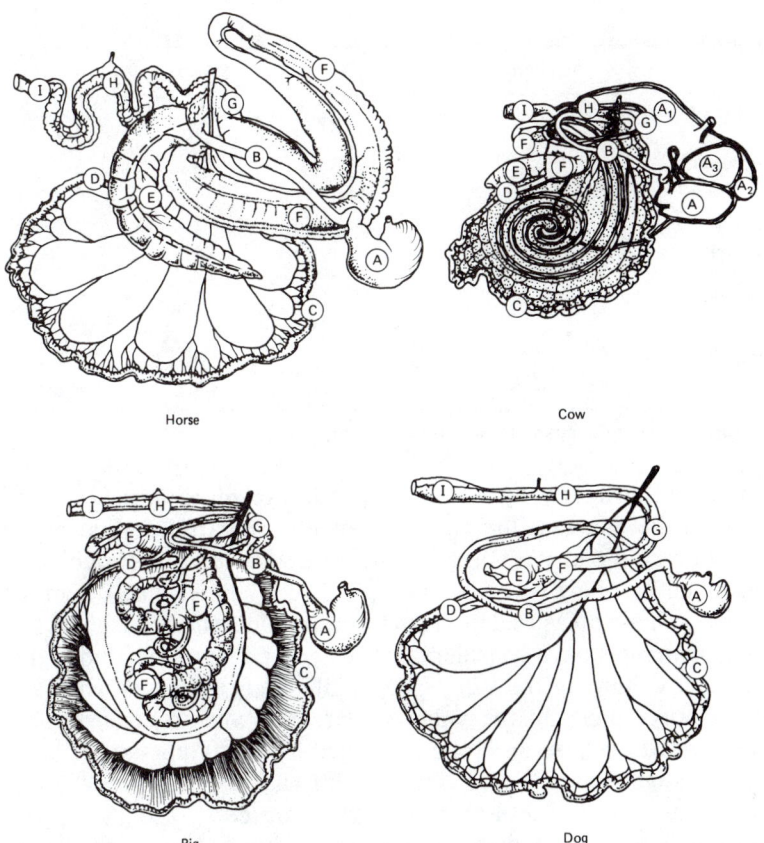

Fig. 3.1 Digestive tracts of horse, cow, pig and dog. A, abomasum (true stomach); A_1, rumen; A_2, reticulum; A_3, omasum; B, duodenum; C, jejunum; D, ileum; E, cecum; F, ascending colon; G, decending colon; H, transverse colon; I, rectum. *Adapted from R. Nickel, A. Schummer and E. Seiferle, Lehrbuch der Anatomi der Haustiere, Paul Parey, Berlin, 1963.*

tive tract. The esophagus empties directly into the crop where the food is stored and soaked. From the crop the food passes through the proventriculus, an enlarged glandular portion where digestive juices are copiously secreted and mixed with food as it passes to the gizzard, a very muscular organ which normally contains stones or grit that aid in grinding hard seeds and grains before they move into the small intestine.

Stomach. The data in Table 3.1 demonstrate that the horse has a relatively small stomach. The stomach of a 190-kg pig has about the same capacity as that of a 450-kg horse and the abomasum or true stomach of a 575-kg cow is almost twice as large as that of the horse. (Incidently the values from the ruminoreticulum are combined because food passes freely back and forth between rumen and reticulum.) Of course, the size and position of the stomach as

TABLE 3.1. Approximate Capacity of the Digestive Tracts of Certain Animals (volume in liters)[a]

	Human	Pig	Horse	Sheep	Cattle
Body weight, kg	75	190	450	80	575
Ruminoreticulum		...	...	17	125
Omasum		...	...	1	20
Abomasum	1	8	8	2	15
Total stomach	1	8	8	20	160
Small intestine	4	9	27	6	65
Cecum	...	1	14	1	10
Large intestine	1	9	41	3	25
Total digestive tract	6	27	90	30	260

[a]These data were selected from various sources to represent physiological fill.

well as of all parts of the digestive tract vary depending on physiological state of animal and type of diet because the digestive organs are capable of considerable stretching. For example, Warner and Flatt[2] reported that earlier published volumes determined by filling with water were 1.5 to 1.7 times larger than the maximum ingesta volume measured immediately after cattle had consumed hay to capacity. Thus, animals have considerable ability to adapt to the kind of diet they consume. The volume of the stomach and intestines increases when animals consume a bulky diet and the volume decreases when a concentrated, high-energy diet is fed because less food is needed to satisfy appetite.

The extent of digestive action in the stomach varies with species. In all species studied hydrochloric acid and pepsin are secreted by the gastric mucosa. These products help break down protein into polypeptides. In the young calf, lamb, and kid, the gastric mucosa secretes rennen which coagulates milk proteins. Gastric lipase is present in low concentration in the gastric juice of carnivores. The amount and type of microbial action in the stomach varies greatly among species, but lactic acid and volatile fatty acids which indicate microbial fermentation have been found in the stomachs of pigs, rabbits, guinea pigs, and horses.

Small Intestine. The small intestine is the principal site of absorption of amino acids, vitamins, minerals, and lipids, and in nonruminants, of soluble carbohydrates. The primary enzymes of the small intestine and other organs of the digestive tract are shown in Table 3.2. Triglycerides and other lipids are hydrolyzed to monoglycerides, glycerol, and free fatty acids (Chap. 7). As shown in Table 3.2, several enzymes function to degrade the different proteins to amino acids. Other aspects of protein digestion and absorption are summarized in Chap. 8. Glucose and other monosaccharides in foods are completely dissolved and absorbed without digestive action. Disaccharides usually are readily hydrolyzed by the enzymes listed by all animal species, but there are some important exceptions. Certain individuals do not secrete effective amounts of lactase and therefore cannot tolerate significant amounts of lactose in the diet. This problem has been studied extensively in humans. Non-Caucasians appear to have a high incidence of lactose intolerance and it has been suggested the lack of lactose may be an autosomal recessive trait. Mature animals in most species have a lower lactose tolerance than the young. It is also interesting that animals that do not normally consume milk may have lactase activity. For example, lactase activity has been reported in the digestive tract of fish.

Ruminants do not secrete significant amounts of sucrase. The digestibility of starch may vary greatly depending upon the source of starch and the animal species and age of animal concerned. In the dog and mink raw cornstarch is very poorly digested but it is well utilized after it is cooked. In the pig raw cornstarch is about as well utilized as the cooked product. Wheat starch, on the other hand, is well digested by the dog, mink, and pig even in the uncooked form. Young calves have very little pancreatic amylase and even in animals 2 years old no increase in blood glucose was noted when starch was put in the abomasum. However, it has been suggested that the starch leaves the abomasum slowly and therefore, even if amylase were present, blood glucose changes might not take place because of the slow release of glucose from the intestine.

TABLE 3.2. Primary Enzymes of the Digestive Tract

Food source (substrate)	Enzyme	Origin	Product of digestion
Carbohydrates			
Starch, glycogen, dextrin	Amylase	Saliva, pancreas	Maltose, glucose
Maltose	Maltase	Small intestine	Glucose
Lactose	Lactase	Small intestine	Glucose, galactose
Sucrose	Sucrase	Small intestine	Glucose, fructose
Fats and oils			
Lipids	Lipase	Gastric mucosa, pancreas	Monoglycerides Glycerol, fatty acids
Proteins			
Milk proteins	Rennin	Gastric mucosa (young calf)	Coagulates milk proteins
Proteins	Pepsin	Gastric mucosa	Polypeptides
Protein breakdown products	Trypsin	Pancreas	Peptides, proteoses
	Chymotrypsin	Pancreas	Peptides
	Carboxypeptidase	Pancreas	Peptides, amino acids
	Aminopeptidase	Small intestine	Peptides, amino acids
	Dipeptidase	Small intestine	Amino acids
Nucleoproteins	Nucleotidase	Small intestine	Nucleotides, nucleosides
	Nucleosidase	Small intestine	Purines, phosphoric acid

Large Intestine.   In the pig and other omnivores there is an enlargement of the cecum and the colon. Some herbivorous animals show great extension of the size of the cecum and colon as compared with other parts of the tract. In the horse and rabbit about 60 percent of the capacity of the digestive tract is in the cecum and colon. There are large differences in the ceca of birds. Some birds such as the parrot, hummingbird, and swift do not have any ceca. Some such as the heron only have one cecum, and some such as the chicken and turkey have two ceca but they do not have much nutritional significance. On the other hand, some birds such as the ostrich, grouse, and ptarmigan have two large ceca with significant bacterial populations. Until recently the large intestine of the ruminant received little attention from nutritionists, but Ulyatt et al.[3] concluded that between 5 and 30 percent of the digestible cellulose is digested in the large intestine. Furthermore, they suggested the amount of hemicellulose digestion is greater than that of cellulose and that digestion in the large intestine of the ruminant accounts for between 4 to 26 percent of the digestible energy.

The microbial digestion of fiber, although not as efficient as in the rumen, is extensive in the horse, guinea pig, beaver, chinchilla, elephant, sea cow, grouse, ptarmigan, and many other nonruminant herbivores (Table 3.3). There is relatively little fiber digestion in the large intestine of many nonruminants such as the dog, chicken, cat, mink, raccoon, and human. But even in these animals some volatile fatty acids are produced, absorbed, and utilized. The factors affecting microbial digestion of fiber in ruminants and nonruminants are discussed in Sec. 3.3.

The nutritional significance of protein and vitamins produced by the microflora of the lower gut is still unresolved. Large amounts of bacterial protein and vitamins are synthesized in the large intestine, but the extent of the absorption of these nutrients is not known.

The large intestine is the major site of sodium absorption and an important site of chloride absorption. Potassium, phosphorus, and magnesium can also be absorbed from the large intestine.

3.2 Microbial Digestion of Carbohydrates

Bacteria are the most important of the symbiotic organisms which break down the higher carbohydrates. This symbiotic relationship is developed to the highest degree in ruminants since the rumen provides both the capacity and other factors which are most favorable to its activity. In addition to digesting the higher carbohydrates, the organisms synthesize essential nutrients, specifically amino acids and the B-vitamins. While this microbial activity is of practical importance primarily in the case of ruminants, it occurs throughout the animal kingdom and is also significant for the horse, guinea pig, rabbit, and other herbivorous animals and even insects such as the termite.

The acids and gases which are formed by microbial action in the rumen are the end products of various intermediary reactions. Cellulose, pentosans, and starch are hydrolyzed to monosaccharides and then fermented. The proportion of the acids formed varies with the nature of the ration, the organisms present, and other factors. Acetic acid makes up from two-thirds to three-fourths or more of the total. Propionic acid stands next in order of amount, followed by butyric acid. Small amounts of other acids have been reported. The actual amounts produced per unit of dry-matter intake, and their relative proportions, depend on the nature of the ration and also vary somewhat between species. Various methods have been used to study the rates of production of these acids. The isotope-dilution technique, introduced by Sheppard and coworkers,[4] with either a single injection or continuous infusion of isotopes has been used extensively. Leng and Brett[5] developed a technique to measure simultaneously the rate of production of the three acids and the rate of production of branched-chain acids and methane.[6] Rates of interconversion are shown. Conversion from acetic to butyric acid accounted for 40 to 50 percent of the butyric acid produced. For 11 sheep on the four rations, the average molar percentages of total volatile fatty acids (VFA) produced were approximately: acetic, 65; propionic, 20; and butyric, 9. However, the ratios of the VFA and methane production can be altered by dietary manipulations. For example, feeding high grain diets or adding monensin, an antibiotic, results in a relatively greater proportion of propionate. Fermentations favoring propionate and reducing methane production usually result in improved feed-to-gain ratios in growing animals. Depressed acetate production may result in milk with a lowered percentage of fat (Sec. 17.14). Methods of manipulating fermentation in order to increase efficiency of food production are currently being studied extensively.

It cannot be concluded that all of the volatile acids found in the rumen arise directly from carbohydrate fermentation. Some may come from the action of microorganisms on protein or other nitrogenous compounds (Chap. 8). There is also some interconversion of acids after their production. The gases formed are methane and carbon dioxide and, under some conditions, hydrogen. The magnitude of microbial digestion is indicated by studies with cattle showing that 40 to 80 percent of the dry-matter intake disappears in the rumen and reticulum, some 80 percent of which is carbohydrate.

Of the fibrous constituents of the feed, lignin is resistant to microbial attack. Cellulose is much more readily broken down, and the hemicelluloses, as a group, are the most digestible of the three. Starch and sugars are readily converted to acids and gases, and in the ruminant this is the chief pathway of their digestion. The growth of bacteria in the rumen results in the synthesis of certain bacterial polysaccharides which may be digested farther on in the tract.

While the rumen is the principal seat of the bacterial breakdown of carbohydrates, this breakdown occurs in significant amounts in the large intestine of some herbivora. This is particularly true for the horse, and in this species there is also considerable activity in the cecum. The process has also been shown to occur to a limited extent in the pig. In all species studied the same three acids have been identified as the principal products, although in some species, such as the rabbit, the relative amount of butyic acid is usually greater than that of propionic acid. Furthermore, the biochemical pathways of fermentation of the large intestine have not been studied as extensively as those of the rumen and it should not be assumed that all information obtained from rumen studies can be applied to the large intestine.

3.3   Factors Governing Microbial Digestion of Fiber

Since in herbivora 20 percent or more of the ration may consist of substances that can be digested only by the action of microorganisms, the factors which may govern this activity are obviously of large importance. The quantitative relations involved in the microbiotic decomposition of carbohydrates vary according to the kind and number of the microorganisms present, which in turn are under the influence of the character of the food. It has been shown, for example, that the addition of easily digestible carbohydrates such as starch, cane sugar, or molasses to the ration of cattle reduces the digestibility of the fiber, and this observation has been explained on the ground that the bacteria attack the simpler carbohydrates by preference. It is clear that such a shift in substance attacked would lower the nutritive value of the entire carbohydrate portion of the ration, in that less fiber would be digested and more of the absorbable sugar would be lost as gases. There is also evidence that the character of the roughage, irrespective of its fiber content, has an influence on the nature of the bacterial flora and on their activity. The replacement of poor-quality hay by alfalfa has been shown to stimulate microbic activity, suggesting that the latter hay may supply specific vitamins or other factors needed for the best growth of the bacteria in question. The higher protein content of the better hay might be the explanation, however, since it has been shown that protein-rich feeds promote the microbial breakdown of fiber. As an example, Williams and coworkers[7] reported that with low-protein diets increases in starch intake reduced the concentration of microorganisms in the rumen of sheep and changed their type. With higher protein levels no such effect occurred. At all starch levels protein addition increased the digestibility of dry matter.

The differences in the extent of fiber digestion in the various animal species are partially explainable on the basis of the varying opportunities presented for the action of microorganisms. Ruminants are able to digest at least 50 percent of the fiber of most feeds and other herbivora can do nearly as well, in contrast to the omnivora, which have only a limited ability to digest the complex polysaccharides. Anatomical differences in the digestive tract explain why hay and other roughages can form such a large part of the rations of cattle, sheep, and horses and yet be tolerated in only limited amounts by pigs and chickens.

For a given species and animal, there are differences in the degree of the breakdown of fiber from different sources which are intimately associated with its chemical and physical nature. The complex polysaccharides of mature plants are less well digested than they are in young, growing plants. The fiber of growing pasture grass, fresh or dried, is more digestible than that of hay. Early-cut hay is more digestible than hay cut in late bloom or in seed. The difference is due to both chemical and physical structure and particularly to the presence of certain substances, notably lignin, which are deposited in the cell wall with age. The lignin is not only indigestible itself, but it also lowers the digestibility of the cellulose and other complex carbohydrates.

It is often stated that the lignin effect on carbohydrate digestion is a result of lignin "encrustation." Van Soest and McQueen[8] state that this theory is not correct and the probable cause of the reduced digestibility is an increased rate of passage.

Much information is being gained regarding microbial digestion in the in vitro studies using artificial rumens in which the natural conditions are simulated. Increasing attention is being given to the bacteriology of the rumen, as is indicated by the reviews by Bryant[9] and Wolin.[10] As we learn more about the conditions which promote the maximum activity of rumen organisms we shall also learn how to take fuller advantage of the rumen processes on which the digestion of the higher carbohydrates depends. It is evident that research on these microbial processes is important to enable us to make the most effective use of roughage in feeding cattle and sheep, particularly the low-grade roughage which is available on many farms.

3.4 Bulk

Fiber content is an important factor governing the bulk of a ration, whatever may be the significance implied in the rather variable usage of this term. As used in connection with a grain mixture the term refers to the weight of a given volume of the feed. For example, oats, which weigh approximately 0.45 kg per liter, are bulky in contrast to corn meal, which weighs 0.70 kg on the same basis. The bulky concentrates are in general those which are high in fiber, although the air spaces between the particles also contribute to bulk. For the ration as a whole, increasing the roughage portion with its high-fiber content increases its bulk. The importance of making up a concentrate mixture so that it will have a certain amount of bulk is stressed by many authorities in order to avoid the formation of a doughlike mass in the stomach which is not readily attacked by the digestive juices. The physiological evidence for this point of view, however, is not entirely conclusive.

Bulk is also considered important from the standpoint that a certain distention of the digestive tract is desirable for the tract's most effective functioning, particularly in the elimination of the feed residue. Of course this distention can be brought about by a large intake of any kind of food, but it is particularly accomplished for the tract as a whole by indigestible material such as fiber. In fact, in human nutrition the term roughage is used synonymously with bulk to denote the indigestible portion of the diet.

Bulk is promoted by ability to absorb water. Some fibrous materials, such as agar, absorb large quantities of water, while others, such as regenerated cellulose, do not. Linseed oil meal, which is much lower in fiber than wheat bran, absorbs three times as much water and thus, in this sense, is a more bulky feed in the digestive tract.

The influence of bulk in promoting the elimination of feed residues is essentially a laxative effect. It is recognized that feeds high in fiber tend to be laxative and that a fiber which readily absorbs water and swells is more laxative than one which does not, at least for certain species. A nonfibrous feed which absorbs a large amount of water is less effective, because it is largely digested and thus does not reach the portion of the tract occupied primarily by feed residues. Of course, bulk is not the sole cause of laxative effect, for many feeds are laxative because of specific chemical substances contained in them which promote peristalsis.

The degree of bulk which is desirable naturally depends upon the species, in view of their variability as regards size and anatomy of their digestive tracts. It is also dependent upon the level of production sought. Too much bulk lessens the consumption of digestible nutrients, and thus the intake of bulky material of low digestibility must be limited. Though alfalfa may be ideal as a sole ration from the standpoint of promoting the normal activity of the digestive tract, high-producing cows cannot consume enough of it to meet their needs for nutrients. On the other hand, a high intake of a ration too low in bulk may result in indigestion and in the animal going "off feed."

Matrone and coworkers[11] reported that lambs grew normally on low-fiber purified diets when sodium and potassium bicarbonate were included to serve as buffers, suggesting that bulk as such is less important than formerly believed. Later studies by various researchers have shown that all-concentration rations consisting of rolled barley or ground ear corn supplemented with protein, minerals, and vitamins give satisfactory rates and efficiency of gain in finishing beef cattle, but injury to the rumen epithelium (*parakeratosis*) and liver abscesses occur in some animals. Adding long hay prevents these pathological changes but does not improve the rate of gain. Buffers have been beneficial in some studies but not in others. The exact role of bulk in ruminant rations has not been fully explained. In certain species, including humans, large intakes of fiber cause intestinal irritation and other gastrointestinal troubles.

But fiber may have beneficial effects for humans. The addition of fiber to low-fiber diets decreases the incidence of diverticulitis and constipation. Reilly and Kersner[12] concluded that there appears to be a correlation between low fiber intake and "diseases of civilization" (appendicitis, cancer of the colon, heart disease) but nothing definitive can be said about cause and effect as yet.

3.5 Rumen Action on Lipids and Proteins

There is no microbial digestion of fats and protein comparable to that previously discussed for carbohydrates. A partial breakdown of fats occurs in the rumen (Sec. 7.10). An extensive protein metabolism takes place which is primarily of a synthetic nature, rather than a digestive process. The details of this action are presented in Secs. 8.13 and 8.14. Some species of nonruminants have microflora that digest wax. For example, the honey guide (a bird) has *miccococcus cerolytius* that break down bees wax.

3.6 Digestive Ability of Different Animals

There are large differences in the ability of different animal species to digest bulky feeds. The average digestion coefficients given in Table 3.3 for alfalfa hay, which contained 86.1 percent of dry matter, 16.2 percent of protein, 1.6 percent of ether extract, and 26.9 percent of crude fiber, show that ruminants

TABLE 3.3. Digestibility of Alfalfa Hay by Certain Animals

Animal species	Organic matter	Crude protein	Ether extract	Crude fiber	Nitrogen-free extract	TDN[a]
	Apparent digestion coefficients, %					
Cattle	61	70	35	44	71	48
Sheep	61	72	31	45	69	48
Goat	59	74	19	41	69	46
Horse	59	75	10	41	68	46
Pig	37	47	14	22	49	30
Rabbit	39	57	21	14	51	30
Guinea pig	52	58	12	33	65	40

[a]Total digestible nutrients (Sec. 9.6).

digested appreciably more of the nutrients than the pig and rabbit. The horse was only slightly less efficient than cattle or sheep in digesting high-quality alfalfa hay, but the difference is considerably greater for low-quality, high-fiber feeds.

Because of extensive bacterial fermentation in the rumen, ruminants are able to digest cellulose and other complex carbohydrates more completely than nonruminant herbivora. Most nonruminants cannot grow and reproduce normally on bulky diets of forages alone because they cannot absorb adequate amounts of energy from such feeds. When high-energy diets are fed, however, nonruminants exhibit digestive ability equal to the ruminants.

Data in Table 3.4 illustrate the comparative digestive abilities of humans, rats, guinea pigs, sheep, and swine to digest the same low-fiber ration. The diet consisted of 68 percent wheat, 7 percent sucrose, 13 percent dried skim milk, 11 percent cream, and 1 percent salt baked and fed as a biscuit along with water. In

TABLE 3.4. Apparent Digestibility of the Same Diet by Several Animal Species

Nutrient	Humans	Rats	Guinea pigs	Sheep	Swine	Average
	Digestion coefficients, %					
Dry matter	90	88	85	79	91	87
Energy	90	87	83	75	91	86
Protein	89	79	76	76	92	82
Ether extract	84	76	35	90	71	71
Carbohydrates	92	90	100	87	93	92
N-free extract	94	92	100	89	95	94

Source: E. W. Crampton and coworkers. The apparent digestibility of essentially similar diets by rats, guinea pigs, sheep, swine and by human subjects, *J. Nutrition*, **43**:541–550, 1951.

the case of sheep the total diet included hay, which had been studied previously, and the digestion coefficients were determined by the difference method. In this comparison most nutrients were slightly better digested by humans and swine than by the other species and sheep tended to be lowest, but the coefficients do not differ greatly for the different species, illustrating that the bacterial fermentation in the rumen of the sheep was not of any particular advantage with the low-fiber diet studied in comparison with enzymic digestion alone.

3.7 Factors Affecting Digestibility

Digestion coefficients are not constants for a given feed or species. They are influenced by several variable factors. Previous discussion (Sec. 3.3) has mentioned various ways in which the breakdown of the higher carbohydrates by rumen bacteria and, in turn, the digestion of other nutrients can be influenced by the nature and relationships of the nutrients fed. Chemical composition of the feed affects digestibility, the stage of maturity at harvest being the most important factor influencing composition of forages. As the plant matures, its cell wall content increases, cell content decreases, and the plant becomes less digestible (Sec. 2.6). Examples of the effect of maturity on composition and digestibility are shown in Table 3.5.

The influence of nutrient relationships in the ration is further illustrated by the results obtained with varying nutritive ratios. As this ratio becomes wider, the digestibility of all nutrients tends to be lower. This is particularly true for

TABLE 3.5. Difference in Composition and Digestibility of Forage Sorghum and Silage Corn with Advancing Maturity

DPE[a]	Dry matter, %	Crude protein, %	CWC,[b] %	Cellulose, %	Lignin, %	NDDM,[c] %
			Forage Sorghum			
70	18	10	45	23	1.9	68
100	31	6	41	20	1.7	66
130	36	5	41	21	2.5	60
180	36	4	51	24	2.8	53
190	39	4	52	28	3.3	48
			Silage Corn			
70	26	4	27	13	1.8	69
100	39	4	30	14	1.9	67
130	43	7	42	20	2.8	65
160	71	5	60	26	3.6	60

[a]Days post emergence.
[b]Cell wall constituents.
[c]In vitro dry matter digestibility.
Source: M. M. Danley and R. L. Vetter, Changes in carbohydrate and nitrogen fractions and digestibility of forages: Maturity and Ensiling, *J. Animal Sci.*, 37:994–1000, 1973.

protein, and the effect here is readily explainable on the basis of output of metabolic nitrogen, since the protein coefficient determined represents the apparent digestibility. Inasmuch as the metabolic nitrogen is governed by total food intake and thus tends to remain constant although the percentage of protein in the food is lowered, the fecal nitrogen as a whole does not decrease proportionally with the decreased protein intake, even though the residual food nitrogen may. Thus the apparent digestibility of protein is lowered with a wide ratio even though the true value may not be. The lowering of the digestibility of nutrients, other than protein, with a ratio having a wide nutritive ratio, is less marked than for protein itself, and published data indicate that it occurs less consistently. In the case of ruminants, however, the addition of protein or of nitrogen compounds utilized by bacteria to a ration having a wide ratio definitely increases the breakdown of the higher carbohydrates, and this in turn makes other nutrients more digestible.

Variations, according to feed sources, in the digestion coefficient for ether extract reflect its variable nature. While in seeds and their by-products this extract consists almost entirely of readily digestible esters of fatty acids, the extract of roughages contains a high proportion of nonsaponifiable constituents as well as nonlipid substances.

Digestibility may be limited by a lack of time for complete digestive action on less easily digestible substances or by a lack of complete absorption. Such an effect is heightened by a rapid passage of the food through the tract. On the other hand, food may move so slowly through the intestines as to be excessively subject to wasteful fermentations.

Many studies have been made on the influence of age which, in general, have shown no marked differences due to this factor. Studies have also been reported of the influence of deficiencies of various minerals and vitamins which, for the most part, have failed to show any marked effects. Some of these deficiencies are reflected, however, in a poorer utilization of the absorbed nutrients, as is discussed later.

It is recognized that the digestibility of a mixture is not necessarily the average of the values for its constituents determined separately or indirectly. Each feed may exert an influence on the digestibility of the others.

3.8 Influence of Level of Intake on Digestibility

Many studies have been made of the influence of level of nutrition upon the digestibility of feeds by various animal species. When feed intakes are reduced below the maintenance level, animals tend to become more efficient in digesting feed and in utilizing the nutrients. The changes may be more of a metabolic effect than in digestive ability alone. During rapid growth nonruminants may consume as much as three times the maintenance level, but this higher intake of feed exerts only a small depressing effect upon the digestibility of the ration. When ruminants are fed roughages alone, level of intake has little influence upon digestibility, but the influence becomes larger as the proportion of concentrate in total ration is increased.

TABLE 3.6. Influence of Level of Intake on Digestibility of Rations by Dairy Cows

Ratios, hay:grain, %	Level of intake[a]	Digestion coefficients, %				
		Dry matter	Crude protein	Ether extract	Carbo-hydrates	TDN
75:25	1.0	69.3	74.7	75.1	71.2	61.3
	2.8	68.7	72.8	72.3	69.7	59.9
50:50	1.0	73.7	75.0	79.9	76.1	64.9
	3.3	70.2	71.7	77.9	71.1	61.5
25:75	1.0	79.9	78.8	86.8	82.7	70.4
	4.0	70.0	68.3	74.5	72.1	61.2

[a]Times maintenance TDN.
Source: D. G. Wagner and J. K. Loosli, Studies on the energy requirements of high producing dairy cows, *Cornell University Agr. Expt. Sta. Memoir* **400**:1–40, 1967.

The data in Table 3.6 show that when dairy cows were fed a ration consisting of 75 percent of alfalfa hay and 25 percent of a concentrate mixture of equal crude protein content there was relatively little change in the digestion coefficients when the intake level was increased from 1.0 to 2.8 times the maintenance level. The depression became greater when the ration contained equal proportions of hay and concentrates. In these cows as the level of intake reached 4.0 times maintenance with concentrates making up 75 percent of the ration, the digestion coefficients declined at least 10.0 percentage units and the total digestible nutrient (TDN) value (Sec. 3.16) of the mixed ration fell from 70.4 to 61.2, a decrease of 12 percent. Greater depressions have been reported at 5.0 to 6.0 times the maintenance intake.

A summary of the effect of level of intake on digestibility in several species is shown in Table 3.7.

3.9 Influence of Feed Preparation on Digestibility

Grinding grain usually does not increase digestibility in those animals which masticate their feed thoroughly, but seeds which escape mastication may remain largely undigested in passing through the tract. This is because their unbroken seed coats resist the action of the digestive enzymes, rather than because of particle size. Sheep masticate their feed so effectively that there is no advantage in grinding grain for them, except in the case of very small and hard seeds. Cattle chew their grain less thoroughly and thus digest it somewhat better when it is ground. As much as 20 percent of the hard corn kernels eaten may pass through into the feces. Grinding helps for very young animals before their teeth are developed and for old animals that have poor teeth. Digestibility in growing swine is only slightly increased by grinding, but the effect is more marked in older animals.

Differing from the case with grains, roughage is chewed by all animals sufficiently to break it up so that the digestive juices can penetrate it. For a

TABLE 3.7. Summary of Effect of Increasing Levels of Intake on Digestibility of Certain Diets by Various Animals

Animal	Kind of diet	Reduction in digestibility?
Chick, growing	Mixed concentrates	No
Hen, laying	Mixed concentrates	No
Rat	Semi-purified	No
Pig	Mixed concentrates	No
Sow	Mixed concentrates	No
Calf, young	Whole milk	No
Rabbit	Hay and concentrates	Yes
Horse	Hay	No
Horse	Hay and concentrates	Yes
Sheep	Forage, long or chopped	No
Sheep	Forage, pelleted-ground	Yes, usually
Sheep	Forage and concentrates	Probably yes
Cattle	Forage, long or chopped	No
Cattle	Forage, pelleted-ground	Yes, usually
Cattle	Forage and concentrates	Yes
Cattle	Corn silage	Yes

Source: J. T. Reid and H. F. Tyrrell, Effect of level of intake on energetic efficiency of animals, *Proc. Cornell Nutrition Conf.*, 1964, pp. 25–38.

given intake, there is no advantage in grinding or chopping hays which are of sufficiently good quality and palatability to be completely consumed without it. Forages harvested at the same stage of maturity and stored in the same way are digested to the same extent by ruminants, whether long or chopped. Several reports have shown that the fine grinding of hay lowers its digestibility, because the ground hay passes through the gut more rapidly. The influence of grinding on voluntary intake and digestibility will depend on the extent to which both the retention time and the rate of breakdown of the food in the gut are altered by the grinding. Pelleting of hay may decrease fiber digestion also because of increased rate of passage. Cooking feeds does not help digestibility in mature farm animals except in the case of a few feeds for swine and poultry, such as soybeans, navy beans, and potatoes. The newborn calf develops the ability to digest uncooked starch rather rapidly. A coefficient of digestibility of 90 percent at four weeks of age has been reported. None of the various processes of fermenting, ''predigesting,'' and malting which have been exploited as methods for getting more nutrient value from roughages and other fibrous feeds have been found to have any advantage when subjected to critical tests.

A special case in which feed preparation influences animal metabolism and performance is represented by the changes in the concentrations of acetic and propionic acid produced in the rumen and the consequent lowering of milk-fat percentage, which results from the feeding of ground roughage and certain concentrates (Sec. 17.14).

3.10   The Significance of Digestion-trial Data

The previous discussion has indicated that a variety of factors influence the nature of the results obtained in a digestion trial. This fact must be borne in mind in interpreting the data and applying them to practice. The most significant data for practical application are obtained where the ration is fed at the level required for satisfactory production. This means both an adequate total intake and also an adequate protein content. A sufficient supply of other nutrients is also desirable because a deficiency of some of them may affect digestion processes even though there is no evident effect on production over the short period of the trial. These various considerations can be met in the case of mixed rations, but not in the case of many individual feeds. The alternative in the latter case is to employ the indirect method, which, however, brings in other possible errors as has been discussed.

Digestion data measure the disappearance of the nutrient in passing through the tract in absorption. In the case of ruminants particularly, coefficients for the higher carbohydrates are always too high as a measure of absorbed nutrients because of the gaseous losses. There are some gaseous losses from nitrogen-free extract also. A crude-fiber coefficient is subject to further question because a part of the undigested residues of this feed component may be sufficiently broken down to appear in the nitrogen-free extract of the feces instead of in the crude-fiber portion. In the case of the usual rations for herbivora, fat-digestion coefficients are subject to rather large errors owing to various causes attributable principally to the ether-extract method. These errors are not of large significance, however, in terms of the digestibility of the ration as a whole because the ether-extract fraction represents such a small part of the total ration.

Despite these various limitations, digestion coefficients remain distinctly useful. A consideration of the limitations serves to emphasize the importance of proper planning of digestion trials and of the matters that should be borne in mind in interpreting the results. Much helpful information along these lines is to be found in the publication by Mitchell.[13] This publication also contains a discussion of the relation between chemical composition and digestibility and considers the usefulness of formulas for calculating digestibility from crude-fiber and protein content. Other studies dealing with this question are cited in the footnotes.

3.11   The Determination of Digestibility

A digestion trial involves a record of the nutrients consumed and of the amounts of them voided in the feces. It is essential that the feces collected represent quantitatively the undigested residue of the measured amount of food consumed. Various methods are employed for this fecal collection. In the case of omnivora and carnivora some indigestible, easily distinguishable substance called a *marker* may be used. The marker is fed just before the beginning of the ingestion of the ration to be tested and again at its close. The feces collection is begun when the first marker appears and is ended with the appearance of the second. A satisfactory marker must be inert physiologically and contain no element under investigation. The less the substance diffuses the better. Carmine, dysprosium, radio cerium, ferric oxide, chromic oxide, barium sulfate, and soot have been employed. No marker can be considered to provide unquestionable accuracy.

In the case of herbivorous animals, with their much larger and more complicated digestive tracts, the use of a marker is a less suitable method. For these species and commonly for other farm animals also, the ration to be tested is fed in constant daily amounts for an extended period. After allowing a certain number of days to elapse as a *preliminary* period to free the digestive tract of any indigestible material coming from the feed consumed prior to the start of the constant intakes of the ration under study, the collection of the feces is begun and continued through the *collection* period. The length of the periods required to obtain reliable results depends upon the species, longer periods being necessary in the case of herbivores, especially ruminants, than of other species. In simple-gutted animals such as the pig, digestion and evacuation are usually complete in 24 hours after the test food is ingested. In ruminants eating a ration largely of roughage, the last residues may not be voided until 150 to 200 hours have elapsed, though 95 percent are usually voided within 140 hours. For pigs and horses preliminary and collection periods of 4 to 6 days each are commonly used. For ruminants these periods must be extended to 8 or 10 days. In general, the longer the period of collection, provided the stated amount of food continues to be consumed regularly and completely, the more accurate the results, since the effect of periodic fluctuations is minimized.

3.12 Methods of Collecting Feces

A digestion trial requires the quantitative collection of the feces uncontaminated by urine. While in the old days this was commonly done manually in the case of farm animals, the modern procedures utilize various types of metabolism cages representing modifications of those earlier designed for laboratory animals. An essential feature of these cages is that the animal must have freedom of movement, particularly regarding lying down and getting up. In one type the bottom is a metal grid through which both the feces and urine pass, the feces being caught on a screen underneath. In the types now more commonly used, the animal is confined so that he cannot turn around, and the length of the cage is adjusted to the size of the animal in such a way that the feces fall into a properly placed container. The feed box is attached to the front, so constructed and placed to prevent scattering.

The construction of such a unit, designed for two steers, is shown in Fig. 3.2 and described in detail by Horn, Ray, and Neumann.[14] The collection boxes are shown detached and a feed box is illustrated separately to give more of the details of construction. The feed boxes are fastened on runners so that they can be moved backward or forward to regulate the length of the unit in accordance

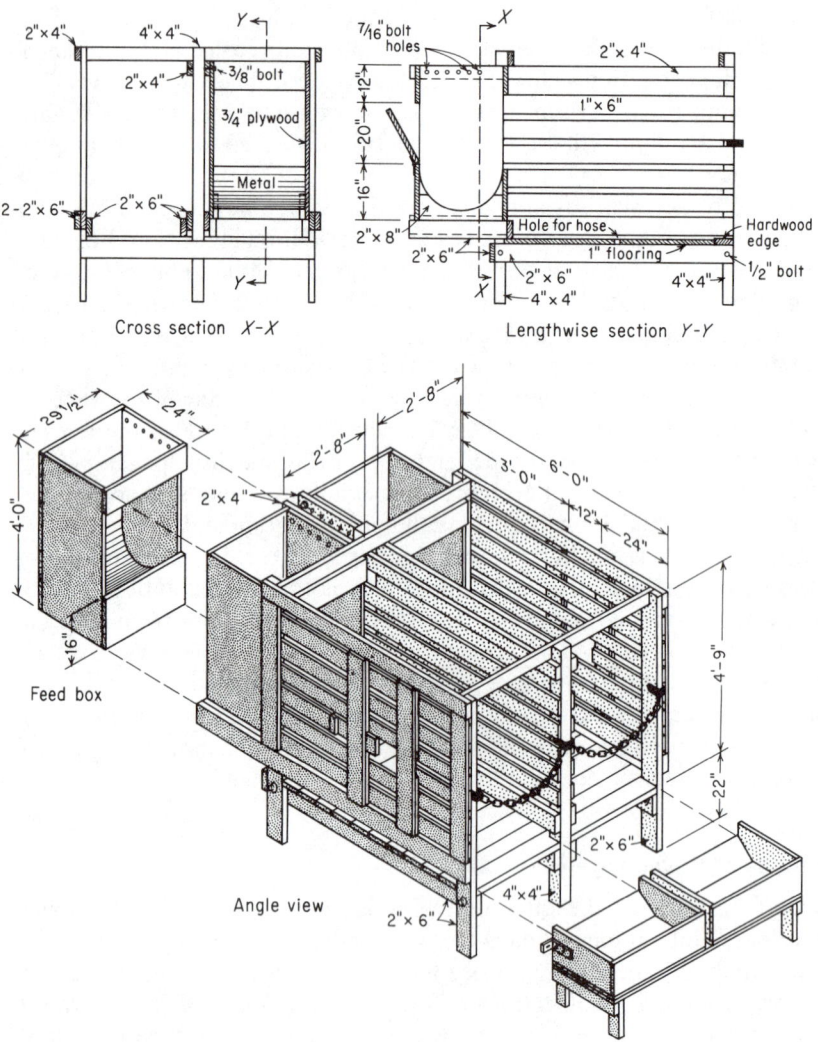

Cross section X-X

Lengthwise section Y-Y

Feed box

Angle view

Fig. 3.2 Digestion and metabolism stall. (*Courtesy of A. L. Neumann, University of Illinois.*)

with the size of the animal. A urinal consisting of a rubber funnel is attached to the belly of the steer by a harness and a tube from it leads down through the floor. This permits the quantitative collection of urine also, as is called for in balance studies.

Obviously the cage described above is suitable only for male animals. Where digestion and metabolism studies are to be carried out with females, as is the case, for example, with dairy cows, a specially designed urine conduit is attached to the animal. While the difficulties here involved are obvious, experience has shown that quantitative feces and urine separations and collections can be obtained by properly designed and attached conduits. The construction and operation of such a conduit is described in the article by Hobbs, Hansard, and Barrick,[15] which also gives the details on a metabolism unit used for heifers. At some experiment stations special buildings are constructed with grates or holes in the floor for the feces and urine to pass through to be collected in receptacles in the lower floor. Schneider and Flatt[16] give details for such buildings.

Feces collections have also been made with steers and wethers by using a collection bag attached to the animal. This method has advantages for certain purposes, notably for the collection of feces from animals on pasture. A bag and harness used by Clanton and Hemstrom of the Nebraska Agriculture Experiment Station is shown in Fig. 3.3. Equipment which has been successfully used by Reid of Australia for collecting both feces and urine on pasture to provide the data needed in a balance experiment (Chap. 9) is shown in Fig. 3.4.

The determination of digestibility in poultry requires a special technique, since the feces and urine are voided together, causing a mixing of the urinary and fecal nitrogen. The two forms of nitrogen can be separated by determining the ammonia and uric acid which represent the urinary output. The determination of digestibility is also carried out by the use of an operative technique which involves the formation of an artificial anus.

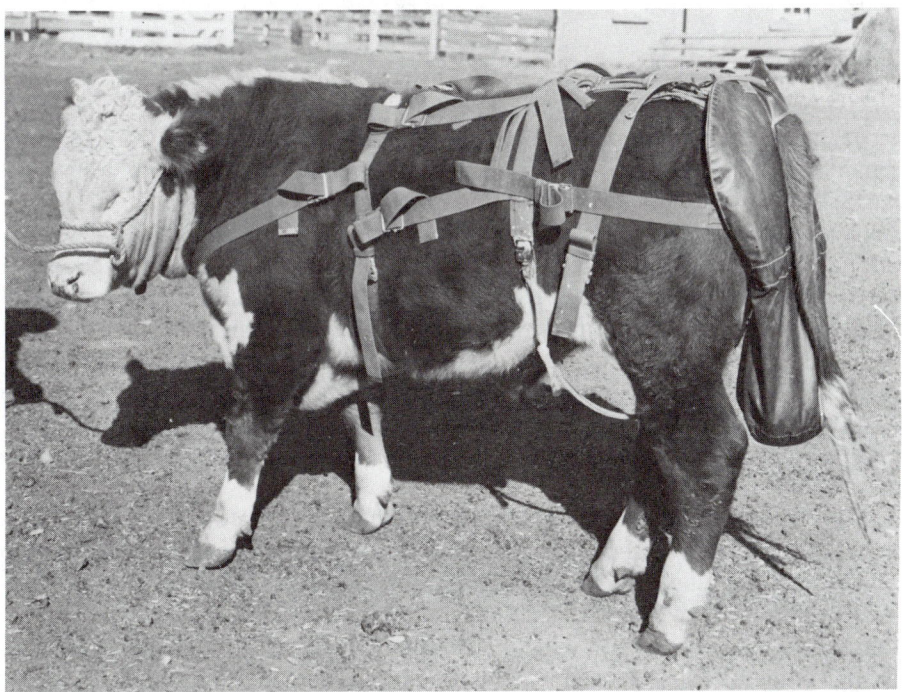

Fig. 3.3 Harness and bag for collecting feces in a digestion trial. (*Courtesy of D. C. Clanton, University of Nebraska.*)

Fig. 3.4 Equipment for collecting excreta of grazing wethers. The roller has the capacity for a 24-hour urine output. (*Courtesy of R. L. Reid, Sheep Biology Laboratory, C.S.I.R.O., Prospect, N.S.W., Australia.*)

3.13 A Digestion Trial

An example of the data obtained in a digestion trial is presented in Table 3.8. In obtaining the data for intake of nutrients, the feed intake was multiplied by figures for its percentage composition as determined by chemical analysis. Similarly, the data for excreted nutrients were calculated, and the digested nutrients obtained by subtraction. The final figures, expressed as percentages, are called *digestion coefficients*. In such a trial, several animals must be used and the results averaged to minimize the factor of individual variability. It has been mentioned (Chap. 8) that the coefficient for protein, determined as indicated in Table 3.8, represents *apparent* digestibility because the feces contain metabolic as well as undigested nitrogen. Since the digested portion is not determined directly in the case of any of the nutrients, the term *coefficient of apparent digestibility* is frequently used.

It is noted that the mineral nutrients are not considered in Table 3.8. The discussion in Chap. 10 shows that some of the absorbed minerals are excreted through the gut, and that the reexcreted portion may be a large as well as a variable part of the whole. Thus data for "digestible ash," which are frequently reported in connection with digestion trials, have no real significance. By the use of isotopes, however, as discussed for calcium and phosphorus, the actually digested or "available" portion can be arrived at.

The digestibility of individual feeds, when fed alone, may be determined in

TABLE 3.8. Digestibility of Dried Grass by a Dairy Cow

	Crude protein	Carbohydrates		Ether extract
		Fiber	NFE	
Intake of 44,684 g dry matter, containing, g	10,216	8,255	20,823	1,697
Output of 11,609 g fecal dry matter, containing, g	2,559	2,158	4,042	783
Digested nutrients, g	7,657	6,097	16,781	914
Digested nutrients, %	75	73.9	80.6	53.9

Source: J. A. Newlander and C. H. Jones, The digestibility of artificially dried grass, *Vermont Agr. Expt. Sta. Bull.* 348, 1932.

so far as they provide a satisfactory ration for the period of the test. The digestibility of concentrates by ruminants cannot be determined in this way because they do not furnish sufficient bulk; their coefficients can be obtained only by difference. In this procedure the digestibility of a roughage as a basal ration is first determined, and then the concentrate is added to the roughage for a second test. By a consideration of differences between the figures obtained for the roughage alone and for the combination, coefficients for the digestibility of the concentrate are calculated. Such figures represent the net effect of the addition of the concentrate to the roughage, but they may not be exact for the concentrate because its addition may have influenced the digestibility of the basal ration. The frequent occurrence of impossible coefficients in data thus obtained testifies to this fact.

3.14 The Indicator Method of Determining Digestibility

The conduct of a digestion trial as previously outlined is obviously a laborious and time-consuming procedure. For many years investigators have sought an indirect method of assessing digestibility. Various workers have proposed formulas for calculating digestible organic matter from crude-fiber or fiber and protein content, but these formulas have been found to have only limited usefulness for the purpose. An indirect method of comparatively recent development, the accuracy and usefulness of which have become definitely established, involves the use of an "inert reference substance" as an indicator. The ideal specifications for such a substance are that it should be totally indigestible and unabsorbable, have no pharmacological action on the digestive tract, pass through the tract at a uniform rate, be readily determined chemically, and preferably be a natural constituent of the feed under test. By determining the ratio of the concentration of the reference substance to that of a given nutrient in the feed and the same ratio in the feces resulting from the feed, the digestibility of the nutrient can be obtained without measuring either the food intake or feces output. The calculation is made as follows:

$$\text{Digestibility} = 100 - \left(100 \frac{\% \text{ indicator in feed}}{\% \text{ indicator in feces}} \times \frac{\% \text{ nutrient in feces}}{\% \text{ nutrient in feed}}\right)$$

In 1918 Edin, a Swedish scientist, first proposed chromic sesquioxide, Cr_2O_3, as an indigestible substance suitable for use as an indicator, but it did not receive adequate testing until some twenty-five years later. Following the recognition of the indigestibility of lignin and the establishment of methods for its quantitative determination, the use of this compound was proposed as one having the advantage of being a natural feed constituent. Indicators have been extensively studied in experiments with ruminants in which the results have been compared with those obtained by quantitatively measuring feed intake and feces voided. In general there has been close correspondence between the results with the indicator method and the conventional method, with the conclusion that either chromic oxide or lignin gives useful results, with little advantage for one over the others. An extensive trial of indicator methods compared with the conventional technique has been reported by Kane and coworkers[17] for dairy cattle in which the advantages and limitations of each are discussed. Use of acid insoluble ash[18] and magnesium ferrite[19] have also been suggested as indicators.

3.15 Indicators in Pasture and Range Studies

The problem of measuring how much feed an animal obtains from pasture or range and of determining the digestibility of this feed is both an important and a difficult one. For estimating consumption on pasture much use has been made of the system devised by Garrigus and Rusk,[20] involving a calculation of the dry matter consumed in a given period from the feces collected during the same period, using a bag as shown in Fig. 3.4. The calculation is based on a previous determination of the digestibility of the dry matter of the feed cut from the pasture in question. The indicator method can be used to shorten this procedure, the calculation being made as follows:

Dry-matter consumption (g per day)
$$= \frac{\text{units indicator per g dry feces} \times \text{g dry matter in feces per day}}{\text{units indicator per g dry matter of forage}}$$

Digestion trials with pasture and range forage have commonly been carried out by using clippings or manually plucked samples from the area in question. Here, as with dry roughage, the indicator method has been shown to be a useful substitute for the standard procedure. A limitation of the use of clippings or pluckings in estimating the composition and digestibility of the feed obtained from pasture or range lies in the fact that the animal grazes selectively in a way that cannot be duplicated in samplings.

It is possible to determine both the feed intake from pasture and its digestibility by the simultaneous use of two indicators, namely, an external one such as Cr_2O_3 to measure dry-matter intake and a naturally occurring one such as a lignin to measure digestibility. The procedures are the same as previously described (Sec. 3.14). The reports of Reid and associates[21] and Smith and Reid[22] give further details of the methods. Animals with esophageal fistulas have been

TABLE 3.9. Digestion Coefficients

Nutrient	Total nutrients in 100 kg, kg	Digestion coefficients, %	Digestible nutrients, kg
Crude protein	20.11	75.0	15.08
Crude fiber	16.25	73.9	12.01
Nitrogen-free extract	40.99	80.6	33.03
Ether extract	3.34	53.9 (X 2.25)	4.04
Total digestible nutrients			64.16

quite useful in range studies.[23] The samples obtained from the fistula can be submitted to an in vitro digestibility determination (Sec. 3.18).

3.16 Total Digestible Nutrients

As a general measure of the nutritive value of a feed, digestion coefficients are used to compute its content of total digestible nutrients (TDN). The dried grass used in the digestion trial presented in Table 3.8 had the following composition: crude protein, 20.11; crude fiber, 16.25; nitrogen-free extract, 40.99; ether extract, 3.34. The digestible nutrients are obtained from these data by multiplying them by the digestion coefficients given in Table 3.9. The digestible fat is multiplied by the factor 2.25 in arriving at the figure 4.04 because of its higher energy value. The calculation of TDN here shown does not take account of digestion losses only, for reasons explained in Sec. 9.6, where the usefulness of TDN values in evaluating feeds and in formulating ratios is also discussed.

Digestion trials carried out as previously described have provided data for the calculation of *average digestion coefficients* for the protein, fat, and carbohydrates in the various feeds and thus, in turn, for the calculation of their content of TDN. Such data have been compiled by Morrison[24] and by Schneider.[25] This latter compilation gives separate tables of digestion coefficients for cattle, sheep and goats, and swine. These and later digestion trials form the basis of the current National Research Council energy values.[26]

3.17 Nutritive Ratio

This is the ratio of the digestible protein, expressed as unity, to the sum of digestible carbohydrates and fat, the latter being multiplied by 2.25. The second factor of the ratio is calculated as follows:

$$\frac{(\text{Digestible fat} \times 2.25) + \text{digestible NFE} + \text{digestible fiber}}{\text{Digestible protein}}$$

Based upon the data previously given for dried grass this calculation results in the figure 3.2, and the ratio is therefore 1 : 3.2. Such a ratio is called a *narrow* one because of the relatively large amount of protein in relation to the other nutrients; where the reverse is true we have a *wide* ratio.

3.18 Indirect and In Vitro Methods of Estimating Digestibility

Digestion trials are time consuming and costly and they require large feed samples. Therefore much effort has been exerted to develop methods to estimate digestibility indirectly or by in vitro methods. Cell wall, lignin, holocellulose, silica, and protein contents have been used individually or in combination to indicate digestible dry matter.[27] In vitro determinations based on the procedure developed by Tilley and Terry[28] have been widely and successfully used.

NOTES

1. E. R. Ørskov, Effect of processing on digestion and utilization of cereals by ruminants, *Proc. Nutrition Soc.*, **35**:245–252, 1976.

2. R. G. Warner and W. P. Flatt, Anatomical development of the ruminant stomach, in R. W. Dougherty (ed.), Physiology of Digestion in the Ruminant, Butterworth and Co. Ltd., London, 1965, pp. 24–38.

3. M. J. Ulyatt, D. W. Dellow, C. S. W. Reid, and T. Bauchop, Structure and function of the large intestine of ruminants, Proc. Fourth International Symposium on Ruminant Physiology, Univ. New England, Armidale, Australia.

4. A. J. Sheppard, R. M. Forbes, and B. Conner Johnson, Rate of acetic acid production in the rumen determined by isotope dilution, *Proc. Soc. Exptl. Biol. Med.* **101**:715–717, 1959. In 1884 Tappeiner in Germany showed that large quantities of volatile fatty acids, notably acetic, were produced from the in vitro fermentation of cellulose by bacteria from the rumen of the ox. Quantitative knowledge of this rumen activity and of its role in nutrition has developed particularly through the use of the permanent rumen fistula technique described in 1886 by the French physiologist Colin. The modern developments, which have dealt both with the production of acids in the rumen and their fate in metabolism, were stimulated by the English physiologist Barcroft and grew out of a large cooperative research project inaugurated in the Unit of Animal Physiology at Cambridge University during World War II. The pioneer work has been reviewed by Elsden and Phillipson in *Ann. Rev. Biochem.*, **17**:705–726, 1948.

5. R. A. Leng and D. J. Brett, Simultaneous measurements of the rates of production of acetic, propionic and butyric acids in the rumen of sheep on different diets and the correlation between production rates and concentrations of these acids in the rumen, *Brit. J. Nutrition*, **20**:541–552, 1966.

6. R. M. Murray, A. M. Bryant, and R. A. Leng, Rates of production of methane in the rumen and large intestines of sheep, *Brit. J. Nutrition*, **36**:1–8, 1976.

7. V. J. Williams and coworkers, Ruminal flora studies in the sheep, Part IV: The influence of varying dietary levels of protein and starch upon digestibility, nitrogen retention, and free microorganisms of the rumen, *Australian J. Biol. Sci.*, **6**:142–151, 1953.

8. P. J. Van Soest and R. W. McQueen, The chemistry and estimation of fiber, *Proc. Nutrition Soc.,* **32:**123–129, 1973.

9. M. P. Bryant, Microbiology of the rumen, in Dukes Physiology of Domestic Animals, Cornell Univ. Press, Ithaca, 1977.

10. M. J. Wolin, Interactions between the bacterial species of the rumen, *Proc. Fourth Internal. Symposium on Ruminant Physiology,* Univ. New England, Armidale, Australia, 1975.

11. G. Matrone, H. A. Ramsey, and G. H. Wise, Effect of volatile fatty acids, sodium and potassium bicarbonate in purified diets for ruminants, *Proc. Soc. Exptl. Biol. Med.,* **100:**8–11, 1959.

12. R. W. Reilly and J. B. Kersner, Fiber deficiency and caloric disorders, *Am. J. Clinical Nutrition,* **28:**293–296, 1975.

13. H. H. Mitchell, The evaluation of feeds on the basis of digestible and metabolizable nutrients, *J. Animal Sci.,* **1:**159–173, 1942.

14. L. H. Horn, Jr., M. L. Ray, and A. L. Neumann, Digestion and nutrient-balance stalls for steers, *J. Animal Sci.,* **13:**20–24, 1954.

15. C. S. Hobbs, Sam L. Hansard, and E. R. Barrick, Simplified methods and equipment used in separation of urine from feces eliminated by heifers and by steers, *J. Animal Sci.,* **9:**565–570, 1950.

16. B. H. Schneider and W. P. Flatt, The Evaluation of Feeds Through Digestibility Experiments, Univ. of Georgia Press, Athens, Ga., 1975.

17. E. A. Kane and coworkers, A comparison of various digestion trial techniques with dairy cattle, *J. Dairy Sci.,* **36:**325–333, 1953.

18. J. F. McCarthy, F. X. Aherne, and D. B. Okai, Use of HCl insoluble ash as an index material for determining apparent digestibility with pigs, *J. Animal Sci.,* **54:**107–111, 1974.

19. H. Neumark, et al., Assay and use of magnesium ferrite as a reference in absorption trials with cattle, *J. Dairy Sci.,* **58:**1476–1481, 1974.

20. W. P. Garrigus and H. P. Rusk, Some effects of the species and stage of maturity of plants on the forage consumption of grazing steers of various weights, *Illinois Agr. Expt. Sta. Bull.* 454, 1939.

21. J. T. Reid and associates, A procedure for measuring the digestibility of pasture forage under grazing conditions, *J. Nutrition,* **46:**255–269, 1952.

22. A. M. Smith and J. T. Reid, Use of chromic oxide as an indicator of fecal output for the purpose of determining the intake of pasture herbage by grazing cows, *J. Dairy Sci.,* **37:**515–524, 1955.

23. G. M. Van Dyne and D. T. Torell, Development and use of the esophageal fistula: A review, *J. Range Management,* **17:**7–10, 1964.

24. F. B. Morrison, Feeds and Feeding, 22d ed., Morrison Publishing Co., Ithaca, N.Y., 1956, Appendix Table 1.

25. B. H. Schneider, Feeds of the world—their digestibility and composition, *West Virginia Agr. Exp. Sta.,* 1947.

26. National Research Council, Atlas of nutritional data on United States and Cana-
     dian feeds, National Academy of Sciences, Washington, D.C., 1971.

27. P. J. Van Soest, The chemical basis for the nutritive evaluation of forages, *Proc.
     Conf. on Forage Quality,* Univ. Nebraska, Lincoln, 1970.

28. J. M. A. Tilley and R. A. Terry, A two-stage technique for the in vitro digestion of
     forages, *J. British Grassland Soc.,* **18:**104–109, 1963.

Chapter 4

The Role and Requirement of Water

Water comprises about 70 percent of the lean adult animal body, and many tissues contain 70 to 90 percent. In fact, one may consider the living elements of the body as water inhabitants even as are true aquatic species. This water is not merely an inert material or solvent but is an active and structural constituent. If a frog's egg weighing a few milligrams is placed in sterile filtered water, a tadpole weighing several grams results. The tadpole contains less dry matter than the original egg, for a part of it has been used to furnish energy for the developmental process. The increase in weight is due to the taking up of water which has become an essential part of the organism. The vital role of water in the body is indicated by Rubner's[1] observation that the body can lose practically all of its fat and over half of its protein and yet live, while a loss of one-tenth of its water results in death. Adolph[2] in an extensive review has pointed out that water ranks far above every other substance in the body as regards rate of turnover. Since it is the nutrient required in largest amounts by man and many animals its consideration provides an appropriate introduction to nutritional physiology.

4.1 Properties and Functions of Water

Water is the ideal dispersing medium because of its solvent and ionizing powers which facilitate cell reactions and because of its high specific heat which enables it to absorb the heat of these reactions with a minimum rise in temperature. Cannon[3] has pointed out that, "The heat produced in maximal muscular effort continued for 20 minutes would be so great that if it were not promptly dissipated it would cause albuminous substances of the body to become stiff like a hard boiled egg." The latent heat of vaporization of water also plays an important role in regulating body temperature. Other properties of large significance in physiology are the high surface tension, the tendency to form hydrates, and the high dielectric constant of water. The functions of water in digestion, in the transport of metabolic products, and in excretion are obvious. Its functions are far more diverse and basic, however, than represented by its roles as a solvent and as a substrate for body reactions. It actually takes part in these reactions, as illustrated by the hydrolysis of proteins, fats, and carbohydrates which takes place in digestion and inside the body and by the many anabolic or catabolic changes in intermediary metabolism which require the chemical addition or release of water. Many illustrations of these functions of water are given in following chapters which discuss the metabolism of the other nutrients.

Water plays many special roles. As synovial fluid, it lubricates the joints, and as cerebrospinal fluid, it acts as a water cushion for the nervous system. In the ear, it transports sounds, and in the eye, it is concerned with sight. An excellent detailed discussion of the functions of water in the body, electrolyte-water interrelationships, water equilibrium in health and disease, and related topics is to be found in a series of articles reporting a symposium held by the Nutrition Society of Great Britain.[4] In view of the variety of its functions and the magnitude of its requirements, water can be considered the most important essential nutrient.

4.2 The Determination of Water

The water present in a biological material is commonly determined by drying it to constant weight at the temperature of boiling water; this is a satisfactory procedure for most routine analysis. However, not all the water, at least in certain materials, is removed by this procedure. The unremoved portion represents water existing in films which has a very low vapor pressure even at 100°C. Thus a refined method involves drying in partial vacuum. Strictly speaking, no figures for moisture content can be considered as absolute values, since their magnitude is influenced by the three variables involved in the determination—temperature, pressure, and time. When the dried material is to be used for subsequent determinations with which previous oxidation might interfere, the drying is carried out in an inert gas such as nitrogen. Certain materials must be dried below 100°C to avoid alterations in some of their constituents. Because acetic acid and other volatile materials are lost when silage and certain other fermented feeds are dried by the usual methods, direct determination of water is carried out by distillation over toluene. Other special procedures are required for certain products. Various methods for the determination of water were assessed in a report of a symposium chaired by Stillman.[5] Methods of measuring water content of the animal body were reviewed in Sec. 2.5.

4.3 Metabolic or Oxidation Water

Most of the water which is utilized by the animal body is ingested, either as such or as a component of the food. There is a further available source which is provided by metabolic processes and which is thus called *metabolic* or *oxidation water*. When the carbohydrate glucose is oxidized to furnish energy for body processes, carbon dioxide and water result:

$$C_6H_{12}O_6 + 6O_2 \longrightarrow 6CO_2 + 6H_2O$$

By calculations from this equation, it can be shown that the metabolism of glucose yields 60 percent of its weight as water and protein produces approximately 40 percent of its weight as water, while in the case of fat the figure is over 100 percent. Metabolic water is also produced by the dehydration synthesis of body proteins, fats, and carbohydrates. Under certain physiological conditions, metabolic water plays an important role in the animal economy. It suffices to meet the needs of hibernating animals. These animals metabolize their reserves of carbohydrate and fat to provide energy for their vital processes, and this metabolism produces enough water to balance that lost by respiration and evaporation. The situation is quite different in the very dry atmosphere of desert regions. The increased ventilation of the lungs by inhalation of relatively dry air to supply the oxygen for fat oxidation causes a greater water loss in the saturated expired air than the metabolic water produced. The relationship between metabolic water produced and oxygen needed is shown by the following data:[6]

	Oxygen per gram of		Metabolic Water per	
	Food, liter	Water formed, liter	Gram of food, g	kcal, g
Starch	0.83	1.49	0.56	0.14
Protein	0.97	2.44	0.40	0.10
Fat	2.02	1.88	1.07	0.12

Starch requires less oxygen per gram of metabolic water formed than fat and much less than protein. Schmidt-Nielsen estimated that the increased respiration needed to oxidize 1.06 kg of body fat by a camel in a dry atmosphere would result in an increased loss of water by evaporation from the lungs of 1.8 kg of water while only 1.1 kg were produced from the fat oxidation. Thus there is no net water gain under these conditions. Metabolic water comprises only 5 to 10 percent of total water intake of domestic animals and varies only with the metabolic rate.

4.4 Effects of Water Deprivation

Animals differ markedly in their ability to conserve water and to withstand water deprivation. When man is deprived of drinking water in a hot, dry environment, he soon exhibits thirst. When the water deficit approaches 4 to 5 percent of body weight, there is discomfort and anorexia. As the water loss increases from 6 to 10 percent there is headache, his movements lack coordination, speech becomes indistinct, and dyspnea and cyanosis is noted. At a deficit of 12 to 14 percent the eyes become sunken, the skin becomes shriveled, there is inability to swallow, and delirium occurs. As water is lost from the blood, its viscosity and the resistance to flow increases. When increasing heart rate is no longer able to circulate the blood fast enough to dissipate the heat from the deeper body parts, a fatal increase in body temperature occurs. In a hot environment dehydration of about 12 percent is fatal to man. Some desert animals such as the camel and gerbil can tolerate a more severe water dehydration than man or the dog without suffering an explosive heat rise. Animals deprived of water refuse to consume food in early stages of dehydration and will not eat dry feed until after they have had a drink of water.

4.5 Water Absorption and Excretion

It is convenient to divide body water into fluid compartments as an aid in understanding its absorption and role in metabolism. Water contained within the cells is considered *intracellular* fluid, that outside of cells *extracellular* fluid, which consists of blood *plasma,* that within the walls of the vascular system, and *interstitial* fluid. Water in the erythrocytes is intracellular, of course. The intracellular fluid accounts for about 50 percent of body weight, interstitial fluid about 15 percent, and blood plasma 5 percent.

Water is being lost from the body constantly in the respired air and evaporation from the skin and periodically through the feces and urine. The water is replaced to satisfy thirst and through foods containing a high percentage of water. Water molecules easily move through cell membranes to maintain osmotic and hydrostatic equilibrium in relation to transfer of mineral elements, nutrients, and waste products. Water absorbed from the intestinal tract enters the extracellular fluid in the blood, the volume of which is largely regulated by the body sodium, the major cation in this fluid compartment. The volume and osmoconcentration of the extracellular fluid are also regulated by thirst—ADH and salt appetite—aldosterone systems working together. Variations in water intake and excretion largely control osmoconcentration. Water gradually moves from the extracellular fluid compartments into the intracellular fluid to maintain osmoequilibrium.

Water losses are related to body size, and in rats a voluntary intake of 800 ml daily per square meter of body surface has been noted. But the losses are conditioned by body processes and are thus highly variable according to the diet, nature of the metabolic end products, and other factors. The losses through the gut vary with the nature of the diet. They increase with the level of roughage intake and with the intakes of other feeds which have laxative qualities. In general, the larger the proportion of undigested material, the greater the loss. In cattle the fecal material contains about 80 percent of water. The feces are much drier in the case of sheep on the same ration, illustrating the fact that

that there are species differences in water loss through the gut. In all species and under all normal conditions, the losses are very small compared with the very large amount of water secreted into the tract in the digestive juices. Actually, the total daily volume of digestive secretions is several times greater than the plasma volume of the body. Normally, almost all the water thus secreted is reabsorbed. In diarrhea large losses occur, resulting in dehydration, and serious consequences if the trouble persists.

The amount of water excreted in the urine is highly variable, depending upon many factors. The kidneys regulate the volume and composition of body fluids, excreting more or less water depending upon intake, outgo through other channels, and the amounts of catabolic products, viz., minerals and nitrogenous end products such as urea, for which water must serve as a solvent. Through its powers of filtration and then of concentration of the filtrate by the reabsorption of water, the kidney can reduce this loss to a minimum. There are large differences in animals in ability to conserve urinary water losses as illustrated by the following comparison showing much higher concentrations of electrolytes in the urine of the camel and kangaroo rat than for humans.

| Species | Urinary concentration | | |
	Urea, mM/liter	Electrolytes, mEq/liter	Osmotic, osm/liter
Human	792	460	1.43
Camel	229	1068	2.8
Kangaroo rat	3840	1200	5.5

Schmidt-Nielsen cited the case of a camel that lost 20 percent of its body weight before any decrease in appetite was noted. After 7 days exposure to direct summer sun in air temperatures of 40°C or more without water and losing 27 percent of its body weight, it was able to drink almost enough water within a few minutes to restore its body weight loss with no apparent ill effects. The additional fact that the camel eats mostly carbohydrates during desert travel further minimizes water used for excretion of waste products. In certain pathological conditions excess water is retained in the tissues, body cavities, and elsewhere, causing *edema*.

There are marked species differences in water excretion according to the nature of the nitrogenous end products. In mammals the principal end product of protein catabolism is urea, which is soluble in water and toxic to the tissues in concentrated solution. Thus much water is required to dilute it to a harmless concentration, remove it from the tissues, and excrete it. Uric acid, the principal nitrogenous end product in birds, is excreted in nearly solid form with minimum loss of water. Further, the breakdown of protein to uric acid provides more metabolic water than does its catabolism to urea. Thus, other conditions being equal, birds have a lower water requirement than mammals and are much less sensitive to the temporary deprivation of it. Mammals will live longer without food than without water, and the consumption of food, especially protein food, without water hastens death as the result of the accumulation of toxic end products. Birds, snakes, and insects survive much longer under these conditions. Clothes moths, which contain 50 percent of water in their bodies, live throughout their cycle on food containing 10 percent or less of this compound. They excrete uric acid, and thus the small amount of water obtained as a component of their food, plus their metabolic water, suffices.

4.6 Water in Body Temperature Control

In all warm-blooded animals the maintenance of a constant temperature is a factor affecting heat production and heat outgo. Since the temperature of the body is normally above its environment, the heat constantly being produced serves in the maintenance of this temperature. The environmental temperature and the amount of heat being produced within the body are the factors that determine the extent to which this heat must be conserved. The amount that is allowed to escape from the body is subject to control, which is referred to as physical regulation. This control is brought about by an adjustment of the blood flow to the skin, by the perspiration mechanism, and by changing the breathing rate to alter evaporation from the lungs. If the conditions call for the dissipation of body heat, and blood flow to the surface is increased as a result of a dilation of the capillaries, which facilitates the escape of heat by radiation, conduction, and convection, and the pores are opened, which allows for a loss of heat through evaporation. These processes are reversed when there is need for the conservation of body heat.

In animals without sweat glands—sheep, swine, and dogs—the relative loss through increased blood flow to the skin and evaporation from the lungs varies with the species. Contrary to the situation with cattle, sheep rely mainly on respiratory evaporative heat loss, according to Brockway, McDonald, and Pullar.[7] Other devices or activities for regulating body heat are changes in the amount of thermal insulation, such as subcutaneous fat, hair, wool and feathers; changes in the exposed body surface by huddling in a cold environment and the reverse in hot weather; and locating a more favorable environment, such as shade or sun, and shelter from or exposure to wind.

The low point of a range is referred to as the "critical temperature." Below it, chemical regulation must come into play. As the environmental temperature continues to drop, there comes a point where the homeothermic mechanisms prove insufficient and the tissues freeze. As the environmental temperature rises above the high point of a range, physical regulation operates without any increase in metabolism until this regulation becomes insufficient to cool the body. At this point a supernormal body temperature ensues as a result of increased metabolism, referred to as the *hyperthermal rise*, which leads eventually to death. Fig. 4.1 illustrates the approximate temperature ranges where extra heat is produced to maintain body temperature and the zone where increasing evaporation of water dissipates excess heat.

Expired air is saturated with water and, even for an animal at rest in a cool

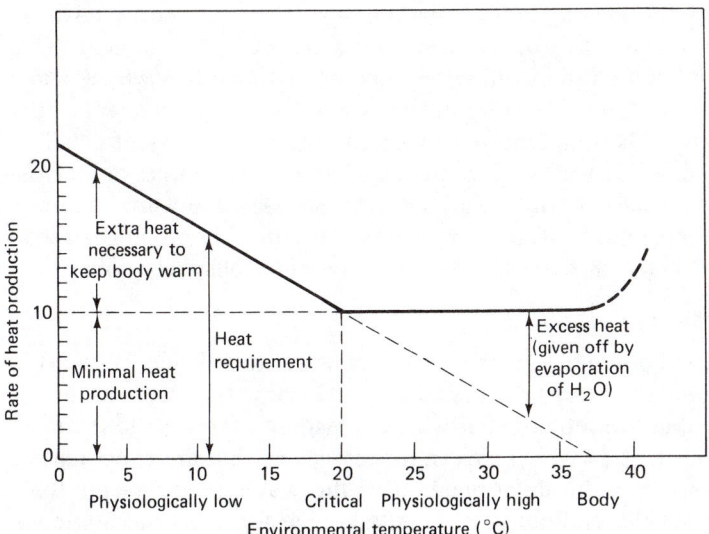

Fig. 4.1 Schematic diagram showing animal heat production versus environmental temperature. Reproduced from Kleiber's, *The Fire of Life,* 1961. (*Courtesy of John Wiley & Sons.*)

environment, results in a substantial loss of water. This loss is greatly increased by physical activity and other factors which step up pulmonary exchange. The magnitude of the water loss in respiration is illustrated by the report of Zervanos[8] that 20-kg peccaries panting from heat stress at 35 to 40°C lost 45 to 49 ml of water per hour. There are also substantial constant losses through the skin via the sweat glands. Perspiration losses represent an evaporation of water for the dissipation of heat in the regulation of body temperature. They increase with muscle activity and temperature. In most animals the sweat glands are few or absent. Under these conditions, the lungs play an important role in the dissipation of heat. The constantly occurring excretion of water vapor through the lungs and skin is largely responsible for the body losses referred to as *insensible perspiration,* which represents the difference between the gaseous intake and outgo. These losses are of considerable magnitude and are related to the environmental temperature and humidity and to the nature of the metabolism, including water intake. When the latter is limited, they may exceed the losses in the urine.

Animals differ greatly in their use of water to prevent elevated body temperatures in warm climates. The small desert rodents have a practice similar to *hibernation,* called *aestivation,* in that they lapse into a dormant state in their deep burrows when the ambient temperature rises too high. There the body temperature is kept below the critical point by heat conduction to the cool ground without evaporation of water by respiration, so they do not become dehydrated. Among the larger animals the camel has exceptional ability to tolerate dehydration to conserve body water. Schmidt-Nielson[9] reported that a camel lost only 1.0 percent of its body weight from water evaporation during exposure to the sun for one day in the Sahara desert in the summer. Under the same conditions a donkey lost 4.5 percent and a man lost 7.0 percent. This difference in loss by perspiration and respiration helps to explain why a camel can survive so much longer without water than a man. An additional fact is that the camel allows its body temperature to vary as much as 6°C (34.2 to 40.7°C) between night and daytime without experiencing daytime heat stress or increasing evaporative water loss or without increasing metabolism to warm the body during the cool desert nights. Water evaporates from the skin rather than wetting the hair as occurs with horses, thus providing more efficient cooling.

4.7 Water Requirements

It is evident that the body must receive sufficient water to balance its losses in addition to the amount required for the formation of new tissues or products, but it is also clear that the requirement will vary widely according to the magnitude of the various factors which govern the losses and that there are marked species differences. Thus the determination of the water requirements for a given species and set of conditions is of limited value for any general recommendations. Leitch and Thomson[10] reviewed factors governing water intake and the requirements of various classes of stock and gave data on water balance for a 500-kg Holstein cow as shown in Table 4.1.

All feeds supply some water, and highly succulant ones, such as silage and green forage, make a substantial contribution to the water requirement. Other conditions being equal, young animals have higher water needs per unit of body size than do mature ones.

The average daily water requirements of various farm animals in a temperate climate is summarized in Table 4.2.

The ambient temperature has a marked effect on water intake. *Bos taurus* cattle weighing 450 kg eating 10 kg dry feed per day drank 28, 41, and 66 liters of water a day when the temperature was 4, 21, and 32°C, respectively. In a cold climate grazing sheep do not need drinking water if they have access to snow

TABLE 4.1. Daily Water Balance of Holstein Cows Eating Alfalfa Hay

	Nonlactating, liters	Lactating, liters
Intake		
Drinking water	26	51
Water in food	1	2
Metoblic water	2	3
Total	29	56
Output		
In feces	12	19
In urine	7	11
Vaporized	10	14
In milk	0	12
Total	29	56

TABLE 4.2. Average Water Requirement of Farm Animals

Animal	Water, liters	Animal	Water, liters
Beef cow, lactating	60	Swine, growing,	
		30 kg	6
Dairy cow, lactating	90	60–100 kg	8
maintenance	60	Lactating sow	
		25 kg	14
Horses		Sheep	
Medium work	40	Lactating ewe	6
Lactating	50	Fattening lamb	4
Poultry, hen	0.5		

for two or three hours each day. Butcher[11] observed that 40-kg sheep confined and fed alfalfa hay drank 1.76 kg of water compared with 1.66 kg of snow consumed by other similar animals. Feed intake was similar, 1.0 kg of alfalfa hay per day, and there was no difference in digestibility of the feed. The camel needs only one-seventh as much water per unit of body weight as man in the summer.

Fortunately, except under pathological conditions, there are no deleterious effects from excessive consumption of water. Thus the requirements can best be taken care of in practice by making sure that the animals have the opportunity to consume all they desire at frequent intervals.

It has been estimated that of the world's water, 97.3 percent is in oceans, 2.05 percent is frozen, and 0.65 percent is fresh. Of the fresh water some 97.6 percent is ground water, half of which is more than one-half mile deep; 1.9 percent flows in streams, rivers, and lakes; 0.28 percent is surface soil moisture; and 0.22 percent is atmospheric vapor moisture.

4.8 Nutrients and Toxic Elements in Water

Although water is sometimes not listed among the dietary essential nutrients for animals it is the most critical nutrient. Animals can survive longer in the absence of any other essential nutrient than without water. Because of its properties as a solvent, water may carry many of the essential elements as well as toxic materials. All types of materials may be suspended in surface water and much of the water available for man and animals is contaminated with sediment, harmful amounts of mineral elements, parasites, and disease-producing microorganisms. A shortage of water for agriculture and industry renders large areas of the world's land nonproductive. Animals, especially ruminants, are able to utilize semidesert areas to produce food for man where water is too limited to produce food crops.

Total dissolved solids (TDS) or salinity is a measure of the usefulness of water for animals or for crop irrigation. One classification follows:

Description	TDS, mg/liter
Slightly saline	1,000–3,000
Moderately saline	3,000–10,000
Very saline	10,000–35,000
Brine	>35,000

All species of animals can tolerate water containing 1,000 to 3,000 mg per liter of total dissolved salts, but there may be mild temporary diarrhea when animals are changed suddenly from salt-free water to this level of salinity. Water containing 3,000 to 5,000 mg per liter is unsatisfactory for poultry, but it is tolerated by other animals. Levels of 5,000 to 7,000 mg per liter of salts may be refused at first. After adaptation they give satisfactory performance of dairy and beef cattle, sheep, swine, and horses. Older ruminants and horses not subject to excess heat or stress may be able to adapt and subsist on water containing 7,000 to 10,000 mg per liter of salts. However, difficulties and death losses may occur in younger or pregnant or lactating animals. Higher levels should always be avoided. Sheep are more tolerant than cattle, and cattle more than pigs. The most abundant salt in saline water is sodium with calcium and magnesium in lower concentrations occurring as carbonates, bicarbonates, chlorides, and sulfates. Sulfates are more harmful than chlorides. Magnesium chloride is more injurious than the calcium or sodium salt. Injury appears to arise because of osmotic effects rather than from any particular ion. If fresh water is also available, animals will not drink water containing injurious levels of salts. Much of the earth's water is too salty to serve as drinking water. The oceans and seas contain more than 30,000 mg per liter of salts and the high salt lakes contain up to ten times that level. The National Academy of Sciences report[12] summarized recommended upper limits of some toxic substances in drinking water. The values are shown below:

Substance	Safe Upper Limit, mg/liter	Substance	Safe Upper Limit, mg/liter
Arsenic	0.2	Mercury	0.01
Cadmium	0.05	Nickel	1.0
Chromium	1.0	Nitrate	100
Cobalt	1.0	Nitrite	10
Copper	0.5	Vanadium	0.1
Fluoride	2.0	Zinc	25.0
Lead	0.1		

A number of other mineral elements are toxic if consumed in sufficiently high levels, but the tolerance level has not been defined.

Table 4.3 shows the mean, maximum, and minimum values for 14 essential mineral elements in surface water samples in the United States. Over the 12-year period more than 80,000 analyses of water samples were made at 1400

TABLE 4.3. Composition of United States Surface Water, 1957–1973

Substance	Mean	Maximum	Minimum
Phosphorus (mg/l)	0.087	5.0	0.001
Calcium (mg/l)	57.1	173.0	11.0
Magnesium (mg/l)	14.3	137.0	8.5
Sodium (mg/l)	55.1	7,500.0	0.2
Potassium (mg/l)	4.3	370.0	0.06
Chloride (mg/l)	478.0	19,000.0	0.0
Sulfate (mg/l)	135.9	3,383.0	0.0
Copper (μg/l)	13.8	280.0	0.8
Iron (μg/l)	43.9	4,600.0	0.10
Manganese (μg/l)	29.4	3,230.0	0.20
Zinc (μg/l)	51.8	1,183.0	1.0
Selenium (μg/l)	0.016	1.0	0.01
Iodine (μg/l)	46.1	336.0	4.0
Cobalt (μg/l)	1.0	5.0	0

Source: National Academy of Sciences, 1974.

different locations. It is clear that some of the samples were too high in sodium, sulfate, and chloride to be acceptable as drinking water for humans or animals. The major water sources, including the Great Lakes and the Mississippi, Missouri, Ohio, Colorado, Columbia, and Kansas rivers contain the following levels (in micrograms per liter): sodium, 1 to 30; copper, 1 to 100; iron, 2 to 300; and manganese, 3 to 800. Some special sources of water could supply a significant proportion of the requirements of calcium, magnesium, potassium, iron, zinc, sodium, and chloride but usual fresh-water sources cannot be relied upon to provide useful amounts of essential mineral elements for animals. Sea water contains (in milligrams per liter): calcium, 410; chloride, 19,440; magnesium, 1,303; sodium, 10,813; and sulfate, 2,713, while water from the Dead Sea contains (in milligrams per liter): calcium, 23,630; chloride, 238,800; magnesium, 56,610; potassium, 6,010; sodium, 19,530; and sulfate, 8,170.

NOTES

1. Max Rubner (1854–1932), a pupil of Voit, served for over forty years at the University of Berlin, first as professor of hygiene and later of physiology. He made many pioneer contributions to the science of nutrition, particularly in the field of energy metabolism, as later discussions show.

2. E. F. Adolph, The metabolism and distribution of water in body and tissues, *Physiol. Rev.*, **13**:336–371, 1933.

3. Walter B. Cannon was professor of physiology at Harvard Medical School. The quotation is from his book, The Wisdom of the Body, W. W. Norton & Company, Inc., New York, 1932.

4. R. A. McCance (symposium chairman), Man's need for water, *Proc. Nutrition Soc.*, **16:**103–134, 1957.

5. J. W. Stillman (symposium chairman), Methods for the determination of water, *Anal. Chem.*, **23:**1058–1080, 1951.

6. K. Schmidt-Nielson, Desert Animals, Oxford Press, New York, 1964.

7. J. M. Brockway, J. D. McDonald, and J. D. Pullar, Evaporative heat loss mechanisms in sheep, *J. Physiol.*, **179:**554–568, 1965.

8. S. M. Zervanos, Seasonal effects of temperature on the respiratory metabolism of the collard peccary (*Tayassu tajacu*), *Comp. Biochem. Physiol. A*, **50:**365–371, 1975.

9. K. Schmidt-Nielson, Comparative physiology of desert mammals, *Missouri Agr. Exp. Sta.*, Special Rept. **21:**1–21, 1962.

10. I. Leitch and J. S. Thomson, The water economy of farm animals, *Nutrition Abstr. & Revs.*, **14:**197–223, 1944.

11. J. E. Butcher, Snow as the only water source for sheep, *J. Animal Sci.*, **25:**590, 1966.

12. R. L. Shirley et al., Nutrients and toxic substances in water for livestock and poultry, National Academy of Sciences, Washington, D.C., 1974.

SUPPLEMENTARY LITERATURE

Asplund, J. M., and W. H. Pfander: Effects of water restriction on nutrient digestibility in sheep receiving fixed water: Feed intakes, *J. Animal Sci.*, **35:**1271–1274, 1972.

Castle, M. E., and T. P. Thomas: The water intake of British Friesian cows on rations containing various forages, *Animal Prod.*, **20:**181–189, 1973.

Clark, R., and J. I. Quin: Studies on the water requirements of farm animals in South Africa: I. The effect of intermittent watering on Merino sheep, *Onderstepoort J. Vet. Sci. Animal Ind.*, **22:**335–343, 1949. II. The relation between water consumption, food consumption and atmospheric temperature as studied on Merino sheep, *Ibid.*, **22:**345–356, 1949.

Fonnesbeck, P. V.: Consumption and excretion of water by horses receiving all hay and hay-grain diets, *J. Animal Sci.*, **27:**1350–1356, 1968.

Harbin, R., F. G. Harbaugh, K. L. Neeley, and N. C. Fine: Effect of natural combinations of ambient temperature and relative humidity on the water intake of lactating and non-lactation cows, *J. Dairy Sci.*, **41:**1621–1627, 1958.

Hoffman, M. P., and H. L. Self: Factors affecting water consumption by feedlot cattle, *J. Animal Sci.*, **35:**871–877, 1972.

Medway, W., and M. R. Care: The effect of excess salt when water intake is restricted, Report N.Y. State Veterinary College, Cornell Univ., 1956–1957, p. 23.

Mitchell, H. H.: The water requirements for maintenance, in Comparative Nutrition of Man and Domestic Animals, Vol. 1, Academic Press, New York, 1962, pp. 192–224.

More water for arid lands, National Academy of Sciences, Washington, D.C., 1974.

Pande, A.: Handbook of moisture determination and control: Principles, techniques and applications, *Vols. 1-4,* Marcel Dekker, New York, 1975.

Robinson, J. R., and R. A. McCance: Water metabolism, *Annual Rev. Physiol.,* **14:**115–142, 1952.

Utley, P. R., N. W. Bradley, and J. A. Boling: Effect of water restriction on nitrogen metabolism in bovines fed two levels of nitrogen, *J. Nutrition,* **100,** 551–556, 1970.

Weeth, H. J., D. S. Sawhney, and A. L. Lesperance: Changes in body fluids, excreta and kidney function of cattle deprived of water, *J. Animal Sci.,* **26:**418–423, 1967.

Winchester, C. F. and M. J. Morris: Water intake rates of cattle, *J. Animal Sci.,* **15:**722–740, 1956.

Chapter 5

Bioenergetics

Since all processes that occur in the animal body when food is ingested and metabolized involve some energy change, an understanding of the principles of bioenergetics is basic to a study of these processes. The word *energy* is derived from the Greek: *en* meaning in and *ergon* meaning work. The term describes a property of matter and reflects the ability of that matter to "do work." The 'work" of the cell is to contract, to actively transport molecules and ions, or to synthesize macromolecules from simpler ones. Its only source of energy to do that work is provided by the chemical energy stored in the food. One of the miracles of life is that cells can effectively extract that energy from food, giving up so little waste to their surroundings.

5.1 Heat Energy

Since, according to the first law of thermodynamics, all forms of energy can be converted quantitatively into heat (all molecules have an inherent "heat content"), the heat energy represented by the constituents of the diet provides a convenient starting point for discussing nutritional energetics. It should be understood, however, that the body is not a heat engine. Life processes operate at constant pressure, and for a heat engine to function at constant pressure, heat must pass from a point of high temperature to a point of lower temperature, an

60

activity which does not exist at the cellular level. The heat produced in life processes is an end product only. It is useful to keep the body warm, as this provides for creature comforts and for a temperature at which chemical reactions can readily occur. Other than that, heat production by the body is a waste of food energy.

The basic unit of heat energy is the calorie (cal), defined as the amount of heat required to raise the temperature of 1 g of water 1°C, measured from 14.5 to 15.5°C. This unit is too small for most convenient use in nutrition. Thus the kilocalorie (kcal)[1] or 1,000 calories, is more commonly used and the one we refer to when considering our own diet. Where larger values are involved, the megacalorie (megacal or Mcal) is employed. There is some discussion for returning to the international unit of work and energy, the joule,[2] but the International Union of Nutritional Sciences has recommended the calorie concept prevail. In the United Kingdom, the kilojoule has been adopted; 1 kcal = 4.184 kJ or 1 kJ = 0.239 kcal.

5.2   Heat of Combustion. Gross Energy

When a substance is completely burned to its ultimate oxidation products, viz., carbon dioxide, water, and other gases, the heat given off is considered as its gross energy, or heat of combustion. This measure is the starting point in determining the energy value of foods for body use. The determination is carried out in a calorimeter.

The *bomb calorimeter* consists of a bomb, in which the food is burned, enclosed in an insulated jacket containing water which surrounds the bomb and which thus provides the means of measuring the heat produced. The construction of the Parr oxygen calorimeter is illustrated in Fig. 5.1. In part (A), the bomb is shown in cross section. A weighed amount of substance to be tested is placed in the cup (a) of the bomb, and the magnesium fuse wire (b) connecting the two terminals is put in place. The cover is screwed on and the bomb charged with 25 to 30 atmospheres of oxygen. The bomb is then placed in the calorimeter jacket as shown in part (B), surrounded by a known volume of water. The stirrer is started, and when the temperature becomes constant, the charge is ignited by the fuse wire and readings are taken on the thermometer to ascertain the maximum rise. This value multiplied by the water equivalent of *that* instrument, gives the number of calories produced by the burning of the sample. The water equivalent is the number of calories of heat input necessary in *that* instrument to change the temperature of the standard amount of water used (usually 2000 g) 1°C and is determined by previously calibrating the instrument with a compound of known caloric content such as benzoic acid.

In the adiabatic instrument in Fig. 5.1, an outside water jacket (a) is maintained at a temperature identical to the water surrounding the bomb by electronically controlled infusions of hot water. Since heat is not transmitted between surfaces of identical temperatures, the temperature rise in the water around the bomb reflects exactly the heat produced. Various corrections, e.g., for the fuse wire that was burned and for the nonphysiological oxidation of N and S to $^-NO_3$

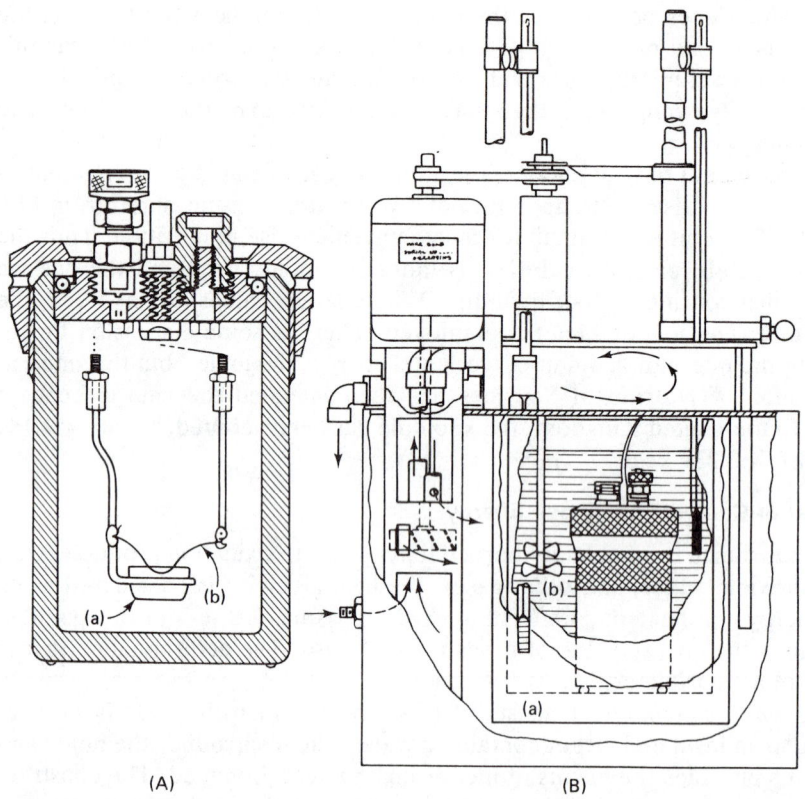

Fig. 5.1 The Parr bomb calorimeter. A, Cross section of bomb: (a) is the cup containing the sample in the bomb, and (b) is the fuse wire. B, Cross section of the adiabatic calorimeter with bomb in place: (a) outside water jacket, (b) inside water jacket surrounding the bomb. (*Permission of the Parr Instrument Co., Moline, Ill.*)

and SO_4^- under high O_2 tension, are subtracted from the calories produced to arrive at the final figure for the caloric value of the food.

The biochemist expresses the heat of combustion of organic compounds as ΔH in kcalories per mole, for example:

	ΔH, kcal/mole
Glucose	673
Glycine	234
Palmitic acid	2380

From the standpoint of nutrition, however, it is more useful to express the values in kcalories per gram, as is done in Table 5.1 for various pure substances and feeds.

TABLE 5.1. Gross Energy Values, or Heats of Combustion (*dry-matter basis*)

Pure substances	kcal/g	Pure substances	kcal/g
Glucose	3.76	Hippuric acid	5.65
Sucrose	3.96	Urea	2.53
Starch	4.23	Uric acid	2.74
Butterfat	9.21	Creatine	4.24
Lard	9.48	Creatinine	4.60
Seed fat (ether extract)	9.33	Methane	13.25
Palmitic acid	9.35		
Stearic acid	9.53	Feeds	
Oleic acid	9.50		
Glycerol	4.30	Corn meal	4.43
Casein	5.86	Oats	4.68
Gliadin	5.74	Soybeans	5.52
Globulin (wheat)	5.36	Wheat bran	4.54
Glycine	3.11	Linseed oil meal	5.12
Alanine	4.35	Timothy hay	4.51
Tyrosine	5.92	Clover hay	4.47
Ethyl alcohol	7.07	Corn stover	4.33
Acetic acid	3.49	Oat straw	4.43
Propionic acid	4.96	Alfalfa hay	4.37
Butyric acid	5.35	Ryegrass	4.54

It should be noted that, as a group, the fats have approximately twice the energy value of the carbohydrates and that the proteins occupy an intermediate position. These differences are governed by their elemental composition (Table 5.2), especially by the relative amount of oxygen contained within the molecule, since heat is produced only by the oxidation resulting from the union of carbon or hydrogen with oxygen from without. The oxygen within has previously liberated heat (or energy) during chemical formation of the compound. In the case of carbohydrates, there is enough oxygen present in the molecule to take care of all the hydrogen present, and thus heat arises only from the oxidation of the carbon. In the case of fat, however, there is relatively much less oxygen present and relatively more atoms requiring oxygen from without, and the combustion involves the oxidation of hydrogen as well as carbon. The burning of 1 g of hydrogen produces over four times as much heat (34.5 kcal per gram) as is the case for carbon (8 kcal per gram). These facts explain the much

TABLE 5.2. Relative Composition of Representative Proteins, Fats, and Carbohydrates (in percent)

	Carbon	Hydrogen	Oxygen	Nitrogen
Fat	77	12	11	—
Protein	52	6	22	16
Carbohydrate	44	6	50	—

greater gross energy values per gram for the fats compared with the carbohydrates. The heat produced in the burning of protein comes from the oxidation of both carbon and hydrogen, but the nitrogen present gives rise to no heat at all because it is set free as such in its gaseous form. No oxidation of it has taken place, and thus no heat is produced, contrary to what occurs in the bomb, for which partial corrections are made.

These differences in elemental composition of the three classes of nutrients carry over and are reflected in the energy content of various feeds (Table 5.1). For example, the energy value of corn, a high-carbohydrate feed, is less than linseed meal, a protein-rich feed, which is less than the soybean, which contains protein and a substantial level of fat. We refer to these in more detail later in the text.

5.3 Chemical Energy

The form of energy primarily concerned in the metabolic processes of plants and animals is chemical. Each bonding between atoms and molecules represents a potential source of chemical energy which is released when the bonds are broken. Conversely, the formation of chemical compounds from simpler units requires energy for the process. Not all of the energy available in chemical bonding is available for life processes, which leads us to contemplate some basic concepts of thermodynamics.

The first law of thermodynamics states that the total energy of a system *and* its surroundings must be a constant, i.e., the law of conservation of energy:

$$\Delta E = E_B - E_A$$

where ΔE is the *change* in *energy* content of the system, E_B is the energy content at the end of a process, and E_A is the energy content at the beginning. Rewritten, it is

$$\Delta E = Q - W$$

where Q is the heat *absorbed* by the system and W is the work *done* by the system on the surroundings. Thus, any change in energy status of a system must be accounted for. A further extension of this law is that as an energy exchange occurs from the system to its surroundings or the reverse, the final state is independent of the path of that transformation. In other words, the energy transformation of a compound like glucose to CO_2 and H_2O is the same whether by ignition or metabolism. Such a concept gives some confidence in studying the energetic processes in situ (within the body), since measurement of all stepwise energy transformations would be enormously difficult.

The second law of thermodynamics states that the energy status of the universe is moving from an organized one to a more disorganized one, one in which molecules are highly random, of a lower energy level, and thus more stable. This must be true. Consider the tremendous daily influx of energy to the earth. If the second law were not so, would we not burn up as the heat content of our planet increased? It is thus with chemical compounds. As they are transformed from one energy level to another, a portion of their energy may be released for useful work (free energy) and a portion contributes to the disorganization (randomness) of the surroundings. The concept is explained by the equation

$$\Delta G = \Delta H - T\,\Delta S$$

where ΔG is the change in free energy, ΔH is the change in enthalpy (heat content of the system), T is the absolute temperature, and ΔS is the change in entropy (degree of randomness). It is important to note that one cannot state that a compound has so much free energy or so much entropy. It is the change in these contents and those of the surroundings that is significant biologically and is measurable. Thus, if the energy status of a system becomes more random (ΔS increases), the free energy of the system is negative and that reaction can occur spontaneously. If, however, entropy decreases (e.g., less randomness, two molecules join together, etc.), ΔG becomes positive, and free energy must be added to the system to make it work. Diagrammatically, it can be visualized as follows when A is transformed to B:

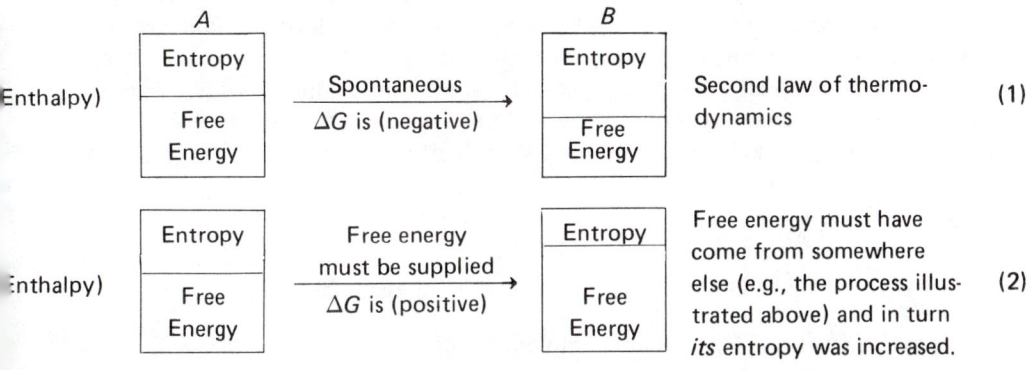

All body reactions which proceed spontaneously are *exergonic* (energy out) reactions and provide free energy to other reactions (Eq. (1), above). Those which require energy, that is, *endergonic* reactions (energy in), cannot take place unless needed free energy is simultaneously released from some exergonic reaction (Eq. (2), above). These two reactions are thus *coupled*. For example, the transformation (endergonic) of glucose to glycogen requires free energy. A part of this energy is provided by the transformation of adenosine triphosphate (ATP) to adenosine diphosphate (ADP) (exergonic), and these reactions are thus coupled:

$$ATP + H_2O - \Delta G \longrightarrow ADP + H_3PO_4 \quad (exergonic)$$

$$glucose + \Delta G \longrightarrow glycogen \qquad (endergonic)$$

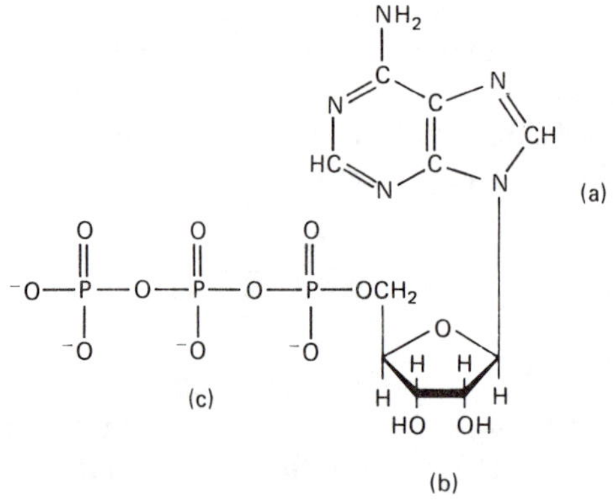

Fig. 5.2 Adenosine triphosphate contains adenine (a), ribose (b) and a triphosphate unit (c). ADP contains one less phosphate and AMP, two less.

5.4 Central Role of ATP in Energy Exchange

At the heart of all biological energy transformations sits ATP (adenosine triphosphate). Its central role was first identified by Lipmann and Kalckar in 1941 and so is relatively recent to our concepts of biochemistry. The formula is shown in Fig. 5.2. Note it consists of adenine (a nitrogen base), ribose (a 5-carbon sugar), and a triphosphate unit. ATP is an "energy-rich" compound, because of the lability (high group-transfer potential) conferred by the two phosphoanhydride bonds. When ATP is hydrolyzed to adenosine diphosphate (ADP) and orthophosphate (Pi), or when it is hydrolyzed to adenosine monophosphate (AMP) and pyrophosphate (PPi), a large amount of free energy is liberated. It is generally considered that these transformations provide about -7.3 kcal/mol under standard conditions, but within the cell it may be closer to -12 kcal/mol.

$$ATP + H_2O \longrightarrow ADP + Pi + H^+; \Delta G = -7.3 \text{ kcal/mol}$$

$$ATP + H_2O \longrightarrow AMP + PPi + H^+; \Delta G = -14.6 \text{ kcal } (-7.3 \text{ kcal/mol})$$

In its active form, ATP is usually complexed with Mg^{++} or Mn^{++}, thus implicating metallic ions in energy transfer. While ATP serves as the major energy source for a myriad of reactions, there are a number of compounds that also possess this high energy transfer capability (Table 5.3) and in fact contribute more energy when hydrolyzed than ATP. The intermediate status of ATP in energy potential thus allows it to give and receive energy and serve as a carrier of phosphoryl groups from high-energy to low-energy compounds. There are other compounds analogous to ATP but which contain another nitrogen base, e.g., guanosine triphosphate (GTP), uridine triphosphate (UTP), and cytidine triphosphate (CTP).

5.5 Biological Oxidation-Reduction

Chemotrophs derive their energy from the oxidation of food molecules such as glucose, fatty acids, and in many instances, amino acids. Fortunately, this

TABLE 5.3. Standard Free Energies of Hydrolysis of Some Compounds Found in Animal Tissues

Compound	$\Delta G^{\circ a}$
Phosphoenol pyruvate	−14.8
Carbamoyl phosphate	−12.3
Acetyl phosphate	−10.3
Creatine phosphate	−10.3
Pyrophosphate	−8.0
ATP to ADP	−7.3
Glucose-1-phosphate	−5.0
Glucose-6-phosphate	−3.3
Glycerol-3-phosphate	−2.2

^{a}All reactants at 1 M concentration; pH 7.0; temperature 25°C.

process is not explosive, as in the bomb, but is gradual with so-called high-energy bonds such as within ATP being generated en route. The process is not 100 percent efficient, with only part being conserved in the form of high-energy bonds, the rest being evolved as heat.

Oxidation-reduction processes are fundamentally the same whether in living or nonliving systems. Oxidation of a molecule involves the combination with oxygen or the removal of hydrogen, or specifically the loss of electrons. Therefore, during oxidation, electrons are transferred from the substance being oxidized to the substance being reduced. The ultimate result, when oxidation is complete in animal tissue, is $CO_2 + H_2O$.

The transfer of electrons in the tissue is not a direct process but effected by special acceptors or electron carriers. They contain specific vitamins as an integral part of their structure, bringing us face to face with the realities of diet and energy metabolism. Three of the most important ones are NAD, NADP, and FAD. The structures of these are shown in Fig. 5.3, with their active sites labeled. Note that all are nucleotides (adenine, ribose, phosphate), with nicotinamide central to NAD and NADP and a riboflavin moiety (dimethyl isoalloxazine) central to FAD.

Nicotinamide adenine dinucleotide (NAD) accepts a hydrogen ion and two electrons (formally a hydride ion) to yield the reduced form, NADH:

It functions in many intracellular reactions of the empirical form:

$$NAD^+ + R\overset{\overset{\displaystyle H}{|}}{\underset{\underset{\displaystyle OH}{|}}{C}}R' \rightleftharpoons NADH + R\overset{}{\underset{\underset{\displaystyle O}{\|}}{C}}R' + H^+$$

One of the hydrogens is accepted directly by the nucleotide, while the other, H^+, remains in the solvent. Thus, the reduced form is usually written $NADH + (H^+)$ or less accurately as $NADH_2$. The flavin adenine nucleotide accepts two hydrogens on the isoalloxazine ring (Fig. 5.3), as indicated by the following empirical equation:

$$FAD + R\overset{\overset{\displaystyle H \quad H}{| \quad |}}{\underset{\underset{\displaystyle H \quad H}{| \quad |}}{C\text{--}C}}R' \rightleftharpoons FADH_2 + R\overset{\overset{\displaystyle H}{|}}{\underset{\underset{\displaystyle H}{|}}{C=C}}R'$$

Both reduced NAD and FAD are central to the production of ATP in the oxidative phosphorylation system (Sec. 6.8). On the other hand, reduced NADP (Fig. 5.3) is almost exclusively an electron donor for important reductive biosyntheses (e.g., fatty acid synthesis) which occur throughout the body.

In addition to the electron carriers, there are other active carrier molecules that exist which have high-energy, group-transfer potential. Whereas ATP carries phosphorus, others carry molecules such as acyl radicals with sufficient free energy to force reactions to proceed. One of the most important is coenzyme A (CoA), described by Lipmann[3] (Fig. 5.4).

When combined with acetate as acetyl CoA, it has a ΔG^0 of -7.5 kcal/mol, much as does ATP. Again it is a nucleotide with the vitamin *pantothenic acid* at its core. A number of carrier molecules are shown in Table 5.4, a few of which have already been discussed.

TABLE 5.4. A Few Activated Carriers Important in Metabolism

Carrier molecule	Component carried in an activated form
ATP	Phosphoryl
NADH and NADPH	Hydrogen
$FADH_2$	Hydrogen
Coenzyme A	Acyl
Lipoamide	Acyl
Thiamin pyrophosphate	Aldehyde
Biotin	CO_2
Tetrahydrofolate	One carbon units
Uridine diphosphate glucose	Glucose
Cytidine diphosphate diacylglycerol	Phosphatidate

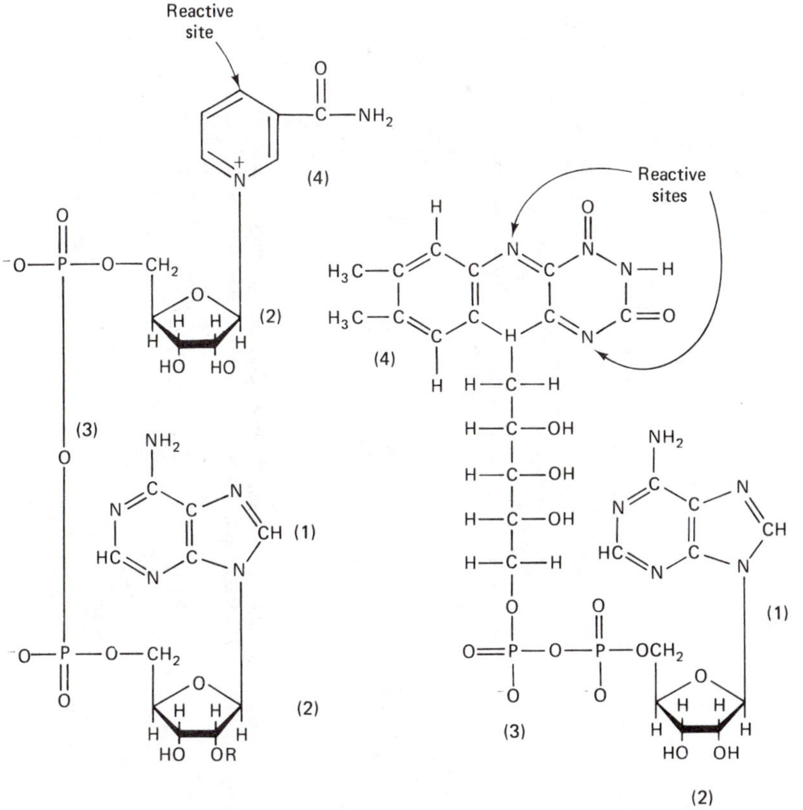

Fig. 5.3 Structures of the oxidized form of (a) nicotinamide adenine dinucleotide containing (1) adenine, (2) ribose, (3) diphosphate and (4) nicotinamide, and (b) flavin adenine dinucleotide containing (1) adenine, (2) ribose, (3) diphosphate and (4) dimethyl isoalloxazine. In NAD^+, R = H; in $NADP^+$, R = a phosphate moiety.

Biochemically the carriers mentioned above, as well as many other compounds in the body, are relatively stable unless a specific catalyst is present. We know these catalysts to be enzymes[4] which activate and hasten the reaction. The substrate on which the enzyme acts is converted into its product(s) in two stages. In the first, the enzyme combines with the substrate and the reaction then proceeds; in the second, the enzyme is liberated and thus free to combine with another molecule of substrate. As a result, large amounts of substrate can be turned over by a minute amount of enzyme. For example, a single molecule of enzyme carbonic anhydrase, which catalyzes the hydration of CO_2 to form carbonic acid, can hydrate 10^5 moles of CO_2 *per second*. This is one of the more spectacular ones, others being as low as 0.5 moles per second. Enzymes are substrate-specific, i.e., they react with only certain compounds and no other.

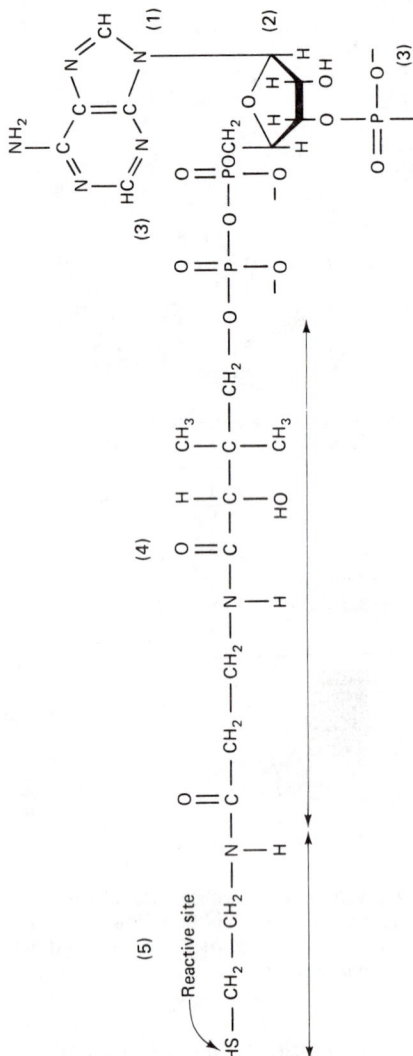

Fig. 5.4 Coenzyme A (CoA) contains (1) adenine, (2) ribose, (3) three phosphate units, (4) pantothenic acid and (5) a β-mercaptoethylamine unit. The HS is the reactive site which carries the acyl radicals.

Each one also has specific limits as to pH and temperature, outside of which they do not function. For example, pepsin, the enzyme responsible for protein degradation in the stomach, functions most effectively at the pH of 1.8 to 2.0 normally found in the stomach but is totally ineffective in the neutral to alkaline pH of the small intestine.

Most enzymes consist of a specific protein moiety called the *apoenzyme* and a prosthetic group called a *cofactor* or *coenzyme* which is essential for the activity of the total enzyme. These prosthetic groups may consist of an organic compound, e.g., a vitamin derivative, or a metallic ion, e.g., Zn^{++}, as functions in carboxypeptidase.

While it is generally stated in the chemical literature that catalysts, and hence enzymes, by definition are not consumed or destroyed as a part of their catalytic activity, such is not true in the animal kingdom. If it were so, a megavitamin pill at birth would last a lifetime and vitamins would not be dietary essentials. Unfortunately, "nothing is ever always" in biology, and they do in fact eventually become destroyed, metabolized, excreted and must be replaced. Hence, while the enzymes and their prosthetic groups do operate very efficiently and catalyze the reactions of many moles of substrate, they are not immortal and must in due time be resynthesized. Hence, we need the diet as a source of renewal.

5.6 Final Common Pathway of Energy Metabolism

We shall be dealing in more detail with the specific steps of energy extraction in the chapters on the nutrients which follow. However, it is probably useful to examine first the "big picture" and then the details. If the student is a bit rusty on the nomenclature and functions of an animal cell, e.g., mitochondrion, endoplasmic reticulum, cell membrane, etc., it may be well to reexamine your freshman biology. The terms will be used frequently as we proceed.

Energy metabolism begins with mastication and digestion of food (Sec. 3.1) so that the cell is presented with a simple molecule. To this point no energy has been extracted. The cell with its thousands of different organic compounds and enzymes is designed to remove the energy of the chemical bonding and generate ATP and other ancillary compounds.

There are three basic cycles through which energy-rich compounds pass: the glycolytic (Embden-Meyerhof); tricarboxylic acid (TCA) (Krebs[5] or citric acid); and oxidative phosphorylation (cytochrome system). In simplest terms, a compound such as glucose enters the glycolytic cycle, passes through to the TCA cycle, and finally through the oxidative phosphorylation sequence. En route ATP is generated, CO_2 is liberated, and finally, H_2O is formed. A simple diagram of these processes is shown in Fig. 5.5.

Note the points of entry for the molecules from different food sources which have been indicated. The fact that such diverse compounds can enter establishes these cycles as a true final common pathway for energy-supplying nutrients. The relative proportions of these nutrients entering at each point will depend on the kind of animal, the diet, and the physiological state of the animal.

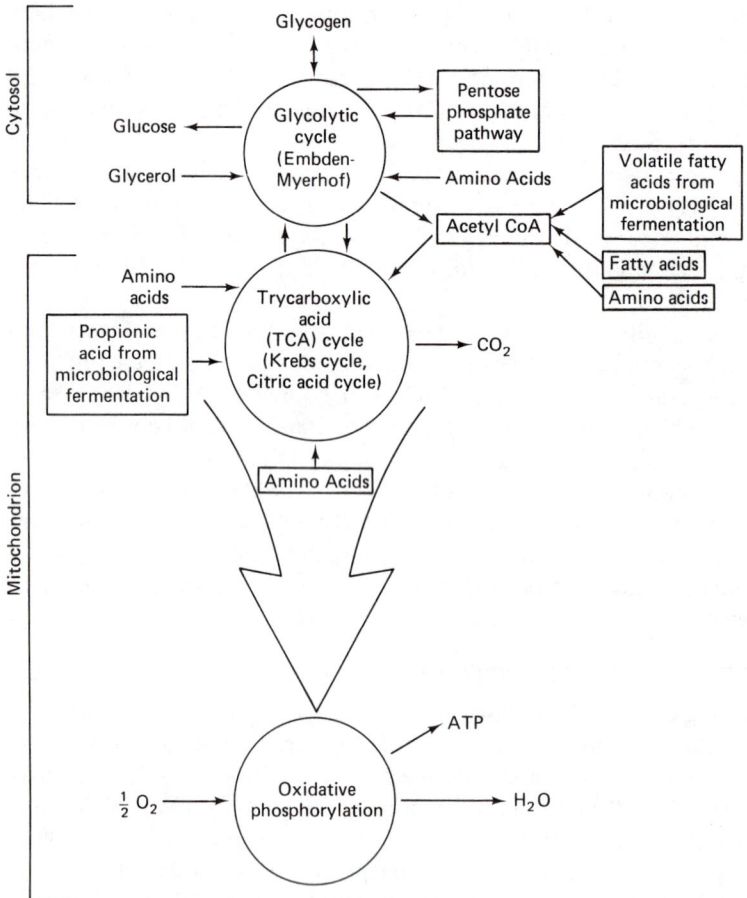

Fig. 5.5 Final common pathway of energy metabolism. A simplified diagram to show the routes most energy-yielding molecules take in the cell as energy is extracted. Note the three primary cycles (glycolytic, TCA and oxidative phosphorylation) and their location within the cell structure.

For example, a rat on a high-fat diet versus a high-carbohydrate diet, a cow on a high-grain diet versus a high-hay diet, a well-fed animal versus one on a subsistance diet will all have different patterns of nutrient flow into these cycles. One of the fundamental principles to be learned about nutrition is how the nutrients and the animal interact at these cycles as they maintain life processes.

NOTES

1. Older literature will often refer to Calories (C.) with the letter c capitalized. This notation is equal to kilocalorie but is becoming obsolete.

2. One joule is the work done by a force of 1 newton exerted through a distance of 1 m. One newton is the force that will give a mass of 1 kg an acceleration of 1 m/sec^2.

3. F. Lipmann (chairman), Symposium on Chemistry and Function of Coenzyme A, *Federation Proc.*, **12**:673–715, 1953.

4. The modern knowledge of enzymes dates from 1926, when Sumner isolated the enzyme urease in crystalline form and proved the crystals to be protein. Dr. James B. Sumner (1887–1955) was professor of biochemistry and director of the laboratory of enzyme chemistry at Cornell University, where he had served for forty-one years as a teacher and a tireless investigator in the field of enzymes. For his outstanding accomplishments, notably for his pioneer work in isolating urease and establishing its protein nature, he was awarded the Scheele Medal at Stockholm, Sweden, in 1937 and the Nobel Prize for Chemistry in 1946.

5. Hans Adolph Krebs, a world-renowned biochemist, was born in 1900 in Hildesheim, Lower Saxony, Germany. He studied at the University of Göttingen, Freiburg, and in Munich and Berlin, and later became an assistant at the Kaiser Wilhelm Institute in Berlin. In 1933 he went to England and later became a British subject. He joined the staff of Sheffield University in 1935 and was appointed professor of biochemistry in 1945. In 1954 he became Whitley professor of biochemistry at Oxford University. For his discoveries of the tricarboxylic and urea cycles he received the Nobel Prize for Physiology in 1953.

SUPPLEMENTARY LITERATURE

Lehninger, A. L.: Biochemistry, 2d ed., Worth Publishing Co., New York, 1975.

Pike, R. L., and M. L. Brown: Nutrition: An Integrated Approach, 2d ed., John Wiley & Sons, New York, 1975.

Stryer, L.: Biochemistry, W. H. Freeman and Company, San Francisco, 1975.

Chapter 6

The Carbohydrates
and Their Metabolism

The group of nutrients called carbohydrates includes the sugars, starch, cellulose, gums, and related substances. Though few of these occur in animal tissues, as such (glucose and glycogen are exceptions), the carbohydrates form the largest part of the animal's food supply. Carbohydrates make up 75 percent of the dry weight of the plant world, upon which animal life primarily depends.

The carbohydrates in plants are produced by means of photosynthesis, the most important chemical reaction in nature. Radiant energy from the sun is captured by chlorophyll and changed to chemical energy, which in turn supports the formation of glucose from carbon dioxide and water—an endergonic reaction. There are many intermediary reactions but the overall process may be represented as follows:

$$6CO_2 + 6H_2O + 673 \text{ kcal} \longrightarrow C_6H_{12}O_6 + 6O_2$$

In this way, plants store the energy of the sun in products that can be used by animals as a source of energy for their life processes. Thus all animal life is dependent upon the process of photosynthesis.

6.1 Classification of Carbohydrates

The carbohydrates owe their name to the fact that they contain carbon combined with hydrogen and oxygen often in the same ratio as in water. While a reasonable generalization, this is a rather imprecise definition. Chemically, carbohydrates are polyhydroxy aldehydes and ketones, or substances which yield them on hydrolysis. An abbreviated classification, which includes the members about which we are particularly interested in nutrition, is presented in Table 6.1.

In nature, carbohydrates are designed to foster plant existence; their use as an animal feed is purely incidental. The more soluble forms serve in plant systems in energy transformation and for tissue synthesis, the less soluble ones such as starch serve as reserve energy, while the relatively insoluble fractions (cellulose, hemicellulose) form plant structural entities. The latter group of compounds is generally irretrievable by the plant. It should be noted from Table 6.1 that the various members of a given subgroup have the same empirical

TABLE 6.1. Classification of Carbohydrates

I. Monosaccharides (single glycose unit)	III. Polysaccharides (Glycan, >10 glycose units)
A. Trioses ($C_3H_6O_3$)	A. Homoglycan (single glycose units)
1. Glyceraldehyde	1. Pentosans ($C_5H_8O_4)_n$
2. Dihydroxyacetone	a. Arabans
B. Tetroses ($C_4H_8O_4$)	b. Xylans
1. Erythrose	2. Hexosans ($C_6H_{10}O_5)_n$
C. Pentoses ($C_5H_{10}O_5$)	a. Glucans
1. Ribose	(1) Starch, α-linked
2. Arabinose	(2) Dextrins, α-linked
3. Xylose	(3) Glycogen, α-linked
4. Xylulose	(4) Cellulose, β-linked
D. Hexoses ($C_6H_{12}O_6$)	b. Fructans
1. Glucose	(1) Inulin
2. Galactose	(2) Levan
3. Mannose	c. Galactans
4. Fructose	d. Mannans
II. Oligosaccharides (2 to 10 glycose units)	B. Heteroglycan
A. Disaccharides ($C_{12}H_{22}O_{11}$)	(2–6 different kinds of glycose units)
1. Sucrose	1. Pectins (α-linked)
2. Maltose	2. Hemicelluloses (β-linked)
3. Cellobiose	3. Gums and mucilages
4. Lactose	4. Mucopolysaccharides
B. Trisaccharides ($C_{18}H_{32}O_{16}$)	IV. Specialized compounds
1. Raffinose	A. Chitin
C. Tetrasaccharides ($C_{24}H_{42}O_{21}$)	B. Lignin (not a carbohydrate)
1. Stachyose	
D. Pentasaccharides ($C_{30}H_{52}O_{26}$)	
1. Verbascose	

formula. They have, however, different structural formulas and exhibit different physical properties. While carbohydrates tend to be a highly heterogeneous set of compounds, American and British chemists have adopted a stringent set of rules for their nomenclature.[1]

CHEMISTRY OF THE CARBOHYDRATES

6.2 Monosaccharides

The monosaccharide is the fundamental unit from which all carbohydrates are derived. It is called a glycose unit and serves as the root word for identifying compounds such as glycosides or glyconic acids. They are further characterized by the number of carbon atoms they contain, C_3, C_4, C_5, etc., and by their structural configuration (aldose or ketose). Only two of the monosaccharides, glucose and fructose, occur to any extent in free form in nature. Most must be obtained by hydrolysis of more complex plant constituents. Monosaccharides are often referred to as simple sugars and are soluble in water and sweet to the taste (Table 6.2).

   Trioses (C_3) and Tetroses (C_4).   Two trioses, glyceraldehyde (glycerose) and dihydroxyacetone are crucial intermediates in the metabolism of glucose in the glycolytic cycle (Sec. 6.8). In establishing proper configurations for monosaccharides, D-glycerose and L-glycerose are used as a reference as shown below. The significant feature is the rotation of the H and OH on the

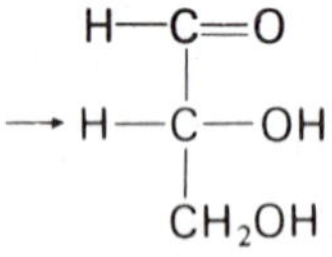

D-glycerose 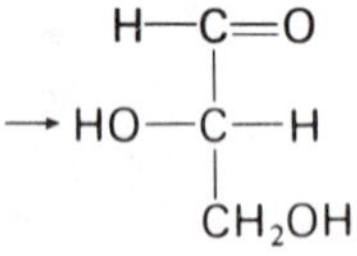   L-glycerose 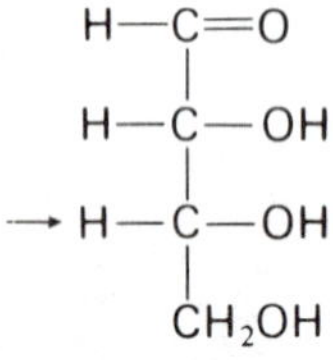   D-erythrose

penultimate ("almost last") carbon atom ($\rightarrow$). The tetrose, erythrose, is a reactant in the phosphogluconate oxidative (pentose) pathway (Sec. 6.8).

TABLE 6.2. Relative Sweetness of Some Organic Compounds

Sucrose	1.0	Maltose	0.45
D-Fructose	1.35	Galactose	0.32
D-Glucose	0.74	Lactose	0.16
Xylose	0.67	Saccharin	200–700
Sorbitol	0.54		

Source: O. C. Dermer, The science of taste, *Proc. Oklahoma Academy Sci.,* **27**:9–20, 1947.

Pentoses (C_5). The pentose, D-ribose, is found in every living animal cell. It occurs in a number of compounds which play crucial roles in metabolism, e.g., ATP, ADP, riboflavin, and RNA (Sec. 5.4). Its reduced form 2-deoxy D-ribose is found in DNA. In addition, D-ribose (aldose) and D-xylulose (ketose) participate in the phosphogluconate pathway. Small amounts of D-xylose are present free in plums, cherries, and grapes, and free arabinose is found in the heartwood of conifers. In general, however, these pentoses exist as polymers called pentosans (Table 6.1). Arabans and xylans make up a significant part of hemicellulose (Sec. 6.4). Xylose can be produced by the hydrolysis of hay, straw, oat hulls, many woods, and especially corncobs. Arabinose is also found in gums such as gum arabic.

Free pentoses constitute a very small part of the animal diet, but nonetheless can be used to some degree in the phosphogluconate pathway. In the polymerized form, as found in hemicellulose, pentose constitutes a significant part of the energy supply for animals if fermented by gastrointestinal microorganisms (Chap. 3).

The following are linear structural formulas of several pentoses together with one source in nature.

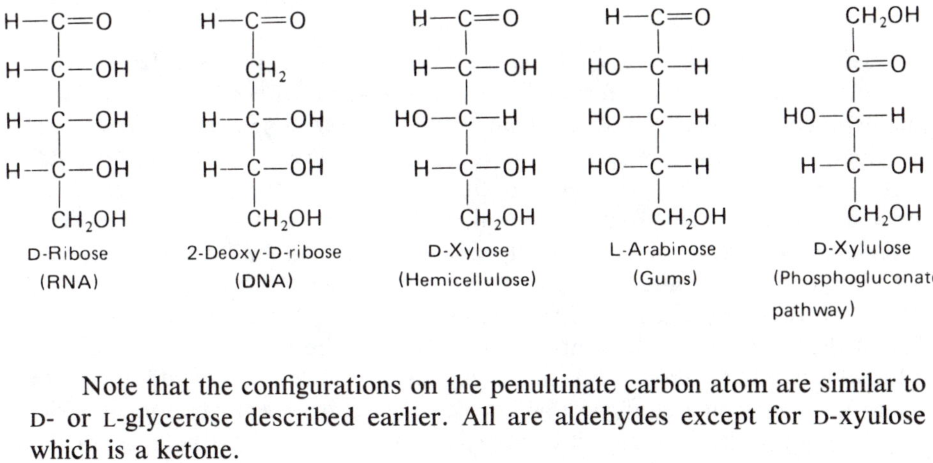

D-Ribose	2-Deoxy-D-ribose	D-Xylose	L-Arabinose	D-Xylulose
(RNA)	(DNA)	(Hemicellulose)	(Gums)	(Phosphogluconate pathway)

Note that the configurations on the penultinate carbon atom are similar to D- or L-glycerose described earlier. All are aldehydes except for D-xyulose which is a ketone.

Hexoses (C_6). The hexoses comprise a large group of sugars, several of which play a significant role in nutrition, either as components of foods or as products of metabolism in the body. Glucose, fructose, and galactose are ones of special importance in nutrition. The hexoses, like the pentoses, are divided into aldoses or ketoses according to whether they contain an aldehyde or ketone group. The presence of these *reducing groups* classifies them as *reducing sugars* and for many years was the basis for their analysis. In that assay the cupric sulfate was reduced to the cuprous ion yielding a brick red precipitate. This method has been largely replaced by enzymatic techniques. The configurations of four important hexoses are shown below.

(1)	CHO	CH$_2$OH	CHO	CHO
(2)	*HCOH	CO=O	*HCOH	*HOCH
(3)	*HOCH	*HOCH	*HOCH	*HOCH
(4)	*HCOH	*HCOH	*HOCH	*HCOH
(5)	*HCOH	*HCOH	*HCOH	*HCOH
(6)	CH$_2$OH	CH$_2$OH	CH$_2$OH	CH$_2$OH
	D-glucose	D-fructose	D-galactose	D-mannose

Note that all are aldehydes except fructose which is a ketone. Because of the presence of asymmetric carbon atoms (labeled with an asterisk) a number of stereoisomers are possible. Only a few occur in nature. The properties of D-glucose indicate that in solution it exists in several stereoisomeric forms which are in equilibrium. It gives a dextro or (+) (right) rotation in polarized light which accounts for its common name *dextrose*. On the other hand, D-fructose gives a levo (−) (left) rotation and is often known as *levulose*.

Glucose.   Glucose occurs in nature as the D-form. It exists widely in fruits and plant sap and constitutes 40 percent of the sugar of honey. It is the basic molecule for the synthesis of starch and cellulose (Sec. 6.4) and is commercially produced by the hydrolysis of cornstarch. It is of central importance in nutrition, as it is the major end product of digestion of carbohydrates by nonruminants and is the primary form to be utilized for energy in which these nutrients circulate in the blood of all mammals. Glucose exists in the form of a pyranose ring and is more accurately depicted in the Haworth perspective below. Note there are two forms, α and β, depending upon the position of the H and OH about the number 1 carbon atom. The α form is the precursor of starch and glycogen, while the β is the form for cellulose. The other configuration is the so-called "chair-like configuration" which is free from any bond angle strain and is in all probability how it exists spatially in nature. This latter form also shows the α and β configuration.

α-D-glucopyranose   β-D-glucopyranose
Haworth perspective

α-D-glucopyranose   β-D-glucopyranose

Chair-like configuration

Galactose.   Galactose is an aldohexose which occurs in milk as a component of the milk sugar, lactose. Certain compounds of galactose, the galactosides, occur in the brain and nervous tissue. While we often think of it as a sugar synthesized by animals, it is found in greens and a wide range of gums and as a component of the galactolipids found in plant leaves.

Mannose.   Mannose is an aldohexose occurring in mannans, a group of polysaccharides widely distributed in plants. It also is found in blood serum globulins and certain egg proteins.

Fructose.   This is the only important ketohexose in nature and is the sweetest of the carbohydrates (Table 6.2). It occurs free along with glucose and sucrose in fruits and honey (Table 6.3). Sucrose is one-half fructose. The polymer of fructose, inulin, is found in a number of plants such as the Jerusalem artichoke, burdock, dandelion, and goldenrod, where it serves as a reserve polysaccharide.

TABLE 6.3. Distribution of Sugars in Plant Materials as a Percent of the Total Solids

Material	Solids, %	Sugars, % of dry matter		
		Glucose	Fructose	Sucrose
Apple	16.0	7.3	37.8	23.6
Peach	12.8	7.1	9.2	54.1
Pear	13.6	6.9	49.8	11.8
Sour cherry	15.1	28.5	21.7	2.6
Table beet	11.2	1.6	1.4	54.6
Carrot	12.0	7.0	7.0	35.3
Lettuce	5.0	5.0	9.2	2.0
Watermelon	9.6	12.3	36.9	24.5
Honey	82.8	43.0	52.0	1.0
Maple syrup	66	–	–	98.0
Sugar beets	16.4	–	–	87.2
Sugar cane	23.2	–	–	65.5

Source: C. Y. Lee, R. S. Shallenberger, and M. T. Vitum, Free sugars in fruits and vegetables, *New York Food and Life Science Bull.*, 1:1–12, 1970. F. B. Morrison, Feeds and Feeding, Morrison Publishing Co., New York, 1956.

Hexose Compounds. Glucose and a number of other hexoses have the capacity to form amino sugars, several acids (depending upon the location of carboxyl group), and alcohols. There are a number which are constituents of blood or body tissues and play some important role in body processes. Below are the linear structural forms of several glucose derivatives which are important in nutrition and metabolism.

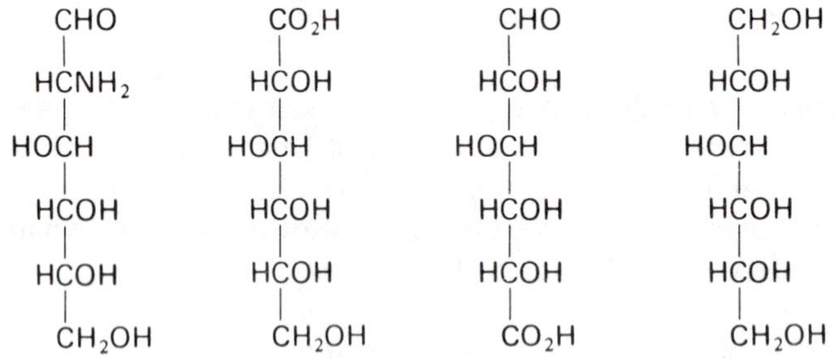

CHO	CO_2H	CHO	CH_2OH
$HCNH_2$	HCOH	HCOH	HCOH
HOCH	HOCH	HOCH	HOCH
HCOH	HCOH	HCOH	HCOH
HCOH	HCOH	HCOH	HCOH
CH_2OH	CH_2OH	CO_2H	CH_2OH

Class of compounds

Amino sugar	Aldonic acid	Uronic acid	Alditol

Compound name

Glucose amine	D-gluconic acid	D-glucuronic acid	D-glucitol (sorbitol)

Glucosamine. Glucosamine is an amino sugar in which one of the OH groups is replaced by NH_2. It occurs in chitin, the shell covering of invertebrates, and is also present in saliva and gastric juice. Galactosamine occurs along with *glucuronic acid* in chondroitin sulfate which, when combined in glycoprotein (Sec. 8.2) is a major component of cartilage. D-*gluconic* acid, in which the aldehyde has been oxidized, is, in its phosphorylated form, 6-phosphogluconate, a key intermediate in the phosphogluconate pathway (Sec. 6.4). Glucuronic acid results from the oxidation of the carbon with the primary hydroxyl group of glucose. It can form an ester or harmless glycoside (glucuronide) with a number of toxic materials such as phenols as well as with a number of hormones such as androgens and cortisone. These can then be excreted in the urine and feces. The bile pigments bilirubin and biliverdin are glucuronides.

Sorbitol. Sorbitol is an alcohol intermediate, produced in the formation of fructose from glucose. It is thus important as a part of the pathway which provides the large amount of fructose normally found in semen. Similar compounds are possible from other hexoses as well.

Perhaps there is no more significant property of the hexoses than their ability to combine with phosphoric acid and generate a high-energy bond (phosphorylation). Glucose-1-phosphate and glucose-6-phosphate are important intermediates in the metabolism of carbohydrates for energy (Sec. 6.8). The formulas for the two compounds are shown below.

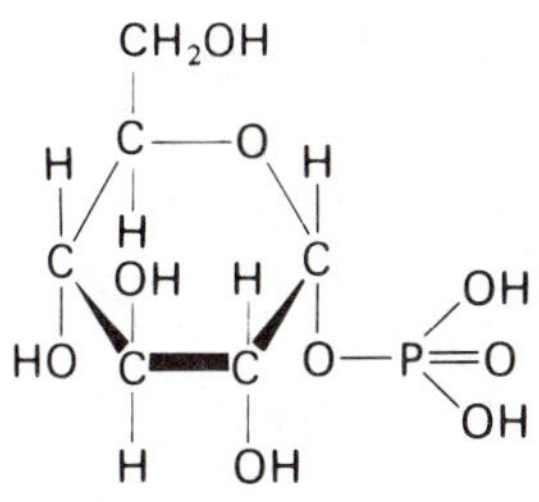

Glucose-1-phosphate

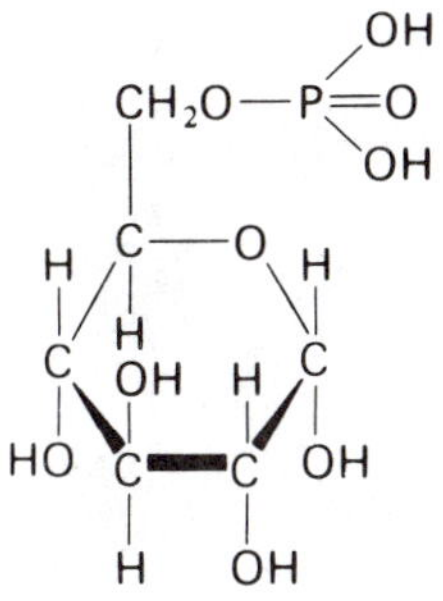

Glucose-6-phosphate

6.3 Oligosaccharides

The oligosaccharides contain sugars with from two to six glycose units.

Disaccharides. The disaccharides derive their name from the fact that they are a combination of two molecules of monosaccharide. Their general formula, $C_{12}H_{22}O_{11}$, indicates that one molecule of water has been eliminated as two monosaccharides combine. They are soluble in water, though to varying degrees.

Sucrose. Sucrose is made up of a combination of one molecule of D-glucose and one molecule of D-fructose. It occurs in sugar cane and sugar beets, sources of commercial sugar. It also occurs in ripe fruits, in tree sap (maple sugar), and in many fruits and vegetables (Table 6.3). Sucrose is dextrorotatary but it is not a reducing sugar as it has no free aldehyde or ketone moiety. When hydrolyzed by dilute acid or the enzyme sucrase, sucrose is split into its two constituent monosaccharides. The resulting sugar is levorotatory. Since the hydrolysis thus results in a change from dextrorotation to levorotation, the process is called inversion and the mixture of glucose and fructose is often called *invert sugar*. Such a process is the way by which honey bees convert the sucrose of plant nectar to honey (Table 6.3).

Maltose. Maltose consists of two molecules of α-D-glucose joined together in an α-1,4 linkage shown below. The position of H on the number 1 carbon atom of molecule (*a*) is in the α position. Note that the number 6 carbon atoms are in a cis configuration. We shall see that this is the fumdamental linkage for the starch molecule and is readily split by mammalian enzymes. Maltose derives its name from the fact that it is produced commercially from starch by the action of malt, obtained from germinating barley which contains a starch hydrolyzing enzyme *diastase*.

Cellobiose. Cellobiose consists of two molecules of β-D-glucose joined together in a β-1,4 linkage shown above. The position of the H on the number 1 carbon atom of molecule (*a*) is in the β position. Note that the number 6 carbon

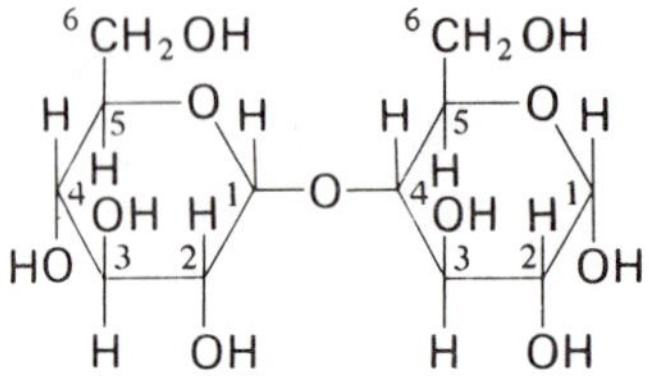

$^6CH_2OH$ $^6CH_2OH$

H $_5$ O H   H $_5$ O H
H   H
$_4$OH H^1 —O— $_4$OH H^1
HO $_3$ $_2$ $_3$ $_2$ OH
H OH   H OH

(a)   (b)

Maltose

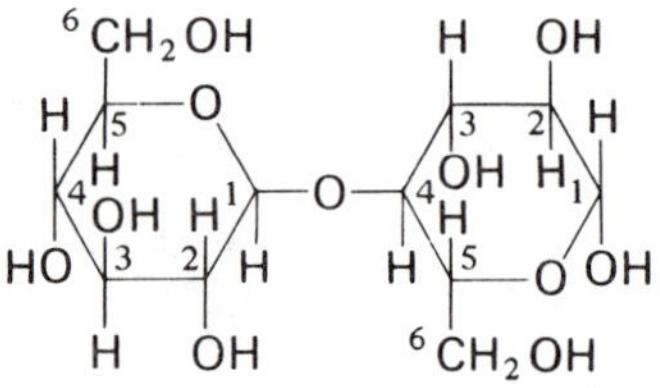

$^6CH_2OH$ H OH

H $_5$ O $_3$ $_2$ H
H   OH H$_1$
$_4$OH H^1 —O— $_4$H
HO $_3$ $_2$ H   H $_5$ O OH
H OH $^6CH_2OH$

(a)   (b)

Cellobiose

H OH $^6CH_2OH$

H^3 $_2'$H   H $_5$ O OH
OH H   H
$_4'$ $_1$ —O— $_4$OH H^1
HO $_{5'}$ $_3$ $_2$ H
O
H $_6$ CH$_2$OH H OH

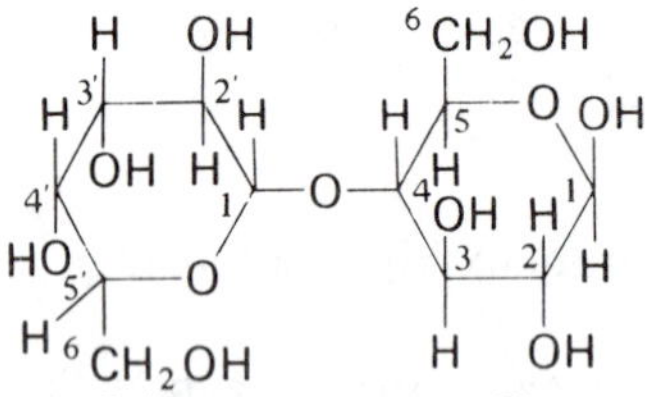

β-D-Galactose β-D-Glucose

Lactose (β form)

atoms are in the trans position. This linkage is the fundamental one for the cellulose molecule and cannot be split by mammalian enzymes. It can be split, however, by microbial and fungal enzymes or acid. Cellobiose does not occur in free form in nature but only as a component of glucose polymers.

Lactose. Lactose is the sugar of milk and consists of one molecule of α-D-glucose and one molecule of β-D-galactose, joined in a linkage as shown above. They can be separated by the enzyme lactase or by the action of acid. It is a reducing sugar and is only one-sixth as sweet as sucrose (Table 6.2). Lactose is of special interest in nutrition, because it makes up nearly half of the solids of milk and because it does not occur in nature except as a product of the mammary gland.

Tri-, Tetra-, and Pentasaccharides. Raffinose is a trisaccharide and is the most widely distributed oligosaccharide in nature except for sucrose. It contains three monosaccharides: D-galactose, D-glucose, and D-fructose. Upon hydrolysis fructose is removed first, leaving the disaccharide *melibiose* which contains galactose and glucose. Melibiose is thus an isomer of lactose. Raffinose with one more molecule of D-galactose becomes *stachyose,* a tetrasaccharide, and with two more becomes *verbascose,* a pentasaccharide. All three of the galatose molecules of these oligosaccharides are linked in an α linkage and are found in substantial quantities in leguminous seeds. Since there is no enzyme in the intestine of animals capable of splitting the α galactosidic linkage (recall lactose is a β linkage), these carbohydrates cannot be digested, are too large to be absorbed, and thus pass into the large intestine relatively unscathed. The microflora then ferment them and, especially for the latter two, produce a large amount of gas, predominantly H_2 and CO_2.[2]

6.4 Polysaccharides

The polysaccharides are polymerized anhydrides of a large number of simple sugars, as their empirical formulas indicate. The various subgroups are difficult to classify with precision but as more is learned of their composition, the pattern becomes clearer.

They are of high molecular weight, and most of them are insoluble in water. Upon hydrolysis by acids or enzymes, they are broken down into various intermediate products and finally into their constituent monosaccharides. Quantitatively, they are the most important nutrients in feeds of plant origin.

For convenience they are divided into homoglycans, those having only one glycose unit, and heteroglycans, those having two to six different glycose units (Table 6.1). In addition it is convenient to assemble those which yield pentoses (pentosans) and those which yield hexoses (hexosans) into subgroups. Some of the more important polysaccharides in animal nutrition are discussed below.

Pentosans yield pentoses (Sec. 6.2) such as xylose and arabinose upon complete hydrolysis with acid. They are α-linked and thus not degradable by mammalian enzymes, but they are rapidly hydrolyzed by microbial and fungal enzymes. Pentosans make up a significant amount of the heteroglycan, hemicellulose. When pentosans are boiled with hydrochloric acid, an aldehyde, furfural, is produced. This reaction is the basis of the quantitative determination of pentosans and is used in the commercial production of furfural from oat hulls and corn cobs.

Starch. Quantitatively hexosans provide the greatest amount of energy for animal needs of any of the carbohydrates in nature. The reserve material of most plants consists of starch, which is found in the tubers, rhizomes, and seeds. However, it is found in small quantities in all parts of the plant such as shoots, stems, and leaves. In ripening fruits there is a conversion of starch to sugars. Two forms of starch exist: amylose and amylopectin. The former is soluble in hot water, has a molecular weight of 10,000 to 100,000, consists of straight chains of D-glucose units in an α-1,4 linkage, and for most plants comprises about 25 to 30 percent of the total starch moeity. The amount of amylose is controlled genetically and increases with maturity. Amylopectin is insoluble in hot water and is a branched polymer of D-glucose with chains of α-1,4-linked glucose connected by an α-1,6 cross linkage (see structure shown in Fig. 6.1). Each side chain contains about 19 to 26 glucose units. In the plant, starch exists as small granules which are characteristic in appearance for each plant. This property furnishes a means of microscopic identification. Because of a high degree of hydrogen bonding on its surface, some granules are quite resistant to rupture. Tuber starch, such as found in the potato, is extremely so and must be cooked before being utilized by species such as pigs or chickens. Such moist heat results in a rupture of the starch granule and an irreversible change in the crystalline structure of the molecule (gelatinization). Enzymatic attack is thus facilitated. For some starches gelatinization is not necessary or desirable. Phos-

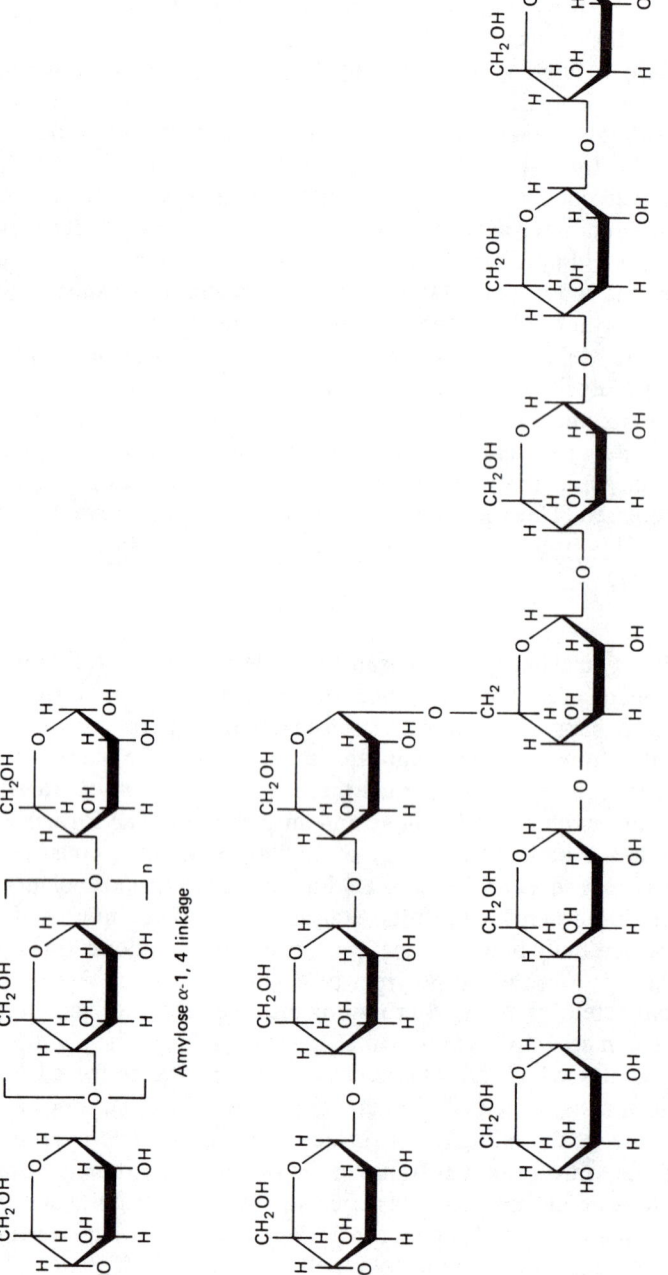

Fig. 6.1 Structure of amylose showing the α-1,4 linkage and amylopectin with the α-1,6 linkage as well (Haworth projection).

phorus is a normal though small component of amylopectin, being about one part per 400 glucose units.

Amylose, because of the cis configuration of the number 6 carbon atoms, tends to exist in a helical coil. In this form it can retain a large amount of iodine in its core and is responsible for the blue color reaction of iodine with starch. Less iodine is retained by amylopectin because helix formation is restricted by the branching and a red color ensues. Thus with the iodine test, the bluer the color, the more amylose.

Upon hydrolysis by either acid or enzymes, starch is changed into dextrin, maltose, and finally glucose. The amylopectin molecule is perceived to be one of the largest in nature, having a molecular weight of over 1,000,000.

Dextrins.   This is an ill-defined group of intermediates resulting from the hydrolysis and digestion of starch as well as from the action of heat on starch. They are a series of lower molecular weight compounds and are highly branched. In animals they result from the removal of glucose from amylopectin leaving branched residues, the α limit dextrin, composed of eight to ten glucose units. Dextrins are also the intermediates, when starch in germinating seeds is converted to glucose for energy to be used by the seedling. Dry heat treatment of starch results in its hydrolysis, depolymerization, rearrangement, and re-polymerization. Of particular interest in nutrition, dextrin is a favored substrate for acidophilic organisms in the digestive tract and, when incorporated in the diet, B-vitamin synthesis in the gut is improved such that dietary levels can be reduced.

Glycogen.   The small amount of carbohydrate reserve in the animal body exists in the liver and muscles in the form of glycogen, which resembles starch in certain properties as well as in function. It is therefore frequently called "animal starch." Glycogen is present in lower as well as in higher animal life. It may make up 10 percent of the wet weight in liver and, toward the end of the larval period, it makes up 33 percent of the dry weight of bee larvae. It is also present in yeasts and certain other fungi. Glycogen is a polymer of D-glucose combined in an α-1,4 linkage and an α-1,6 cross linkage. It resembles amylopectin in molecular weight, but its side chains consist of only 12 D-glucose units. Glycogen is water-soluble, gives a red color with iodine (no helical core), and yields glucose as the sole end product of hydrolysis.

Cellulose.   Cellulose is the most abundant substance in the plant kingdom and is the major structural component of plant cell walls (Sec. 6.5). Chemically it is a polymer of β-1,4-linked D-glucose units. As such, the six carbon atoms are in the trans position which results in cellulose being a flat, band-like microfibril. Recall that with starch the number 6 carbon atoms exist in a cis position, and a helical form results. Since in cellulose the glucose units exist in a chair-like configuration, as well as being in the β linkage, it is highly stable internally and, further, the microfibrils are held firmly to each other by hydrogen bonding. This configuration makes cellulose essentially insoluble and extremely resistant to enzymatic degradation. It can be hydrolyzed to glucose by strong acids. It may or may not be combined with lignin. While not acted upon by any mammalian enzyme, it can be degraded by fungi and bacteria. If a hydroxy-methyl group is added to carbon 6, a much more soluble and hydrolyzable compound (carboxymethyl cellulose) is formed. There are many different forms of cellulose, depending upon the species of plant. The number of polymerized glucose molecules may range from 900 to 2,000. Cotton is one of the purest forms.

Hemicellulose. Hemicellulose is *not,* as the name implies, one-half a cellulose. It is a complex, heterogeneous mixture of a number of different polymers of monosaccharides including glucose, xylose, mannose, arabinose, and galactose. It is a principal component of plant cell walls. A predominant hemicellulosic molecule is xyloglucan, which as its name implies, consists of a chain of β-1,4-linked D-glucose units with terminal branches of α-1,6-linked xylose units. This molecule is linked covalently to the pectic fraction of the cell wall and by hydrogen bonding to the cellulose microfibrils, thus adding significant strength to plant cells. Hemicellulose may also contain xylans, glucomannans, and galactoglucomannans.

Hemicellulose is much less resistant to chemical degradation than cellulose and is defined as a carbohydrate soluble in mild alkali. It can also be hydrolyzed by a relatively mild acid treatment. It is the cell wall fraction most closely allied with lignin, which is discussed below.

Pectin. Pectin is found primarily in the spaces between plant cell walls (middle lamella) but also infiltrates the cell wall itself. It consists of a backbone polymer of α-1,4-linked D-galacturonic acid units interspersed with 1,2-linked rhamnose units. It can be extracted with hot or cold water and will form a gel, as any jelly expert can attest. While the linkage is the α form, there is no mammalian enzyme capable of hydrolyzing pectins, and their digestibility relies entirely on microbial action. In spite of this fact, it is highly digestible by most species, including humans. Because of its water-holding (gel) capacity, it often is used to reduce diarrhea in infants and calves.

Lignin. Lignin is a class of noncarbohydrate compounds which give structural support to plant cell walls and as such is appropriately discussed with carbohydrates. True lignin is a high molecular weight, amorphous polymer of phenyl-propane derivatives. Its specific structure is not well described and its form can vary widely from plant to plant. Fundamentally it is a complex structure consisting of carbon-to-carbon bonds and ether linkages resistant to acid and alkali. Lignin is found in the woody parts of plants such as cobs, hulls, and the fibrous portion of roots, stems, and leaves. Hardwoods contain the most lignin of any plant. Its content increases steadily as the growing plant matures and its chemical linkage, especially with hemicellulose and cellulose, markedly reduces the digestibility of the latter. In grass forage this bond is an ester, while in legume forages it is an ether, neither of which is attacked by mammalian or microbial enzymes elaborated by anerobic organisms. Aerobic organisms and fungi can break the linkage, resulting in the rotting of forage and wood that we observe in nature. Alkali treatment of highly lignified grass such as straw breaks the hemicellulose-lignin bond, improving digestibility of the hemicellulose but not destroying the lignin.

Plant Gums. Plant gums are formed at the site of injury or by a deliberate incision and are viscous fluids which become hard when dry. They are used commercially as thickening agents or stabilizers for emulsions. They are complex, highly branched residues containing D-glucuronic and D-galacturonic acids along with other simple sugars such as arabinose and rhamnose. *Gum arabic* is a well-known commercial gum.

Mucilages. Mucilages are found in seeds and bark and appear to function as a water-holding compound which protects against dessication. They are highly variable in composition. The mucilage of alfalfa seed, for example, is predominantly a D-galacto-D-mannan. Others contain D-gluco-D-mannan and L-arabino-D-xylan.

Mucopolysaccharides. Mucopolysaccharides are widespread in animal connective tissue, are highly viscous, and readily form gels. They are synthesized primarily from the amino sugars D-glucosamine and D-galactosamine together with several uronic acids. Their physical-chemical properties provide some rigidity to animal tissues.

Chitin. Chitin is the principal component of the exoskeleton of insects and crustaceans and is composed of N-acetyl-D-glucosamine.

6.5 Development of Plant Cell Walls

Teleologically plant cell walls evolve to provide the plant with sufficient structural stability to obtain proper light and nutrition and to protect their reproductive parts and seeds from premature destruction. It also provides a barrier to the invasion of plant diseases and insects. While these features insure plant survival and propagation, they do impede their utilization by animals which consume them. From a pragmatic viewpoint, greater good no doubt results from rigid cell walls than if they were more maleable. To nutritionists interested in evaluating and utilizing plant materials for feed and food, cell-wall development is not an insignificant event.

Plant cells have traditionally been divided into three components, the middle lamella, the primary wall, and the secondary wall. The middle lamella is that space between the walls of two adjoining cells. The primary wall forms first in the growing plant and is the most dynamic of the structures. It is the exterior wall of the cell, provides its shape, and elongates as the plant grows. The secondary cell wall forms within the primary wall, provides strength to the cell, and is largely synthesized after cell elongation is complete. The lumen of the cell is centrally surrounded by the secondary cell wall and contains the active protoplasm of the cell. It is generally believed that there is a structural protein, high in the amino acid hydroxyproline, in plant cell walls as well. The physical and chemical distinction between the walls is not abrupt, there being a gradual transition between one and the other. The middle lamella consists primarily of pectin which also infiltrates and is bonded to the cellulose and hemicellulose of the primary wall. The primary wall consists of loosely bound cellulose fibrils, while the secondary cell wall consists of highly organized layers of cellulose fibrils lying in several directions. Hemicellulose is in greatest quantity in the secondary cell wall but infiltrates the primary wall, even to the middle lamella. When plant growth nears completion, lignin is formed in the secondary cell wall combining with hemicellulose and cellulose to give final rigidity to the cell. The plant cell dies when lignification is complete. A cross section of the quantitative relationship of these compounds as they might appear in cell wall is shown in Fig. 6.2.[3]

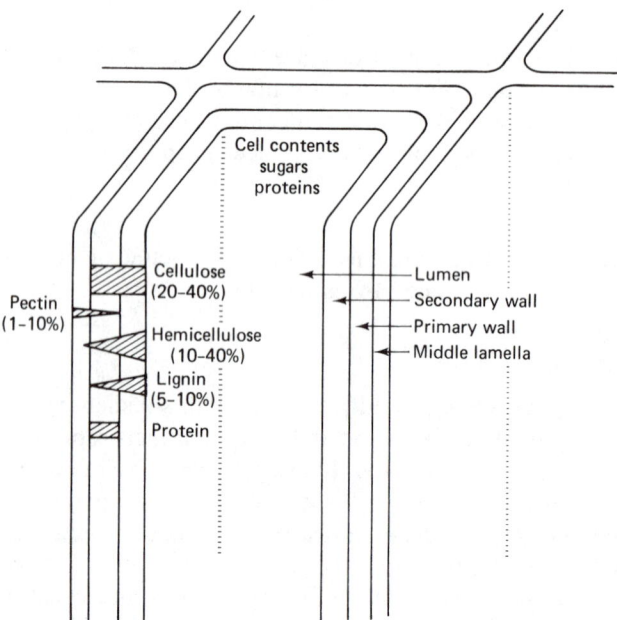

Fig. 6.2 Schematic representation of the structure of a forage plant cell showing the component layers. The relative amounts of each of the carbohydrate fractions in the respective layers are depicted by the shaded areas, e.g., hemicellulose largely in the secondary wall; pectin largely in the middle lamella. The figures in parentheses are amounts often found in forage dry matter.

6.6   The Determination of Carbohydrates for Nutritional Purposes

The analysis of feeds for all of the individual carbohydrates would be a tremendous task, especially since some of the procedures are difficult and time consuming. Thus there have been attempts to assay carbohydrates in groups: those well digested and those less well digested. A method was proposed over 100 years ago by Henneberg in Germany, known as the Weende method, after the experiment station where it was developed. It separated carbohydrate into two groups: *crude fiber* and *nitrogen-free extract* (NFE). The former was determined by alternately boiling the sample in weak acid and then alkali. The remaining residue was free of soluble components such as fat, protein, sugars, hemicellulose, and starch and contained the less soluble carbohydrate fraction, such as ligno-cellulose, and cellulose. The loss upon ignition represented the crude fiber. The NFE was determined by difference, i.e., that remaining after

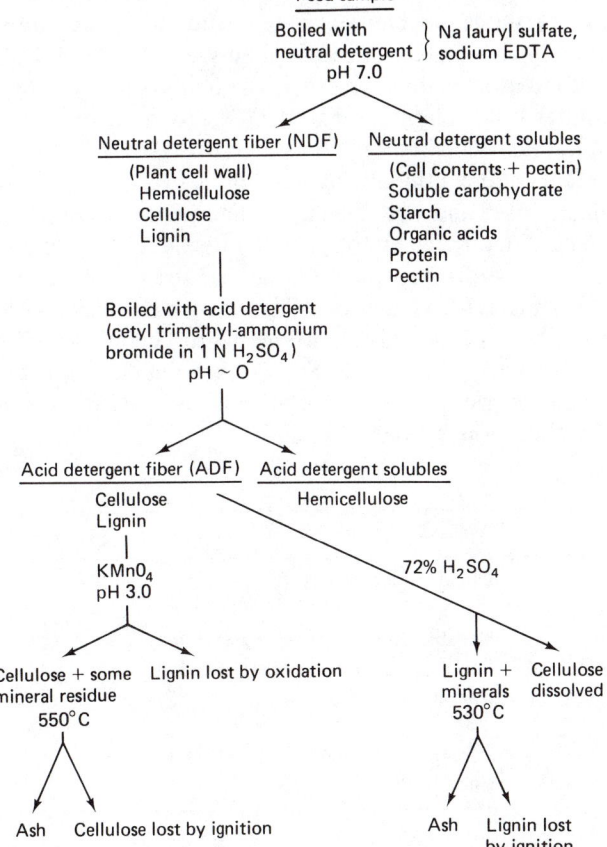

Fig. 6.3   The Van Soest method of partitioning fiber in feeds. (*Source: H. K. Goering and P. J. Van Soest, Forage Fiber Analysis. Agr. Handbook No. 379., A.R.S., U.S.D.A., 1975.*)

subtracting all the assayed values, e.g., crude fiber, protein, ether extract, and mineral matter (ash). This method has served the nutritionist well for over a century because it makes it possible to separate feeds into broad categories such as roughages with high crude fiber and concentrates with low crude fiber. Probably more than any other assay, it established the basis for merchandizing livestock feeds in which the feeder could be sure it contained quality ingredients and not undue amounts of coarse, undigestible fiber.

However, the crude-fiber–NFE assay has serious flaws. Since NFE is determined by difference, it accumulates all of the errors in the other assays. In addition, it is not the result of an assay itself, but just "what's left." Crude fiber also is a tenuous descriptive term and it has been shown in a number of cases that the digestibility of crude fiber is higher than the NFE in the same feed.[4] Hence crude fiber does not separate carbohydrates into readily digestible and less readily digestible fractions.

Recognizing this problem, Van Soest and his associates have devised a rapid method of partitioning feed carbohydrate into fractions based on nutritional availability. In these methods he has separated feedstuffs into those components that are (1) mostly available, (2) incompletely available, and (3) mostly unavailable. The method utilizes detergents which complex with protein to render it soluble and utilizes a chelating agent (EDTA) to remove heavy metal and alkaline earth contamination. Boiling the sample with a neutral detergent solubilizes the cell contents and pectin leaving behind the cell-wall fraction containing cellulose, hemicellulose, and lignin (neutral-detergent fiber, NDF). Subsequent boiling with an acid detergent hydrolyzes the hemicellulose that is free and that which is combined with lignin, leaving behind the cellulose and lignin as the acid detergent fiber (ADF). Oxidation of lignin with $KMnO_4$ leaves cellulose and ash, which, upon ignition, gives the value for cellulose. The procedure is graphically described in Fig. 6.3. Such a scheme fractionates carbohydrates into more realistic groups according to their utilization by animals and their resident microbial populations.

Fraction	Availability
Cell contents	Largely available
NDF	
Hemicellulose	Variable depending upon lignification
Cellulose	Variable depending upon lignification
Lignin	Unavailable
ADF	
Cellulose	Variable depending upon lignification
Lignin	Unavailable

This procedure should in time replace the less useful crude-fiber–NFE technique as a means of describing the carbohydrate portion of feedstuffs.

DIGESTION, ABSORPTION, AND METABOLISM OF CARBOHYDRATES BY NONRUMINANTS

From the standpoint of nutrition, two processes are essential for life: the assimilation of food for use in bodily functions, and the removal of waste products. There are three basic actions of the body which effect these changes: (1) *digestion* or the preparation of food for entrance into the bloodstream, (2) *absorption* into the bloodstream of the molecules derived from digestion, and (3) *metabolism* of the absorbed nutrients for use by the body and/or subsequent excretion.

6.7   Digestion and Absorption

The principles of digestion and the gastrointestinal organs involved have been described briefly in Chapter 3. Fig. 6.4 shows the several digestible carbohydrates and their respective enzymes and end products. The α-amylase, elaborated by the saliva and pancreas, attacks the internal α-1,4 linkages, hydrolyzing amylose to maltose and maltotriose (two and three glucose polymers, respectively). The hydrolysis of the branched amylopectin yields in addition α limit dextrins made up of eight to ten glucose molecules each with an α-1,6 branching linkage in addition to the α-1,4 linkages. Maltase, secreted at the intestinal brush border, promptly hydrolyzes maltose and maltotriose to glucose while an α-dextrinase (isomaltase) hydrolyzes the α limit dextrin to glucose and maltose.[5] The brush border also secretes lactase and sucrase to yield

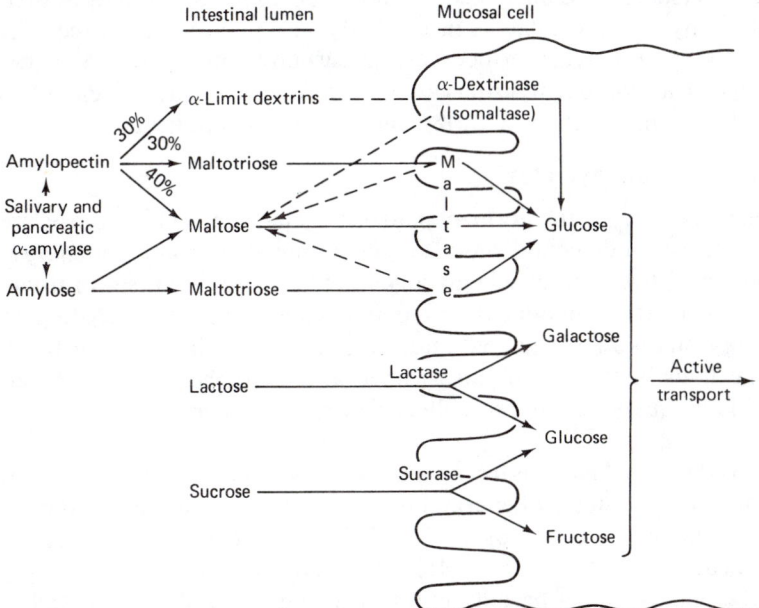

Fig. 6.4 Outline of carbohydrate digestion and absorption. (*Adapted from G. M. Gray, Federation Proc.* **26**:*1415–1419, 1967.*)

galactose, glucose, and fructose, as shown in Fig. 6.4. It has also been shown that glucose can be excreted back into the lumen to be absorbed farther on down the tract.

The absorption of glucose, galactose, and fructose is an active process utilizing a specific carrier protein that translocates the molecules across the brush border membrane. Fructose is transported less rapidly than the others suggesting a somewhat different mechanism. Energy is required and Na^+ and K^+ ions are involved.[6]

Aberrations of carbohydrate digestion and absorption are not uncommon and can compromise adequate nutrition even with a good diet. Diseases such as tropical sprue, or protein-calorie malnutrition (kwashiorkor or marasmus) cause sufficient damage to the intestinal wall that many of the dietary carbohydrates are not digested and hence not absorbed. Genetic anomalies such as a low level of intestinal lactase can prevent proper utilization of lactose, the valuable carbohydrate of milk. In all these syndromes, the osmotic effect of the undigested carbohydrate attracts water into the lumen of the intestine (hydrogogue effect) with diarrhea a prominant symptom. Bacterial degradation of the carbohydrate in the lower bowel often produces excessive gas to add to the person's discomfort.

Some non-Caucasians show a degree of lactose malabsorption due to a low level of intestinal lactase. Latham's[7] studies show that most of those who are labeled lactose malabsorbers can readily consume 200 ml of milk on an empty stomach without discomfort. Thus, while milk *ad libitum* may be contraindicated for these persons, judiciously used it should be an important part of their diet. As with many enzyme systems in the body, the presence or absence of substrate results in an increase or decrease in carbohydrates in the intestine. Such situations as starvation, semistarvation, nature of the diet, and circadian rythms alter their concentration in varied and not always predictable ways.[6]

6.8 Metabolism of Carbohydrates

Absorbed carbohydrate in the form of glucose, galactose, and fructose are metabolized in three fundamental ways: (1) as an immediate source of energy, (2) as a precursor of liver and nuscle glycogen, and (3) as a precursor of tissue triglyceride. The relative amount diverted into each pathway is contingent upon the energy status of the animal and the amount of carbohydrate being absorbed. The liver is the first organ to have access to the newly acquired nutrients and is uniquely adapted to collect and process them.

Glucose, Galactose, Fructose for Energy.   As was discussed in Sec. 5.3, the ultimate fate of the chemical energy bonded in the carbohydrate is for the production of high-energy bonds useful for the "work of the body" with the concomitant production of CO_2 and H_2O. The long tortuous pathway from glucose to CO_2 and H_2O was broadly outlined in Fig. 5.5 and is discussed in more detail in this chapter. It is the author's view that a thorough understanding of the biochemical interrelationship of nutrients is paramount to talking intelligently about nutrition. However, it is possible to understand the complexity without spending undue time on the minutia. Hence, while biochemical formulas are critical to the ultimate understanding of a chemical process, we have chosen to leave those details to courses in biochemistry and instead will stress a broader view of the processes involved. As students we recall how quickly formulas leave our consciousness and, having once been exposed to them, how quickly they can be retrieved by opening the proper textbook.

Fig. 6.5 outlines the pathway of energy processing for all carbohydrates and lipids. Lipids together with proteins will be discussed in more detail in Chaps. 7 and 8, respectively.

Note that there are three basic groupings of reactions: (1) glycolytic cycle (Embden-Meyerhof), (2) tricarboxylic acid cycle (TCA, or Krebs cycle), and (3) oxidative phosphorylation (cytochrome system). The first is anaerobic while the latter two require oxygen and are thus aerobic. Note further that (1) exists in the cytosol while (2) and (3) are within the mitochondrion. In this diagram the number of carbon atoms in each compound and the changes the reactions have on the amount of high-energy bonds have been recorded.

The first step in the oxidation of glucose to CO_2 and H_2O is the phosphorylation of glucose to glucose-6-phosphate by the enzyme hexokinase utilizing a phosphoryl group from ATP. A second ATP plus an isomerization reaction yields fructose-1,6-diphosphate. Because fructose is a C_6 molecule, it results in the production of two moles of glyceraldehyde-3-phosphate which proceeds to pyruvate. Glycolysis thus eventually results in the conversion of one mole of glucose to two moles of pyruvate with a utilization of 2ATP, a production of 4ATP and the reduction of NAD to $NADH_2$. This latter compound, after entering into the oxidative phosphorylation cycle, yields 3ATP per mole. However, since glycolysis occurs in the cytosol and oxidative phosphorylation is within the mitochondrion, an ATP is used when $NADH_2$ enters the mitochondrion. Thus, the gain is only 2 ATP/$NADH_2$. As a result, the overall ATP generation for one mole of glucose during glycolysis is (-1, -1, $+4$, $+2$, $+2 = 6$).

Pyruvate can freely enter the mitochondrion and is oxidatively decarboxylated in the presence of thiamin diphosphate. The CO_2 is lost and the $NADH_2$ passes into the oxidative phosphorylation reaction to yield 3ATP. The acetyl CoA then condenses with oxalacetate and is oxidized to CO_2 and H_2O via the TCA and oxidative phosphorylation pathways. Note the points of production of CO_2 and $NADH_2$ and note further that, in the transformation of succinate to fumarate, $FADH_2$ is produced. This coenzyme yields only 2ATP through the oxidative phosphorylation cycle. Totaling the ATPs produced from pyruvate (recall two moles), one gets 15 ATP per mole (4$NADH_2$, 1$FADH_2$, 1ATP directly), or a total of 30. The net gain from 1 mole of glucose is thus 36 ATPs (including 6 from glycolysis). If a value of 7.3 kcal per mole of ATP is used, the efficiency of energy conservation of glucose (673 kcal per mole) is about 39 percent with the loss appearing as waste heat.

Note in Fig. 6.5 the routes of entry of galactose and fructose into the glycolytic cycle. For all practical purposes these monosaccharides are as func-

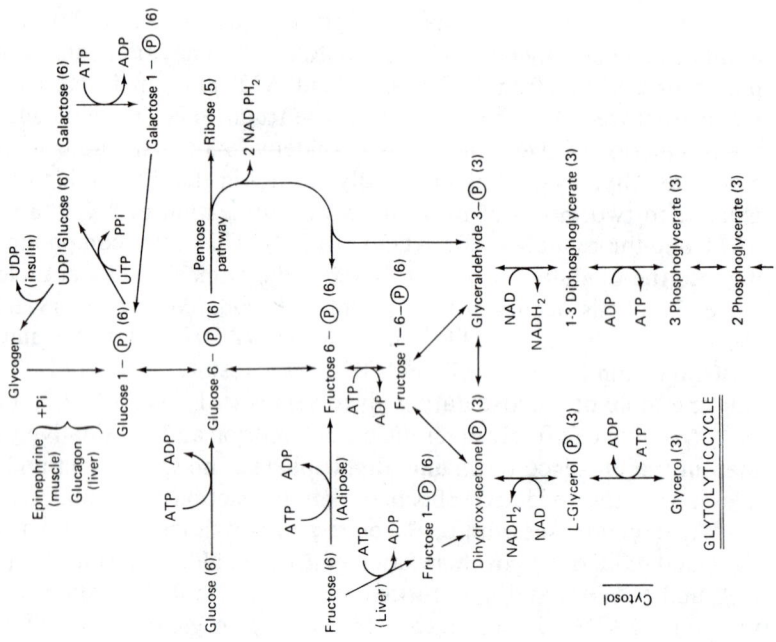

GLYTOLYTIC CYCLE

Cytosol

94

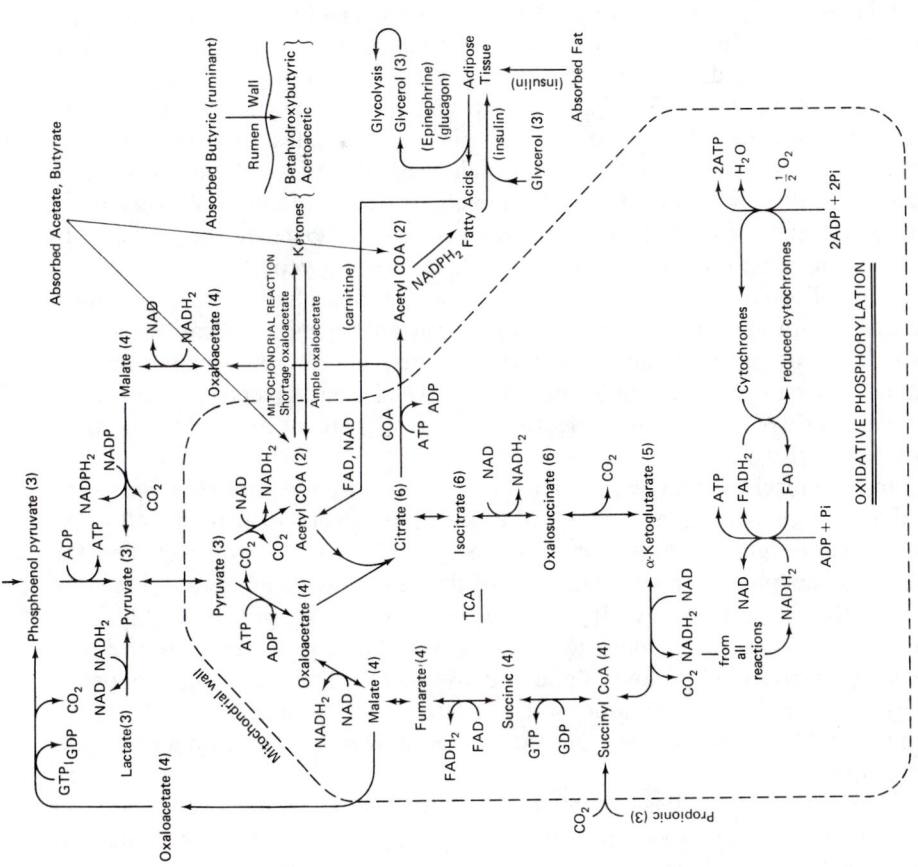

Fig. 6.5 Metabolic pathways used as carbohydrate and fat are converted to energy.

tional as glucose. A serious malady develops in infants who do not have the proper enzyme (galactose-1-phosphate uridyl transferase) to convert galactose-1-phosphate to glucose-1-phosphate. A galactosemia develops followed by vomiting, diarrhea, and jaundice. Many suffering with this affliction are mentally retarded. The only treatment is a galactose-free diet. While it cures the outward signs, the mental retardation is not reversible.

Glucose to Glycogen.   Although the body has need for energy on a continuing basis, most animals consume food periodically and in amounts exceeding their immediate need. The excess is stored as either liver or muscle glycogen or as fat for future needs. In Fig. 6.5, the pathway followed from glucose to glycogen is indicated. When blood glucose is elevated following a meal containing carbohydrate, insulin is secreted which triggers the formation of glycogen. A high-energy compound uridine triphosphate (UTP) (Table 5.4) is used for synthesis. Conversion of glycogen back to glucose-1-phosphate is by a separate pathway, utilizing the enzyme phosphorylase. Since glycogen is branched, in order for all of it to be converted to glucose, a debranching enzyme and a transferase are also required in addition to phosphorylase. Overall these processes use only one high-energy bond to elevate glucose-6-phosphate to glycogen and back to glucose-6-phosphate again. Thus using glycogen as a storage element, besides being handy teleologically, is 97 percent as efficient as using glucose directly.

In the muscles, glycogen, under the influence of epinephrine, is converted to glucose-6-phosphate and enters the glycolytic cycle to provide ATP. As muscle has no glucose-6-phosphatase (to split glucose-6-phosphate), and the glucose-6-phosphate cannot diffuse out of the cells, this metabolite can only be used at the cell for energy. In the liver, however, glycogenolysis (glycogen hydrolysis) occurs in response to the hormone glucagon secreted by the pancreas in response to a low blood glucose level. The resulting glucose-6-phosphate is hydrolyzed by glucose-6-phosphatase present in liver cells releasing glucose to the circulation where it can be utilized by the brain and muscle for energy.

Thus, while each tissue may have a small store of glycogen, the liver contains the main supply which, when thrown into the general circulation as sugar, becomes available for use by any tissue in the body (Table 6.4). The glycogenolytic function of the liver thus provides a mechanism whereby the blood-sugar level may be held within the comparatively narrow limits compatible with normal metabolism. The blood of cattle and sheep contains from 40 to 60 mg of sugar per 100 ml, a lower level than in nonruminants of 80 to 120 mg. Newborn ruminants have a level comparable to other species, which decreases as the animal matures. Birds have higher blood-sugar values than do mammals, but cold-blooded animals show very low figures, such as 20 mg per 100 ml commonly found for the frog.

The temporary storage of glycogen following carbohydrate absorption prevents *hyperglycemia,* i.e., a blood-sugar level above the normal range; and the

TABLE 6.4. Available Metabolic Fuels in a Normal 70-kg Male

	Weight of Fuel, kg	Caloric equivalent, kcal
Triglycerides	15	141,000
Proteins (muscle)	6	24,000
Glycogen (muscle)	0.150	600
Glycogen (liver)	0.075	300
Circulating fuels	0.023	100
		166,000

Source: G. F. Cahill, Jr., Starvation in man, *N. Eng. J. Med.,* **282**:668–675, 1970.

later release of this glycogen as glucose to balance the withdrawal of sugar from the blood by the tissues prevents the opposite, *hypoglycemia.* In spite of the reserve of glycogen, a 24-hour fast will reduce the levels to nearly zero. The storage is thus short-lived unless constantly replenished.

Glucose to Fat.   The ability of the liver and other tissues to store sugar as glycogen is limited, and thus, when the carbohydrate intake regularly exceeds the current need of the body for energy purposes, sugar is transformed into fat (Table 6.4). This process takes place on a large scale in the fattening of animals, since their food consists primarily of carbohydrates. This formation of body fat from carbohydrate food was first demonstrated by Lawes and Gilbert[8] by means of a slaughter experiment over a century ago. They chose pigs from the same litter and of the same size. Some of these animals were slaughtered at the start and analyzed as controls, while the others were killed after being fed for an extended period on a low-fat ration of known composition. The data obtained from the analysis of these animals, compared with the data from the controls, showed that the pigs had stored more fat than could have resulted from all of the fat and protein fed and, therefore, a part of their fat must have been formed from carbohydrates. Today this empirical observation is clearly supported biochemically.

The formation of fat from glucose involves the synthesis of the two components, fatty acids and glycerol. In both cases the glucose is first broken down via the glycolytic pathway (Fig. 6.5). Dihydroxyacetone serves as a precursor for glycerol utilizing one ATP. Pyruvate is decarboxylated to acetyl CoA which is the precursor of the fatty acids. The mechanism of fatty acid synthesis is explained in detail in Sec. 7.12.

Phosphogluconate Oxidative Pathway (Pentose Shunt).   Not all of the glucose-6-phosphate need pass through the glycolytic cycle; a portion is diverted into the phosphogluconate oxidative pathway for the express purpose of synthesizing 5-carbon sugars (ribose) and the reduced coenzyme $NADPH_2$. Ribose is used in the synthesis of DNA, RNA, ATP, etc., and the coenzyme $NADPH_2$, which provides reducing power for the synthesis of fatty acids. Fig.

6.5 shows the pathway in general form and Fig. 6.6 in detail. Note that within the pathway there exist a number of intermediates, e.g., glyceraldehyde-3-phosphate, fructose-6-phosphate and fructose-1,6-diphosphate, all of which could be a part of the glycolytic cycle. Thus the reaction can be tilted towards ribose, or $NADPH_2$ with the "leftovers" returned to the glycolytic cycle. As one might expect, this pathway is more active in adipose tissue "where the action is." Note that in the TCA cycle (Fig. 6.5) the reaction of malate to pyruvate also generates $NADPH_2$.

Oxygen Debt. A Way to Prolong Strenuous Exercise.   In cases of extreme exertion the need for ATP exceeds the ability of the animal to provide oxygen to the tissues to keep the oxidative phosphorylation and TCA cycles operative. As a result glucose metabolism stops at pyruvate and pyruvate concentration increases (Fig. 6.5). However, in the presence of lactate dehydrogenase, pyruvate is promptly reduced to lactate and $NADH_2$, which was produced in the reaction glyceraldehyde-3-phosphate to 3-phosphoglycerate, is oxidized back to NAD^+. This oxidized cofactor can then be returned to react with glyceraldehyde-3-phosphate and allow glycolysis to continue. This process, from glucose to pyruvate, results in a net production of 2ATP (-1, -1, $+2$, $+2$) (note $NADH_2$ yields no ATP in this situation), together with a buildup of muscle lactate. Lactate promptly diffuses from the muscle cell into the blood, is carried to the liver where it is oxidized to pyruvate, and to glucose via a special pathway known as the Cori cycle. Glucose then diffuses back into the circulation to return to the muscle to be reduced again to lactate via the glycolytic cycle. In this way, ATP continues to be generated to a limited degree in spite of

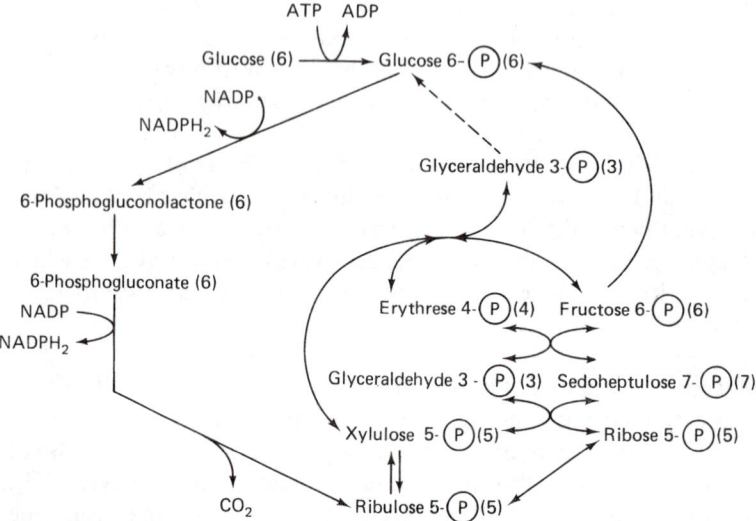

Fig. 6.6 Phosphogluconate oxidative pathway.

a dearth of oxygen. When the strenuous exertion ceases and tissue oxygen increases, the TCA and oxidative phosphorylation cycles become operative again and hence ATP generation returns to normal. Blood and tissue lactate levels then decline to normal levels as well. The organism has thus been able to function in spite of a limited tissue oxygen supply.

DIGESTION, ABSORPTION, AND METABOLISM OF CARBOHYDRATES BY RUMINANTS

The impact of the gastrointestinal architecture of the ruminant on the digestive processes of carbohydrates was discussed in Chapter 3. The net effect is that while nonruminants absorb primarily monosaccharides from carbohydrates, the ruminant absorbs primarily volatile fatty acids and little if any monosaccharides.

6.9 Digestion and Absorption

A probable system of the chemical degradation of carbohydrates is given in Fig. 6.7.[9] You will note that initially all dietary carbohydrate is converted to glucose. However, glucose is present only transiently and is promptly converted via pyruvate to volatile fatty acids. Methanogenic bacteria then utilize formate, H_2, and CO_2 to produce methane. The relative proportions of the volatile fatty acids vary with diet (Sec. 3.2). In general the molar percentages on an all hay diet are acetic 65, propionic 20, butyric 12, and others such as valeric, isovaleric, and isobutyric 1. Increasing the level of grain to 70 percent can change acetate and propionate to about 40 and 37 molar percent, respectively.

The shift in acid percentages is not a chance event but the end result of a complicated adjustment of the biomass in the rumen. Such a dramatic change in diet (i.e., microbial substrate) has a marked impact on the numbers and kinds of microorganisms in the rumen contents. Numbers of cellulolytic organisms decline while starch fermenters increase. Again, the degree of change of each is the result of a fantastically complicated interaction among microbial species, amount and kind of substrate, end products produced, absorption of products into the blood, and passage of materials down the tract. In the final analysis it is axiomatic that the oxidation-reduction potential of the biomass must be in balance for the reactions to proceed, and in the rumen—proceed they do.

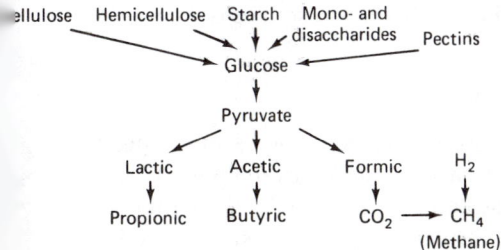

Fig. 6.7 Suggested metabolic pathways for the fermentation of carbohydrates by the rumen microorganisms.

Van Soest[10] has adapted the concepts reported by Wolin[11] and described the kinds of equations that exist in the fermentation of glucose to the principal volatile fatty acids.

1. Acetate = $C_6H_{12}O_6 + 2H_2O \longrightarrow 2C_2H_4O_2 + 2CO_2 + 8H$
2. Propionate = $C_6H_{12}O_6 \longrightarrow 2C_3H_6O_2 + 2[0]$ (acrylate pathway)
3. Butyrate = $C_6H_{12}O_6 \longrightarrow C_4H_8O_2 + 2CO_2 + 4H$

These equations in effect tell us that if these compounds are produced, there will be an excess of 8H per mole acetate, 4H per mole butyrate, and a deficiency of 4H for each mole of propionate formed via the acrylate pathway. Thus in a normal fermentation there is an excess of [H] and the rumen contents constitute a highly reduced medium. To stay in balance, extra hydrogen must be removed. Methane is the hydrogen sink which keeps this process in balance (Fig. 6.7). As methane cannot be metabolized by the animal, it is thus a net loss of feed energy. Extensive efforts have been made to depress methane production and thus to divert its energy to compounds metabolizable by the animal. One of the most effective has been Monensin (Sec. 12.3). Because less methane is produced and the oxidation-reduction balance must remain, a shift in the molar percentages of volatile fatty acids must occur. In fact, there is significant increase in propionate, the VFA deficient in [H], and a reduction in acetate, the VFA providing excess [H].[12] In short, in spite of its complexity, the rumen fermentation is a finely tuned biochemical process. It does and will represent a significant challenge to the ruminant biochemist, perhaps for a long time.

As a result of the pioneer studies of Barcroft and associates and later investigations by others, it has become clearly established that the fatty acids produced by microbial action are directly absorbed from the rumen, reticulum, omasum, and large intestine. The absorption from the rumen is prompt, and elevated levels in the portal blood have been noted within 10 minutes after eating.[13] The process appears to be by simple diffusion with the undissociated acid passing through at the normal pH of the rumen of about 6.7. The rumen epithelium is not a simple sieve but has the capacity to metabolize volatile fatty acids as they are absorbed. It is believed 80 to 90 percent of the butyrate is converted to ketone bodies (acetoacetic acid and β-hydroxybutyric acid). Thus portal and systemic blood levels of butyrate are extremely low. Up to 50 percent of the propionate may be metabolized to lactate and pyruvate during absorption. Relatively little acetate is used other than as a source of energy by ruminal epithelium and muscle.

6.10 Metabolism of Volatile Fatty Acids (VFA)

The liver is presented with a variety of carbohydrate metabolites in the portal blood, none of which is glucose, as in the case of the nonruminant. Acetate largely passes through to enter the bloodstream. It is the only VFA found in appreciable quantities in the peripheral circulation. It is phosphorylated to acetyl CoA and enters the TCA cycle. It requires two high-energy bonds to become activated and yields 12 moles of ATP upon oxidation (Fig. 6.5, $3NADH_2$, $1FADH_2$, 1ATP directly). Thus there is a net gain of 10 ATP per mole of acetic acid absorbed. Acetate can also be used directly for the synthesis of milk fat, especially the short chain acids (Sec. 17.12).

Propionic acid is largely removed from portal blood by the liver where it is converted to glucose. In fact, it is the primary source of glucose for the ruminant and, in the case of a cow producing 40 kg of milk per day, has been estimated to provide as much as 60 percent of that needed.[14] Propionate is converted to glucose by first entering the TCA cycle as succinyl CoA (Fig. 6.5) in the following manner:

$$\text{Propionate (3)} + \text{CoA} \xrightarrow{\quad \text{ATP} \quad \text{AMP} \quad} \text{Propionyl CoA (3)}$$

$$\xrightarrow[\text{ATP} \quad \text{AMP}]{\quad CO_2 \text{(Biotin)} \quad} \text{Methylmalonyl CoA (4)} \xrightarrow{\quad (\text{Vitamin B}_{12}) \quad} \text{succinyl CoA (4)}$$

The reactions require three high-energy bonds and involve two vitamins, biotin and vitamin B_{12}. Since the oxaloacetate-phosphoenolpyruvate reactions are not reversible, succinyl CoA must bypass these reactions on the way to phosphoenolpyruvate and "up" to glucose. Since oxaloacetate cannot move through the mitochondrial membrane, while malate can, succinyl CoA is converted to malate, which passes through the membrane and is converted to oxaloacetate and then into phosphoenolpyruvate. Phosphoenolpyruvate then moves in the reverse of glycolysis to yield glucose. If propionate is converted to glucose and then metabolized to CO_2 and H_2O there is a net gain of 17 ATPs per mole or 34 per glucose equivalent (C_6). This contrasts with 36 from glucose alone.

Butyric acid is absorbed as a ketone body, being eventually metabolized as acetyl CoA as discussed in Sec. 7.25. The net ATP production is 25 per mole.

NOTES

1. W. Pigman and D. Horton, The Carbohydrates—Chemistry and Biochemistry, *Vol. 11B,* Academic Press, New York, 1970, chap. 46, pp. 809–834.

2. E. Cristofaro et al., Involvement of the raffinose family of ologosaccharides in flatulence, in H. L. Sipple and K. W. McNutt (eds.), Sugars in Nutrition, Academic Press, New York, pp. 314–336.

3. D. F. Bateman and H. G. Basham, Degradation of plant cell walls and membranes by microbial enzymes, in P. Heitfuss and P. H. Williams (eds.), Encyclopedia of Plant Physiology, New Series No. 4, Springer-Verlag, New York, 1976, pp. 316–355.

4. P. J. Van Soest, Physico-chemical aspects of fibre digestion, in I. W. MacDonald and A. C. I. Warner (eds.), Digestion and Metabolism in the Ruminant, The Univ. New England Publishing Unit, Armidale, Australia, 1975, pp. 351–365.

5. L. J. Filer, Jr., Digestion absorption and metabolism of starch, in L. M. Hood et al. (eds.), Carbohydrates and Health, AVI Publishing Co., Westport, Conn., 1977, pp. 39–43.

6. S. Rieser, Digestion and absorption of dietary carbohydrates, in C. D. Berdanier (ed.), Carbohydrate Metabolism, Halsted Press, New York, 1976, pp. 45–78.

7. M. C. Lathman, Public health importance of milk intolerance, *Nutrition News,* **40:**13–31, Dec. 1977.

8. J. B. Lawes and J. H. Gilbert, Experimental enquiry into the composition of the animals fed and slaughtered as human food, *Trans. Roy. Soc (London)*, **149:**493–680, 1859.

9. Adapted from R. A. Leng, Formation and production of volatile fatty acids in the rumen, in A. T. Phillipson (ed.), Physiology of Digestion and Metabolism in the Ruminant, Oriel Press, 1970.

10. P. J. Van Soest, Personal communication, 1977.

11. M. J. Wolin, Interactions between bacterial species of the rumen, in I. W. MacDonald and A. C. I. Warner (eds.), Digestion and Metabolism in the Ruminant, Univ. of New England Publishing Unit, Australia, 1974.

12. M. L. Thonney, Use of Monensin sodium in feeding cattle: A review, *Proc. Cornell Nutrition Conf.,* 1977, pp. 104–112.

13. L. E. Chase, P. J. Wangsness, and R. J. Martin, Portal blood insulin and metabolite changes with spontaneous feeding in steers, *J. Dairy Sci.,* **60:** 410–415, 1977.

14. J. M. Elliot, Glucose economy of the lactating dairy cow, *Proc. Cornell Nutrition Conf.,* 1976, pp. 59–66.

SUPPLEMENTARY LITERATURE

Berdanier, C. D.: Carbohydrate Metabolism, Regulation and Physiological Role, Halsted Press, New York, 1976.

Bergen, W. G., and M. T. Yokoyama: Productive limits to rumen fermentation, *J. Animal Sci.,* **46:**573–584, 1977.

Hood, L. M., E. K. Wardrip, and G. N. Bollenback: Carbohydrates and Health, AVI Publishing Co., Inc., Westport, Conn., 1977.

Lehninger, A. L.: Biochemistry, Worth Publishers, Inc., New York, 1975.

McLaren, G. C., et al.: Level of readily fermentable carbohydrates and adaptation of lambs to all-urea supplemented rations, *J. Nutrition,* **87:**331–336, 1965.

Moir, K. W.: Accuracy and precision of two methods for determining acid-detergent fibre, *Lab. Pract.,* **25:**521–584, 1976.

Pigman, W., and D. Horton: The Carbohydrates, Chemistry and Biochemistry, Vol. IA, IB, IIB, Academic Press, New York, 1970–1972.

Putnam, P. A., and R. E. Davis: Post-ruminal fiber digestibility, *J. Animal Sci.,* **24:** 826–829, 1965.

Sipple, H. H., and K. W. McNutt (eds.): Sugars in Nutrition, Academic Press, New York, 1974.

Teague, H. S., and L. E. Hanson: The effect of feeding different levels of a cellulosic material to swine, *J. Animal Sci.,* **13:**206–214, 1954.

Weinberg, S., and A. L. Sheffner: Buffers in Ruminant Physiology and Metabolism, Church and Dwight Company, New York, 1975.

Chapter 7

The Lipids
and Their Metabolism

Plant and animal materials contain a group of substances, insoluble in water but soluble in ether, chloroform, and benzene, which are most commonly referred to as lipids. The group includes the fats and a number of closely related compounds (Table 7.1). From the standpoint of the amounts present in the animal body and its food, the fats are by far the most important members of the group, but several of the other lipids play very significant roles in nutrition and physiology. As an example, one may cite cholesterol, which is the precursor of vitamin D and sex hormones on the one hand, and the infamous component of atheromatous plaques of cardiovascular disease on the other. In the body, the lipids serve as a condensed reserve of energy, as structural elements of the tissues, and as essentials for various reactions in intermediary metabolism.

7.1 Classification of Lipids

Table 7.1 is a suggested classification of the lipids which, while based on their chemical similarity, includes most of those which have nutritional or paranutritional significance.

7.2 Fatty Acids

Since the fatty acids are constituents of most of the other lipids, it is helpful to take them up first. While over 100 have been isolated from nature, Table 7.2 lists those which commonly occur in plant and animal fats and are of especial

TABLE 7.1. A Classification of the Lipids

Saponifiable		Nonsaponifiable
Simple	Compound	Terpenes
		Steroids
Fats	Glycolipids	Prostaglandins
Waxes	Phospholipids	

interest in nutrition. The common names are given, while names in parentheses represent the modern chemical nomenclature. The suffix denotes the state of saturation: *-anoic,* saturated; *-enoic,* one double bond; *-dienoic,* two double bonds; *-trienoic,* three double bonds; *-tetrenoic,* four double bonds; etc. These double bonds are reflected in the formulas for the acids in that they have a smaller number of hydrogen atoms relative to the carbon atoms present. The symbols also reflect the number of carbon atoms as well as the number and location of the double bonds. The term "polyunsaturated fatty acids" is frequently applied to those having more than one double bond and they have especial significance in nutrition as we shall see later. Note that the melting point of the fatty acid increases as the chain length increases, but that adding a double bond lowers it for a given chain length; two double bonds lower it even

TABLE 7.2. Fatty Acids Commonly Found in Lipids

Acids	Empirical formula	Symbols		Melting point, °C
Saturated				
Butyric (butanoic)	$C_4H_8O_2$	$C4:0^a$		−4.3
Caproic (hexanoic)	$C_6H_{12}O_2$	$C6:0$		−2
Caprylic (octanoic)	$C_8H_{16}O_2$	$C8:0$		16.5
Capric (decanoic)	$C_{10}H_{20}O_2$	$C10:0$		31.4
Lauric (dodecanoic)	$C_{12}H_{24}O_2$	$C12:0$		44
Myristic (tetradecanoic)	$C_{14}H_{28}O_2$	$C14:0$		58
Palmitic (hexadecanoic)	$C_{16}H_{32}O_2$	$C16:0$		63
Stearic (octadecanoic)	$C_{18}H_{36}O_2$	$C18;0$		71.5
Unsaturated				
Palmitoleic (hexadecenoic)	$C_{16}H_{30}O_2$	$C16:1^{\Delta 9\,b}$		1.5
Oleic (octadecenoic)	$C_{18}H_{34}O_2$	$C18:1^{\Delta 9}$	$C18:1\omega 9^c$	16.3
Linoleic (octadecadienoic)	$C_{18}H_{32}O_2$	$C18:2^{\Delta 9,12}$	$C18:2\omega 6$	− 5.0
Linolenic (octadecatrienoic)	$C_{18}H_{30}O_2$	$C18:3^{\Delta 9,12,15}$	$C18:3\omega 3$	−11.3
Arachidonic (eicosatetraenoic)	$C_{20}H_{32}O_2$	$C20:4^{\Delta 5,8,11,14}$	$C20:4\omega 6$	−49.5

[a] $C4:0$; four carbon atoms; 0 double bonds.
[b] Denotes that the fatty acid has one double bond which starts at the ninth carbon atom beginning with the carboxyl carbon.
[c] A designation used for unsaturated fatty acids. The ω denotes the first carbon atom of the double bond counting from the terminal methyl. Similar ω positions group unsaturated fatty acids into families and indicate, for example, $C20:4\omega 6$ was synthesized from $C18:2\omega 6$.

more. Thus the melting point of C18:0 > C18:1 > C18:2. This characteristic is important as it influences the properties of the fats of which they are a part.

The general formulas for four of the more common unsaturated fatty acids are shown below. With the double bonds they can exist in the cis or trans form; in nature, most are in the cis form. The double bonds are highly reactive and are especially susceptible to oxidation (Sec. 7.3).

$$\overset{\omega 9}{CH_3}-(CH_2)_7-\overset{\Delta 9}{CH}=CH-(CH_2)_7-COOH$$
Oleic acid

$$CH_3(CH_2)_4-\overset{\omega 6}{CH}=CH-CH_2-\overset{\Delta 9}{CH}=CH-(CH_2)_7-COOH$$
Linoleic acid

$$CH_3-CH_2-\overset{\omega 3}{CH}=CH-CH_2-CH=CH-CH_2-\overset{\Delta 9}{CH}=CH-(CH_2)_7-COOH$$
Linolenic acid

$$CH_3(CH_2)_4\overset{\omega 6}{CH}=CHCH_2CH=CHCH_2CH=CHCH_2\overset{\Delta 5}{CH}=CH(CH_2)_3COOH$$
Arachidonic acid

All of the acids listed in Table 7.2 have straight chains and an even number of carbon atoms. This is a characteristic of almost all natural fatty acids. Microbial lipids, however, are often branched and have chains of an odd number of carbon atoms. Hence the fats of the ruminant body and milk characteristically contain these ''aberrant'' forms. The first five listed are classed as volatile fatty acids (VFA) since they can be distilled with steam. By having a carboxyl group at one end, they can react with Ca^{++} or Mg^{++} to produce soap.

7.3 Fats

Fats are defined as the esters formed by the union of a trihydroxy alcohol *glycerol* and three moles of fatty acids. The basic reaction is a dehydration shown below, with water being removed.

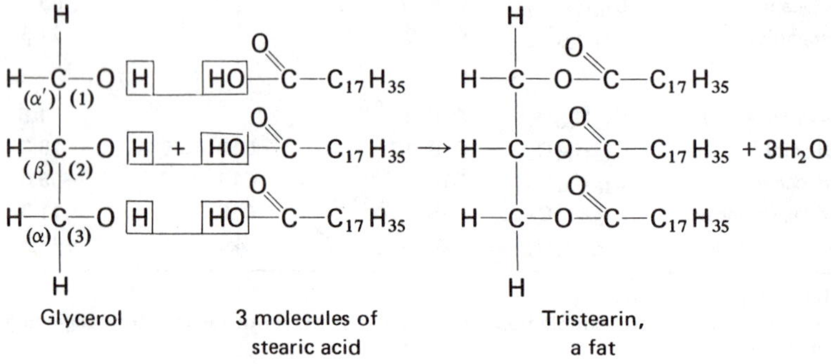

Glycerol   3 molecules of stearic acid   Tristearin, a fat

In this reaction three moles of stearic acid (C18:0) have been used to form the fat, tristearin. If the fatty acids had been palmitic, oleic, and stearic, the molecule would have been called palmitoleostearin. In common parlance this compound is referred to as a *neutral fat,* a *triglyceride* (TG), or more correctly a *triacylglycerol.* The kinds and proportions of fatty acids in the molecule determine the chemical and physical properties of the fat, which may vary considerably, as shown in Table 7.3. For example, the vegetable triglycerides, such as corn oil, are composed largely of unsaturated fatty acids, especially polyunsaturated acids, and melt below room temperature. Tallow, on the other hand, has substantially fewer unsaturated fatty acids, about 40 percent, and is solid at room temperature (melting point, 42°C). *Oils* are generally those liquid at room temperature (20°C) while *fats* are solid.

Note that butter has a low level of unsaturated fatty acids but still has a rather low melting point. This is due to its unique character of having substantial levels of short-chain fatty acids, which in turn have low melting points. Coconut fat is the only vegetable fat which has similar qualities and is one reason it has been popular as a nondairy fat substitute. Lard and poultry fat have

TABLE 7.3. Physical and Chemical Characteristics of Several Fats and Oils

	Corn[a]	Soy[a]	Safflower[b]	Coconut[a]	Grass pasture[c]	Butter[a]	Tallow[a]	Lard[a]	Egg[d]
Saturated acids									
Butyric C4:0						3.2			
Caproic C6:0				0.2		1.8			
Caprylic C8:0				8.2		0.8			
Capric C10:0				7.4		1.4			
Lauric C12:0				47.5		3.8			
Myristic C14:0			0.2	18.0	1.0	8.3	3		0.3
Palmitic C16:0	7.0	8.5	12.3	8.0	16.0	27.0	27	32.2	22.1
Stearic C18:0	2.4	3.5	1.8	2.8	2.0	12.5	21	7.8	7.7
Total	9.4	12.0	14.3	92.8	21.1	58.8	51	40.0	30.1
Unsaturated acids									
Palmitoleic C16:1					2.0				3.3
Oleic C18:1	45.6	17.0	11.2	5.6	3.0	35.0	40	48.0	36.6
Linoleic C18:2	45.0	54.4	74.3	1.6	13.0	3.0	2	11.0	11.1
Linolenic C18:3		7.1			61.0	0.8	0.5	0.6	0.3
Arachidonic C20:4									0.8
Total	90.6	79.5	85.5	7.2	79.0	38.8	42.5	59.6	52.1
Melting point °C	<20	<20	<20	20–35		28–36	36–45	35–45	
Iodine no.	105–125	130–137		8–10		26–38	46–66	40–70	
Saponification no.	87–93	190–194		250–260		220–241	193–200	193–220	
Reichert-Meissl no.				6–8		23–33			

[a]Most of these data were taken from a compilation by Vera R. Goddard and Louise Goodall, Agr. Res. Service, U.S.D.A., 1959.
[b]M. C. Nesheim (personal communication).
[c]G. A. Garton, Fatty acid composition of the lipids of pasture grasses, *Nature,* **187:**511–512, 1960.
[d]Agriculture Handbook 8-1, Composition of Foods: Dairy and Egg Products, Raw and Prepared, U.S.D.A., 1976.

more linoleic acid than either tallow or butter and therefore are somewhat "softer" at room temperature.

It is obvious that the assembling of the fatty acids by the plant or animal cell is not a random event but dictated by genetics and, as we shall see in the case of animals, to some extent by diet. Positioning of specific fatty acids on the glycerol moiety is also not completely random. Lard, for example, has all of its palmitic acid on the 2 position while in plant oils, all of the saturated acids are esterified to the 1 and 3 positions.[1]

Fat Analysis and Characterization. In routine feed analysis, the lipids are determined as *ether extract*. The feed is dried to a moisture-free basis and then extracted for 16 hours with anhydrous ethyl ether. The extract is weighed after the evaporation of the ether. In addition to lipids, ether extracts plant pigments, such as chlorophyll, xanthophyll, and carotene, and traces of various other substances. Ether also removes certain *essential oils* which are nonlipid products consisting primarily of aromatic esters, aldehydes, and ethers. Thus the use of the term ether extract as a synonym for fat, in speaking of the nutrient composition of feeds and rations, is not strictly accurate. In certain leafy materials, the amount of ether extract other than esters of fatty acids may represent 25 to 40 percent of the total. In those foods, which we recognize to be the chief sources of dietary fat, viz., seeds and animal products, the ether extract consists very largely of triglycerides. While a useful procedure, the limitations of ether extract must be considered.

Modern techniques are now available using both gas-liquid and/or thin-layer chromatography to estimate the amounts of the specific fatty acids as well as their location on the glycerol molecule.[2]

There are a number of standard assays which are also helpful in characterizing fat. The *melting point* (or solidifying point) is one discussed above which gives a useful measure of the hardness of the fat and indirectly reflects the carbon chain length or degree of unsaturation. A more specific measure of the degree of unsaturation is the *iodine number*. Iodine readily unites with the unsaturated fat at the double bond, two moles of iodine per double bond. It is defined as the number of grams of iodine absorbed by 100 g of fat. Table 7.3 indicates the marked difference between soybean oil (130) and butter fat (30).

When a fat is boiled with alkali such as potassium hydroxide, it is split into glycerol and the alkali salt of the fatty acids. These alkali salts are called soaps and the process is called saponification. The number of milligrams of alkali (KOH) required to saponify 1.0 g of fat is called the *saponification number*. Since one mole of K^+ reacts with each fatty acid, the larger the saponification number, the smaller the average chain length. As an illustration, corn oil has a much smaller value than butter fat (Table 7.3).

The determination of the amount of water-soluble, steam-volatile fatty acids is a useful measure of the character of butter fat and is used for detecting its adulteration, since the high percentages of these acids in butter fat are its distinctive feature. This measure is called the *Reichert-Meissl* number. In Table

7.3 it can be seen that coconut fat is the only one, aside from butter, which has a value of any size, and they can be easily differentiated.

Rancidity.   Fats, in feeds and foods, are susceptible to two kinds of rancidity, *hydrolytic* and *oxidative*. The former is the result of the hydrolytic splitting of the fatty acids from the glycerol, the opposite of triglyceride (TG) synthesis. It is produced by the action of the enzyme lipase (Sec. 7.9) which occurs in molds and bacteria and in some fresh foods. It results in the production of mono- and diglycerides (one or two fatty acids split off the TG) and free fatty acids. In situations where the free fatty acid has a specific taste or odor, the result can be offensive although not altering its nutritional value. A case in point is very aged butter, in which butyric acid has been released which accounts for its unmistakable, offensive, rancid odor.

Oxidative rancidity occurs in fats having substantial levels of polyunsaturated fatty acids and, in addition to producing abnormal flavors and odors, may seriously alter the nutritional value of the feed. Polyunsaturated fatty acids are of the methylene-interrupted type, i.e., a CH_2 group exists between the two double bonds. The polyunsaturated acids are of this type. This is in contrast to the conjugated type in which only a single C—C bond is between the double bonds. Under proper conditions of temperature, moisture, and often catalytic amounts of copper, a hydrogen atom is abstracted from the methylene carbon (C) leaving a free radical. The double bond shifts to the conjugated form and the free radical can then unite with oxygen (O) to form a peroxide and then with hydrogen (H) to form a hydroperoxide as shown below. The hydroperoxide can

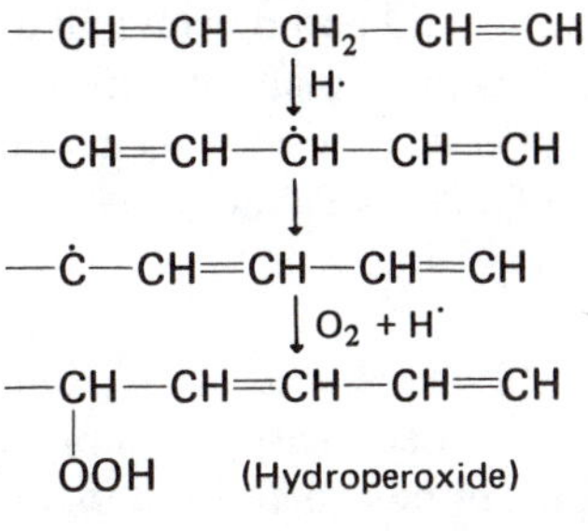

$$—CH\!\!=\!\!CH—CH_2—CH\!\!=\!\!CH$$
$$\downarrow H\cdot$$
$$—CH\!\!=\!\!CH—\overset{\cdot}{C}H—CH\!\!=\!\!CH$$
$$\downarrow$$
$$—\overset{\cdot}{C}—CH\!\!=\!\!CH—CH\!\!=\!\!CH$$
$$\downarrow O_2 + H\dot{}$$
$$—CH—CH\!\!=\!\!CH—CH\!\!=\!\!CH$$
$$\underset{OOH}{|} \quad \text{(Hydroperoxide)}$$

Autooxidation of methylene-interrupted unsaturated fatty acid

polymerize with other hydroperoxides to form a film as in the drying of linseed oil paint. It can also break down into aldehydes and ketones which are odorous. The oxidation is autocatalytic which means that after it once gets started it continues at an increasing rate. The result is the production of undesirable end products and the destruction of essential fatty acids (EFA) since the peroxidized fats are no longer methylene interrupted (Sec. 7.18).

The reaction can be stopped if antioxidants are present which promptly replace the H once it has been abstracted. Vitamin E is a natural feed ingredient which does this (Sec. 11.18) but is used up in the process, posing the problem of producing a vitamin E deficient feed if oxidation is extensive. Today commercial antioxidants are available such as hydroquinone or ethoxyquin which can supply the H and "spare" the natural vitamin E plus prevent the rancidity from occurring. Oxidized flavors of milk stem from this reaction and can also be prevented if the vitamin E level of milk is increased. Many vegetable oils have natural antioxidants present, while animal fats, particularly when refined, have little such activity. Vitamin A and carotene, in addition to vitamin E, are subject to destruction by rancid fats.

Hydrogenation. Double bonds of fatty acids will take on hydrogen much as they do iodine. A catalyst is required. This process of hydrogenation produces a saturated, and thus a hard, fat from an unsaturated, soft one. The saturation of the double bonds makes the fat less reactive and tends to prevent the oxidative changes of rancidity. Thus hydrogenation is used for improving the keeping qualities of certain fats, especially vegetable oils, used for food; this process produces solid cooking fats. The oils are not completely hydrogenated because, if this were done, the products would be too hard for convenient use. Completely hydrogenated cottonseed oil melts around 62°C and gives no iodine number, whereas the partially hydrogenated products used in cooking melt between 35 and 43°C and have an iodine number of 60 to 75.

7.4 Waxes

Waxes are lipids resulting from the combination of fatty acids and higher monohydroxy or dihydroxy alcohols. They have high melting points and are sufficiently difficult to saponify that they are not readily digested by animals. A common example is beeswax, a combination of palmitic acid and myricyl alcohol ($C_{30}H_{61}OH$). They occur as secretions and excretions in animals and insects and as protective coatings for plants, e.g., apples. They are removed in the ether extract and thus often inflate the nutritive significance of that assay.

7.5 Compound Lipids

Fats and waxes as described above are essentially hydrophobic, i.e., completely immiscible with water. In biological systems it is necessary that lipids, especially the fatty acids, often be a contiguous part of, or be transported in, an aqueous medium. A number of compound lipids which contain a hydrophobic (nonpolar) end (fatty acids) and a hydrophylic (polar) end (nitrogen base, sugars, etc.) are thus important in the animal and plant system. They have the ability to form a lipid bilayer and thus form the principal structural component of the cell membranes in animals.

Phospholipids (phosphatides) consist of glycerol in which the 1 and 2 positions are esterified with long-chain fatty acids. Often a saturated form is at the 1 position and an unsaturated one at the 2 position. Position 3 is esterified to phosphoric acid and in turn to a nitrogen base. If the nitrogen base is choline (Sec. 11.47) the phospholipid is *lecithin;* if it is ethanolamine it is *cephalin.* The formula is shown below. They both are components of animal cell membranes and of the lipid transport moieties in the plasma (Sec. 7.9), e.g., *chylomicrons*

$$\text{R}_2-\overset{\displaystyle O}{\overset{\|}{C}}-\text{O}-\underset{\underset{\displaystyle H}{|}}{\overset{\overset{\displaystyle H}{|}}{\text{CH}}}$$

$$\overset{\overset{\displaystyle H}{|}}{\underset{\underset{\displaystyle H}{|}}{\text{HC}}}-\text{O}-\overset{\displaystyle O}{\overset{\|}{\text{C}}}-\text{R}_1$$

$$\overset{\overset{\displaystyle H}{|}}{\underset{\underset{\displaystyle H}{|}}{\text{HC}}}-\text{O}-\overset{\overset{\displaystyle O}{\|}}{\underset{\underset{\displaystyle \text{O}^-}{|}}{\text{P}}}-\text{O}-\text{CH}_2\text{CH}_2\overset{+}{\text{N}}(\text{CH}_3)_3$$

(Choline)

Lecithin

$$\text{HO}-\text{CH}_2-\text{CH}_2-\text{NH}_3^+$$

(Ethanolamine)

and the *lipoproteins*. They are essential for proper digestion and absorption of fat, especially in the ruminant (Sec. 7.10). Lecithin is also found in the lipid of the soybean seed from which it is produced commercially. It is an excellent emulsifying agent and is widely used as a smoothing agent in foods and to improve the utilization of fat in milk replacers by the calf.[3]

Sphingolipids do not contain glycerol but consist of the amino alcohol sphingosine to which is added a fatty acid, phosphate, and either choline or ethanolamine. In animals they are found predominantly in brain and nerve tissue and called *sphingomyelin*. One special species of sphingolipid has the choline replaced by glucose and is called a *cerebroside*, as shown below.

$$\text{CH}_3(\text{CH}_3)_{12}-\text{CH}=\text{CH}-\overset{\overset{\displaystyle H}{|}}{\underset{\underset{\displaystyle H}{|}}{\text{C}}}-\overset{\overset{\displaystyle }{|}}{\underset{\underset{\displaystyle NH}{|}}{\text{C}}}-\text{CH}_2-\text{O}-\overset{\overset{\displaystyle O}{\|}}{\underset{\underset{\displaystyle O^-}{|}}{\text{P}}}-\text{CH}_2-\text{CH}_2\overset{+}{\text{N}}(\text{CH}_3)_3$$

$$\overset{|}{\underset{\underset{\displaystyle R_1(FA)}{|}}{O=C}}$$

(Sphingosine) R_1(FA)   (Choline)

Sphingomyelin

Glycolipids contain glycerol, with two polyunsaturated fatty acids (largely linoleic) joined to the 2 and 3 position. On the first carbon atom one or two moles of galactose are attached. Whereas triglyceride is the major lipid of seeds, galactolipid is the predominant one found in the leaves of plants. It therefore represents a major source of lipid for animals consuming forage alone.

7.6 Prostaglandins

Prostaglandins are a large family of compounds containing 20 carbon atoms and a cyclic structure between the eighth and twelfth carbon atom. They are synthesized by nearly all mammalian tissues from arachidonic acid which in turn is synthesized from linoleic acid (Sec. 7.18). Their synthesis is dependent on an intact pituitary, as removal of this gland precipitates an essential fatty acid deficiency which can be cured by prostaglandins but not by arachidonic acid. The prostaglandins modulate the action of a number of hormones on the synthesis of cyclic adenosine monophosphate (AMP). They are involved in a tremendously wide range of actions including lipolysis of adipose tissue and a number of key reactions in the reproductive process.[4]

7.7 Steroids

Steroids are a large group of compounds in plants and animals all derived from the basic ring structure shown below. They include the sterols, cholesterol and

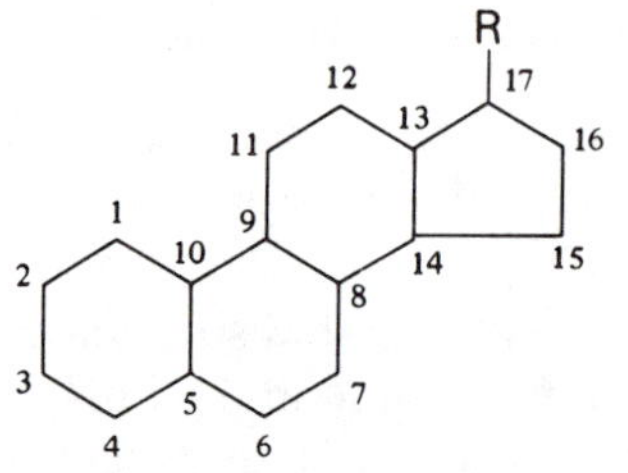

ergosterol, the bile acids, and the adrenal and sex hormones. Modification of the side chain on carbon atom 17 and rearrangement of the double bonds on the nucleus results in a wide range of compounds and biological activity.

Cholesterol is one of the most important animal sterols. It is insoluble in water and chemically unreactive, which is useful since it is a structural component of cells. It occurs free or in combination with fatty acids in all cells, in blood, and in wool grease lanolin. It is synthesized from acetate (C_2) in the liver and thus is not a dietary essential. The level of cholesterol is monitored by a negative feedback mechanism, i.e., when dietary intake increases, tissue synthesis declines. It serves as a precursor of a number of compounds (below) and is an essential component of lipid-carrying moieties in the blood (Table 7.4). 7-dehydrocholesterol is the animal-derived precursor of vitamin D_3 (Sec. 11.9).

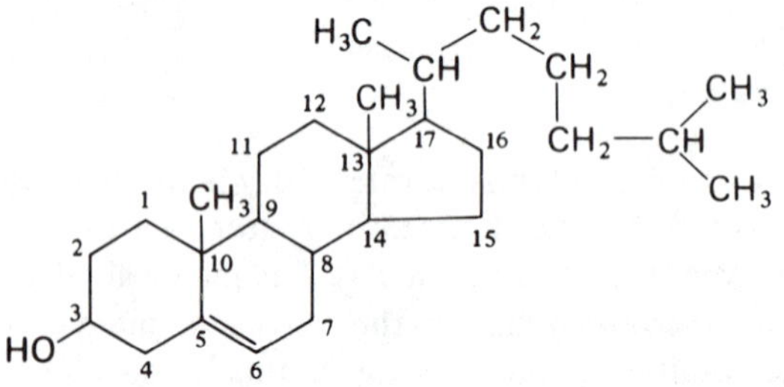

Cholesterol

Ergosterol is the principal plant sterol of importance in animal nutrition. It is present in high concentrations in yeast and upon irradiation is converted to vitamin D_2 (Sec. 11.14). All plant leaves contain ergosterol and thus contribute some vitamin D_2 when irradiated after harvesting. Few other plant sterols are absorbed by animals.

Bile acids are polar derivatives of cholesterol synthesized in the liver and either stored in the gall bladder to be released after eating, or released continuously for animals having no gall baldder (e.g., rats and horses). They are major components of bile and aid in the emulsification of fat in the small intestine

(Sec. 7.9). The polar component is attached to C_{17} and includes either glycine (glycocholic acid) or taurine (taurocholic acid).

Steroid hormones are synthesized from cholesterol primarily at the sites of secretion, viz., androgens, testis; estrogens, ovary; glucocorticoids and mineral corticoids, adrenal cortex. All of the syntheses require a supply of $NADPH_2$ from carbohydrate metabolism in order to proceed.

7.8 Terpenes

Extraction of feed samples will result in the removal of the terpenes which have nutritional, nonnutritional, or paranutritional significance. They are substances which will yield an isoprene moiety upon degradation. The carotenoids (Sec. 11.4) are vitamin A precursors; xanthophyll is not a nutrient but gives a yellow color to poultry skin; the phytol portion of chlorophyll and the essential oils have no nutritive value. They are only a small part of the ether extract of most feeds and provide no energy to the animal.

LIPID DIGESTION, ABSORPTION, AND METABOLISM

Following the nursing stage, the lipids make up only a small part of the diet of most animals, except man and carnivores. However, the metabolism of lipids is of great importance in nutrition, both because of the vital roles played by specific lipids and also because of the extensive fat synthesis which occurs in the body during fat deposition and the secretion of milk. Lipids occur as essential constituents in every cell in the body. While the depot fat serves primarily as a source of energy, that deposited under the skin serves also as a nonconducting layer which prevents the too rapid escape of body heat, and that around the viscera and certain other organs performs a supporting function.

7.9 Lipid Digestion and Absorption in Nonruminants

The primary object of lipid digestion and absorption is to arrange the lipid in a form that is water miscible since the microvilli of the small intestine are often considered to be covered with an unstirred aqueous layer.

Of the lipids being ingested, fat and cholesterol are essentially nonpolar and thus immiscible with water. Phospholipids on the other hand have a polar compound on the number 3 carbon atom and are much more miscible. In fact they help emulsify fat globules.

Under the influence of the peristaltic action of the stomach and duodenum, fat exists in the duodenum as a coarse emulsion. Little if any hydrolysis has occurred, although in the calf a pregastric esterase secreted at the base of the tongue does partially hydrolyze milk fat. In the presence of bile, the pancreatic lipase and colipase hydrolyze the triglyceride droplets into fatty acids and monoglycerides and reduce the lipid to a finer and finer emulsion. The fatty acids are preferentially removed from the 1 and 3 positions leaving a 2-monoglyceride which, having a polar (glycerol) and nonpolar (fatty acid) end, is itself an excellent emulsifying agent. The bile, free fatty acids, and monoglyceride become oriented into a mixed micelle (Fig. 7.1) containing a lipid core and a polar exterior. Whereas the fat droplet probably exceeded 5,000 Å, the micelle has a diameter of 30 to 100 Å.

The micelle migrates to the brush border where it is disrupted. All but the bile is absorbed into the mucosa; the bile remains in the lumen and eventually moves down the intestine to be absorbed and recirculated through the liver. Most of the triglyceride is absorbed by the time the ingesta reaches the mid-jejunem. Within the mucosa, the fatty acids and monoglycerides are resynthe-

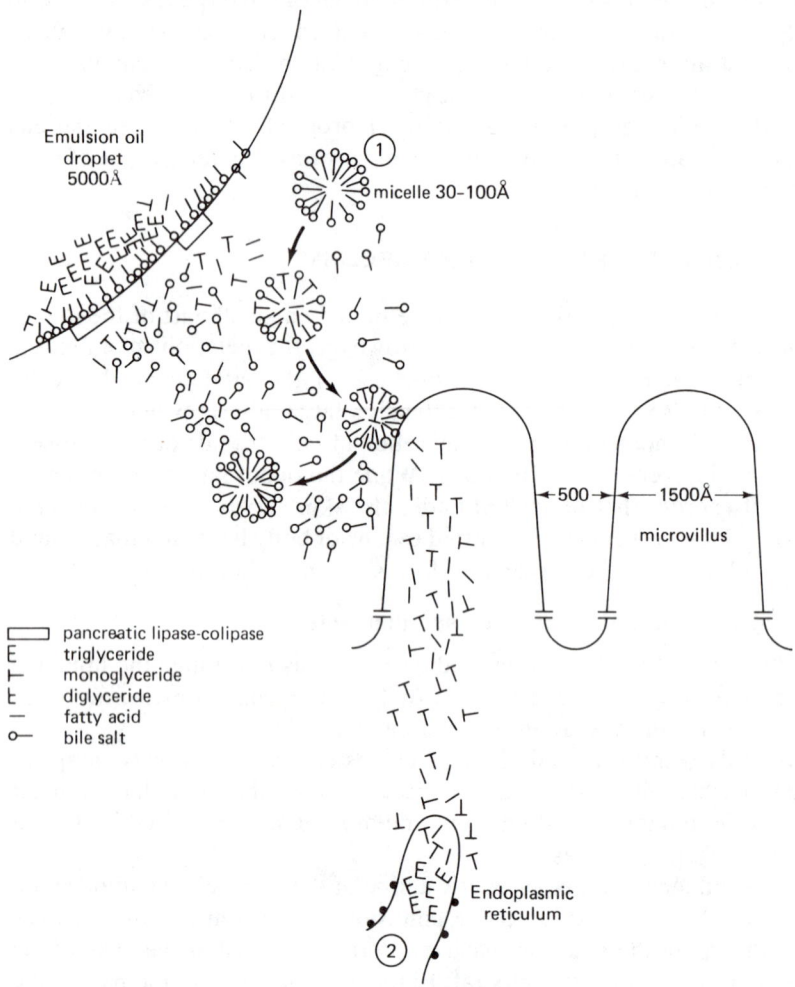

Fig. 7.1 Schematic diagram of triglyceride absorption. (1) Bile is secreted as a micellar solution containing bile salts, phospholipids and cholesterol in the ratio of 12.5 to 2.5 to 1. (2) Resynthesized triglyceride coalisce to form an oil droplet stabilized by phospholipid and protein and discharged into the central lacteal. (*Courtesy of A. Bensadoun, Cornell University.*)

sized into triglycerides, combined with cholesterol and phospholipid, "encased" in a thin layer of protein and secreted into the central lacteal of the villus as either chylomicron or very low density lipoprotein (VLDL) particles. The composition of these and other lipid transport particles are shown in Table 7.4. The central lacteal drains into the lymph vessels and enters the general blood circulation via the thoracic duct at the right atrium. On a high-fat meal, the lymph drainage appears milky due to the turbidity of the VLDL and chylomicrons.

The hydrolysis and resynthesis of triglycerides during the process of digestion and absorption produces similar but not identical triglyceride molecules in the lymph. For example, 78 percent of the glycerol is from the diet, 22 percent is synthesized de novo. About 88 percent of the fatty acids on the 1 and 3 carbon atoms and 75 percent on the 2 carbon atom were on the same one in the diet.[5] The minor shifts result from translocation to and from the other positions. Fatty acids of chain lengths below C_{12} do not appear in the lymph but are absorbed directly into the hepatic portal circulation.

Both phospholipid and cholesterol are secreted in the bile in substantial quantities; in fact, in the case of man, in amounts equivalent to 5 or 6 eggs per day. Phospholipid (largely lecithin) is preferentially hydrolyzed at the 2 position by a pancreatic phospholase to yield a 1-lysolecithin and a free fatty acid. They become part of the micelles and are absorbed with them to be resynthesized into phospholipid by the enterocyte (mucosal cell). The phospholipid is utilized to form the chylomicrons or VLDL with the excess being synthesized into triglycerides. About 20 percent of the phospholipid is synthesized de novo.

TABLE 7.4. Characteristics of Lipid Transport Particles in Human Plasma[a]

	Chylomicron	Very low density lipoprotein (VLDL)	Low density lipoprotein (LDL)	High density lipoprotein (HDL)	Albumin FFA[b] complex
Density, g/ml	<0.95	0.95–1.006	1.006–1.063	1.063–1.21	—
Diameter, Å	>800	300–800	200–300	75–200	—
Protein, %	2	6	10	41	92
Phospholipid, %	8	12	20	28	—
Total cholesterol, %	3	22	40	25	—
Triglyceride, %	86	58	29	3	—
Free fatty acids, %	1	2	1	3	8
Plasma concentration, mg/100 ml	30	129 ± 122	439 ± 99	300 ± 83	4000

[a]Data are an order of magnitude rather than a number as separation of the particles is by a density gradient procedure. It is clear, however, that protein increases and triglyceride decreases as density increases. Further, LDL is the prime carrier of cholesterol.
[b]Free fatty acid.
Adapted from R. M. S. Smellie (ed.), Plasma Lipoproteins. Academic Press, New York, 1971; S. Eisenberg and R. I. Levy, Lipoprotein metabolism, *Adv. Lipid Res.*, 13:2–89, 1975; J. D. Morrissett, R. L. Jackson, and A. M. Gotto, *Annual Rev. Biochem.*, 44:183–207, 1975.

Cholesterol esters are hydrolyzed by a pancreatic cholesterol esterase; the cholesterol becomes dissolved in the lipid core of the mixed micelle and is absorbed with it. It appears in the lymph as a part of the chylomicra or VLDL. Level of dietary fat has been shown to increase the amount of cholesterol absorbed. It makes little difference whether the fat is saturated or unsaturated although a higher proportion of absorbed triglycerides appears in VLDL when the fat is saturated. Obviously cholesterol absorption and metabolism are extremely active areas of investigation, and should be (Sec. 7.20).

Alterations in the type of dietary lipid can significantly change the amount that is digested and absorbed by animals. In general, absorbability (digestion and absorption) is superior with (1) shorter chain length fatty acids, (2) more unsaturated fatty acids, (3) as triglycerides rather than free fatty acids. In the calf, absorbability of butter > lard > tallow which is evidence that the above postulates are valid. These have been well worked out for the chick. In all probability the ability to participate in micelle formation is the reason for the greater absorbability.

7.10 Lipid Digestion and Absorption in Ruminants

Assimilation of milk fat by the preruminant animal is considered to be essentially the same as that of a nonruminant. The presence of a pregastric esterase (before stomach), however, does give the calf and the lamb a head start on lipid hydrolysis. The adult ruminant is another matter. The adult's diet consists of a high proportion of unsaturated fatty acids found in the galactolipids of forage and in the triglycerides of the cereal grains. However, the lipid of the rumen ingesta becomes uniquely different from the feed. The rumen microbial population promptly hydrolyzes the triglyceride and galactolipid, releasing free fatty acids (FFA) and allowing the glycerol and galactose to be fermented to volatile fatty acids (VFA). Because the intraruminal environment is a highly reduced one (Sec. 6.9), the unsaturated fatty acids are rapidly hydrogenated to a saturated form with stearic acid predominating. The bacteria and protozoa are also capable of synthesizing a number of odd chain fatty acids from propionate (C3) and branched-chain fatty acids from the carbon skeletons of the amino acids valine, leucine, and isoleucine. Table 7.5 shows the relative composition of forage lipid and the lipid of rumen ingesta.

Whereas most lipid in the nonruminant enters the duodenum as a coarse emulsion of triglycerides, in the ruminant it is largely present in the form of thin layers of free fatty acids on the surface of the feed particles. In addition, the contents of the duodenum and upper jejunum are more acidic than that of the nonruminant, not becoming alkaline until about three quarters of the way down the jejunum. These conditions mean that there is little triglyceride available to be converted to monoglyceride (a potent emulsifying agent) and further that pancreatic lipase will be less active in the duodenum and upper jejunum. Nonetheless, active micelle formation of the fatty acids does occur in the upper tract under the influence of bile salts (especially larger amounts of taurocholate which is more active in an acid medium) and biliary and ingesta phospholipid,

TABLE 7.5. Composition of Fatty Acids in Hay and Rumen Ingesta Lipids[a]

Fatty acid	Hay, %	Ingesta, %
14:0br[b]	—	0.6
14:0	0.9	1.6
14:1	0.8	—
15:0br	—	2.5
15:0	0.8	2.3
16:0br	—	1.0
16:0	33.9	30.0
16:1	1.2	—
17:0	—	2.4
18:0	3.8	41.4
18:1	3.0[c]	7.0[d]
18:2	24.0	3.9
18:3	31.0	6.0

[a] I. Katz and M. Keeney, Characterization of the octadecanoic acids in rumen digesta and rumen bacteria, *J. Dairy Sci.*, **49**:962–966, 1966.
[b] Branched chain.
[c] Cis form.
[d] Trans form.

largely lecithin. Although absorption of lipid does occur in the upper jejunum (15 to 26 percent), the larger percentage is absorbed in the lower three quarters of the jejunum where the pancreatic phospholipase has had a chance to hydrolyze lecithin into a fatty acid and the highly polar lysolecithin, which further enhances micelle formation.

Because of the continuous nature of digestion in the ruminant, the lymph draining the intestine is invariably milky. The particles are made up of about 75 percent in the VLDL fraction and 25 percent of chylomicrons about the reverse for the nonruminant. They contain about 70 percent triglycerides and 20 percent phospholipid with only limited amounts of cholesterol; most of the unsaturated fatty acids are carried on the phospholipid.

As with the nonruminant the rate of absorption of fatty acids is less for the saturated, long chain than for the shorter chain or the unsaturated acids (C18:0 < C16:0 < C18:1). However, stearic is apparently better utilized by ruminants, in part due to its high degree of dispersion on entering the duodenum. The student is referred to several excellent reviews on the subject which is at best complicated and changing rapidly.[6,7,8]

7.11 Transport and Storage

The lipid in the form of chylomicrons and VLDL is carried by the capillaries to the adipose tissue. Under the influence of a lipoprotein lipase, the triglycerides are hydrolyzed at the capillary wall into diglycerides and free fatty acids which remain in the blood while the diglyceride is transported across the capillary wall to be hydrolyzed further. Glycerol is then released back to the blood stream and the free fatty acids resynthesized into triglycerides within the adipose tissue cell utilizing de novo glycerol derived from the glycolytic cycle. The free fatty acids, glycerol, and cholesterol esters released by the disintegration of the chylomicrons and the very low density lipids are carried to the liver to be metabolized. Insulin increases the activity of lipoprotein lipase which is consistent with its function to foster storage of energy.[9]

7.12 Lipid Synthesis

Most animals consume food in excess of their caloric needs, usually coming from dietary carbohydrates and going preferentially to form liver and muscle glycogen. When these stores are full, fat is synthesized (Sec. 6.8). In the case of the nonruminant, the primary substrate for fat synthesis is glucose which enters the glycolytic cycle and eventually becomes pyruvate (Fig. 6.5). During a surfeit of food, there is ample oxaloacetate, and pyruvate is diverted to acetyl CoA which is used for fat snythesis rather than for energy. Acetyl CoA cannot penetrate the mitochondrial wall, whereas citrate can. Thus CoA condenses with oxaloacetate to form citrate, which then passes into the cytosol, where oxaloacetate is removed, leaving acetyl CoA available for fatty acid synthesis. The oxaloacetate is converted to malate and to pyruvate and returns to the TCA cycle.

Acetate is synthesized into palmitic acid (C16:0) in the manner shown below; acetyl CoA is converted to malonyl CoA by the addition of CO_2 with biotin as a cofactor; both acetyl CoA and malonyl CoA combine with a carrier protein (ACP; acyl carrier protein) and condense to form acetoacetyl-ACP. The latter product is reduced, utilizing 2 moles of $NADPH_2$ to form butyryl-ACP. Elongation of butyryl-ACP is effected by sequentially combining with a mole of malonyl-ACP until palmitic acid is synthesized. The enzyme responsible for

$$\text{Acetyl CoA(2)} + \text{ATP} + \text{HCO}_3^- \xrightarrow{\text{Biotin}} \text{Malonyl CoA(3)} + \text{ADP}$$

$$\text{Acetyl CoA(2)} + \text{ACP} \longrightarrow \text{Acetyl-ACP(2)} + \text{CoA}$$

$$\text{Malonyl CoA(3)} + \text{ACP} \longrightarrow \text{Malonyl-ACP(3)} + \text{CoA}$$

$$\text{Acetyl-ACP(2)} + \text{Malonyl-ACP(3)} \longrightarrow \text{Acetoacetyl-ACP(4)} + CO_2$$

$$\text{Acetyl-ACP(4)} + 2NADPH_2 \longrightarrow \text{Butyrl-ACP(4)} + \text{NADP} + H_2O$$
Synthesis of Fat From Acetate

the synthesis is the multienzyme complex, *fatty acid synthetase*. The significance of reduced NADP in fatty acid synthesis is clear from the above equations and emphasizes the importance of the phosphogluconate pathway (and also the malate pyruvate pathway) in maintaining this reaction.

In the case of the ruminant, excess energy from the rumen exists primarily as acetate and butyrate. While propionate is available, the glucose economy is usually sufficiently low that it is preferentially diverted to that compound. The ruminant is unique in that it *cannot* convert glucose to fat. From a teleological point of view this is good, conserving glucose for more vital functions. The reason it cannot use glucose for fat is that it does not have two key enzymes, ATP citrate lyase (splitting citrate into oxaloacetate and malate) and NADP-malate dehydrogenase (converting malate to pyruvate, Fig. 6.5). Thus the mechanism described above for the nonruminant is inoperable and the ruminant must rely entirely on acetate or butyrate for fat synthesis. Interestingly enough, it has been shown that, in the ruminant, the conversion of isocitrate to α-ketoglutarate yields reduced NADP in amounts much larger than that found in the rat.[10] In fact, it appears to contribute about as much reduced NADP needed by the bovine mammary gland as the phosphogluconate pathway. Thus the absence of the malate → pyruvate source is of little consequence.

Palmitic acid can be elongated to C18:0 via a mitochondrial enzyme system and desaturation to C18:1 or C16:1 is effected by a microsomal enzyme. In animals the mechanism for synthesizing palmitic, stearic, and oleic acids from acetate or glucose are thus present. Synthesis of linoleic or linolenic is not possible since mammals have no enzymes capable of introducing a double bond beyond Δ 9. Since these acids are required for satisfactory nutrition, they are referred to as the essential fatty acids (EFA) (Sec. 7.18).

Not all tissues are equally effective in synthesizing fat. In nonlactating animals, the liver and adipose tissue are the primary sites as shown in Table 7.6 for several species. This table is a clear example of why interspecies comparisons in nutrition, even between nonruminants are hazardous.

7.13 Fat Storage

It is evident from the previous section that the adipose tissue is an integral part of energy storage, whether it be derived from carbohydrate, lipid or, as we shall see, from amino acids. Approximately 50 percent of the adipose tissue is found under the skin, i.e., subcutaneous fat. The balance is located around certain organs, notably the kidneys, in the membranes surrounding the intestines, in the muscles, and elsewhere. Adipose tissue is not entirely inert. It has a blood

TABLE 7.6. Tissue Site of Excess Energy Conversion to Fatty Acids in Various Species[a]

	Fatty acid synthesis	
	Tissue	Substrate
Ruminant		
Sheep	Adipose	Acetate, from rumen fermentation
Cow	Adipose	
Nonruminant		
Pig	Adipose	Glucose, from carbohydrate digestion
Rat	Liver and adipose	
Chick	Liver	
Human	Liver	

[a]D. E. Bauman, Fat metabolism in ruminants, *Proc. Cornell Nutrition Conf.*, 1974, 69–73.

and nerve supply and is extremely dynamic. The modern discoveries made by the use of isotopes have shown that the fats in the body are in a state of flux. Fatty acids from the depots are being constantly mobilized and transported. Absorbed fatty acids merge with these from the depots. Some of the acids of this pool are constantly being converted into others. Some are degraded, while others are combined with glycerol and transported back to the depots. All of these reactions are so balanced that mixtures of fatty acids in the depots, blood, and organs tend to remain qualitatively and quantitatively constant for a given species.

The adipose tissue is not solely lipid but also contains water and nitrogen. The Missouri workers[11] reported an extensive study of the fat deposited in various parts of the steer. The data show that different fat deposits vary in water content (4.5 to 14.4 percent), in nitrogen content (0.18 to 0.62 percent), and in character, as shown by different physical constants. The nitrogen in fat deposits occurs principally in connective tissue.

Since adipose tissue always contains some water, it is evident that fat deposition involves a deposition of water also. With a ration rich in fat, there is some retention of water in all tissues, including the blood. Fat deposits are considered to be water-in-oil emulsions, in which albumin, lecithin, or soaps act as the emulsifying agent. When the depots are called upon to furnish energy, there may be a retention of water in place of fat. This has been clearly shown for the human subject by Newburgh and Johnston.[12] By taking account of the water intake and outgo, as well as the energy metabolism, they found that obese individuals frequently maintained or even increased their weight temporarily on a reducing diet, because water was being stored despite the fact that depot fat was being used up. Particularly striking is the observation of Trowbridge[13] that the kidney fat of a steer on a submaintenance ration for 11 months contained 81.4 percent water, 9.6 percent protein, and only 4.6 percent fat, respectively. These observations illustrate the limitations of the weight measure as the sole criterion of nutritive state in maintenance or in fattening. The fat-water relationships in adipose tissue may have a bearing on the amount of "shrink" in animals rapidly fattened for market.

7.14   Food Source and Nature of Body Fat in Nonruminants

The fact that depot fat arises from both carbohydrates and also directly from various fatty acids in the bloodstream explains why the nature of the fat deposited can be markedly affected by the character of its food source. Large differences can result in the degree of hardness of the fat, which is a considerable factor in the market value of the carcass of pigs as indicated by the "soft pork" problem. The influence of the kind of fat fed upon the character of the body fat is strikingly shown by Table 7.7 obtained by Anderson and Mendel with rats in which the oils listed furnish 60 percent of the energy intake.

Anderson and Mendel found that the iodine numbers of body fat deposited from various carbohydrates and proteins fell within the range 55 to 70. This range, representing fat synthesized within the body, was thus considered to

TABLE 7.7. Iodine Number in Food Fat and Body Fat

Food fat	Iodine number of food fat	Iodine number of body fat
Soybean oil	132	123
Corn oil	124	114
Cottonseed oil	108	107
Peanut oil	102	98
Lard	63	72
Butterfat	36	56
Coconut oil	8	35

Source: W. E. Anderson and L. B. Mendel, The relation of diet to the quality of the fat produced in the animal body, *J. Biol. Chem.*, **76**:729–747, 1928.

typify the normal depot fat of the rat. Taking this range as a base line, the data for the various oils given above show the striking influence of large intakes of fats differing widely as regards degree of saturation from that normally deposited. It is noted, however, that the extremes exhibited by the food fats are never reached by the body fats, reflecting the capacity of the organism, in depositing ingested fat, to modify the latter where it is widely different from the normal deposit. The data given illustrate the fact that carbohydrates produce a less unsaturated and thus a harder body fat than do most fats found in feeds of vegetable origin.

Many experiments have shown that fat deposition in the hog follows the sample principles illustrated above for the rat. The data in Table 7.8 from the classical work of Ellis and Isbell demonstrate the effect of adding various oils to a ration of corn and tankage, which by itself produces a firm fat in hogs. The percentage of fat in these rations is much smaller than that used in obtaining the data previously cited for the rat, and thus the effects are less marked. It is noted, however, that the ration containing cottonseed oil, having the lowest iodine number of the oils used, was the only one which produced a carcass graded as hard. This grade is reflected in a high melting point, a relatively low iodine number, and the highest percentage of saturated fatty acids. All the other oils, when added at the same level as the cottonseed, produced medium-soft carcasses, readily explainable by the constants and fatty-acid distribution of their fats. Increasing the level of corn oil produced an oily carcass with fat which melted at room temperature and which consisted of unsaturated acids to the extent of 73 percent. Such a "soft" carcass would meet with packer disapproval and be objectionable to the consumer.

The data in Table 7.8 explain why soybeans and peanuts, feeds rich in highly unsaturated fats, can be used in only a limited way for fattening hogs without producing soft pork (Fig. 7.2).

Deposits of soft fat can be modified by a change in diet. When, after a period on feeds rich in unsaturated fat, a ration which will produce a hard fat is given, the deposited fat gradually becomes harder. Ellis[14] described such a

TABLE 7.8. Influence on Hog Carcass of Adding Various Oils to a Basal Ration of Corn and Tankage

Oil supplement	Firmness grade[a]	Melting point, °C	Iodine number	Fatty acids, %		
				Oleic	Linoleic	Total saturated
Peanut oil, 4.1%	MS	34.3	72.4	47.9	13.8	32.5
Cottonseed oil, 4.1%	H	45.3	64.4	35.9	15.7	43.0
Soybean oil, 4.1%	MS	31.2	75.7	43.3	18.6	33.8
Corn oil, 4.1%	MS	36.3	76.3	45.0	16.8	33.0
Corn oil 11.5%	O	24.5	97.2	41.4	31.4	23.1

[a]H = hard, MS = medium soft, O = oily.
Source: N. R. Ellis and H. S. Isbell, Soft pork studies. II. The influence of the character of the ration upon the composition of the body fat of hogs. III. The effect of food fat upon body fat, as shown by the separation of the individual fatty acids of the body fat, *J. Biol. Chem.,* **69**:219–248, 1926.

change in hogs, which results when a ration containing peanuts is followed by corn and nonsoftening supplements. The process is called "hardening off" and is taken advantage of in feeding practice in finishing hogs for market. Anderson and Mendel[15] showed that the process takes place more rapidly where the animal is fasted for a period before the hardening ration is given. The recognition that adipose tissue is in a dynamic state explains why this process occurs. While experiments have shown that the feeding of tocopherol, an antioxidant, to hogs makes their fat somewhat less susceptible to oxidative rancidity, the practical importance of this finding seems doubtful.

In a summary of recent studies, Bitman[16] concludes that the adipose tissue of both swine and poultry reacts to vagaries of saturation in the diet, while egg lipid can be altered readily only towards more unsaturation.

Fig. 7.2 Lard from hard, soft and oily carcasses. (*Taken from O. G. Hankins, N. R. Ellis, and J. H. Zeller, Some results of soft pork investigations II. U.S. Dept. Agr. Bull. 1492, 1928.*)

7.15 Food Fat and Nature of Milk and Body Fat in Ruminants

The intense biohydrogenation system in the rumen suggests that any attempt to increase the level of unsaturated fatty acids in the milk fat or adipose tissue by feeding them at high levels would be unrewarding. In a quantitative way this is true. However, nothing is rarely 100 percent in biology. The classical studies of Maynard, McCay, and Madsen[17] demonstrated a significant and prompt change in the iodine number of milk fat when the diet fat was changed from a high to a low degree of unsaturation (Fig. 7.3). In this experiment, a ration containing approximately 3.5 percent of fat on a dry-matter basis was used. In the first and third periods the grain mixture was selected to have a high iodine number, primarily by the inclusion of ground flaxseed, while in the middle period the iodine number was reduced to a low value by the omission of the flaxseed and the inclusion of coconut-oil meal. The curve shows that a change in the iodine number of the food fat from 107 to 43 resulted, in the first 24 hours, in a drop in the value for milk fat from 38 to 32, with a later drop to approximately 26 as the minimum. The restoration of the food fat of high iodine number resulted in a quick rise in the milk-fat value to its level during the first period. These rapid and large changes with a ration which contained only about 3.5 percent of fat are striking indeed. Yet from a quantitative point of view the iodine number of the milk fat was still within the normal range and somewhat more saturated than lard (Table 7.3). More recent studies summarized by Bitman[16] showed that the modest increase in unsaturation was probably due to an increase in oleic acid (C18:1) with little increase in linoleic acid (C18:2). The rumen microflora are apparently rapidly able to hydrogenate only one double bond. Attempts to alter the fatty acid composition of ruminant body fat have been generally unfruitful.

In recent years the concept of feeding "protected lipids" has emerged. Australian workers[18] developed a procedure of encapsulating small droplets of lipid in a thin layer of protein. By treating the protein with formaldehyde the droplet avoids attack by rumen microorganisms but is released by the acidic and proteolytic conditions of the abomasum. The lipid is thus available for digestion

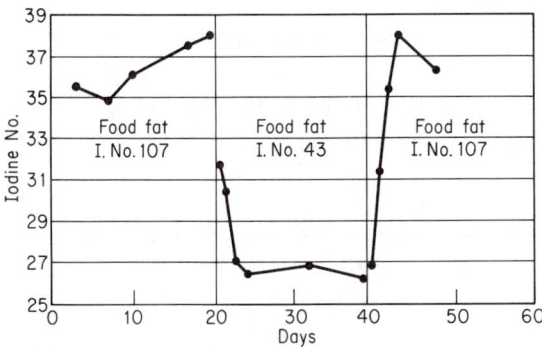

Fig. 7.3 The iodine number of milk fat as influenced by food fats of high and low degrees of unsaturation.

and absorption as with the nonruminant. In spite of the conditions of the duodenum which mediate against fat hydrolysis (Sec. 7.10), digestion and absorption of protected lipid is highly efficient. Following feeding of unsaturated fats in a protected form there is a prompt rise in the degree of unsaturation of the serum lipids, tissue fat, and milk fat. With light-weight steers, incorporation in the adipose tissue is much higher than with heavier steers, suggesting a somewhat reduced rate of lipid turnover in the older animal. Milk fat promptly shows an increase of from about 2 percent C18:2 as a percent of total fatty acids to as much as 30 percent. This represents an efficiency of transfer of as much as 40 percent. In many cases milk-fat yield is increased. Unfortunately the unsaturated milk fat is susceptible to oxidation and thus the shelf life is poor. The problem could be easily remedied by the addition of antioxidants, a technique patently illegal at the moment.

Clearly, the means for altering the fatty acid composition of animal fats, ruminant and nonruminant, are available should it be economically feasible to do so.

7.16 Catabolism of Fats and Fatty Acids

The end result of the catabolism of fats and fatty acids is the production of ATP, CO_2, and H_2O with the liberation of excess heat. The initial degradation of fat leads to the formation of glycerol and acetyl CoA. In the special case of the ruminant, absorbed acetate, butyrate, and ketone bodies are also available for immediate catabolism.

Glycerol oxidation is shown below. In the process, 2 moles of glycerol become 1 mole of glucose which, when totally catabolized to CO_2 and water, yields a net of 38 ATP or 19 per mole.

	ATP
2 glycerol(3) $\longrightarrow$ 2L-glycerol Ⓟ(3)	-2
2L-glycerol Ⓟ(3) $\longrightarrow$ 2 dihydroxyacetone Ⓟ(3)	+4
2 dihydroxyacetone Ⓟ(3) $\longrightarrow$ 1 glucose(6)	spontaneous
1 glucose $\longrightarrow$ $6CO_2 + 6H_2O$	36
Net	38 (19/mole)

Acetate oxidation is via the TCA cycle and yields 10 ATP per mole. *Butyrate oxidation* is via acetyl CoA as follows:

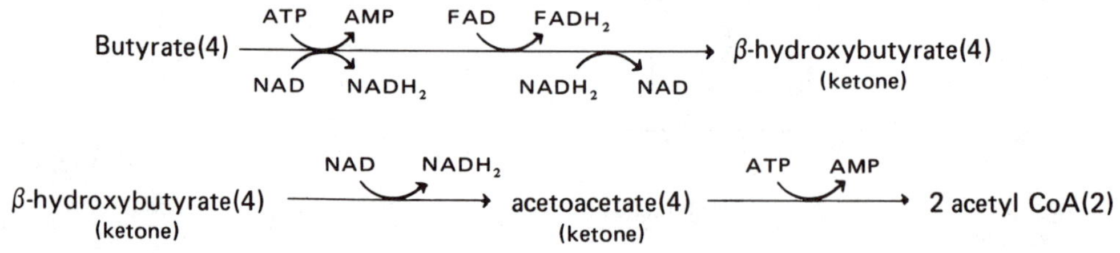

Butyrate(4) $\xrightarrow{\quad\quad\quad\quad\quad}$ β-hydroxybutyrate(4) (ketone)

ATP   AMP   FAD   FADH$_2$
NAD   NADH$_2$ NADH$_2$ NAD

β-hydroxybutyrate(4) (ketone) $\xrightarrow{\quad NAD \quad NADH_2 \quad}$ acetoacetate(4) (ketone) $\xrightarrow{\quad ATP \quad AMP \quad}$ 2 acetyl CoA(2)

The tally for ATP production is

	ATP
Butyrate $\longrightarrow$ hydroxybutyrate	0
hydroxybutyrate $\longrightarrow$ 2 acetyl CoA	+1
2 Acetyl CoA $\longrightarrow$ $4CO_2 + 4H_2O$	+24
Total	+25/mole

Mobilization and Oxidation of Fat.   Except for a short time following a meal, the utilization of fat occurs after existing adipose reserves have been mobilized. The release of triglycerides from the adipose tissue is under hormonal control which activates adenyl cyclase which in turn causes the synthesis of cyclic AMP. Hydrolysis of the triglyceride to glycerol and fatty acids follows with their release into the circulation. The fatty acids are combined with an albumin (Table 7.4) and circulate as an albumin–fatty-acid complex. There is a marked species difference in the hormones reacting at the cell membrane. Epinephrine is especially active for rats, dogs and humans; glucagon has a marked effect in rabbits and avian species. The case of the ruminant is unclear with glucagon a possibility according to Bauman and Davis.[10] Prostaglandins also are active in stimulating lypolysis.

It is well known that between meals the blood free fatty acids are elevated as a result of the mobilization just discussed. Upon eating, they drop promptly, an example of the finely controlled homeostatic mechanism related to energy needs. It has been demonstrated in newborn lambs that the blood free fatty acids rise within 2 to 5 hours after birth; lambs that were lethargic and did not survive did not show the elevated level. Is this a case of a mechanism that failed?[19]

Oxidation of long-chain fatty acids takes place in the mitochondrion. However, they cannot pass into the mitochondrion alone. They need a special carrier mechanism in the form of carnitine to transverse the wall. For example, palmitic acid (C14) is first activated to acyl CoA as follows:

$$RCOO^- + CoA + ATP \longrightarrow acyl\ CoA + AMP + PPi$$

Carnitine replaces the CoA to form acyl-carnitine which together with the CoA diffuse into the mitochondrion (Fig. 6.5). Acyl CoA reforms and carnitine diffuses back to the cytosol. An activated form of palmitic acid is now oxidized by removing acetyl CoA generating 5 ATPs (Fig. 7.4) and contributing acetyl CoA to the TCA cycle. The remaining activated C14 is again oxidized to yield another acetyl CoA. The process continues until all eight acetates are removed. It is known as β-oxidation as each cleavage of the fatty acid occurs at the β-carbon atom. The ATP production tally is as follows:

	ATP
Palmitic(16) $\longrightarrow$ palmitic CoA(14)	-2
Palmitic CoA(16) $\longrightarrow$ 8 acetyl CoA (7 cleavages $\times$ 5)	$+35$
8 acetyl CoA $\longrightarrow$ $16H_2O + 16CO_2$ (8 $\times$ 12)	$+96$
Total	$+129$

Since the heat of combustion of palmitic acid is 2338 kcal per mole, the efficiency of the oxidation to ATP is about 40 percent, in the same ball park as glucose [(129 $\times$ 7.3)/2338 $\times$ 100 = 40 percent]. Apparently all long-chain fatty acids are oxidized by β oxidation.

Brown fat is a unique type of adipose tissue found in newborn goats and lambs and in a number of rodents. It is located near and around the spinal cord, thoracic organs, and kidneys and has a reddish-brown color. The color is due to an intense vascularity, and the presence of a high concentration of cytochrome and flavin compounds. Histologically the adipose cells are multivacuolar, have a central nucleus, and many mitochondria with large numbers of crista. Normal adipose cells are white and have single vacuoles and the nuclei are oriented peripherally.

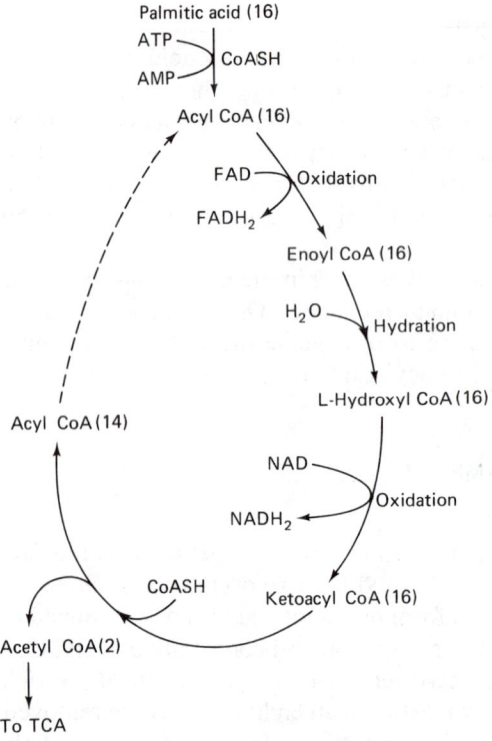

Fig. 7.4 β-oxidation of palmitic acid.

Brown fat is specifically involved in the process of nonshivering thermogenesis. During a cold stress, the sympathetic nervous system releases norepinephrine into the brown fat which causes a release of fatty acids. Unlike other adipose tissue, the fatty acids are metabolized within the brown fat tissue itself in what has been described as a "poorly understood loosening" of the coupling of oxidative phosphorylation.[20] The effect is to have fatty acid oxidation with little ATP formed so that substantially more of the energy is released as heat. The proximity of the tissue to the major blood vessels and the receptors involved in shivering provides the animal with a means to maintain body temperature without shivering. The advantage to the young born in a hostile environment is obvious. Brown fat increases as animals become acclimatized to the cold and is important to hibernating animals during spring arousal and other periods of thermal stress.[21]

Fat to Glucose. The age-old controversy of whether fatty acids can be converted to glucose has finally been laid to rest. While the mechanism exists in plants, it is not possible in animals. Note in Fig. 6.5 that acetyl CoA, the basic unit of fatty acid oxidation, contributes two moles of CO_2 by the time the cycle reaches succinyl CoA. Thus 2 carbons are in, 2 carbons are out long before glucose could be formed. It is, therefore, impossible biochemically for acetate to contribute any carbon and thus give a net synthesis of glucose. Since, however, the fatty acid has been used for energy it has been used in place of glucose and thus has *spared* glucose. But do not forget that the glycerol moiety of fat can yield glucose as noted earlier and thus "fat" can contribute somewhat to the glucose pool.

7.17   Ketosis

As can be seen in Fig. 6.5 the entrance of acetyl CoA into the TCA cycle is contingent upon an adequate supply of oxaloacetate. Under conditions of nutritional and/or physiological stress, where energy demand exceeds the available tissue glucose, fat is mobilized to acetyl CoA; it cannot be converted to energy via the TCA cycle and is thus converted to ketones as shown in Fig. 7.5. Such conditions exist: in cows just after parturition (heavy glucose demand for milk production), in sheep just before parturition (heavy glucose drain to the fetus),

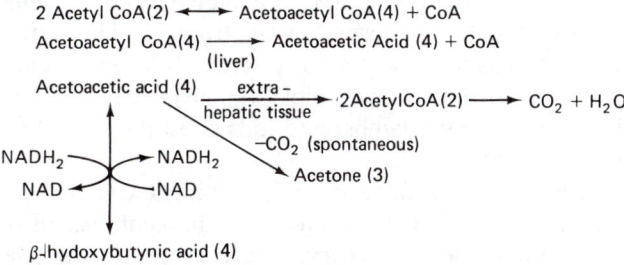

Fig. 7.5 Pathways of ketone production and use.

in humans after a long siege of vomiting (limited food [carbohydrate] consumption). Under these conditions the level of ketones in the blood increases (*acetonemia*) and since two of them (acetoacetic acid and β-hydroxy butyric acid) are highly acidic, an acidosis develops. The ketones appear in the urine; *acetonuria* and simple urine tests can be used to verify a diagnosis. Acetone is exhaled via the lungs and in severe cases is readily detectable as a sweet odor. Other signs of ketosis include anorexia, a marked drop in blood glucose and in cows a spectacular drop in milk production.

Treatment consists of making some form of glucose available to the tissues. Glucose injection is an effective clinical treatment although in cattle it may be short lived and require repeating. Cortisone and/or ACTH injections are effective as they mobilize tissue amino acids, some of which can be converted to glucose (Sec. 8.21). Feeding of readily fermentable carbohydrate is not effective in ruminants as it is rapidly converted to volatile fatty acids, only one of which (propionic) yields glucose. However, highly successful recoveries follow feeding of sodium propionate or propylene glycol. They are both glucogenic. The latter is preferred as it is absorbed directly.

The fundamental reason(s) why ketosis develops in ruminants is still unclear. General recommendations for prevention include the following: (1) maintain an adequate feed intake to insure a reasonable level of rumen propionate, and (2) don't have the females too fat. The reader is referred to two reviews which treat the subje in great detail.[22]

7.18   Essential Fatty Acids (EFA)

Despite the fact that certain lipids are essential constituents of animal tissues, the knowledge that carbohydrates are readily changed into fat and that such essential lipid constituents as phospholipids and cholesterol can be made in the body naturally led to the view that lipids as such are not required in the diet. This viewpoint was changed some forty years ago by the studies of Burr and Burr.[23] These investigators found that, with a diet almost entirely devoid of fat, rats developed a scaly condition of the skin and a necrosis of the tail, accompanied by failure of growth and eventual death. Harmful effects on reproduction and lactation performance were also noted. The addition of small amounts of pure linoleic acid was strikingly effective in preventing or curing these conditions, but saturated acids were ineffective. Later, arachidonic and linolenic acids were each found partially effective in correcting the troubles. These two, along with linoleic, are referred to as the essential fatty acids. Deficiency symptoms have been reported in mice, dogs, pigs, chicks, guinea pigs, fish, and infants. Suggestive evidence that calves, lambs, and kids need these acids in their rations has also been reported.

We generally speak of three essential fatty acids (EFA): linoleic (C18:2ω6), linolenic (C18:2ω3), and arachidonic (C20:4ω6). None can be synthesized by mammalian tissue but linoleic and linolenic can serve as precursors of a number of highly unsaturated fatty acids which can be grouped into families as follows:

linoleic series

$\quad$ C18:2ω6 $\rightarrow$ 18:3ω6 $\rightarrow$ 20:3ω6 $\rightarrow$ <u>20:4ω6</u> $\rightarrow$ 22:4ω6

linolenic series

$\quad$ C18:3ω3 $\rightarrow$ 18:4ω3 $\rightarrow$ 20:4ω3 $\rightarrow$ 20:5ω3 $\rightarrow$ 22:5ω3

Note that linoleic acid is a precursor of arachidonic acid (underlined) which in turn has been shown to be the precursor of prostaglandins (Sec. 7.6). In most species linoleic is the most effective EFA, but in fact arachidonic is the tissue level EFA. Cold water fish are an exception in that they preferentially require linolenic acid and their tissues contain high levels of the 4 and 5ω3 compounds shown above.

The complete biochemical function of the EFA is not known other than in relation to prostaglandins. However, they are widely distributed in phospholipids and cholesterol esters which in turn constitute a significant part of all cell membranes and lipid transport moieties. Clearly they are needed.

On an EFA deficient diet the tissues attempt to produce unsaturated acids from oleic acid (C18:1) which results in the family shown below. The result is

oleic series

$\quad$ 18:1ω9 $\rightarrow$ 18:2ω9 $\rightarrow$ 20:2ω9 $\rightarrow$ <u>20:3ω9</u> $\rightarrow$ 22:3ω9

an accumulation of the trienoic acid underlined above (eicosatrienoic acid). In studies with rats, Holman[24] found that on a fat-free diet there was a high triene-tetraene ratio in the tissue lipids which was progressively reversed as linoleate was added to the ration in increasing amounts. This reversal was also accompanied by a remission or disappearance of the skin symptoms. Holman proposed the triene-tetraene ratio as a measure of EFA requirement and stated that, in the case of the rat, a ratio of less than approximately 0.4 in the tissues studied indicated that the minimum requirement of linoleate (1 percent of kcal) has been met. Workers in the same laboratory, Hill and associates,[25] found in a study with pigs which had received a fat-free diet that as linoleic acid was added, the triene-tetraene ratio in the heart and liver lipids dropped from around 9 at zero intake to below 0.5 at 2 percent intake. The triene-tetraene ratio has been used widely as an index of EFA deficiency in spite of a lack of visible signs of deficiency.

A number of species differences cast an enigmatic cloak over the importance of the EFA. There is doubt that the pig requires EFA.[26] Although the calf has been shown to require EFA, the adult ruminant apparently grows normally on a lipid-free diet.[27] The microbial population appear to provide enough EFA to meet the requirements. On the other hand it has been shown that the lamb and calf utilize linoleic acid much more efficiently than nonruminants. One could speculate that this ability has evolved in response to the hydrogenating activity of the rumen continually suppressing the effective levels of dietary EFA consumed.

EFA are widely distributed among feed fats. For example, corn oil, soybean oil, cottonseed oil, peanut oil, and certain others are excellent sources (Table 7.3). If farm animals actually have a dietary need for EFA, it seems probable that they are adequately supplied by the commonly fed rations, but the previous discussion has indicated that there are questions involved here that require further study.

There are nutritional disadvantages from excessive intakes of EFA. These readily oxidized acids increase the requirement for vitamin E which appears to serve primarily as an antioxidant in the body (Sec. 11.17). Several experiments have shown that levels of the vitamin which were normally sufficient to prevent such E-deficiency symptoms as muscular dystrophy and encephalomalacia proved inadequate as the intakes of EFA were increased. Some years ago it was reported that when calves were fed certain vegetable oils as supplements to skim milk poor results were obtained compared with those where butterfat was used. The results were interpreted to mean that butterfat, nature's fat for sucklings, was superior to vegetable oils for the young. The real explanation came later from experiments by Adams et al.[28] When "filled milk" included corn oil as the fat, muscular dystrophy resulted, but no such troubles were found in groups receiving hydrogenated vegetable oils, or corn oil plus a large supplement of vitamin E.

7.19 Special Nutritive Value of Fats as a Group

As far as we know at present, lipids are not specifically required in the diet except as a source of the essential fatty acids and of choline. They do have, however, other important dietary qualities. They serve as carriers of certain nonfat nutrients, notably the fat-soluble vitamins A, D, E, and K. There are other reasons why fat is a useful constituent of the diet. It promotes the absorption of both vitamin A and carotene, particularly the latter. Russell and co-workers[29] found that the absorption of carotene by hens, for example, was about 60 percent of the intake in a ration containing approximately 4 percent of fat, in contrast to a 20 percent absorption on a ration extracted to 0.07 percent of this nutrient. In practical feed formulation, fat is added to reduce dustiness and thus improve palatability and feed consumption.

By far the most important reason to include fat in animal diets is that it is a concentrated form of energy containing two and one-quarter times as much energy as carbohydrates. Thus the higher the fat content of the diet, the greater the energy value per kilo. This makes sound commercial sense if the cost per unit of energy is competitive. However, the addition of fat must be done with some clear recognition of the implications on a number of fronts, viz., acceptability and feed intake, availability of the fat, concentration of other nutrients in the diet, and effect on feed utilization.

Fat is readily digested by all species exceeding 90 percent, even for beef steers fed tallow. In general, additions of 5 to 10 percent are permitted for poultry while intake is often depressed if it exceeds 5 percent in beef cattle rations. Protected fats up to 30 percent of the concentrate are possible for milk cows.[30] Carnivores like the mink and dog relish fat, and for mink, diets of 20 to

25 percent of fat are common and superior to those with lower fat. Veal calves need a milk replacer diet of 20 percent or more in order to have an acceptably finished carcass. While it is generally agreed that animals eat to meet a caloric need, the addition of fat often increases energy intake to some extent although total dry-matter consumption declines.

The overall utilization of the gross energy from fat appears to be superior to that supplied from carbohydrates. In reviewing the various experiments with rats, Mendel and Anderson[31] stated that an abundant intake of fat is much more effective for storing fat than an equicaloric intake of carbohydrates. It has also been shown that, within limits, increasing the fat content of the ration of growing pigs and fattening steers increases the feed efficiency beyond that which could be accounted for by the additional energy thus provided. In the case of chickens, Carew and Hill,[32] by studies of tissue energy in birds receiving equal intakes of metabolizable energy, found that corn oil increased the metabolic efficiency of energy utilization. In later studies the authors obtained similar results with beef tallow and soybean oil.

According to early studies by Forbes and associates at Pennsylvania State University, the greater than expected value of fat as a source of energy can be explained on the basis that, with equicaloric diets, increasing the fat component decreases the heat increment. With fewer calories thus lost as heat, relatively more are available for production. Data illustrating this point are shown in Fig. 7.6 for rats fed high- and low-fat diets of equal calorie content. This chart is reproduced from a publication by Swift,[33] which reviews the Pennsylvania studies and discusses other aspects of the role of fat in the diet.

Fat additions have implications for other nutrients as well. Fatty acids can combine with Ca^{++} and Mg^{++} to form insoluble soaps which pass out in the feces, thus wasting minerals and energy. Often calcium levels are increased in beef cattle diets when fat is added.

Since added fat increases the energy content of the diet without a commen-

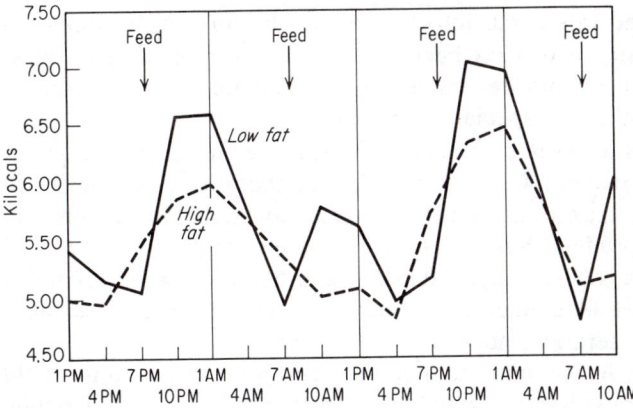

Fig. 7.6 Curves of heat production per 3-hour period of rats on diets of high and low fat content.

surate increase in dry matter, the proportions of nutrients to energy are reduced with each fat addition. For example, a 5 percent addition of fat to a poultry diet increases the metabolizable energy value by 200 kcal per kilogram of diet. In order to maintain a constant protein/energy ratio, the *percent* protein in the diet must therefore be increased. In practice, poultry diets are often formulated with an eye to the nutrient/energy ratio rather than to a strict percent of the diet.

7.20 Animal Lipids in Human Nutrition

By far the principal market for animal products is the human diet and thus their values for this purpose are of prime importance to the animal industry. Here, there are some important considerations regarding animal fat versus vegetable fats which are cheaper sources of energy. As carriers of other nutrients, milk fat and egg fat are outstanding sources of vitamin A, but animal fats in general are low in vitamin E and EFA compared with vegetable fats. Of much greater significance is the controversy now going on, both among investigators and in the public press, regarding the association of certain dietary lipids with the onset or severity of heart disease. To date, this controversy has unduly hurt the market of animal fats. The animal industry should have a keen interest in the status of the research in this field.

The lipids most directly concerned are cholesterol and the polyunsaturated fatty acids. Cholesterol is the principal constituent of the plaques which form in the walls of the arteries causing them to narrow and restrict the blood flow, a condition called *atherosclerosis,* which may lead to the formation of a blood clot and thus a heart attack. An elevated blood cholesterol is one of the factors associated with the occurrence of heart attacks. There appears to be a failure of the liver to exert its normal control over the blood cholesterol level. Here the dietary intake comes into question, and it is frequently recommended that the intakes of animal fats, particularly eggs and butter, be restricted in favor of vegetable fats.

The polyunsaturated fats come into the picture because high intakes of them relative to other fatty acids have been shown to decrease elevated cholesterol levels. Here again dietary recommendations call for a substitution of vegetable fats for animal fats, especially ruminant fats.

Despite many years of research on heart disease, the importance of diet in its control remains ambivalent. There is no evidence that dietary changes will lessen the risk of heart attack although maintaining a lowered blood cholesterol level would seem to be prudent. Since there are many other risk factors reputed to be involved, smoking, obesity, hypertension, and exercise, so diet remains only one of the avenues of concern. As one noted cardiovascular scientist stated "we are after moderation, not martyrdom."[34]

The animal industry has an important stake in further research in this field to get all the facts. It also has a justifiable interest in combating the public concern about animal fats which is promoted on the basis of wrong or questionable data.

NOTES

1. F. H. Mattson, Fat 24–32, in Present Knowledge in Nutrition, 4th ed., Nutrition Foundation, Inc., New York, 1976.

2. M. Kates, Techniques of Lipidology, American Elsevier, New York, 1972.

3. D. T. Hopkins, R. G. Warner, and J. K. Loosli, Fat digestibility by dairy calves, *J. Dairy Sci.*, **42**:1815–1820, 1959.

4. V. J. Goldberg and P. W. Ramwell, Role of prostaglandins in reproduction, *Physiol. Rev.*, **55**:325–351, 1975.

5. F. H. Mattson and P. A. Volpenheim, The digestion and absorption of tri-glycerides, *J. Biol Chem.*, **239**:2772–2777, 1964.

6. W. M. F. Leat and F. A. Harrison, Digestion absorption and transport of lipids in sheep, in I. W. MacDonald and A. C. I. Warner (eds.), Digestion and Metabolism in the Ruminant, Univ. of New England Publishing Unit, Armidale, NSW, Australia, 1975, pp. 481–495.

7. A. L. Lough, Aspects of lipid digestion and absorption in the ruminant, in E. K. Rommel, H. Goebell, and R. Böhmer (eds.), Lipid Absorption: Biochemical and Clinical Aspects, Univ. Park Press, Baltimore, Md., 1976, pp. 255–279.

8. G. A. Garton, The digestion and assimilation of lipids, in R. W. Dougherty et al. (ed.), Physiology of Digestion in the Ruminant, Butterworths, Washington, D.C., 1965, pp. 390–398.

9. R. L. Jackson, J. D. Morisett, and A. M. Gotto, Jr., Lipoprotein structure and metabolism, *Phys. Rev.*, **56**:259–316, 1976.

10. D. E. Bauman and C. L. Davis, Regulation of lipid metabolism, in I. W. MacDonald and A. C. I. Warner (eds.), Digestion and Metabolism in the Ruminant, Univ. of New England Publishing Unit, Armidale, NSW, Australia, 1975, pp. 496–509.

11. P. F. Trowbridge, C. R. Moulton, and L. D. Haigh, Composition of the beef animal and energy cost of fattening, *Missouri Agr. Expt. Sta. Bull.* 30, 1919.

12. L. H. Newburgh and M. W. Johnston, Endogenous obesity—a misconception, *J. Am. Dietet. Assoc.*, **5**:275–285, 1930.

13. P. F. Trowbridge, The resorption of fat, *Proc. Am. Soc. Animal Nutrition,* 1910, pp. 13–20.

14. N. R. Ellis, Changes in quantity and composition of fat in hogs fed a peanut ration followed by a corn ration, *U.S. Dept. Agr. Tech. Bull.* 368, 1933. (See also earlier papers cited.)

15. W. E. Anderson and L. B. Mendel, The relation of diet to the quality of the fat produced in the animal body, *J. Biol. Chem.*, **76**:729–747, 1928.

16. J. Bitman, Status report on the alteration of fatty acid and sterol composition in lipids in meat, milk and eggs, *Proc. Symposium on Fat Content and Composition of Animal Products,* N.A.C. Publication, Washington, D.C., 1976, pp. 200–237.

17. L. A. Maynard, C. M. McCay, and L. L. Madsen, The influence of food fat of varying degrees of unsaturation upon blood lipids and milk fat, *J. Dairy Sci.,* **19**:49–53, 1936.

18. T. W. Scott and L. J. Cook, Effect of dietary fat on lipid metabolism in ruminants, in I. W. MacDonald and A. C. I. Warner (eds.), Digestion and Metabolism in the Ruminant, Univ. of New England Publishing Unit, Armidale, NSW, Australia, 1975, pp. 510–523.

19. J. H. Moore and R. C. Noble, Foetal and neonatal lipid metabolism, in I. W. MacDonald and A. C. I. Warner (eds.), Digestion and Metabolism in the Ruminant, Univ. of New England Publishing Unit, Armidale, NSW, Australia, 1975, pp. 465–480.

20. J. Himms-Hagen, Lipid metabolism during cold exposure and during cold acclimation, *Lipids*, 7:310–323, 1972.

21. R. E. Smith and B. A. Horwitz. Brown fat and thermogenesis, *Physiol. Rev.*, 49:330–425, 1969.

22. L. H. Schultz, Management and nutritional aspects of ketosis, *J. Dairy Sci.*, 54:962–973, 1971; Ketosis, in B. L. Larson and V. R. Smith (eds.), Lactation, Vol. 2, Biosynthesis and Secretion of Milk-Diseases, Academic Press, New York, 1974, pp. 318–350.

23. G. O. Burr and M. M. Burr, A new deficiency disease produced by the rigid exclusion of fat from the diet, *J. Biol. Chem.*, 82:345–367, 1929; On the nature and role of the fatty acids essential in nutrition, *ibid.*, 86:587–621, 1930.

24. R. T. Holman, The ratio of trienoic: tetraenoic acids in tissue lipids as a measure of essential fatty acid requirements, *J. Nutrition*, 70:405–410, 1960.

25. E. G. Hill and coworkers, Essential fatty acid nutrition in swine. I. Linoleate requirement estimated from triene-tetraene ratio of tissue lipids, *J. Nutrition*, 74:335–341, 1961.

26. Nutrient requirements of swine, National Academy of Sciences, Washington, D.C., 1978.

27. R. R. Oltjen and P. P. Williams, Microbial populations and metabolic parameters of ruminants fed a purified diet with and without dietary lipids, *J. Animal Sci.*, 38:915, 1974.

28. R. S. Adams et al., Some effects of feeding various filled milks to dairy calves (four subtitles), *J. Dairy Sci.*, 42:1552–1592, 1959.

29. W. C. Russell et al., The absorption and retention of carotene and vitamin A by hens on normal and low fat rations, *J. Nutrition*, 24:199–211, 1942.

30. W. L. Dunkley, N. E. Smith, and A. A. Franke, Effects of feeding protected tallow on composition of milk and milk fat, *J. Dairy Sci.*, 60:1863–1869, 1977.

31. L. B. Mendel and W. E. Anderson, Some relations of diet to fat deposition in the body, *Yale J. Biol. Med.*, 3:107–137, 1930.

32. L. B. Carew and F. W. Hill, Effect of corn oil on metabolic efficiency in chickens, *J. Nutrition*, 83:293–299, 1964.

33. R. W. Swift, The importance of fat in the diet, *Ann. N.Y. Acad. Sci.*, 56:4–15, 1952.

34. D. Kritchevsky, Diet, blood lipids and cardiovascular disease. Diet variations, *Proc. Symposium Am. Dairy Sci. Assoc.*, 1972, pp. 11–17.

SUPPLEMENTARY LITERATURE

Bensadoun, A., and J. T. Reid: Effect of physical form, composition and level of intake of diet on the fatty acid composition of the sheep carcass, *J. Nutrition,* **87:**239–244, 1965.

Cunningham, H. M., and J. K. Loosli: The effect of fat-free diets on lambs and goats, *J. Animal Sci.,* **13:**265–273, 1954.

Goldstein, J. L., and M. S. Brown: The low density lipoprotein pathway and its relation to atherosclerosis, *Annu. Rev. Biochem.,* **46:**897–930, 1977.

Hathaway, H. D.: The composition of feed fats as related to digestibility and metabolizable energy, *Proc. Annu. Meet. AFMA (Am. Feed Manufacturers Ass.) Nutr. Council,* **37:**19–22, May 1977.

Lambert, M. R., et al.: Lipid deficiency in the calf, *J. Nutrition,* **52:**259–272, 1954.

Lech, J. J.: Control of endogenous triglyceride metabolism in adipose tissue and muscle: Perspectives, *Fed. Proc.,* **36:**1984–1985, 1977.

Palmquist, D. L.: Ruminant adipose tissue metabolism—Symposium. *Fed. Proc.,* **35:**2300–2318, 1976.

Patton, S., and E. M. Kesler: Saturation in milk and meat fats, *Science,* **156:**1365–1366, 1967.

Poskitt, E. M. E., and T. J. Cole: Do fat babies stay fat? *British Med. J.,* **1:**7–9, 1977.

Snyder, F., (ed.): Lipid Metabolism in Mammals, Plenum Press, New York, 1977.

Wiggers, K. D., et al.: Type and amount of dietary fat affect relative concentration of cholesterol in blood and other tissues of calves, *Lipids,* **12:**586–590, 1977.

Chapter 8

The Proteins and Their Metabolism

Since protein is the principal constituent of the organs and soft structures of the animal body, a liberal and continuous supply is needed in the food throughout life for growth and repair, and thus the transformation of food protein into body protein is a very important part of the nutrition process. The term *protein* is a collective one which embraces an enormous group of closely related but physiologically distinct members. Plant proteins differ from each other and from animal proteins; each animal species has its own specific proteins, and a given animal contains many different ones in its organs, fluids, and other tissues. In fact no two proteins seem to be exactly alike in their physiological behavior. From the standpoint of nutrition the important distinguishing feature of the various proteins is their amino acid makeup (Sec. 8.2).

8.1 Elementary Composition of Proteins

In common with the fats and carbohydrates, the proteins contain carbon, hydrogen, and oxygen. In addition they contain a large and fairly constant percentage of nitrogen. In practical terms, the figure commonly used is 16 percent. Most proteins also contain sulfur, and a few contain phosphorus and iron. They are complex substances, colloidal in nature and of high molecular weight. The range of *elementary composition* of the more typical proteins is as follows:

	Percent
Carbon	51.0–55.0
Hydrogen	6.5– 7.3
Nitrogen	15.5–18.0
Oxygen	21.5–23.5
Sulfur	0.5– 2.0
Phosphorus	0.0– 1.5

8.2 Amino Acids

Proteins are polymers of amino acids which vary in relative amounts and kind from protein to protein. These amino acids are obtained as hydrolytic end products, when proteins are heated with strong acids or when they are acted upon by certain enzymes. They are also the end products of protein digestion, the building stones from which body protein is made, and the degradation products in protein catabolism. Thus our study of protein nutrition deals primarily with amino acids. There are some 20 to 22 different amino acids present in proteins, although in nature over 150 other amino acids exist that never are part of proteins.

The amino acids are derivatives of the short-chain fatty acids and contain a basic amino group ($-NH_2$) and an acidic carboxyl group ($-COOH$). The amino group is attached to the α carbon atom (α-amino acids), and in nature the amino acids generally assume the L configuration, when compared with L-glycerose (Sec. 6.2). Most are soluble in water and all but glycine are optically active. Because they have both an amino group and a carboxyl group, they are amphoteric, assuming acidic or basic properties depending upon the pH of the medium. In acid pH, the amino acid is a cation; in basic pH it is an anion; at the pH at which it is electrically neutral it is dipolar and called a "zwitterion." At this point the pH is termed the isoelectric point for that amino acid.

$$
\begin{array}{ccc}
\text{CHO} & \text{COOH} & \text{COO}^- \\
| & | & | \\
\text{HO}-\text{C}-\text{H} & \text{H}_2\text{N}-\text{CH}-(\alpha\text{ carbon}) & ^+\text{HN}-\text{CH} \\
| & | & | \\
\text{CH}_2\text{OH} & \text{R} & \text{CH}_2 \\
\text{L-glycerose} & \text{L-amino acid} & \text{L-alanine} \\
& & \text{(Zwitter ion)}
\end{array}
$$

The amino acids which have been identified in plant and animal proteins are classified according to the series of organic compounds in which they belong, and those in the aliphatic series are further classified according to the number of amino groups and carboxyl groups present. The formulas of those which have received special attention in nutrition studies are shown with the other amino acids. An asterisk has been placed in front of the names of those which have a nonpolar side chain (R group) which is significant in the development of protein structure.

The relative distribution of 12 of these amino acids in the body protein of different species, as determined by Williams and coworkers, is presented in Table 8.1. It is noted that the figures are very similar for the three species, tryptophan being the lowest for all and arginine, leucine, and lysine ranking at the top. What does this tell us about the relative tissue needs of different species for amino acids?

I. Aliphatic amino acids

A. Monoamino-monocarboxylic acids (neutral)

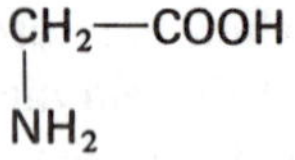

$$CH_2-COOH$$
$$|$$
$$NH_2$$

Glycine, $C_2H_5NO_2$
Amino-acetic acid

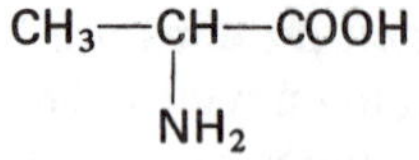

$$CH_3-CH-COOH$$
$$|$$
$$NH_2$$

*Alanine, $C_3H_7NO_2$
α-Amino-propionic acid

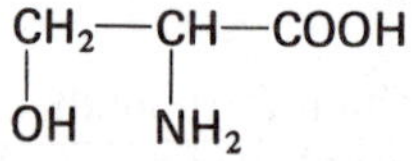

$$CH_2-CH-COOH$$
$$|\quad\ |$$
$$OH\ \ NH_2$$

Serine, $C_3H_7NO_3$
α-Amino-β-hydroxypropionic acid

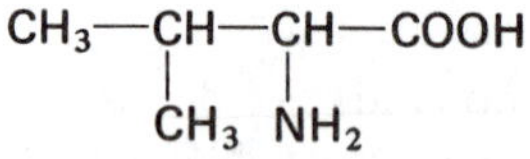

$$CH_3-CH-CH-COOH$$
$$|\quad\ |$$
$$CH_3\ NH_2$$

*Valine, $C_5H_{11}NO_2$
α-Amino-β-methyl-butyric acid

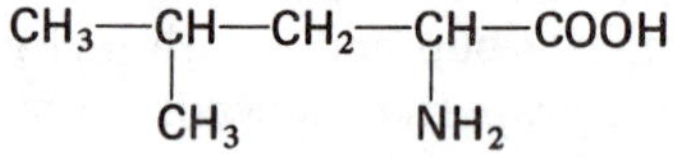

$$CH_3-CH-CH_2-CH-COOH$$
$$|\qquad\qquad |$$
$$CH_3\qquad\ \ NH_2$$

*Leucine, $C_6H_{13}NO_2$
α-Amino-γ-methyl-valeric acid

$$CH_3-CH_2-CH-CH-COOH$$
$$|\quad\ |$$
$$CH_3\ NH_2$$

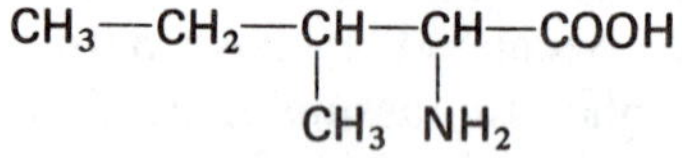

*Isoleucine, $C_6H_{13}NO_2$
α-Amino-β-methyl-valeric acid

$$CH_3-CH-CH-COOH$$
$$|\quad\ |$$
$$OH\ \ NH_2$$

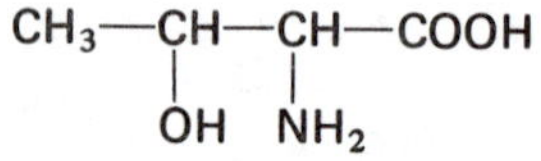

Threonine, $C_4H_9NO_3$
α-Amino-β-hydroxybutyric acid

B. Monoamino-dicarboxylic acids (acidic)

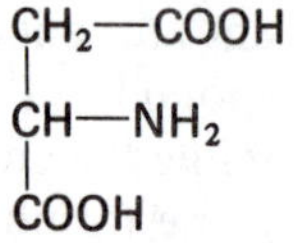

$$CH_2-COOH$$
$$|$$
$$CH-NH_2$$
$$|$$
$$COOH$$

Aspartic acid, $C_4H_7NO_4$
Amino-succinic acid

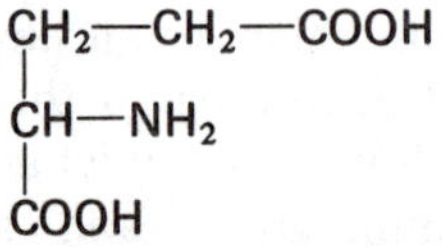

$$CH_2-CH_2-COOH$$
$$|$$
$$CH-NH_2$$
$$|$$
$$COOH$$

Glutamic acid, $C_5H_9NO_4$
α-Amino-glutaric acid

C. Diamino-monocarboxylic acids (basic)

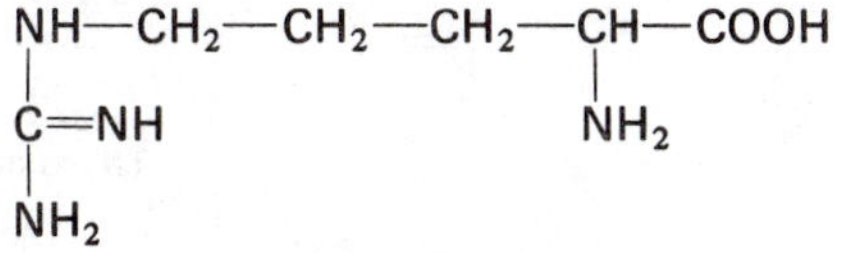

$$NH—CH_2—CH_2—CH_2—CH—COOH$$
$$C=NH \qquad NH_2$$
$$NH_2$$

Arginine, $C_6H_{14}N_4O_2$
α-Amino-δ-guanidine-valeric acid

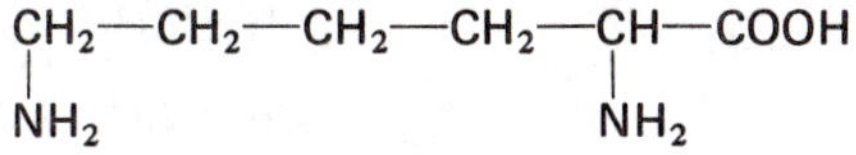

$$CH_2—CH_2—CH_2—CH_2—CH—COOH$$
$$NH_2 \qquad\qquad NH_2$$

Lysine, $C_6H_{14}N_2O_2$
α-ϵ-Diamino-caproic acid

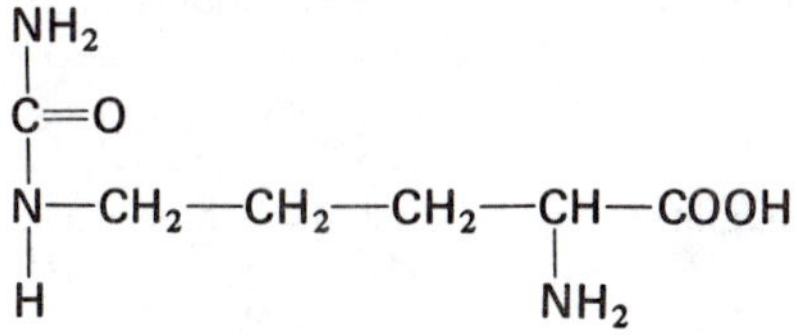

$$NH_2$$
$$C=O$$
$$N—CH_2—CH_2—CH_2—CH—COOH$$
$$H \qquad\qquad NH_2$$

Citrulline, $C_6H_{13}O_3N_3$
δ-Carbamido-α-amino-valeric acid

D. Sulfur-containing amino acids

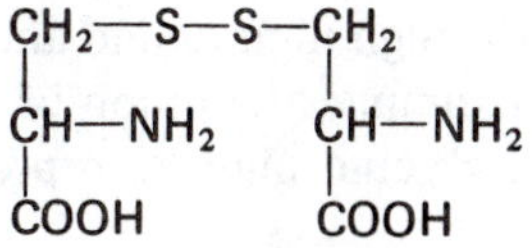

$$CH_2—S—S—CH_2 \qquad\qquad H—S—CH_2$$
$$CH—NH_2 \quad CH—NH_2 \qquad\qquad CH—NH_2$$
$$COOH \qquad COOH \qquad\qquad COOH$$

Cystine, $C_6H_{12}N_2O_4S_2$ Cysteine, $C_3H_6NO_2S$
Di(α-Amino-β-thio-propionic acid) α-Amino-β-thio-propionic acid

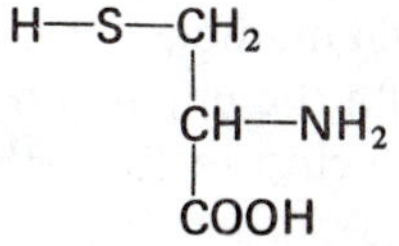

$$CH_3—S—CH_2—CH_2—CH—COOH$$
$$NH_2$$

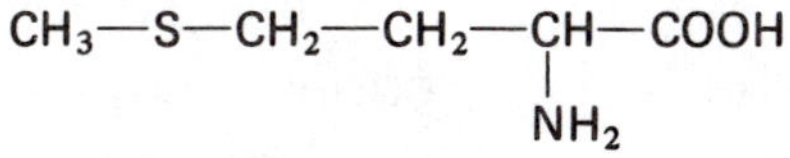

*Methionine, $C_5H_{11}NO_2S$
α-Amino-γ-methylthio-butyric acid

II. Aromatic amino acids

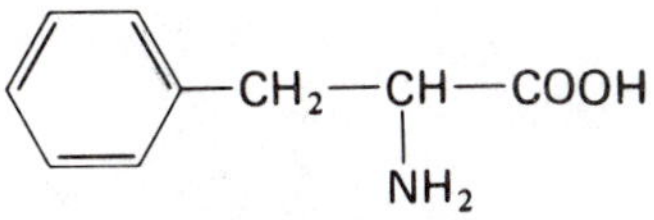

$$—CH_2—CH—COOH$$
$$NH_2$$

*Phenylalanine, $C_9H_{11}NO_2$
α-Amino-β-phenyl-propionic acid

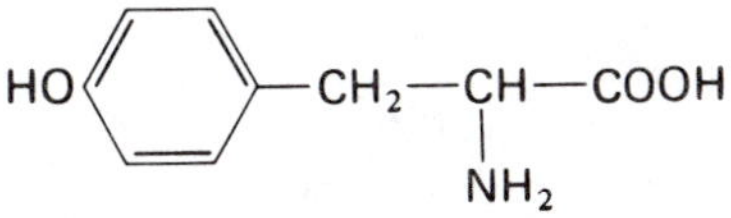

$$HO—\quad—CH_2—CH—COOH$$
$$NH_2$$

Tyrosine, $C_9H_{11}NO_3$
α-Amino-β-parahydroxy-
phenyl-propionic acid

Diiodotyrosine   Thyroxine

III. Heterocyclic amino acids

Histidine, $C_6H_9N_3O_2$
α-Amino-β-imidazole-propionic acid

*Proline, $C_5H_9NO_2$
Pyrrolidine-α-carboxylic acid

Hydroxyproline, $C_5H_9NO_6$
Hydroxypyrrolidine-α-carboxylic acid

*Tryptophan, $C_{11}H_{12}N_2O_2$
α-Amino-β-indolepropionic acid

8.3 The Structure of the Protein Molecule

Our knowledge of the structure of proteins began with the work of Emil Fischer who devised methods for uniting amino acids through their amino and carboxyl groups, with the elimination of water. As an example, the union of two molecules of glycine to form the dipeptide glycyl-glycine may be represented as follows:

Glycine   Glycine   Glycyl-glycine

It was found that the principal linkage existing between the amino acids in the protein molecule is through the amino group of one acid and the carboxyl group of another. This type of union is called a *peptide bond* or *peptide linkage*. The

amino acids so joined are referred to as amino acid residues. The sequential

TABLE 8.1. Amino Acid Content of Body Protein (g/16 g of nitrogen)

Amino acid	Rat	Chicken	Pig
Arginine	5.89	6.71	7.12
Histidine	2.16	1.96	2.65
Isoleucine	3.49	4.12	3.84
Leucine	6.46	6.63	7.14
Lysine	7.61	7.46	8.55
Methionine	1.71	1.76	1.77
Phenylalanine	3.69	3.95	3.77
Threonine	3.87	4.02	3.79
Tryptophan	0.76	0.77	0.74
Valine	5.51	6.72	6.00
Tyrosine	2.88	2.49	2.59
Cystine	1.49	1.75	1.01

Source: Harold H. Williams and coworkers, Estimation of growth requirements for amino acids by assay of the carcass, *J. Biol. Chem.,* **208:**277–286, 1954.

addition of up to several hundred amino acid residues by this covalent, peptide bond results in the formation of a long-chain polypeptide and is referred to as the *primary structure* of protein. The formation of protein, however, is much more than the synthesis of polypeptides. Because of the diversity of the order and kinds of amino acids along the chain, the polypeptides can further arrange themselves into forms which are known as secondary, tertiary, and quaternary structure. In *secondary structure* the polypeptides align themselves along one dimension in the form of a right-turning α helix. The helix is stabilized by hydrogen bonding between the carbonyl (CO) and imido (NH) groups. The α helixes can then intertwine with each other to form a supercoil, much like a rope. Secondary structure is characteristic of fibrous protein and is extremely stable.

In *tertiary structure,* the polypeptides are folded and tightly coiled into a globular form. They are stabilized by several mechanisms, viz., (1) the nonpolar ends of a number of amino acids become a hydrophobic center of the polypeptide coil, (2) two cysteine molecules in apposition to each other forming a disulfide linkage (–S–S–) to form a mole of cystine, (3) acidic and basic amino acids linked by a salt bridge, and (4) hydrogen bonding. Obviously the form, secondary or tertiary, depends upon the kind and location of amino acids on the polypeptide chain. For example, the amino acid proline is always at a fold.

Quaternary structure refers to the alignment of several tertiary structures into one protein. Hemoglobin, for example, consists of four single-strand tertiary forms compactly associated into a single globular protein.

It is now well known that a protein is what it is by virtue of the kinds and sequences of amino acids in its polypeptide chains. Each protein is precisely built to serve its specific function in the body and has a structure which is determined genetically. Further, the protein molecules in each species are characteristic for the species, and many can be distinguished immunologically. These untold numbers of proteins vary dramatically in the number of polypeptide chains and molecular weight. For example, myoglobin has one chain and a molecular weight of 16,900. Bovine β-lactoglobulin has two chains and a molecular weight of 35,000, whereas tobacco mosaic virus has a molecular weight of 40 million with 2,100 chains.

The proteins occurring in nature are built from their constituent amino acids. In the case of plants, including the lower forms such as yeast and bacteria, nitrates and ammonium salts are used as the initial nitrogenous compounds for protein synthesis. In the case of animals, however, the constituent amino acids must be available with the exception of some which can be synthesized in the body from simpler compounds, as is discussed later.

The complete amino acid sequences of over 100 proteins are now known. In 1955 the work of Sanger[1] and associates culminated in the determination of the complete primary structure of the hormone polypeptide insulin, molecular weight 6,000, a brilliant achievement for which Sanger received the Nobel prize. Today's understanding of the structure of proteins and how they are biosynthesized (Sec. 8.19) is a result of the amazing developments in the field of molecular biology during the last twenty years.

8.4 Properties of Proteins

Proteins are amphoteric and can combine chemically with both acids and bases. This occurs by virtue of their having free and reactive amino and carboxyl groups. Those involved in the peptide linkage are nonfunctional in this context. Each protein has its characteristic isoelectric point at which the tendencies to acidic and basic dissociations are equal, and at this point the protein is least soluble, most readily precipitated and, as such, can be washed. This procedure is taken advantage of in the separation and purification of proteins, e.g., vitamin-free casein. Proteins precipitated in this fashion retain their original structure and chemical properties when resolubilized, by carefully changing the pH.

Proteins vary widely with respect to their solubility in water and various aqueous solutions, from insoluble (hair) to soluble (albumin). None are soluble in fat solvents such as ether and petroleum ether. In addition to forming chemical combinations, proteins in solution have colloidal properties. They do not pass through the membranes or gels which are used as the criteria for separating out colloidal particles, and many proteins can bind ions physically by adsorption as well as by uniting with them chemically. Many proteins can also be precipitated from solution by neutral salts such as ammonium or sodium sulfate in a process called salting out, a mechanism that is not clearly understood. By varying the ionic strength, separation of a number of proteins is possible. The protein maintains its biological and chemical properties upon being resolubilized. Some less soluble proteins can be made more soluble by "salting in," with the addition of certain salts to the aqueous solution.

When the protein, particularly a globular one, irreversibly unfolds due to the breaking of the bonds which hold it in a globular form, the protein is said to be denatured. Heating of egg white is a common example. Other reagents such as nonpolar solvents (alcohol), salts of heavy metals, strong acid or alkali, and trichloracetic acid result in denaturation. In this case biological activity, e.g., enzyme function is destroyed. Digestive action on some kinds of protein is improved if the protein is cooked before eating.

8.5 Classification of Proteins

The various proteins cannot be identified or distinguished from each other by any simple chemical method and thus their classification is based primarily on form, physical properties, and chemical configuration. While not an official classification, the following breakdown allows one to visualize proteins as they relate to animal nutrition and metabolism. Note that the classifications are not mutually exclusive but overlap to some degree.

1 *Simple Proteins.* This group includes those that yield only amino acids or their derivatives on hydrolysis.

2 *Conjugated Proteins.* Here are included those in which simple proteins are combined with a nonprotein radical (prosthetic group). Examples of these types with a source and prosthetic group, respectively, include *nucleoproteins* (ribosomes; RNA), *phosphoproteins* (casein of milk; phosphate), *metalloproteins* (cytochrome oxidase; iron and copper), *lipoproteins* (VLDL; phospholipid, fat, cholesterol), *flavoproteins* (succinic dehydrogenase; FAD), *glycoproteins* (γ globulin; galactose, mannose, hexosamine).

Proteins can also be distinguished by their structural conformation as follows.

Fibrous proteins are long chains of polypeptides as noted above and are found in the animal in the form of *collagen, elastin,* and *keratin. Collagen* is the most abundant single protein in mammals and is a principal component of the cornea and of connective tissue; it makes the meat of older animals less tender. It is insoluble and indigestible, but upon heating or treating with acid, it becomes a soluble, easily digested mixture of polypeptides, called gelatin. Since it has no (–S–S–) cross linkages as a protein, it contains no cysteine or cystine. It is also devoid of tryptophan. It contains substantial hydroxyproline, an amino acid rarely found elsewhere. It can be used to identify the presence of large amounts of collagen in feed mixtures. It is considered a *very* poor quality protein (Sec. 8.10). *Elastin* is a part of tendons, arteries, and other elastic tissue and unlike most fibrous proteins can stretch in two directions and is very poorly digested. Its synthesis requires copper as an obligate cofactor (Sec. 10.32) as evidenced by ruptured aortas in copper-deficient fowl and guinea pigs. *Keratins* in the α helix form are found in hair, horn, and wool. They are high in cystine (22 percent) and thus have many disulfide linkages. These linkages make hair quite malleable and permit it to stretch when moist heat is applied and then contract when cooled and dried. β-keratins have no disulfide linkages and do not stretch, e.g., the beaks of birds.

Globular proteins include the enzymes, protein hormones, and oxygen-carrying proteins. They are soluble in water or aqueous mixtures of acids, bases, or alcohol. *Albumins* are water soluble and are a significant part of serum protein and egg protein. *Globulins* are insoluble in water but solubility increases with changes in neutral salt concentration. Immune globulins and hemoglobin in blood, lactoglobulins of milk, and myoglobins in muscle are examples. Many plant seeds contain globulins. Some proteins fall between fibrous and globular, having a fibrous structure but a solubility of the globular form. Myosin of muscle and fibrinogen of blood are examples.

8.6 Purine and Pyrimidine Compounds

Purines and pyrimidines are nitrogen bases which exist at the core of a number of compounds central to energy metabolism and the perpetuation of genetic information. Their basic structure is shown below. The principal purines are

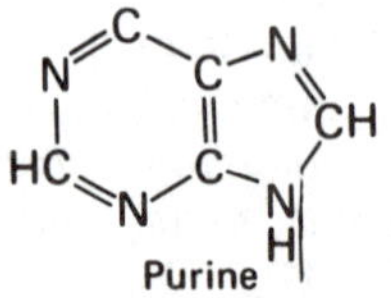

Purine 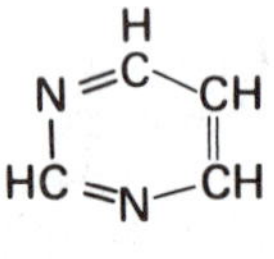   Pyrimidine

adenine and guanine while the major pyrimidines are cytosine, uracil, and thymine and are distinguished by the configuration of their side chains. When condensed with a pentose, either ribose or deoxyribose (Sec. 6.2), they become nucleosides and take the name of their nitrogen base, i.e., the major ribose nucleosides are adenosine, guanosine, uridine, and cytidine. The deoxyribo-nucleosides are deoxyadenosine, deoxyguanosine, deoxythymidine, and deoxy-cytidine. When each becomes phosphorylated, they are termed nucleotides. ATP and its hydrolytic products, ADP and AMP (Fig. 5.2, Table 5.4), are adenine nucleotides. Adenine nucleotide is also a major component of the coenzymes NAD (nicotinamide adenine dinucleotide), FAD, and CoA (Figs. 5.3 and 5.4). The condensation of a large number of nucleotides forms a nucleic acid of which the most important are deoxyribonucleic acid (DNA) and ribo-nucleic acid (RNA). In these structures the sugar is either ribose or deoxyribose as indicated. The phosphate and sugar serve as structural components, while the arrangement of the pyrimidine and purine bases identify the genetic code.

DNA is the master key to protein synthesis because it contains the infor-mation which determines the kinds and amounts of the different proteins syn-thesized in each cell. This information is passed on from parent to offspring. The DNA for each type of cell in a given species is a specific one and thus directs the exact reproduction of the species, except when mutations occur. DNA is composed of a double helix, made up of many mononucleotide units containing deoxyribose, phosphate, and a purine or pyrimidine base. The bases represented in DNA are the purines, adenine and guanine, and the pyrimidines, cystosine and thymine. The size of these macromolecules is enormous, the largest known having an estimated molecular weight of around 7 million.

RNA is a polynucleotide containing recurring units of mononucleotides, each consisting of ribose, phosphate, and purine or pyrimidine bases. The purines are adenine and guanine and the pyrimidines are cytosine and uracil (not thymine as in DNA). There are three that play a major role in protein synthesis. One is messenger RNA (*m*RNA) which carries the genetic information from DNA to the ribosome for protein synthesis. The second is ribosomal RNA (*r*RNA) which makes up a major part of the structure of the ribosomes (Sec. 8.19). The third is transfer RNA (*t*RNA) which functions in the transfer of enzyme-activated amino acids into the ribosomes where they can be added to the developing polypeptide chain. There is a specific *t*RNA for each amino acid required for synthesis. A more complete discussion of protein synthesis is outlined in Sec. 8.19.

8.7 Nonprotein Nitrogen Compounds (NPN)

There exist in plants and animals a number of nitrogen-containing compounds which, by definition, are not proteins, i.e., they are not amino acids joined by a peptide bond. They all are classed as nonprotein nitrogen (NPN) compounds and that is all they have in common. They are too diverse in structure and function to be further classified. The purine and pyrimidine compounds just discussed fall into that category and of course are distributed widely in plant and animal tissue.

The other nonprotein compounds occurring in feeds include amides, amino acids, nitrogenous glucosides and fats, alkaloids, ammonium salts, and nitrate. Of these, the amides and the amino acids are the ones which are of major nutritional importance. They are especially abundant where growth is rapid, and thus they make up as much as one third of the total nitrogen in pasture herbage and early cut hay. Fifty percent of the nitrogen in silage crops occurs in this form, due in part to their immaturity at harvest and in part to the fermentation processes during ensiling which hydrolyze protein to amino acids. For example, fresh corn forage contains 10 to 20 percent NPN, while in corn silage it is up to 50 percent. The developing seed is high in NPN at the start, but is less than 5 percent at maturity. Mature hays and the commonly fed concentrate mixtures of seeds and their by-products contain relatively little NPN.

In recent years a number of NPN feed additives have been used as a source of nitrogen for ruminant diets. Such compounds as urea, biuret (2× urea), uric acid, and ammoniated products of various kinds have been used effectively. They are discussed in more detail in Sec. 8.16.

In addition to the NPN compounds which occur in feeds, there are a number which are important in nutrition, either as intermediates or end products of protein metabolism or as essential constituents of various tissues and secretions. Some of these such as asparagine, glutamine, uric acid, urea, and creatine are discussed in later sections.

AMINO ACIDS AND PROTEIN QUALITY

The recognition that the nitrogen present in the body had its origin in nitrogen compounds present in the food dates primarily from the work of Magendie[2] published in 1816. After it became established that proteins were the nitrogen compounds essentially concerned, Magendie produced the first evidence that all proteins were not of equal value. In his famous "gelatin report" published in 1841, he showed that gelatin would not take the place of meat protein in the diet.

This finding stimulated growth and nitrogen balance studies by German, Swiss, and Danish scientists as to why gelatin was inferior. The first satisfactory explanation as to why proteins differ in nutritional quality was proposed around 1870 by L. Hermann, a German physiologist who stated that digestion produces units for building body protein and that all the units of this protein, probably amino acids, were needed in the food. In 1876 Escher, a Swiss physiologist, fed dogs a purified diet containing gelatin, which caused them to lose weight. Weight was maintained when tyrosine was added. Amino acid analysis of proteins by Abderhalden in Germany provided the basis for more meaningful studies by Kauffmann, who showed in 1905 that cystine, in which gelatin was low, was needed as well as tyrosine as a supplement.

Next, work was begun with other proteins by more exact methods. Willcock and Hopkins[3] in England, using purified diets, showed that mice receiving zein as the sole protein ingredient died, while those receiving casein lived. Supplementation of zein with tryptophan, isolated and identified by Hopkins in 1901, helped, but failure eventually ensued. In 1909 Osborne and Mendel[4] in the United States began their classic purified-diet studies with rats using pure proteins prepared and analyzed by Osborne and supplemented with various amino acids. They found that certain proteins which caused nutritive failure could be rendered satisfactory by the addition of missing amino acids. For example, as shown in Fig. 8.1, young rats lost weight on zein as the sole protein, but grew when lysine and tryptophan were added. The second curve

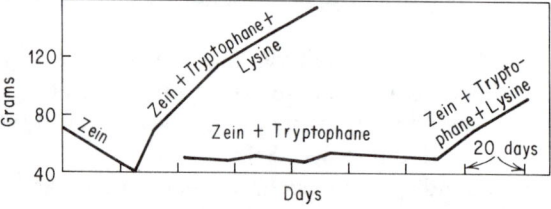

Fig. 8.1 The effect on rat growth of supplementing the protein, zein, with tryptophan and lysine. (*From Osborne and Mendel.*)

implies that lysine is not needed for maintenance. We now know that it actually is, though at a low level. During this same period Abderhalden was conducting nitrogen-balance studies with dogs fed mixtures of 16 amino acids, which were only partially successful, quite probably because not all the needed amino acids had been discovered at that time.

From 1915 on, further developments occurred rapidly, including studies with chickens and pigs, and led to the realization that the value of a given protein in nutrition is governed by its amino acid makeup. The body is unable to synthesize many of the amino acids which are present in its proteins, and thus the protein in the food must be of a nature which will supply them.

8.8 Essential or Indispensable Amino Acids

The modern advances in the field of amino acid nutrition date from 1930, when W. C. Rose[5] of the University of Illinois began a brilliant series of studies, using a new technique which has given us specific information as to the acids which must be present in the food. By the use of semipurified diets designed to be otherwise adequate for the normal growth of rats, in which the sole source of nitrogen was supplied by amino acids, the effect of the addition or removal of each of the acids was studied. Thus the Illinois workers were able to classify 10 as essential dietary constituents and the others nonessential, as shown in Table 8.2 for the rat. Arginine was found to be a special case in that growth occurred in its absence but not at the normal rate. This meant that the body could synthesize this acid but not sufficiently rapidly to meet fully the needs for growth. Rose thus defined an essential amino acid as *one which cannot be synthesized in the body at a rate required for normal growth.* He found that, for maintenance, rats needed the same ones as for growth, with the exception of arginine. We now know that arginine is an end product of the urea cycle (Sec. 8.21) and thus is available as "excess" amino acids are being readied for excretion. This mechanism apparently provides enough for maintenance but not enough for growth.

Following his lead, a number of investigators have laboriously determined the amino acid requirements for a number of nonruminant species using semi-purified diets and crystalline amino acids as suggested by Rose. Their results are shown in Table 8.2. Eight amino acids are required by all species and thirteen are required by one or more species. These thirteen must be included in the diet for at least one species and are clearly dietary *essential* or *indispens-able* amino acids. The remaining are not an obligatory part of the diet as they can be synthesized within the body. They have been called *nonessential* amino acids.

Harper[6] has taken exception to the concept of nonessentiality and suggested that amino acids that can be "synthesized in the body from whatever precursor" be described by the term *dispensable.* In most instances this latter group of amino acids will be synthesized by the cell using *nonspecific* sources of *amino* nitrogen (glutamic acid, diammonium citrate, alanine, essential amino acids, etc.). This nonspecific nitrogen, however, is clearly "essential" to the

TABLE 8.2. Amino Acid Requirements for Several Species

Amino acid	Human[a] Adult	Human[a] Infant	Rat[b]	Chick[b]	Pig[b]	Fish[b]
Essential						
Phenylalanine + tyrosine	16	132	0.8	1.34	0.88	2.1
Isolencine	12	80	0.5	0.80	0.63	0.9
Lysine	12	97	0.7	1.20	0.95	2.0
Leucine	16	128	0.75	1.35	0.75	1.6
Methionine + cystine	10	45	0.6	0.93	0.56	1.6
Threonine	8	63	0.5	0.75	0.56	0.9
Valine	14	89	0.6	0.82	0.63	1.3
Tryptophan	3	19	0.15	0.23	0.15	0.2
Arginine	–	–	0.6[c]	1.44	0.25[c]	2.4
Histidine	0	33	0.3	0.35	0.23	0.7
Glycine + serine	–	–	–	1.5	–	–
Asparagine	–	–	0.4[f]	–	–	–
Proline	–	–	0.4	0.2[g]	–	–
Dispensable						
Cystine	[d]	[d]	[d]	[d]	[d]	[d]
Tyrosine	[e]	[e]	[e]	[e]	[e]	[e]
Glutamic Acid	–	–	–	–	–	–
Alanine	–	–	–	–	–	–
Aspartic Acid	–	–	–	–	–	–
Hydroxyproline	–	–	–	–	–	–
Citrulline	–	–	–	–	–	–
Ornithine	–	–	–	–	–	–
Protein requirement	0.8 g/kg	2.0 g/kg	12%	23%	20%	40%

[a]Milligram per kilogram body weight per day. National Research Council, Recommended Dietary Allowances, 8th ed., National Academy of Sciences, Washington, D.C., 1974.
[b]Percent of air dry (90 percent dry matter) diet. National Research Publication on the nutrient requirements of laboratory animals (1972), Poultry (1977), Swine (1978), Warm Water Fish (1977), National Academy of Sciences, Washington, D.C.
[c]Growth only.
[d]A portion but not all of the methionine requirement can come from cystine.
[e]A portion but not all of the phenylalanine requirement can come from tyrosine.
[f]Required only for 3–8 days on an amino acid diet.
[g]R. E. Austic, personal communication.

animal, since a diet composed solely of essential amino acids, even though in excess, is improved by adding mixtures of nonspecific nitrogen sources. Further, mixtures are superior to a single compound giving strength to the term *nonspecific*.

Two other points are evident in Table 8.2. First, there is a marked species difference not only in the protein requirement but also in the qualitative and quantitative amino acid requirements. The very high protein need of the fish is probably a reflection of its basically carnivorous diet and relative inability to utilize carbohydrate. Arginine would be expected to be higher in the chick as they have no urea cycle to help (Sec. 8.21) and, in addition, feathers are very high in arginine. The failure to detect a histidine requirement for adult humans has been a puzzle but appears to be a fact. It is clear from this table that extrapolating from one species to another would be hazardous. Second, from Table 8.2 one can see there is a marked reduction in the requirements between the growing infant and the adult. This difference is characteristic of all species, e.g., the baby pig versus the older pig.

Biochemically, the essentiality of the several amino acids is a reflection of the cell's inability to synthesize the necessary carbon skeletons (α-keto acids) to which the amino nitrogen can be attached. If this is true, the feeding of these carbon skeletons or their analogs should, in the presence of adequate tissue ammonia, substitute for a dietary amino acid. This is, in fact, the case[7] as the α-keto acids of all but two of the ten essential amino acids for the rat have been shown to be utilized with varying degrees of efficiency as a replacement for their respective amino acids. Neither threonine nor lysine could be replaced, however, indicating that the means of adding ammonia to those keto acids is lacking. Medically, feeding α-keto acids might enable patients with elevated blood ammonia levels, as a result of chronic renal failure (Sec. 8.21), to use this excess nitrogen as a source of essential amino acids and to be fed a very low protein diet. Need for kidney dialysis would thus be reduced. Research in this vein is active today.

In nature, most amino acids occur in the L form and are well utilized. In industry it is possible to synthesize mixtures containing the D and L form economically. When searching for amino acid supplements caution must be exercised, as, except for methionine, the D form of amino acids is utilized much less efficiently than the L form. In humans, however, even D-methionine is poorly utilized. With the commercial availability of the pure L isomers of amino acids, the research worker should use them in preference to the less expensive D-L mixtures, in experimental diets. While more costly, the added precision is worth the added cost.

Can we have normal performance on amino acid diets? If so, it suggests our qualitative and quantitative data on amino acids are reasonably correct. Because of the expense of pure amino acid diets, only the smaller species have been tested. Pleasants[8] obtained reproduction into five generations on a completely soluble diet using pure L-amino acids fed to germ-free mice. Rat growth on an amino acid diet was equal to diets containing casein or casein hydrolysate.[9] It would then appear that a proper balance of the essential amino acids and a mixture of nonspecific nitrogen is all that is required for proper nitrogen nutrition.

8.9 The Chemical Determination of Protein and Amino Acids

The direct determination and identification of the various proteins which are present in a feed or tissue is an impractical procedure. Thus the chemist takes advantage of the fact that nitrogen occurs in the different proteins in a fairly constant percentage, 16 percent on the average (Sec. 8.1). He analyzes for nitrogen and multiplies the result by the factor 6.25. The analytical technique, in use for over a century, is known as the Kjeldahl procedure. In this method, the amino nitrogen ($-NH_2$) is oxidized by sulfuric acid in the presence of catalysts to $(NH_4)_2SO_4$. The ammonium ion is converted to NH_3 by NaOH and collected by distillation. The NH_3 is then quantitatively titrated by one of several techniques and the nitrogen in the sample computed. It is specific for ($-NH_2$) and not nitrate ($-NO_3$).

While the average factor 6.25 is applied to feeds in general, specific factors may be used in the case of products for which the protein and nitrogen relationship is definitely known. For example, it has been found that the combined proteins of milk contain approximately 15.7 percent of nitrogen on the average, and hence the factor 6.38 is used. Wheat-flour protein, on the other hand, contains 17.5 percent nitrogen, and thus the factor is 5.71. Specific factors are also employed for certain animal tissues. Jones[10] has published specific factors for 121 different proteins and foods.

The estimation of protein content from an analysis for nitrogen assumes that all of the nitrogen in the substance analyzed is in the form of protein. This is not strictly true for any feedstuff, as there are certain ones which contain a considerable amount of nitrogen in compounds other than protein, viz., non-protein nitrogen (Sec. 8.7). This fact was recognized by the early workers in animal nutrition, and methods were accordingly proposed for determining the true protein as distinguished from the crude protein obtained by multiplying the total nitrogen by a factor. This distinction has largely lost its significance as evidenced in the previous discussion in which many of the nonprotein nitrogen compounds such as amides and ammonium salts can be effectively used by the body to form dispensable amino acids. In the case of the ruminant, they are, of course, useful to form essential amino acids as well (Sec. 8.16). Actually the nonprotein nitrogen content of nonruminant feeds is relatively low in any event. In view of all these considerations, a distinction between crude and true protein of feeds seems no longer worthwhile. In this book, unless otherwise specifically stated, the term protein is used, without qualifying adjective, to express the value obtained by multiplying the total nitrogen by 6.25 (or some other stated factor). It should be borne in mind that the value so expressed includes other nitrogen compounds besides protein and that a more exact measure of the value of feed nitrogen for protein nutrition must include an estimate of its amino acid composition or some indirect means of assessing its value biologically (Sec. 15.16).

While chemists have long been active in the development of methods for determining the quantitative distribution of amino acids in protein, the recognition of the essentiality of some for proper nutrition greatly stimulated the effort towards a rapid amino acid assay. Early techniques were largely by microbiological procedures. Today rapid separation and quantification of amino acids in many materials is possible by use of ion exchange columns operated automatically. It is first necessary to hydrolyze the peptide bonds yielding a mixture of the constituent amino acids which are then passed through the column.[11] Table 8.3 includes representative values of several animal feeds. With these data it is possible to formulate diets meeting minimum requirements as suggested in Table 8.2.

8.10 Essential Amino Acids and Protein Quality

The discovery that many of the amino acids composing body proteins must be supplied as such by food protein explains why different foods or feeds with the same protein content have different protein values in nutrition, i.e., they differ in *protein quality*. Those proteins whose assortment of amino acids more nearly approximates the needs of the animal are of *high* quality and those which do not are of *low* quality. In a facetious vein, pig protein would be the best protein to feed to pigs: nutritionally sound but economically disastrous.

Examination of Table 8.3 helps to clarify the concept of protein quality. Note that the values are expressed in two ways: percent of the feed and percent of the protein. The first set permits one to combine feeds and obtain proper amino acid levels in a feed mixture. The second is a better measure of the quality of the individual feed protein per se. For example, it is clear that corn protein is very low in lysine, a well known fact that must be accounted for in feed compounding. On the other hand, the proteins of soybean meal, milk solids, whey, or fish meal are much higher in lysine and could be mixed with corn to satisfy a lysine requirement. The reverse situation exists for leucine in which corn protein is richer than the others, whereas relatively little difference exists with respect to valine. It is clear from the table that the proteins from animal sources have no shortages of the essential amino acids (i.e., they are of high quality), whereas the plant proteins can be marginal (low quality). Yet in practice, for a 25-kg pig, a mixture of soybean meal and corn can meet all of the essential amino acid requirements without supplementation. At younger ages, richer sources of protein are required to meet the higher protein and amino acid requirements.

The impact of putting the information about amino acid requirements into practice and its effect on animal performance and efficiency was clearly demonstrated by a very simple experiment conducted by Scott, reported by Baker,[12] and shown in Table 8.4. Treatment 1 had soybean meal alone to provide 15 percent protein. In treatment 2 sufficient essential amino acids were added to insure that all the essential amino acid requirements were met; no account was made of the excesses which existed. With all the essential amino acids supplied, feed intake, gain, and PER all improved. All measures are evidence of a higher quality protein being consumed. In treatment 3 sufficient soybean meal was used to just meet the minimum requirement for that essential amino acid which was in greatest excess *with respect to requirement,* in this case phenylalanine. Sufficient amounts of all the remaining essential amino acids were added such that all essential amino acids were at levels which *just* met the requirement, i.e., a "perfectly" balanced protein for the chick. Neither gain nor feed intake changed much, but PER increased dramatically. Diet 3 did

TABLE 8.3. Amino Acids in Feeds[a]

Amino Acids	% of air dry feed						% of protein					
	Alfalfa	Corn meal	Soybean meal	Nonfat dried milk	Dried whey	Fish meal	Alfalfa	Corn meal	Soybean meal	Nonfat dried milk	Dried whey	Fish meal
Arginine	0.80	0.50	3.68	1.12	0.34	3.79	4.6	5.7	7.5	3.3	2.8	6.6
Glycine	0.90	0.37	2.29	0.27	0.30	4.19	5.1	4.2	4.7	0.8	2.5	6.9
Histidine	0.32	0.20	1.32	0.84	0.18	1.46	1.8	2.3	2.7	2.5	1.5	2.4
Isoleucine	0.84	0.37	2.57	2.15	0.82	2.85	4.8	4.2	5.3	6.4	6.8	4.7
Leucine	1.26	1.10	3.82	3.23	1.19	4.50	7.2	12.5	7.9	9.6	9.9	7.4
Lysine	0.73	0.24	3.18	2.40	0.97	4.83	4.2	2.7	6.6	7.2	8.1	8.0
Methionine	0.23	0.20	0.72	0.93	0.19	1.78	1.3	2.3	1.5	2.8	1.6	2.9
Cystine	0.20	0.15	0.73	0.44	0.30	0.56	1.1	1.7	1.5	1.3	2.5	0.9
Phenylalanine	0.79	0.47	2.11	1.58	0.33	2.48	4.5	5.3	4.4	4.7	2.8	4.0
Tyrosine	0.56	0.45	2.01	1.13	0.25	1.98	3.2	5.1	4.1	3.4	2.1	3.3
Threonine	0.70	0.39	1.91	1.60	0.89	2.50	4.0	4.4	3.9	4.8	7.4	4.1
Tryptophan	0.28	0.09	0.67	0.44	0.19	0.68	1.3	1.0	1.4	1.3	1.6	1.1
Valine	0.84	0.52	2.72	2.30	0.68	3.23	4.8	5.9	5.6	6.9	5.7	5.3
Protein, %	17.5	8.8	48.5	33.5	12.0	60.5						

[a]N.R.C. Nutrient Requirements of Poultry, National Academy of Sciences, Washington, D.C. 1977.

TABLE 8.4. Minimizing Excess Amino Acids for the Chick

Treatment	$N \times 6.25$	Gain, g	Feed intake, g	PER[a]
1. Soybean meal (49.7% protein)	14.7	49	102	3.3
2. As 1 plus deficient amino acids	15.7	65	110	3.8
3. As 1 with all amino acid equal to requirements	13.1	66	112	4.5

[a]Protein efficiency ratio (grams gain per gram protein consumed) (Sec. 15.18).

not serve the needs of the animal any better for growth but since it met the requirements with less protein, there was less wastage and hence it was much more efficiently utilized. If account could have been taken of the absorbability of *each* of the essential amino acids, no doubt PER could have been increased even more.

This experiment is positive proof of the value of the tedious job of ascertaining minimum amino acid requirements, in the interest of conserving feed supplies for use by animals, to say nothing of reducing the cost to the producer.

DIGESTION AND ABSORPTION OF NITROGEN COMPOUNDS BY NONRUMINANTS

A simple chemical analysis of a feed protein, while indicative of its nutritive value, is only part of the answer. It is the kind and quantity of the essential amino acids and nonspecific amino nitrogen that arrive at the individual cell which determines the value of the feed protein. Thus a background of the process of digestion and factors which affect it are essential for "interpreting" protein nutrition.

8.11 Digestion and Absorption

In general, dietary proteins are hydrolyzed to their constituent amino acids, absorbed, and transported to the liver via the hepatic portal vein. Some amino acids appear in the lymph but the amounts are small. An exception to this principle exists with certain neonatal (newborn) mammals who, during the first day(s) of life can absorb intact immune globulins directly into the lymphatic system (thoracic duct). This ability lasts for only 24 hours in the calf, several days in the rat, but does not even exist in the human infant or the guinea pig. Where functional, the intestinal villi of the newborn are able to absorb the globulins by *pinocytosis* (an engulfing phenomenon). This capacity is soon lost in a process referred to as closure. It enables those species which do not normally obtain adequate immune protection through placental transfer to acquire "instant" immunity by ingesting colostrum high in immunoglobulins. Aside from this unique situation, protein must be digested.

The proteolytic enzymes, their source of secretion, substrate requirements, and hydrolytic products are summarized in Table 3.2. The enzymes secreted by the gastric mucosa and pancreas are discharged into the lumen of the stomach and small intestine, respectively. The enzymes of the intestinal mucosa act within the intestinal mucosal cell itself. Two types of enzymes exist, the endoenzymes characterized by pepsin, trypsin, and chymotrypsin and the exoenzymes characterized by the carboxypeptidases and peptidases. The former cleave large molecules into smaller ones by acting within the peptide chain, while the latter attack terminal amino acids yielding free amino acids. The endoenzymes do not cleave at random but are specific for certain peptide bonds, e.g., pepsin cleaves bonds adjacent to an aromatic amino acid.

Protein digestion begins in the stomach with significant denaturation by HCl followed by peptic digestion which is active at a low pH (2.1). It results in the production of large peptides and relatively few amino acids.[13] The ingesta moves to the duodenum where it is acted on by the several pancreatic enzymes and results in the production of a substantial amount of free amino acids (over 60 percent of the ingesta protein) and oligopeptides. These latter compounds are absorbed directly into the intestinal mucosa where they are hydrolyzed by peptidases to amino acids and in turn transported to the portal circulation. No peptides appear in the portal blood, indicating hydrolysis is complete before movement into the systemic circulation.

The rate of absorption of amino acids is not uniform although virtually all occurs in the proximal two-thirds of the small intestine. Absorption is an active type, much as with glucose, in which the transport of sodium is involved. Tripeptides are absorbed more rapidly than dipeptides, which are in turn faster than free amino acids. In addition, there appears to be a competition for absorption within groups of free amino acids, viz., acidic, basic, neutral, and imino acids.[14] There is, however, no competition between groups, suggesting that slightly different mechanisms of transport exist for different chemical configurations. When the amino acids are absorbed as oligopeptides, this competition disappears.

The nitrogen compounds in the chyme are a combination of feed components plus endogenous excretions of the animal itself. It has been estimated that in humans these excretions constitute about 50 g per day for a 70-kg person.[15] They consist of digestive enzymes and mucus (20–30 g), desquammated epithelial cells (30 g), and plasma protein (1–2 g). This amount is about equal to the daily protein requirement. Fortunately most of this excretion is digested and "recycled" into the body to be used as needed elsewhere. Some portion of this digestion is at the hand of microbes since fecal nitrogen is higher in the germ-free animal than the normal one. The small fraction which is not digested, together with intestinal microbes, passes on down the tract to appear in the feces as metabolic fecal nitrogen (MFN).

The reabsorption of the digested protein excretions poses a problem in evaluating the significance of the postprandial (after eating) amino acid increase in the portal circulation. It has been thought that the pattern of amino acids in this blood would reflect the amounts and kinds of amino acids being absorbed from the diet and hence would be useful in evaluating the diet. The leveling

effect of the absorption of the excretions makes this type of evaluation somewhat imprecise. However, by estimating the difference between portal and
arterial levels, the increase due to diet does give a reasonable approximation of
the rate of amino acid absorption. By adding blood flow data, the total absorption with time can be estimated. Only the amounts used by the intestinal tissue
itself cannot be accounted for. Using techniques of this kind it has been shown
that the essential amino acids are not all absorbed with equal efficiency, varying from 54 percent for isoleucine to 89 percent for histidine in the pig consuming fish meal.[16]

In most cases, the contribution of the stomach to the total digestive process is about 20 percent. Removal of peptic digestion is rapidly accommodated
for in the small intestine, but removal of pancreatic digestion seriously reduces
total protein digestibility and absorbability.

For the neonatal animal receiving milk, the stomach is substantially more
important. In the ruminant, rennin causes the casein to clot within two or three
minutes after consumption. In the nonruminant, pepsin serves a similar function. The noncasein protein (whey, lactalbumin, lactoglobulins) is expressed
from the clot and moves slowly into the duodenum. It has been shown that 53
percent of the milk protein is still in the abomasum 12 hours after feeding.[17]
Similar observations exist for the baby pig. Clearly with milk diets, the stomach
acts to digest, but also serves to release protein and other nutrients to the small
intestine more gradually. In this context milk is unique, since when milk protein is replaced by nonmilk proteins, passage from the abomasum is much more
rapid.

True versus Apparent Digestibility. The dietary protein not appearing in
the feces is assumed to be digested. It is referred to as *apparent digestibility*
since it is recognized that a portion of the fecal nitrogen has been derived from
the animal and is not a feed residue. Deducting this metabolic fecal nitrogen
(MFN) from the fecal nitrogen results in the estimation of *true digestibility*
which thus accurately reflects the absorption of the feed nitrogen per se. It has
generally been impossible to separate MFN from feed nitrogen residues chemically. Thus MFN has been determined using nitrogen-free diets, in which all
fecal nitrogen is thus derived from the animal. Since such diets are abnormal,
it was early suggested by Mitchell that a small amount of a highly utilized, well
balanced protein such as 4 percent of defatted egg protein be fed and the fecal
nitrogen measured. The fecal nitrogen was the same for both diets indicating
the egg protein was 100 percent absorbed. Clearly the egg diet provides a much
better experimental regime. A summary of a number of studies has shown that
MFN is proportional to feed intake, i.e., about 2 mg N per gram of dry-matter
intake.[18] Using such a constant, it is possible to convert *apparent* digestibility to
true digestibility.

There are ample data to suggest that the true digestibility of protein is not
affected by level of protein in the diet. However, apparent digestibility increases in a curvilinear fashion as the level of dietary protein increases while
dry matter is held constant. This is true due to the fact that at low levels of

protein intake, the constant excretion of MFN makes up a greater proportion of the fecal nitrogen output which becomes progressively less as feed protein levels and hence total fecal nitrogen residues increase. From a practical point, however, changes in digestibility are quantitatively small, above 12 to 15 percent of dietary protein.

While the distinction between true and apparent digestibility is desirable for certain experimental work, e.g., determination of biological value (Sec. 15.16), it has relatively little significance in feeding practice. The excreted metabolic nitrogen represents a loss which must be taken into account and assessed against some body process. Since it is related to feed intake as a whole and it is a loss which occurs in the course of the digestion of this food, it is more appropriately assessed against digestion than against any other body function. Thus the figures for the digestibility of protein of feeds commonly determined and employed represent apparent digestibility, though generally spoken of without this qualifying term.

8.12   Factors Modifying Protein Digestibility

While the digestibility of protein can be expected to vary between feeds there are at least three factors which can seriously alter protein utilization: (1) age of the animal, (2) presence of protease inhibitors in the feed, and (3) heat damage to protein via the Maillard reaction.

The problem of age is of practical significance primarily in the neonatal calf. Apparent digestibility increases between 1 and 4 weeks with relatively little change thereafter. The difference is more pronounced on a nonmilk, liquid diet.

Raw or inadequately heated soybean meal has long been an anathema to swine, fish, and poultry feeders because of the presence of toxic proteins including trypsin inhibitors and hemoglutinins (lectins). Trypsin inhibitors depress the activity of trypsin and chymotrypsin, which reduces protein digestibility and results in the hypertrophy of the pancreas and a reduction in the available energy of the feed. Growth is depressed, as you might expect. The lectins cause agglutination of erythrocytes in vitro, reduce amylase activity, and also cause a severe growth depression. Soybeans are not the only culprit, as most beans such as navy or kidney beans contain similar "antinutritional" factors. Ruminants are not affected as the rumen fermentation renders the factors ineffective. Fortunately, for nonruminants including humans, cooking or proper heat treatment destroys the factors. In the case of commercial production of soybean meal, proper heat treatment is routine.

A simple test has recently been developed at Cornell[19] for assaying these inhibitors and, because of the high degree of sensitivity, has shown even barley and alfalfa to contain small amounts.[20]

Trypsin inhibitors are not all bad, however. Colostrum contains a powerful one which aids immune globulin absorption in the newborn by suppressing proteolytic activity in the small intestine. Any proteolysis would alter the protein structure and destroy the effectiveness of the immune globulins.

Mild heat treatment of protein is advantageous as noted above and as a means of preliminary denaturation before ingestion. Extensive heating or prolonged storage, however, can result in impairment of protein quality by the well known Maillard or browning reaction. In this reaction, free amino groups of the peptide chain, most usually the ε amino acid from lysine, react with the aldehyde group of reducing sugars such as glucose or lactose to yield an amino-sugar complex that is no longer available to the animal. The basic reaction is shown in Fig. 8.2. Note that the aldose has become a ketose and if the amino acid is lysine the compound formed is fructosyl-lysine. In the case of dried-milk products, a galactose-fructosyl-lysine complex is formed. As a result of the complexing, trypsin can no longer cleave the peptide bond and lysine is not available. Under the influence of the intestinal microflora, a small part may be released and eventually absorbed but is promptly excreted in the urine.[21] Analysis of the feed for lysine shows it to be present in normal amounts; in effect it is the availability not the amount that is the problem. An example of the effect of heat treatment on lysine availability is shown in Table 8.5.

Although the lysine level has not changed significantly, the ease of hydrolysis and value to the rat is severely reduced by heating. As animal assays are time consuming, a simple laboratory test was devised in which fluorodinitrobenzene (FDNB) was mixed with the feed protein.[22] It complexes with the free amino groups and, after hydrolysis of the protein to its constituent amino acids, the FDNB-amino acid compound yields a yellow color. FDNB does not react with amino groups attached to a sugar moiety. Hence, after hydrolysis, a more intense yellow color per gram of protein is produced from an undamaged protein than from a damaged one. Table 8.5 shows the good relationship between the in vitro test and the quality of milk protein evaluated by other means.

Lysine is the amino acid most severely affected. This has particular nutritional significance since it is often the most limiting in many plant feeds (Table 8.3). Free amino acids not part of a protein are also susceptible, because fructosyl complexes of phenylalanine, methionine, tryptophan, and leucine can be produced. Such reactions are of special concern, when purified amino acids

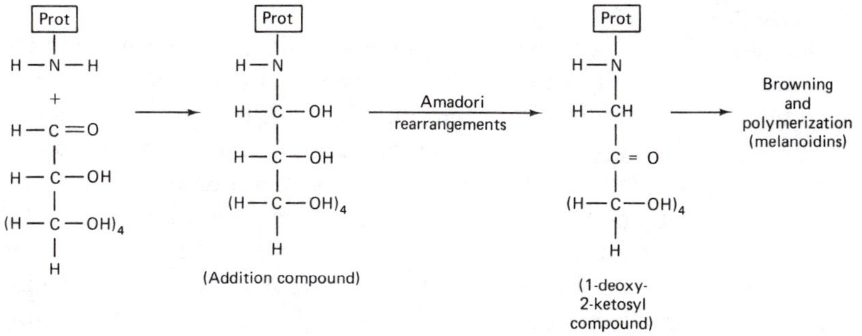

Fig. 8.2 Early phase of Maillard reaction between free epsilon-NH$_2$ groups in amino acids and an aldo-hexose sugar. (*Adapted from J. E. Hodge. J. Agr. Food Chem. 1:928–943, 1953.*)

TABLE 8.5. Lysine levels (g/16 g nitrogen) in Four Samples of Milk Powder Estimated by Four Techniques

Sample	Total analysis	By in vitro digestion	Rat test	Reacting with FDNB
Good quality	8.0	8.3	8.1	8.2
Slightly damaged	7.6	6.2	6.1	6.4
Scorched	6.8	4.5	4.0	3.8
Severely scorched	6.1	2.3	2.0	1.9

Source: F. Mottu and J. Mauron, The differential determination of lysine in heated milk. III: Comparison of the in vitro methods with the biological value, *J. Sci. Food Agr.,* **18:**57–62, 1967.

are added to animal feeds, or in the sterilization of intravenous feeding mixtures containing pure amino acids and glucose for human medicine. The end products of heat damage may not necessarily be harmless, as it has been shown that fructosyl-phenylalanine or its metabolite(s) can depress protein synthesis in the liver.[23]

DIGESTION AND ABSORPTION OF NITROGENOUS COMPOUNDS IN RUMINANTS

8.13   Digestion and Absorption

In the discussion of carbohydrate metabolism, it was pointed out that bacteria and other microorganisms play a large role in the breakdown of complex carbohydrates in the digestive tract, especially in the ruminants. As the bacteria multiply, they synthesize protein to construct their own bodies, obtaining the raw material from the ingested food. For this purpose they can utilize amides, ammonium salts, and even nitrates, as well as protein itself. Bacterial protein so formed in the rumen is digested later in the stomach and intestine, and thus the microorganisms of the rumen play an important role in protein as well as carbohydrate nutrition.

The concept that microorganisms play a useful role in protein metabolism was put forward long before the specific importance of amino acids was appreciated. From a study of cellulose digestion Zuntz,[24] in 1891, expressed the view that rumen bacteria use by preference amides, amino acids, and ammonium salts instead of protein. Other studies led to the belief that the protein supplied by a given ration was augmented as a result of the formation of protein in the bodies of bacteria and protozoa which were later digested. These early observations were followed by many experiments indicating that the protein requirements of animals, especially herbivora, could be met in part by such nonprotein nitrogen compounds as asparagine, urea, and even ammonium salts, and these findings were frequently explained on the basis that microorganisms intervened to transform these simple compounds into protein which was later digested and thus served the body.

The many contradictory findings reported kept the question open for nearly fifty years after the initial observations of Zuntz. In 1937 Fingerling and coworkers[25] produced clear evidence from nitrogen-balance studies with calves that urea can be utilized to supply a part of the protein needs for growth. Shortly thereafter, convincing growth studies with both calves and lambs were published in both Great Britain and the United States, and it was established that urea could be utilized for milk production by the dairy cow. Following these early observations, the subject of protein nutrition by ruminants has received ever increasing study. The student is referred to several extensive reviews presented in two symposia which consolidate much of our current concepts of how nitrogen is used by the ruminant[26,27] in much greater detail than is possible in this text.

The general pathways of nitrogen digestion and absorption are shown in Fig. 8.3. The dotted lines indicate routes which are used but are probably quantitatively small. It is well established that the key to nitrogen metabolism in the ruminant is the ability of the microbial population to utilize ammonia and *in the presence of adequate energy,* to synthesize the proper amino acids needed for their own protein requirements. It has been shown that 80 percent of the bacterial species existing in the rumen can utilize ammonia as the sole source of nitrogen for growth while 26 percent require it absolutely and 55 percent could use either ammonia or amino acids. A few species can use peptides as well. Protozoa cannot use ammonia but derive their nitrogen by consuming bacteria and particulate matter. The ingesta moving into the small intestine thus contains feed protein which has not been degraded as well as bacterial and protozoal bodies. In the small intestine, enzymatic degradation, as described earlier (Sec. 8.11), produces amino acids from the chyme plus the endogenous secretions. These are in turn absorbed via the portal circulation in a manner probably not different from the nonruminant. Although protein digestion and absorption in the nonruminant is primarily in the proximal two-thirds of the small intestine, in the sheep the ileum is an excellent site for digestion and absorption.[28]

The cecum and large intestine receive whatever has not been digested in the small intestine plus urea from the blood, and largely as a result of microbial action, support an active fermentation. Amino acid absorption is essentially nil, but because of the active fermentation, substantial carbohydrate fermentation occurs, resulting in some energy absorption in the form of volatile fatty acids. This process is especially important in the nonruminant herbivore such as the horse. The feces contain the indigestible feed and metabolic nitrogen.

Unique to this system is the movement (flux) of ammonia and urea. If rumen ammonia is produced in excess of the ability of the microbes to use it, it can be absorbed into the portal circulation, transported to the liver, and converted to urea (Sec. 8.21). The urea can then be either excreted by the kidney into the urine or recycled into the rumen by way of the saliva, to get rid of

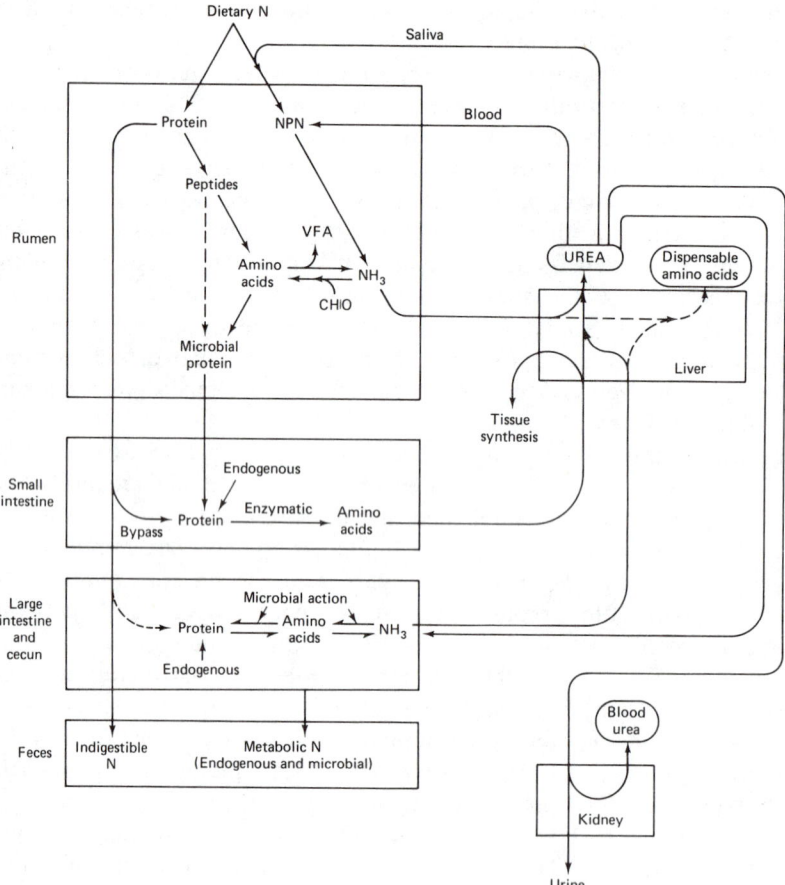

Fig. 8.3 Pathways of digestion, absorption and metabolism of nitrogen in the ruminant.

excess ammonia when needed or to conserve it when the excess in the rumen
is only transitory. On low-protein diets, the kidney reabsorbs a greater quantity
of urea and thus a fair proportion is returned to the blood to be recycled into
the rumen to provide added nitrogen for microbial fermentation.[29] Young growing lambs extract more than mature wethers.[30] While ammonia is toxic in
excess, it can be used as a source of nitrogen for the synthesis of dispensable
amino acids (Sec. 8.18). Quantitatively this pathway is significant, but small.

In terms of overall protein nutrition, the animal is concerned with the
amount and kinds of amino acids absorbed from the small intestine. From Fig.
8.3, it is clear that, aside from level of protein, there are a wide range of factors
which impinge on our effort to predict accurately what and how much the
absorption milieu will be. These factors include (1) percent of true versus
nonprotein nitrogen in the diet, (2) degradability of the dietary protein by the
rumen microbes, (3) amount, digestibility, and quality of the microbial protein,

(4) level of completely indigestible protein in the diet. These factors will be discussed in Sec. 8.14.

8.14   Factors Modifying Efficiency of Protein Utilization in Ruminants

Twenty years ago ruminant nutritionists relaxed with the concept that the rumen microbes could solve their protein quality problems. Feed them enough amino nitrogen and energy and relax. The nonruminant nutritionists, on the other hand, had the problems of adjusting amino acids. However, as more information has been obtained about rumen function, and as pressures build to use feed protein efficiently, the ruminant nutritionists find their problems multiplying. From the perspective of the nonruminant, digestibility or amino acid composition of the diet is not enough. We need to know exactly what is absorbed.

Recently Burroughs[31] and his associates have attempted to do just that. They have suggested a concept of *metabolizable protein,* an attempt to equate the needs of the animal with the ability of the feeds to provide the proper quantity of amino acids to the tissues. It represents a challenging idea and as data become available could well serve as a basis for feeding ruminants in the future. Its accuracy will depend upon recognition of the factors itemized in previous and following sections.

Percent of True versus Nonprotein Nitrogen (NPN).   A characteristic of dietary NPN is its high degree of solubility and as a result its rapid conversion to ammonia. The presence of ample bacterial urease also promptly degrades urea into CO_2 and ammonia as shown below:

$$NH_2\overset{\overset{\displaystyle O}{\|}}{-C}-NH_2 + H_2O \xrightarrow{\text{urease}} 2NH_3 + CO_2$$

Since most NPN compounds, like urea, contain no carbohydrate and hence energy, the bacteria are forced to synthesize amino acids by combining ammonia with carbon skeletons derived from other dietary components. This is a finite reaction, which can proceed only so rapidly, and if ammonia supply exceeds the capability of the bacteria to utilize it, ammonia is lost, as shown in Fig. 8.4. The ability to recycle urea, however, does help to prevent all ammonia which escapes at first from being lost for good. Protein, on the other hand, is metabolized more slowly, releases ammonia more gradually, and provides a source of carbon for amino acid synthesis. The bacteria hence have all the substrate "at hand" to synthesize their own brand of amino acids. One must remember (Sec. 8.7) that the nitrogen of most feeds is not solely protein but some NPN as well. Historically, the judicious combination of true protein and NPN has been a more effective nitrogen combination than diets high in NPN, but no better than predominantly protein diets.

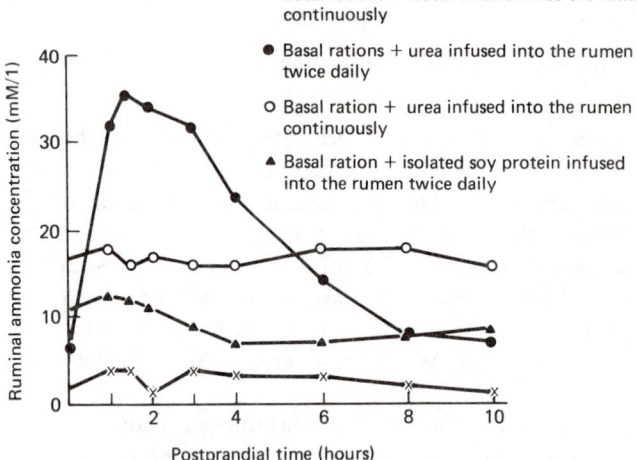

× Basal rations + water infusion into the rumen
  continuously

● Basal rations + urea infused into the rumen
  twice daily

○ Basal ration + urea infused into the rumen
  continuously

▲ Basal ration + isolated soy protein infused
  into the rumen twice daily

Fig. 8.4 Influence of source of supplementary nitrogen on ruminal ammonia pattern. *(From C. L. Streeter et al., Influence of rate of ruminal administration of urea on nitrogen utilization in lambs, J. Animal Sci. 37:796–799, 1973.)*

Degradability of Protein. From Fig. 8.3 it is seen that a proportion of the protein may bypass the rumen to be digested in the small intestine. As such, the amino acid composition of the "bypass protein" is a factor to be considered. While the rumen contains a potent supply of proteases and deaminases, all proteins are not degraded to the same degree (Table 8.6). As expected, urea is 100 percent degraded as is the very soluble casein. Plants are highly variable but corn, with its high content of insoluble *zein,* is only 40 percent degraded. The net effect is to have casein, a very high quality protein, largely coverted to microbial protein, while much of fish meal, also of high quality, largely passes intact. Barley is largely converted to microbial protein.

TABLE 8.6. Degradation of the Protein from Selected Feed Ingredients by Rumen Microbes (in percent)

Urea	100
Casein	90
Barley	80
Cottonseed meal	70
Peanut meal	65
Soybean meal	60
Alfalfa hay	60
Milo	40
Corn	40
Corn silage	40
Fish meal	30

Adapted from W. Chalupa, Rumen bypass and protection of proteins and amino acids, *J. Dairy Sci.,* **58**:1198–1218, 1975.

The practical application of these observations has been to describe proteins as to their relative solubility using buffers[32] and a bacterial protease,[33] and thus to rank them as to their probable degradability in the rumen. In addition, proteins have been treated by mild heat or chemicals such as formaldehyde or tannin to decrease rumen microbial attack and to thus increase the amount of bypass protein[34] which will be digested in the small intestine. Applying these principles of solubility and dietary treatment, it has been shown that the efficiency of dietary protein usage can be improved by presenting to the portal circulation a greater quantity and better assortment of amino acids. The implications of these results are that somehow the synthesis of protein by the rumen microbes is ineffective and/or inefficient. In part that is true.

Yield and Quality of Bacterial Protein.   If rate of microbial synthesis and quality and digestibility of microbial protein were ideal, the ruminant should be able to perform effectively on urea plus a source of energy. In fact they almost can. In 1949 Loosli and associates, using a nearly protein-free purified diet with lambs, produced specific evidence that microbial action in the rumen can synthesize from urea all of the 10 amino acids required for rat growth. The data summarized in Table 8.7 show that three to ten times as much of each amino acid was excreted as was fed. Since the lambs were gaining weight and were in positive nitrogen balance, the excreted acids could not have come from tissue breakdown but must have been synthesized in the rumen. On the basis of analysis of rumen samples, the concentration of the various amino acids in the rumen contents was estimated to be 9 to 20 times greater than in the feed. In Finland, Virtanen[35] reported that a cow produced 4,271 kg of milk when fed a protein-free (urea) diet for 12 months. When 20 percent of the dietary nitrogen came from protein, milk production increased. Depending entirely upon microbial protein is possible but it is not efficient nor practicable.

TABLE 8.7. Average Daily Amino Acid Balance of Sheep (in grams)

Amino acid	Intake	Excreted		
		Urine	Feces	Total
Arginine	0.19	0.48	0.06	0.54
Histidine	0.05	0.18	0.02	0.20
Isoleucine	0.00	0.52	0.06	0.58
Leucine	0.15	0.61	0.08	0.69
Lysine	0.24	0.71	0.12	0.83
Methionine	0.03	0.21	0.02	0.23
Phenylalanine	0.05	0.48	0.04	0.52
Threonine	0.07	0.67	0.06	0.73
Tryptophan	0.01	0.13	0.01	0.14
Valine	0.14	0.69	0.08	0.77

Source: J. K. Loosli et al., Synthesis of amino acids in the rumen, *Science*, **110**:144–145, 1949.

One can readily deduce from the number of biochemical steps involved, that the synthesis of microbial protein from NPN, amino acids, and recycled urea would require more energy than the enzymatic degradation of protein by the small intestine. Further, protein synthesis is a finite process, proceeding only as rapidly as the rumen environment will permit and obviously reaching some limit even when conditions are optimal. Microbial yield of protein is variable but ranges between 90 and 230 g per kilogram of organic matter digested. This amount is adequate to provide the protein for growing animals over about 100 kg and to maintain levels of milk production up to 10 kg per day. Feed protein per se obviously must be provided for high-producing cows.[36]

In addition, a number of characteristics affect the usefulness of microbial protein and therefore must be considered when evaluating the contribution of absorbed microbial amino acids.[37,38,39] For example, (1) diet does not affect the amino acid composition of individual species of bacteria or protozoa per se, (2) protein quality of protozoa and bacteria are quite similar (biological value of 78), but cellulolytic bacteria are higher than noncellulolytic, (3) true digestibility of protozoa is higher (88 percent) than bacteria (66 percent), (4) protozoa numbers are higher on high roughage diets than on concentrate diets, (5) microbial protein contains 20 percent nucleic acid which is probably of little use to the animal. Overall, at the small intestinal level, microbial protein is somewhat inferior to high-quality animal protein, equivalent to soybean and alfalfa leaf protein, and superior to most cereal proteins. The system is obviously complex.

The amalgamation of the principles of protein degradation, microbial synthesis, and protein digestion represents one of the most challenging and potentially rewarding aspects of current ruminant nutrition research. Use of mathematical modeling as suggested by Waldo[40] and isotope dilution studies reported by Nolan[41] should reap immeasurable benefits which will reflect a more efficient animal production.

Heat Damaged Protein.   From Fig. 8.3 it can be seen that a proportion of the diet passes into the feces unscathed. In ruminant diets most of this is the result of heat damage which involves the same basic reaction as noted in Sec. 8.12. In the case of forages, the primary reactive carbohydrate is hemicellulose. Cellulose is essentially unreactive. The polymer eventually produced is completely indigestible, contains 11 percent nitrogen, and has many of the characteristics of lignin. In fact it is often referred to as "artifact lignin." Analysis for it is made by determining the nitrogen content of the acid detergent fiber (Sec. 6.6). It is referred to as acid detergent insoluble nitrogen (ADIN).

Production of ADIN results from a combination of heat, moisture, and time. Optimal conditions for damage are 70 percent moisture, 60°C or higher, and the more time the worse. In the process, hemicellulose is destroyed.[42] In the normal drying or ensiling of forage, conditions often exist where extensive heat damage will occur. Low-moisture silages (haylage) are likely to be most affected. Less than 10 percent of the nitrogen appearing as ADIN in forages is considered normal while values up to 50 percent are not uncommon and are undesirable. Today, programs to evaluate farmer's forages deduct the ADIN before considering the contribution of the forage to the nitrogen of the diet.

8.15 True versus Apparent Digestibility

In ruminants, the principle of true digestibility is the same as that for nonruminants (Sec 8.11). The estimate of metabolic fecal nitrogen, however, is difficult to obtain using a nitrogen-free diet. In 1927 Titus[43] introduced a technique with steers, which involved the plotting of the total nitrogen intake as a function of the total fecal nitrogen excretion, using rations of varying protein content but of constant feed intake. He extrapolated the straight line thus obtained back to the point of zero protein intake and arrived at the estimated metabolic nitrogen excretion for the feed intake in question. As a result of these and other studies a value of between 0.545–0.576 g nitrogen per 100 g food dry matter (about 5 mg per gram) was obtained. This is over twice the value for rats and would appear logical since both microbial residues and tissue desquammation would be expected to be higher in ruminants.

In recent years, Van Soest[44] has reported that the true digestibility of neutral detergent soluble fractions of feeds including protein, exceeds 90 percent and often approaches 100 percent. Thus the only indigestible nitrogen fraction in the feces will be the ADIN fraction. Consequently, it has been suggested that relatively little improvement can be made in the evaluation of feed protein for ruminants other than by knowing the ADIN content of the feed and assuming essentially 100 percent true digestibility for the remainder. Thus recent feeding standards for protein for ruminants have returned to the concept of total dietary rather than digestible protein. As data on potential degradability of feed proteins become more quantitative, they surely will be considered in feeding programs as well.

8.16 Use of NPN for Ruminants (and Nonruminants)

The place of nonprotein nitrogen (NPN) in ruminant nutrition had its origin in 1879 in Germany and entered the United States feeding scene with research at Wisconsin in 1939. While there are a number of NPN compounds (Sec. 8.7), urea, because of cost, availability, and long history of usage, is the one most commonly fed today. It has one of the most voluminous bibliographies of any field of ruminant nutrition and a recent publication[45] should be studied for a key to that literature. Used properly, urea has been shown to be an important and effective feed ingredient.

Proper use of urea must take into account its unique characteristics, viz., it is deficient in all minerals; it contains no methionine and cystine and thus, in contrast to natural protein, is deficient in sulfur; it has no energy value of its own; it is extremely soluble and is converted to ammonia in the rumen quickly;

and fed in large doses, it results in sufficient ammonia release to cause a fatal toxicity. Further, rumen microorganisms must become adapted to dietary urea, a period taking from 2 to 4 weeks.

The first two problems are handled easily by adding suitable quantities of inorganic minerals. A readily available source of energy is necessary for the microbes to utilize the ammonia as it is released. Starch is the most satisfactory, being fermented at a fairly rapid rate. Molasses is somewhat less valuable as it is fermented too rapidly, while cellulose is least valuable being fermented too slowly. A level of 1 kg of starch per 100 g of urea is often suggested as a guideline. Commercial mixtures of urea and dehydrated alfalfa meal or urea and heated starch have been prepared to facilitate urea utilization.

From a theoretical point of view, to be utilized maximally, urea should be metered into the rumen in amounts which will be released at a rate not to exceed the ability of the microbes to assimilate it. Low levels of urea would of course meet this requirement. Feeding frequently, as is usually done with a completely mixed diet fed ad libitum, would also satisfy this requirement. In Fig. 8.4 note that periodic additions of small amounts of urea prevent a high peak of rumen ammonia developing. It is also important that the other ingredients of the diet are not too high in NPN nor particularly soluble (Sec. 8.14). Total ammonia release could be excessive for maximum efficiency. In spite of the logic of this reasoning there is not unanimity of opinion that it always works in practice and further quantitative research is badly needed.

Obviously, adding urea to diets already meeting the protein requirement is an exercise in futility. For dairy cattle it has been suggested that urea is of no benefit in diets containing over 13 percent protein since evidence in vitro indicates the bacteria are unable to use ammonia effectively if rumen concentrations exceed 5 to 8 mg per 100 ml, a level generated by a 13 percent protein diet.[46] The recommendation is difficult to follow for dairy cattle since the protein requirement for high producing cows exceeds 13 percent. Practical feeding trials have recently indicated urea may be well utilized in diets above 13 percent.[47] Further research will no doubt clarify this important controversy and more critically establish when urea can be used most effectively.

Urea toxicity results from a grossly elevated rumen ammonia level (80 mg per 100 ml) which in turn causes blood ammonia to rise. The animal shows signs of nervousness, excessive salivation, muscular tremors, repiratory difficulty, and tetanic spasms. Death occurs within 1/2 to 2 1/2 hours. Neutralization of the ammonia by drenching with glacial acetic acid or cooling the rumen with water may prevent death.

Toxicity should never occur in practice, if feeds are thoroughly mixed and total intakes are moderate. For example, to obtain maximum efficiency, it is recommended that dairy cattle should never be fed more than 1 percent urea in the total diet dry matter. With diets high in NPN like silages, it should be cut by one-half.

Nonprotein nitrogen is of little practical value for nonruminants. It is ineffective for swine, is used to some degree by mature horses on low-protein diets, and can be used for synthesis of dispensable amino acids for hens fed diets well balanced in the essential amino acids. Any value is derived from the ammonia released by the intestinal microflora. It apparently cannot be used to spare essential amino acids.

METABOLISM OF ABSORBED NITROGEN

Few animals are continuous eaters, which means that the influx of nutrients into the body is sporadic rather than uniform. The metabolic machinery must thus be equipped to handle sharp increases in nutrients, arrange for their temporary storage, and then meter them to the tissues during leaner times. Nitrogen absorption and metabolism are no exception. The liver is the key organ in this effort, as it synthesizes proteins, supplies amino acids to the circulation when needed, and processes nitrogen for excretion when in excess. Its proper functioning depends not only on its ability to absorb and retain amino acids, but on its ability to provide a judicious and carefully modulated release of them to the system.[48]

A diagram of the main chain of events in protein metabolism is shown in Fig. 8.5. Often overlooked is the fact that the intestinal wall is not a sieve, but an active metabolic tissue which requires nutrients for its sustenance. For example, glutamic and aspartic acids are extensively utilized and are low in the portal blood while alanine is added to the portal circulation (Sec. 8.17) by transamination. Consequently, the liver receives the algebraic sum of the nutrients absorbed, synthesized, and metabolized by the intestine. The varied routes shown in Fig. 8.5, by which the absorbed amino acids are metabolized, are discussed in the following sections.

8.17 Reactions of Amino Acids

There are many reactions involving amino acids but two main types provide for interconversion, synthesis of dispensable amino acids, utilization of amino acids for energy, and the utilization or excretion of excess ammonia.

Transamination.   This reaction occurs in the mitochondria and cytoplasm of most cells of the body. It permits the exchange of ammonia from an amino acid to the keto moeity of an α-keto acid. The most important transaminase involves glutamic acid and α-ketoglutaric acids, but aspartic-oxaloacetic transaminase is also a significant one. The reaction is reversible, requires vitamin B_6 as a cofactor, and serves to promote the following events: (1) excess amino acids (essential or dispensable) can be relieved of their ammonia (to glutamic) and the keto acid metabolized in the TCA cycle for energy (reaction right), (2) dispensable amino acids can be synthesized from glutamic acid and TCA intermediates (reaction left). Of the essential amino acids only lysine and threonine do not participate in transamination; recall their α-keto acids are not used for essential amino acid synthesis when fed or infused (Sec. 8.8).

```
 R            COOH         COOH      R
 |            |            |         |
 CHNH₂        CH₂          CH₂       C=O
 |            |    Vitamin B₆   |    |
 COOH    +    CH₂  ⇌       CH₂   +   COOH
              |            |
              C=O          CHNH₂
              |            |
              COOH         COOH
```

Amino acid   α-Ketoglutaric acid   Glutamic acid   α-Keto acid

Important Transaminase Reactions

alanine + α ketoglutaric ⟷ glutamic + pyruvic

aspartic + α ketoglutaric ⟷ glutamic + oxaloacetic

Deamination. There are two kinds of deamination reactions: *oxidative* and *nonoxidative*. The former is by far the most important, being widely distributed in the cytosol and mitochondria of most cells. The principal enzyme is L-glutamate dehydrogenase which catalyzes the reversible reaction:

L-glutamate + NAD(NADP) + H_2O ⟷ α-ketoglutarate + NH_4^+ + NADH(NADPH)

Because of this reaction, ammonia can be released to enter the urea cycle and be excreted (Sec. 8.21) and α-ketoglutarate can enter the TCA cycle. It is an exergonic reaction, as it is a dehydrogenation and not a hydrolysis. It goes to the right under the influence of ADP and GDP (indications of an energy shortage) and to the left under the effect of ATP and GTP (indication of an energy surfeit). Note it is a means of utilizing free NH_3 which later can be swept up by transamination. Two important nonoxidative deamination reactions are catalyzed by amino acid dehydratase and also require vitamin B_6, e.g., serine → pyruvate + NH_4^+; threonine → α-ketobutyrate + NH_4^+. In addition to the amination reaction of α-ketoglutarate, glutamic and aspartic acids can be aminated to glutamine and asparagine and they can in turn release ammonia and amino acid by subsequent deamination.

NH_3 + glutamate + ATP ⇌ glutamine + ADP + Pi

NH_3 + aspartate + ATP → asparagine + ADP + Pi

asparagine + H_2O → aspartate + NH_3

The presence of all of the above pathways provides the animal cell with a high degree of flexibility to provide the proper assortment of amino acids needed by the tissues and to excrete the excess efficiently.

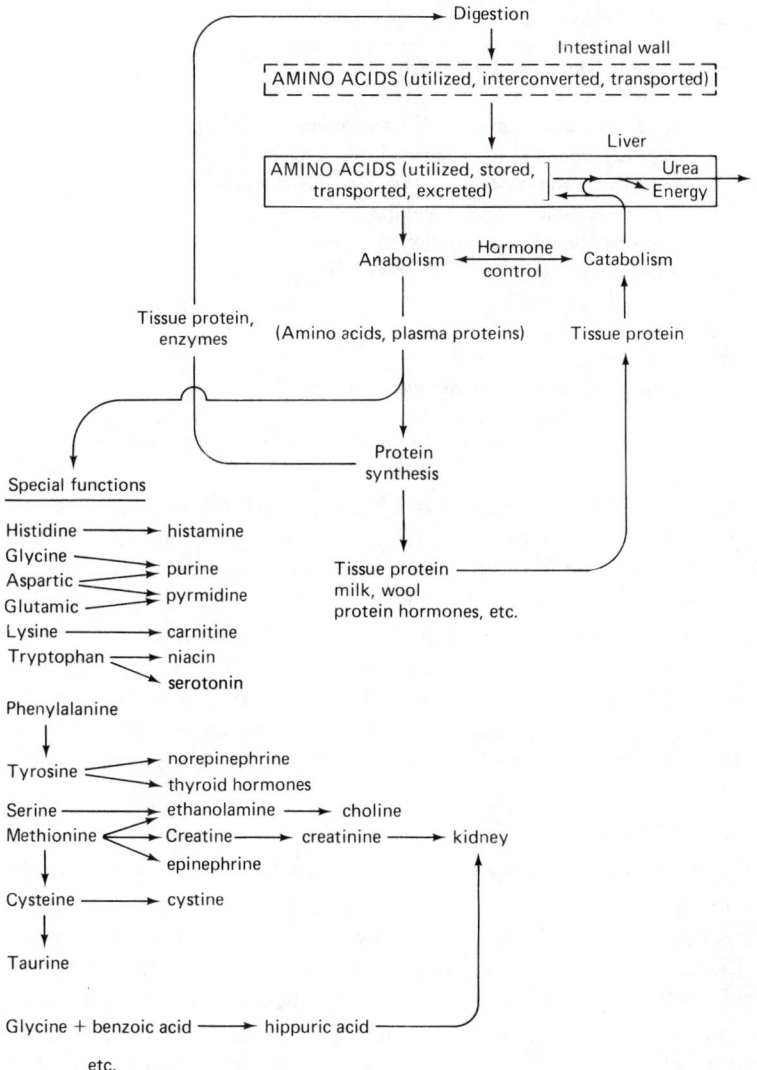

Fig. 8.5 Pathways of nitrogen absorption and metabolism.

8.18 Synthesis of Dispensable Amino Acids

The dispensable amino acids make up almost 40 percent of tissue protein and are obviously of quantitative importance. Not all must be synthesized by the animal as, under normal conditions, the diet will provide many. However, if they are in short supply the cells, with an adequate supply of amino nitrogen and a source carbon and energy, will make up the deficit. A list of some of the precursors and the dispensable amino acids from which they are synthesized

TABLE 8.8. Dispensable Amino Acids and Their Precursors

1. Glutamic (ketoglutarate + NH_4^+) (amination)
2. Glutamine (glutamic + NH_4^+) (amination)
3. Alanine (glutamic + pyruvate) (transamination)
4. Aspartic (glutamate + oxalacetate) (transamination)
5. Serine (3-phosphoglycerate + glutamate) (transamination)
6. Proline (glutamic − H_2O) (dehydration)
7. Hydroxyproline (hydroxylation of proline)
8. Asparagine (aspartic + NH_4^+) (amination)
9. Glycine (CO_2 + NH_4^+)
10. Serine (from glycine and vice versa)
11. Arginine (from urea cycle)
12. Tyrosine (hydroxylation of phenylalanine)
13. Cysteine (methionine + serine)

are shown in Table 8.8. The importance of glutamic acid in these syntheses is obvious, being involved in the first six. This fact perhaps explains why, on amino acid diets, glutamic acid makes such an effective source of nonspecific nitrogen. Note that tyrosine and cystine are derived from the essential amino acids phenylalanine and methionine, respectively. With a deficiency of either tyrosine or cystine, the essential amino acids will have to be used to synthesize them. However, if the diet contains plenty, the levels of phenylalanine and methionine in the diet can be reduced (Table 8.2). In no case, however, can tyrosine or cystine substitute for *all* of the needed phenylalanine or methionine.

8.19 Biosynthesis of Proteins

The manufacture of protein is by far the principal synthetic activity of the cell. It has been estimated that in an adult human over 300 g of protein are synthesized a day; this compares with a requirement of about 50 g per day and in the western world with an intake of often 100 g per day.[49] In 100-g (rapidly growing) rats a total of 713 mg of muscle protein is synthesized per day while net deposition of muscle was only 149 mg.[50] The high rates of synthesis give not only an indication of the degree of synthetic activity but also point to the fact that many amino acids are normally reutilized as a result of rapid turnover and recycling within the body (Sec. 8.21). In fact a major focus of protein synthesis is *not* the deposition of new tissue but the replacement of old. Like all synthetic processes, protein synthesis requires energy which approximates, on a theoretical basis, 8 to 10 high-energy phosphate bonds per mole of amino acid incorporated.[51] Direct attempts to measure the energy cost in the animal have resulted in values two to four times that. The higher "practical" value is probably the result of an increase in the rate of metabolism and an increased rate of amino acid recycling during actual protein accretion.

The mechanics of protein synthesis is one of the most intriguing and exciting stories to emerge from the biochemistry laboratories in the past twenty years. It is covered in detail in all biochemistry and physiology texts and it is therefore unnecessary to recount here other than in summary form.

The nucleus contains the cellular components of DNA in which resides the inherited information necessary to synthesize all of the proteins required by that cell. It also has the capacity to respond to environmental changes in the cell, synthesizing on the one hand, or not synthesizing on the other, as the situation demands.

DNA serves as the template for the synthesis of three RNAs: ribosomal RNA (rRNA), messenger RNA (mRNA), and transfer RNA (tRNA). These RNAs are all synthesized in a process called *transcription* by the action of RNA polymerase in conjunction with DNA. Ribosomal RNA constitutes over two-thirds of the mass of the ribosomes, the sites of polypeptide synthesis in the cytoplasm. Transfer RNA is the smallest RNA. In a reaction driven by ATP the hydroxyl group at one end of the tRNA is attached to the carboxyl group of an amino acid to form aminoacyl-tRNA,

Alanine + ATP + tRNA $\longrightarrow$ Alanyl-tRNA

There are 64 separate tRNAs, 61 attuned to a specific amino acid, and 3 designed to signal peptide chain termination. With 20 amino acids to be incorporated into proteins, several amino acids have more than one tRNA. On the end of the tRNA opposite the amino acid are three purine or pyrimidine bases, which serve as the *anticodon* and identify it as carrying a specific amino acid. After transcription, messenger RNA moves to the ribosome. It is composed of a sequential series of three bases, each called a condon, and when paired with the anticondon of tRNA, identifies the sequence of amino acids to be placed on the polypeptide chain. The peptide linkage is formed in a process called *translation,* the tRNA is released to be used again, and the ribosome moves along the mRNA molecule. Upon reaching a termination codon the polypeptide is released to the cytoplasm to migrate to its proper site of action.

It is obvious from the above description, why a deficiency of just one amino acid would disrupt the orderly process of protein synthesis and thus interfere with a whole myriad of biological activities—growth, digestion, reproduction, or disease prevention. A shortage of total protein would also have the same effect.

Protein synthesis is far from a haphazard event. It has a very small K_M constant, i.e., it will proceed when concentrations of amino acids are very low. It is also under significant hormonal control. For example, growth hormone and insulin stimulate protein synthesis while the glucocorticoids cause protein degradation. Certain specific proteins such as ovalbumin and avidin in the fowl require estrogen and progesterone, respectively. Hormones appear to act at many levels of protein synthesis, e.g., increasing amino acid uptake, making more ribosomes available, increasing numbers and kinds of mRNA by affecting transcription and increasing the speed of movement of ribosomes along the mRNA.[52]

An interesting summary of the effect of a number of environmental and nutritional variables on the synthesis of muscle protein in vitro is shown in Table 8.9. It is not difficult to comprehend that these same variables are active in the whole animal as well.

8.20   Synthesis of Special Compounds

Absorbed nitrogen is used in the synthesis of a number of special compounds as outlined in Fig. 8.5. This list is not complete, but identifies a number that have special significance in nutrition. It also points to the considerable interaction that exists between some of the amino acids. Two of these deserve special comment. Details on the biosynthesis and metabolism of all of these compounds are found in texts on biochemistry and a complete discussion is beyond the scope of this text.

From Fig. 8.5 it is clear that methionine plays more than a casual role. It is significant because it contains sulfur and also has a labile methyl group (Sec. 8.2). As a bearer of sulfur it is a precursor of cysteine (and hence cystine and taurine). The latter compound is well known as a component of bile acids and as an excretory pathway for sulfur. Recently it has been shown that in cats[53] a taurine deficiency results in retinal degeneration of the eye. Further investigation showed that the cat is unable to synthesize adequate levels via the methi-

TABLE 8.9. Affectors of Muscle Biosynthesis as Measured by Polyribosome Activity

Affector	Polyribosome activity, %
Hormones	
Control	100
Castrated + testosterone	209
Hypophysectomized + growth hormone	159
Control + cortisone	74
Infection	
Control (preinfection)	100
S. typhimurium + adequate protein	35
S. typhimurium + low protein	10
Dietary protein quality	
Casein + methionine (control)	100
Barley	104
Soya protein	87
Gelatin	67
Amino acid diet devoid of threonine	49
Dietary protein quantity	
18% casein (control)	100
3% casein	50

Adapted from G. R. Beecher, Some practical applications of protein synthesis and muscle growth, *J. Animal Sci.*, **38**:1071–1078, 1974.

onine-cysteine pathway, as is true for most species. For the cat, then, taurine becomes a required nutrient.

Methionine, which has been activated to S-adenosylmethionine by ATP, serves as the principal methyl donor ($-CH_3$) to as many as 40 different methyl group acceptors. A few of the methylated compounds are creatine, epinephrine, and choline. Since choline is a required nutrient, a deficiency automatically raises the requirement for methionine. Conversely, a slight deficiency of methionine can be overcome by an excess of choline which can methylate homocysteine.

The metabolism of creatine is especially significant both in energy and nitrogen metabolism. In the presence of an excess of ATP, creatine is spontaneously converted to phosphocreatine which contains a single high-energy bond and serves to store reserve energy. When ATP reserves are low, phosphocreatine donates its phosphate to ADP to yield ATP again. Creatine is promptly dehydrated to form creatinine and is excreted in the urine. The amount of creatinine excreted is extremely constant on a day-to-day basis. Its excretion is highly correlated with the lean body mass and is therefore often used to predict the amount of lean tissue in the whole animal. The empirical reactions between methionine, homocysteine and creatine are shown below.

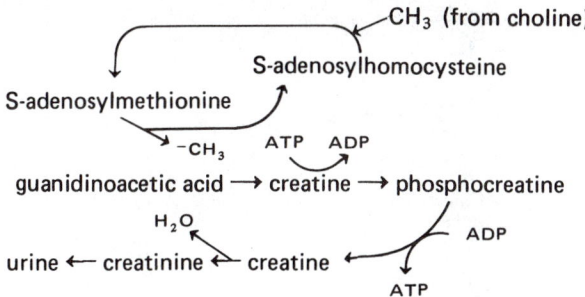

Tryptophan deserves special comment as well. In most species it can be converted to niacin with vitamin B_6 as a cofactor. The niacin required in the diet is thus based in part upon the conversion ratio of available tryptophan to niacin. In the rat, it is estimated that 33 to 40 mg of tryptophan will yield 1 mg of niacin, while the comparable figure for man is about 60 mg. In practice, where minimum dietary protein is an economic necessity, niacin should be provided without regard to the tryptophan in the diet. Strangely enough, neither the cat nor the mink can utilize tryptophan for niacin synthesis.

8.21 Disposal of Excess Amino Acids

The anabolic reactions discussed above provide for the utilization of the amino acids required by the tissues. In the real world, no diet provides precisely the proper amount of protein at any given time for optimal utilization. Good feeding practice will always provide a slight excess. In many instances it is more economical not to worry about excesses, for example, young ruminants grazing alfalfa. But there is the case of adult humans, where excess protein intake is often the rule rather than the exception. The extra amino acids are promptly metabolized, often within hours. The ammonia is excreted and the carbon skeleton is used as a source of energy, glucose, or fat depending upon the amount. In times of low feed intake, tissue protein is mobilized and metabolized for energy (Sec. 8.22).

Most terrestial vertebrates excrete ammonia in the urine as urea; fowl and land-based reptiles, because of limited water intake, excrete it as uric acid which is a solid; most fishes excrete ammonia directly through the gill tissue into their water environment. In fowl, uric acid is synthesized, from glutamine, glycine, and aspartic acid in a very complicated manner under the control of the enzyme xanthine oxidase.

Urea is formed in the liver by means of the urea cycle, first described by Krebs of TCA-cycle fame. It is diagrammed in Fig. 8.6. Note that both NH_3 groups found in the urea molecule came from glutamate, one from the oxidative

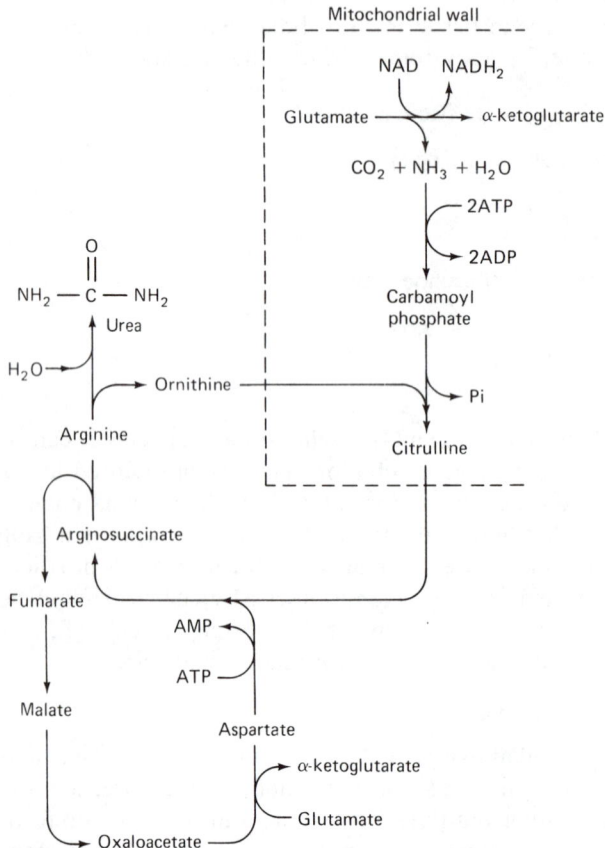

Fig. 8.6 The urea cycle.

deamination and the other from transamination with oxaloacetate. A total of three high-energy bonds are generated and four used up giving a net loss of one per mole of urea synthesized from glutamic acid. This compares with a zero energy requirement for the fish which requires no energy for excretion of nitrogen after the ammonia is formed.

The carbon skeletons of the amino acids which have lost their amino groups are now nothing more than carbohydrate, eligible for entry into the TCA cycle to be used as needed. The pathways of degradation in which the ammonia is removed and a carbon skeleton is produced are complex for many of the amino acids. However, a summary of the pathways of the amino acid residues is helpful in order to give an indication as to their potential for providing glucose, or energy. Fig. 8.7 shows the points of entry of the amino acids into the TCA cycle. All of those which do not enter at either acetyl CoA or acetoacetate have the potential of being converted to glucose in the process of *gluconeogenesis* and are called *glucogenic*. Those entering at acetyl CoA or acetoacetate cannot result in a net synthesis of glucose, but could provide ketones and thus are referred to as *ketogenic*. Leucine and lysine are obligatorily ketogenic.

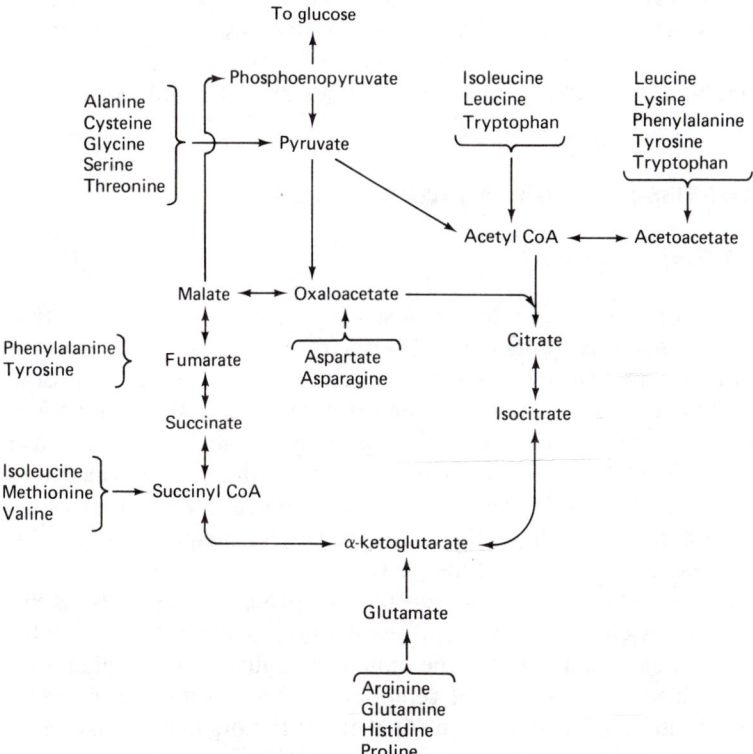

Fig. 8.7   Points of entry for the carbohydrate residues resulting from amino acid catabolism.

The process of gluconeogenesis has special significance for the ruminant.[54] As has been discussed (Sec. 6.9), little starch or glucose bypasses the rumen and therefore little glucose per se is absorbed directly. It must depend upon rumen propionate, amino acids, lactate, pyruvate, or glycerol for its glucose. When feed intake is high, propionate and amino acids will be the primary source although, as was stated earlier (Sec. 6.10), a high-producing cow can derive no more than 60 percent of its glucose from propionate. When feed intake is below the maintenance level, glycerol from mobilized fat (Sec. 7.16) and amino acids from mobilized protein (Sec. 8.22) will become the predominant precursors. About 85 percent of the glucose is produced in the liver and 15 percent in the kidney. Under normal feeding regimes, it is not necessarily true that all the carbon skeletons will result in glucose synthesis; as participants in the TCA cycle, they may just spare glucose as a source of energy.

By referring to Figs. 5.4 and 6.5 we can see how the amino acids can be integrated into the energy-producing cycles with fats and carbohydrates and become a part of the *final common pathway* of energy metabolism.

The purines and pyrimidines absorbed from the intestine can be used for the synthesis of nucleotides and nucleic acids although the efficiency of their use may not be high. In ruminants, where 20 percent of the absorbed microbial nitrogen is in the form of nucleic acids, the efficiency is poor. Purine compounds which are degraded in the tissue are readily reused. When excreted they are converted to uric acid in primates and the Dalmation dog, to allantoin in most mammals, and to urea in fish. Pyrimidines are excreted as urea and ammonia.

CATABOLISM OF TISSUE PROTEIN AND AMINO ACIDS

8.22   Protein. A Dynamic Tissue

The protein mass of the body, like the adipose mass, is in a continuous state of flux, with tissues constantly being catabolized and resynthesized. The overall rate of degradation is precisely regulated in much the same manner as protein synthesis.[55] As amino acids are released they become available to the general amino acid pool and can either be reused for protein synthesis or utilized as a source of energy. Two kinds of catabolism occur, that which is a normal function of tissue maintenance and renewal and that which follows periods of undernutrition. The amino acids have the same options for metabolism in either case, but the needs of the animal dictate the route.

Why should there be a continuous turnover or protein renewal?[56] As protein synthesis requires energy, on the surface it would appear to be wasteful. Yet it represents an excellent way for the animal to adjust to a changing environment. Tissues need to involute or be remodeled, enzymes no longer needed must be removed, and during starvation (or dieting) the organism must selectively feed on its own tissue to survive. With a continuing protein turnover capability, the animal has the means, flexibility, and speed to adapt to gross or subtle changes in its environment which, in the long run, may affect its ability to survive.

Not all tissues have a similar rate of protein turnover. Renewal of intestinal mucosa is extremely rapid, 1 to 3 days in most species.[57] The soluble protein of liver has a half-life of 0.9 days while that in muscle is 10.7 days.[58] Further, the fibrous proteins of muscle, viz., myosin and actin, have half-lives of 35 and 51 days, respectively. The shorter half-life obviously reflects a more intense rate of degradation and resynthesis.

When a nutritional stress such as starvation strikes, tissues lose their protein and release amino acids at different rates as well. In general the integrity of those organs which are crucial to survival such as the brain and kidney are maintained while the liver rapidly loses tissue to support the organism. Muscle is somewhere in between.

During fasting or starvation, the primary need is for energy, particularly glucose. It has been shown that over 50 percent of the glucogenic acid(s) released by the peripheral tissues during fasting are alanine for man[59] and alanine and glutamine for sheep.[60] These are clearly glycogenic, can be derived by transamination, and provide a way for peripheral tissues to metabolize the carbon skeletons of other amino acids for energy and then transport the excess ammonia to the liver on a glycogenic amino acid. This mechanism also explains why the injection of corticosteroid or ACTH hormones which mobilize protein results in an increase in blood sugar in a ketotic dairy cow and is a fairly successful treatment for this disease.

8.23 Endogenous and Exogenous Catabolism

In 1905 Folin[61] put forth the theory that there are two forms of protein catabolism, essentially independent and quite different from each other: a variable one which he called *exogenous,* dependent on the level of protein consumed, and a constant type, the *endogenous catabolism,* related to body size and other body factors. According to this theory the endogenous catabolism represents metabolic processes which are the essential characteristics of living cells, exemplified by the nitrogen excretion on a nitrogen-free, otherwise adequate diet. It reflects metabolic processes which are essential to life, and the end products thus excreted tend to be constant, per unit of body size, unaffected by the character or amount of the food protein. The endogenous nitrogen excreted represents a loss which must be made good by dietary protein in order to maintain the integrity of the nitrogenous tissues of the body. The exogenous catabolism, on the other hand, reflects the breakdown of absorbed dietary nitrogen compounds which are not synthesized into body protein.

This theory of Folin was early challenged by certain workers, and the arguments against its validity have been reviewed by Borsook and Dubnoff.[62] It is at least subject to reinterpretation as a result of the recognition of the dynamic state of body proteins and amino acids. There can be no strict chemical distinction between exogenous and endogenous catabolism as the two are hopelessly intertwined in the body. Yet there clearly is a minimum excretion which results from tissue metabolism and meets Folin's definition of "wear-and-tear" protein catabolism. Thus, while any sharp distinction between endogenous and exogenous nitrogen, as set forth by Folin, may be no longer valid, it seems very doubtful that the dynamic state significantly changes the end results of catabolism.

This question is an important one for students of nutrition because one method of arriving at the maintenance requirement of protein (Sec. 14.15) and also the most widely used method of determining the biological value of protein (Sec. 15.17) assume the reality of Folin's endogenous catabolism. In this connection Mitchell,[65] who has been largely responsible for developing both of these methods, has come to the conclusion, from his own studies and from reviewing the work of others both pro and con, that the Folin concept remains valid, that there are no well-demonstrated findings to the contrary, and that results of isotope studies have clarified but not destroyed the basic concept. His conclusion appears justified. The questions here involved are discussed further in the later sections, referred to above, where the use of endogenous nitrogen measure is described.

It is important to bear in mind that the term endogenous as here used refers to the urinary excretion. The metabolic fecal nitrogen (Sec. 8.15) also comes from body sources rather than from food protein, but it is a loss incident to the digestion process and not to the essential cellular activity for which Folin coined the term endogenous. Since this term is used by some writers to denote the urine losses only and by others to include the fecal losses as well, in this text the specific terms endogenous urinary nitrogen (EUN) and metabolic fecal nitrogen (MFN) are employed unless the context makes clear specifically what is meant.

8.24 Dietary Protein and Amino Acid Anomalies

Protein Deficiency.   Protein is too ubiquitous a part of the animal to show a specific set of deficiency signs. As one could expect, most metabolic activities are depressed. Food intake declines, rumen fermentation is less, growth is impaired, and reproduction and lactation are suboptimal. On a nitrogen-free diet, death will result.

In humans, the classical disease of protein malnutrition of the young is *kwashiorkor.* The sister affliction *marasmus,* is defined as a calorie-deficient state. The problem of separating the two diseases clinically has led to the general term of a *protein/calorie malnutrition* (PCM). In this state, blood protein is low, digestion is poor, the patient is lethargic and has a reduced resistance to infectious disease as evidenced by lowered antibody production and cell-mediated immunity defense mechanisms.[64] There is also evidence that ability to handle the pharmacological effects of foreign chemicals is impaired.[65]

Considerable attention has been directed toward the effect of protein deficiency as a part of general undernutrition and brain development and hence learning.[66] It is clear that rats[67] and pigs[68] show behavioral peculiarities following protein insufficiency during various stages of growth. In a hungry world, research in this area has special meaning to us all.

In growing rats a protein deficiency results in anemia, hypoproteinemia, and in adults an edema due to lowered plasma protein. Chicks or hens on a diet somewhat low in protein have a lower growth rate, reduced egg production, reduced egg size, poor feathering, and a slight overcomsumption of feed. As a result carcass fat is increased.

Amino Acid Deficiency. A deficiency of any one of the essential amino acids is in effect a protein deficiency. Because of the inability to utilize all of the amino acids for protein synthesis, the unused ones are deaminated and urea excretion increases. A unique feature of an amino acid deficient diet is the prompt reduction in food intake. With rats, feed intake is reduced by two-thirds within 24 hours after first consuming the diet. When the amino acid is replaced, intake during the first 20 minutes is three to five times more than that on the deficient diet and returns to normal within 24 hours. Injection of the deficient amino acid into the carotid artery relieved the intake depression while a jugular injection did not, indicating the effect was in the central nervous system. It was later shown that the prepyriform cortex center of the brain is intimately involved in this mechanism. Since force-feeding the deficient diet results in significant pathology, the intake depressing mechanism must serve as a protective mechanism.[69]

Amino Acid Imbalance. An amino acid imbalance "results from the addition to a low-protein diet of one or more amino acids, other than the growth-limiting one, in amounts that are not individually toxic and yet cause depressions in food intake and growth that are readily prevented by a supplement of the growth-limiting amino acid.[70]" Rats prefer a protein-free diet to an imbalanced one and, when corrected, leave the protein-free diet. Feed intake is depressed via the same mechanism described for an amino acid deficiency but involves also the medial amygdala of the brain. The imbalanced diet also interferes with protein synthesis in the liver.

Amino Acid Antagonism. In special cases it has been shown that one amino acid affects the requirement of another by interfering with its metabolism. An example is the lysine-arginine antagonism, wherein dietary lysine increases the requirement for arginine.[71] Lysine acts by competing with arginine for reabsorption in the renal tubules increasing arginine excretion and by increasing renal arginase activity and thus splitting arginine into urea and ornithine. High lysine also suppresses feed intake. In poultry a valine-leucine-isoleucine antagonism also exists. Corn is especially high in leucine (Table 8.3) which could cause an imbalance. Although in practical feeding the above anomalies may be infrequent, persons preparing practical diets should be aware of potential problems.

8.25   Minimum and Optimum Protein Intake

The question whether a level of protein in excess of the minimum required to meet the protein needs of the body is advantageous or disadvantageous has been long debated in the field of human nutrition, and it has implications for animals also. Under the influence of Liebig,[72] who erroneously believed that protein was broken down to furnish the energy for muscular work, the importance of large intakes of proteins was greatly overemphasized for many years following the middle of the last century. Gradually, as a result of research, the pendulum swung to the other extreme marked by the publication in 1904 by Chittenden[73] of experiments supporting the view that minimum intakes favored health and bodily vigor. Chittenden's views were by no means universally accepted, and today most authorities recommend intakes that slightly exceed the minimum requirements, largely as an insurance against underestimating the variation of feed and animal. Some feel that extra protein results in an increase in tissue protein, the so-called *protein reserve* or *deposit protein*. It has been shown in chicks that high-protein diets help to combat the ill effects of nutritional disease and stress through utilization of this protein reserve.[74] However, the Food and Nutrition Board [75] indicates that in healthy individuals there is little need to consume protein in excess of requirements. On the other hand, neither is there any harm in consuming more than the recommended allowances except for cost, as the body has the ability to eliminate excesses.

NOTES

1. F. Sanger, The structure of insulin, in D. E. Green (ed.), Currents in Biochemical Research, Interscience, New York, 1956, pp. 434–459.

2. François Magendie (1783–1855) the great French physiologist, is recognized as the founder of the modern experimental method in animal-feeding experiments. He employed diets of pure carbohydrates and fats to prove that food nitrogen is essential. These studies were published under the title: Sur les propriétés nutritives des substances qui ne contiennent pas d'azote, *Ann. chim. et phys.*, (1) **3**:66–77, 1816.

3. Frederick Gowland Hopkins was born in England in 1861 and was trained as a chemist. His activities in the field of biochemistry began in 1898 at Cambridge University. In addition to his work on tryptophan he showed that histidine is an essential amino acid. In 1921 he discovered glutathione. His most famous work, for which he received jointly with Eijkmann the Nobel Prize in medicine, was in the field of vitamins (Sec. 11.1).

4. Thomas B. Osborne and Lafayette B. Mendel, Amino acids in nutrition and growth, *J. Biol. Chem.*, **17**:325–349, 1914, and later reports.

5. W. C. Rose et al., Comparative growth on diets containing ten and nineteen amino acids, with further observations upon the role of glutamic and aspartic acids, *J. Biol. Chem.*, **176**:753–762, 1948.

6. A. E. Harper, Nonessential amino acids, *J. Nutrition*, **104**:965–967, 1974.

7. K. W. Chow and M. Walser, Substitution of five essential amino acids by their alpha-keto analogues in the diet of rats, *J. Nutrition*, **104**:1208–1214, 1974.

8. J. R. Pleasants, B. S. Reddy, and B. S. Wostmann, Qualitative adequacy of a chemically defined liquid diet for reproducing germ-free mice, *J. Nutrition*, **100**:498–508, 1970.

9. L. H. Breuer, W. G. Pond, R. G. Warner, and J. K. Loosli, The role of dispensable amino acids in the nutrition of the rat, *J. Nutrition*, **82**:499–506, 1964.

10. B. Jones, Factors for converting percentages of nitrogen in foods and feeds into percentages of proteins, *U.S. Dept. Agr. Cir.* 183, 1931 (slightly revised, Feb. 1941).

11. R. J. Simpson, M. R. Neuberger, and T. Y. Liu, Complete amino acid analysis of proteins from a single hydrolysate, *J. Biol. Chem.*, **251**:1936–1940, 1976.

12. D. H. Baker, Amino acid balance and imbalance in the chick, *Feedstuffs*, **46**, no.15, 1974.

13. D. M. Matthews, Protein absorption, *J. Clinical Pathol.*, **24**(suppl. 5):29–40, 1971.

14. B. G. Munck, Amino acid transport, in D. J. A. Cole et al. (eds.), Protein Metabolism and Nutrition, Butterworth, New York, 1976, pp. 73–95.

15. G. Fauconneau, and M. C. Michel, The role of the gastrointestinal tract in the regulation of protein metabolism, in H. N. Munro (ed.), Mammalian Protein Metabolism, Academic Press, New York, 1970.

16. A. Rerat, T. Corring, and J. P. Laplace, Protein digestion and absorption, in D. J. A. Cole et al. (eds.), Protein Metabolism, *op. cit.*, pp. 97–138.

17. J. H. B. Roy and I. J. F. Stobo, Nutrition of the preruminant calf, in D. J. A. Cole et al. (eds.), Protein Metabolism, *op. cit.*, pp. 30–48.

18. B. O. Eggum, A study of certain factors influencing protein utilization in rats and pigs, *Publ.* 406, Institute of Animal Science, Dept. of Animal Physiology and Chemistry, Copenhagen, Denmark, 1973.

19. M. R. Sandholm et al., Determination of antitrypsin activity on agar plates: Relationship between antitrypsin and biological value of soybean for trout, *J. Nutrition*, **106**:761–766, 1976.

20. M. L. Scott, M. Sandholm, and H. W. Hochstetler, Effects of antitrypsin and hemaglutinins in soybeans and other feedstuffs upon feed digestion in chickens, Proc. Cornell Nutrition Conf., 1976, pp. 22–25.

21. H. Erbersdobler, Amino acid availability, in D. J. A. Cole et al. (eds.), Protein Metabolism, *op. cit.*, pp. 139–158.

22. K. J. Carpenter and V. H. Booth, Damage to lysine in food processing: Its measurement and significance, *Nutrition Abst. Rev.*, **43**:424–451, 1973.

23. G. H. Johnson, D. H. Baker, and E. G. Perkins, Nutritional implications of the Maillard reaction: The availability of fructose-phenylalanine to the chick, *J. Nutrition*, **107**:1659–1664, 1977.

24. Nathan Zuntz (1847–1920) was a pioneer in the field of basal metabolism and in respiration studies with farm animals. He developed the first portable respiration apparatus. Trained as a physician, he early forsook medicine to become a teacher and investigator in physiology, first at Bonn and later at Berlin. He devoted himself particularly to work with farm animals and to basic problems related to their nutrition. His publications numbering over 400, deal with a wide variety of physiological problems.

25. G. Fingerling and coworkers, Ersatz des Nährungseiweisses durch Harnstoff beim wachseden Rinde, *Landw. Vers. Sta.* **128:**221–235, 1937.

26. D. J. A. Cole et al. (eds.), Protein Metabolism and Nutrition, Butterworth, New York, 1976.

27. I. W. McDonald and A. C. I. Warner (Eds.), Digestion and Metabolism in the Ruminant, Univ. of New England Publishing Unit, Armidale, NSW, Australia, 1975.

28. D. Ben-Ghedalia and H. Tagari, The ileum as a site of protein digestion, *British J. Nutrition,* **36:**211–217, 1976.

29. B. Schmidt-Nielsen, Excretion in mammals: Role of renal pelvis in the modification of the urinary concentration and composition, *Fed. Proc.* **36:**2493–2503, 1977.

30. S. A. Allen and E. L. Miller, Determination of nitrogen requirement for microbial growth from the effect of urea supplementation of a low N diet on abomasal N flow and N recycling in wethers and lambs, *Brit. J. Nutrition,* **36:**353–368, 1976.

31. W. Burroughs, D. K. Nelson, and D. R. Mertens, Protein physiology and its application in the lactating cow: The metabolizable protein feeding standard, *J. Animal Sci.,* **41:**933–944, 1975.

32. J. E. Wohlt et al., Nitrogen metabolism in wethers as affected by dietary protein solubility and amino acid profile, *J. Animal Sci.,* **42:**1280–1289, 1976.

33. G. Pichard and P. J. Van Soest, Protein solubility of ruminant feeds, Proc. Cornell Nutrition Conf., 1977, pp. 91–98.

34. K. A. Ferguson, The protection of dietary proteins and amino acids against microbial fermentation, in I. W. McDonald and A. C. I. Warner, Digestion and Metabolism, *op. cit.,* pp. 448–464.

35. A. I. Virtanen, Milk production of cows on protein-free feed, *Science,* **153:**1603–1614, 1966.

36. S. W. Chalupa, Protein solubility—Its importance in ruminant rations, Proc. 10th Ruminant Health Nutrition Conf., N.Y.S. Veterinary College, Ithaca, N.Y., 1977.

37. R. H. Smith, Nitrogen metabolism in the rumen and the composition and nutritive value of nitrogen compounds entering the duodenum, In I. W. McDonald and A. C. I. Warner, Digestion and Metabolism, *op. cit.,* pp. 399–415, 1976.

38. W. Chalupa, Metabolic aspects of nonprotein nitrogen utilization in ruminant animals, *Fed. Proc.,* **31:**1152–1164, 1972.

39. D. B. Purser, Nitrogen metabolism in the rumen: Microorganisms as a source of protein for the ruminant animal, *J. Animal Sci.,* **30:**988–1001, 1970.

40. D. R. Waldo, Nitrogen metabolism in the ruminant, *J. Dairy Sci.,* 265–275, 1968.

41. J. V. Nolan, Quantitative models of nitrogen metabolism in sheep, in I. W. McDonald and A. C. I. Warner, Digestion and Metabolism, *op. cit.,* pp. 416–431, 1975.

42. H. K. Goering, P. J. Van Soest, and R. W. Hemken, Relative susceptibility of forages to heat damage as affected by moisture, temperature and pH, *J. Dairy Sci.,* **56:**137–143, 1973.

43. H. W. Titus, The nitrogen metabolism of steers, on rations containing alfalfa as the sole source of the nitrogen, *J. Agr. Research,* **34:**49–58, 1927.

44. P. J. Van Soest, personal communication, 1978.

45. Urea and other nonprotein nitrogen compounds in animal nutrition, National Academy of Sciences, Washington, D.C., 1976.

46. L. D. Satter and R. E. Roffler, Nitrogen requirement and utilization in dairy cattle, *J. Dairy Sci.,* **58:**1219–1237, 1975.

47. K. Kwan et al., Use of urea by early postpartum cows, *J. Dairy Sci.,* **60:**1706–1724, 1977.

48. H. N. Christensen, Implications of the cellular transport step for amino acid metabolism, *Nutrition Rev.,* **35:**129–133, 1977.

49. H. Munro, Eukaryocyte protein synthesis and its control, in D. J. A. Cole et al. (eds.), Protein Metabolism, *op. cit.,* pp. 3–18.

50. P. J. Buttery and K. N. Boorman, The energetic efficiency of amino acid metabolism, in D. J. A. Cole et al. (eds.), Protein Metabolism, *op. cit.,* pp. 197–206.

51. J. Kielanowski, Energy cost of protein deposition, in D. J. A. Cole et al. (eds.), Protein Metabolism, *op. cit.,* pp. 207–215.

52. K. L. Manchester, Hormonal control of protein metabolism, in D. J. A. Cole et al. (eds.), Protein Metabolism, *op. cit.,* pp. 35–47.

53. K. C. Hayes, R. E. Carey, and S. Y. Schmidt, Retinal degeneration associated with taurine deficiency in the cat, *Science,* **188:**949–951, 1975.

54. E. N. Bergman, Glucose metabolism in ruminants, Proc. Third International Conf. on Prod. Diseases in Farm Animals, Agr. Pub. Doc., Wageningen, The Netherlands, 1974, pp. 25–29.

55. A. L. Goldberg and A. C. St. John, Intracellular protein degradation in mammalian and bacterial cells, *Ann. Rev. Biochem.,* **45:**747–803, 1977.

56. R. W. Swick and H. Song, Turnover rates of various muscle proteins, *J. Animal Sci.,* **38:**1150–1157, 1974.

57. W. J. Visek, Effects of urea hydrolysis on cell-life span and metabolism, *Fed. Proc.,* **31:**1178–1193, 1972.

58. D. J. Millward et al., Protein turnover, in D. J. A. Cole et al. (eds.), Protein Metabolism, *op. cit.,* pp. 4–69.

59. P. Felig and J. Wahren, Protein turnover and amino acid metabolism in the regulation of gluconeogenesis, *Fed. Proc.,* **33:**1092–1097, 1974.

60. E. N. Bergman and R. N. Heitmann, Metabolism of amino acids by the gut, liver, kidneys and peripheral tissues, *Fed. Proc.* **37:**1228–1232, 1978.

61. Otto Folin, A theory of protein metabolism, *Am. J. Physiol.,* **13:**117–138, 1905.

62. H. Borsook and J. W. Dubnoff, The metabolism of proteins and amino acids, *Ann. Rev. Biochem.,* **12:**183–189, 1943.

63. H. H. Mitchell, The validity of Folin's concept of dichotomy in protein metabolism, *J. Nutrition,* **55:**193–207, 1955. Harold H. Mitchell (1886–1966) was born in

Illinois and received his bachelor's degree and his doctorate in chemistry at the State University. He was appointed to the faculty of the University in 1913 and rose to professor of animal nutrition and head of the division of animal nutrition, holding these positions until his retirement in 1954. He was an outstanding teacher and investigator. Many of his research contributions in the fields of energy, protein, mineral, and vitamin metabolism are cited in this book. Following his formal retirement, he completed and published (1963) his two-volume work, Comparative Nutrition of Man and Domestic Animals, Academic Press, New York.

64. H. McFarlane, Nutrition and Immunity: Present Knowledge in Nutrition, Nutrition Foundation, 1976, pp. 459–466.

65. T. C. Campbell and J. R. Hayes, The effect of quantity and quality of dietary protein on drug metabolism, *Fed. Proc.*, **35**:2470–2474, 1976.

66. Present Knowledge of the Relationship of Nutrition to Brain Development and Behavior, National Academy Sci., Washington, D.C., 1973.

67. D. A. Levitsky, T. F. Massaro, and R. H. Barnes, Maternal malnutrition and the neonatal environment, *Fed. Proc.*, **34**:1583–1586, 1975.

68. R. H. Barnes et al., Effect of postnatal dietary protein and energy restriction on exploratory behavior in young pigs, *Develop. Psychobiol.* **9**:425–435, 1976.

69. Q. R. Rogers and P. M. B. Leung, The influence of amino acids on the neuroregulation of food intake, *Fed. Proc.*, **32**:1709–1719, 1973.

70. Q. R. Rogers, The nutritional and metabolic effects of amino acid imbalances, in D. J. A. Cole et al. (eds.), Protein Metabolism, *op. cit.*, pp. 279–301.

71. R. E. Austic, Involvement of food intake in the lysine-arginine antagonism in chicks, *J. Nutrition,* **105**:1122–1131, 1975.

72. Justus von Liebig (1803–1873) was the foremost organic chemist of his time and is frequently spoken of as the founder of agricultural chemistry. He was the father of the modern methods of organic analysis, and with him began the accumulation of knowledge regarding the composition of foods, tissues, feces, and urine not available to earlier nutrition workers. He wrote several books dealing with the relations of organic chemistry to agriculture and to animal economy which are well worth reading by the modern student.

73. Russell H. Chittenden (1856–1943) served for forty years as professor of physiological chemistry at Yale University, where he made many outstanding contributions to the modern science of nutrition and inspired a host of students who have continued his work. An account of his protein studies is given in his book Physiological Economy in Nutrition, Frederick A. Stokes Co., Philadelphia, 1904.

74. H. Fisher et al., Protein reserves: evidence for their utilization under nutritional and disease stress conditions, *J. Nutrition,* **83**:165–170, 1964.

75. Recommended Dietary Allowances, National Academy of Sciences, Washington, D.C., 1974.

SUPPLEMENTARY LITERATURE

Allison, J. B., et al.: Influence of dietary proteins in protein synthesis in various tissues, *Fed. Proc.*, **22**:1126–1130, 1963.

Ashworth, A., and A. D. B. Harrower: Protein requirements in tropical countries: Nitrogen losses in sweat and their relation to nitrogen balance, *British J. Nutrition*, **21**:833–843, 1967.

Barnes, R. H., and E. Kwong: Effect of different postnatal periods of protein-energy malnutrition in young rats upon subsequent protein utilization, *J. Nutrition*, **107**:412–419, 1977.

Bergen, W. G.: Postruminal digestion and absorption of nitrogenous compounds, *Fed. Proc.*, **37**:1223–1227, 1978.

Bergen, W. G., and M. T. Yokoyama: Productive limits to rumen fermentation, *J. Animal Sci.*, **46**:573–584, 1977.

Hardy, A. J., J. G. Morris, and Q. R. Rogers: Valine requirement of the growing kitten, *J. Nutrition*, **107**:1308–1312, 1977.

Oltjen, R. R., and P. A. Putnam: Plasma amino acids and nitrogen retention by steers fed purified diets containing urea or isolated soy protein, *J. Nutrition*, **89**:385–391, 1966. Protein metabolism and nutrition, Proc. Second International Symp. Agr. Publ. Doc., Wageningen, The Netherlands, 1977, p. 178.

Robbins, K. A., and D. H. Baker: Phenylalanine requirement of the weanling pig and its relationship to tyrosine, *J. Animal Sci.*, **45**:113–118, 1977.

Rose, W. C., M. J. Coon, and G. F. Lambert: The amino acid requirements of man. VI: The role of the caloric intake, *J. Biol. Chem.*, **210**:331–342, 1954.

Schneider, B. H.: The subdivision of the metabolic nitrogen in the feces of the rat, swine, and man, *J. Biol. Chem.*, **180**:845–856, 1949.

Tagari, H., and E. N. Bergman: Intestinal disappearance and portal blood appearance of amino acids in sheep. *J. Nutrition*, **108**:790–803, 1978.

Vickery, H. B., and C. L. A. Schmidt: The history of the discovery of the amino acids, *Chem. Rev.*, **9**:169–318, 1931.

Yu, Yu, and D. M. Veira: Effect of artificial heating of alfalfa haylage on chemical composition and sheep performance, *J. Animal Sci.*, **44**:1112–1118, 1977.

Chapter 9

Energy Concepts

Food is the source of energy for humans and animals. The carbohydrates, fats, and proteins which food supplies to the body may all be used as energy, for regulating body temperature, and for maintaining the vital functions of growth, activity, reproduction, and production. Depending on the age and species concerned, 70 to 85 percent of the total dry matter consumed is used for energy for these functions. Minerals, vitamins, and enzymes play an important role in the digestion and metabolism which releases and makes the food energy available. The lack of sufficient food or poor quality of food which limits its acceptability and/or digestibility, thus restricting the energy supply, is a serious problem to both human and animal health and productivity in many areas of the world.

The major cost of producing animals is in providing adequate feed supplies. Cost of food also helps to explain the high incidence of hunger and malnutrition among the poor, especially in the underdeveloped countries. Although specific nutritional deficiencies such as protein, vitamins, or minerals are important causes of retarded growth and development, low productivity, poor health, and even death, a shortage of food represents the biggest nutritional problem in animals and humans alike.

Famines have occurred periodically throughout recorded history[1] resulting in widespread malnutrition and death. Chronic shortages of food persist in many areas of the world, but research has made large increases in the yields of food and feed crops possible,[2] and the developed countries have made tremendous progress in increasing the efficiency of utilizing plant and animal resources in providing human food,[3] but continued progress depends upon expanding technological developments into all countries. A recent world food study concluded, "Possibly as many as 450 to 1,000 million persons in the world do not receive enough food. Human beings suffer from several kinds of malnutrition. They can be harmed by eating too little food, or too much food, or by the wrong amounts or balance of particular elements in their diets."[4] The same applies to animals, but animals overeat only when they are managed in confinement and fed unnatural types of diets and not under grazing conditions where a lack of energy or of specific nutrients account for most of their nutritional problems.

It is important to recognize that animals increase the supply of food for humans by consuming resources that otherwise would contribute little or nothing to feeding people. These resources include forages from grasslands and other ranges, plant by-products, cellulosic wastes, crop residues, and browse, materials from which humans are unable to digest any useful energy.

A deficiency of energy results in retardation or failure of growth, weight loss, emaciation, and eventually death if the deficiency is severe and extended. These signs are nonspecific since many other nutritional deficiencies cause the same gross effects under some conditions. Many cases of malnutrition are caused by multiple deficiencies.

9.1 Forms of Energy

The original source of energy, the sun, or *solar energy,* is stored in plants in the form of carbohydrates, lipids, and protein through photosynthesis. This stored *chemical energy* becomes available to man and animals to the extent that they are able to eat and digest the plants. Thus this is the form of energy which has significance in nutrition. Other forms of energy include electrical, thermal, and radiant. Everyone knows of the tremendous power of atomic or nuclear energy, illustrated by the atomic bomb, nuclear-powered generating plants, and nuclear-powered ships. The radioactive tracers C^{14}, C^{45}, Co^{60}, Fe^{59}, I^{131}, P^{32}, and S^{35} are by-products of atomic research that are widely used in nutrition studies. The animal body also absorbs solar energy from sunshine as a source of heat to warm the body, but this cannot be stored in chemical form for later use as plants are able to do.

The physiological processes by which food energy becomes available to animals is presented in Chaps. 5, 6, 7, and 8.

9.2 Models for Energy Metabolism

There is much interest in the design of models for the various physiological body processes and production systems. Models help to understand the various interrelationships of complex systems and define more accurately the phases which require additional definition. Among others, Lucas has presented a model in the form of an energy flow diagram (Fig. 9.1) which shows the flow of energy from the food eaten through the body processes to useful products in ruminants. The place of nutrient pools and energy reserves in maintaining homeostasis is evident and the nature of energy losses during metabolism becomes clear. Later discussions will point out the magnitude of the various fractions of energy as it is moved through the body.

9.3 Body Temperature Control

In all warm-blooded animals the maintenance of a constant temperature is a factor affecting heat production and heat outgo. Since the temperature of the body is normally above its environment, the heat constantly being produced serves in the maintenance of this temperature. The environmental temperature and the amount of heat being produced within the body are the factors that determine the extent to which this heat must be conserved. The amount that is allowed to escape from the body is subject to control, which is referred to as *physical regulation*. This control is brought about by an adjustment of the

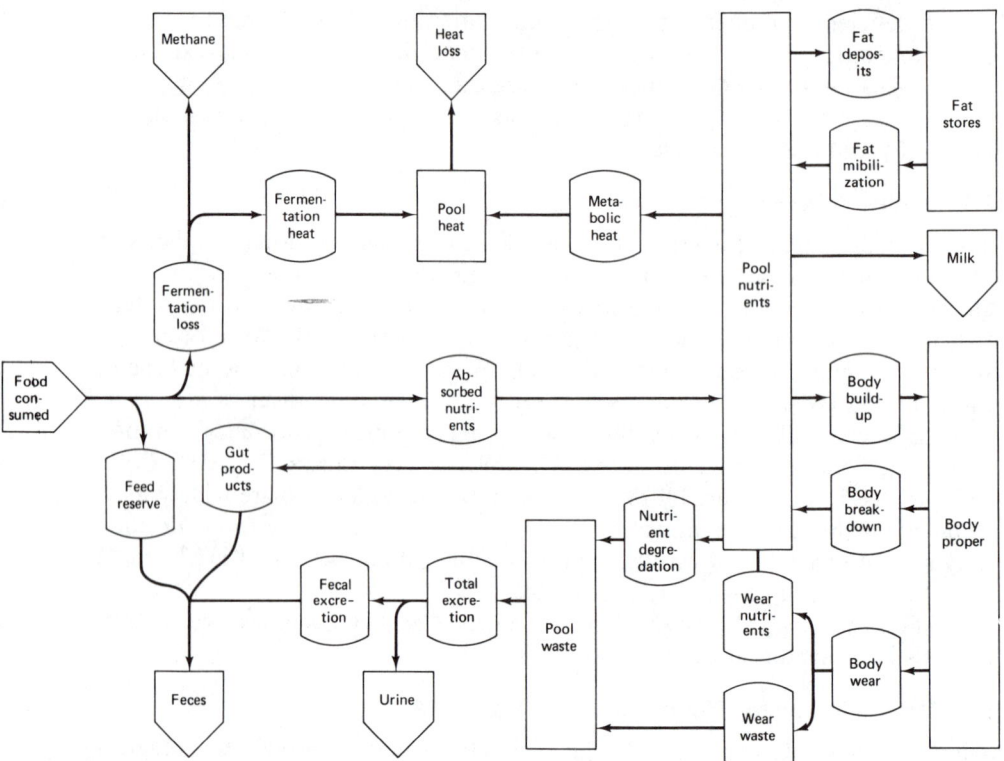

Fig. 9.1 Energy flow diagram. (*Source: H. L. Lucas, Formulation and role of imput-output models in animal production. Symposium, Food Composition, Animal Nutrient Requirements and Computerization of Diets. Utah State University, Logan Utah, 1976*).

blood flow to the skin, by the perspiration mechanism, and by changing the breathing rate to alter evaporation from the lungs. If the conditions call for the dissipation of body heat, the blood flow to the surface is increased as a result of a dilation of the capillaries, which facilitates the escape of heat by radiation, conduction, and convection, and the pores are opened, which allows for a loss of heat through evaporation. These processes are reversed when there is need for the conservation of body heat.

In animals without sweat glands—sheep, swine, and dogs—the relative loss through increased blood flow to the skin and evaporation from the lungs varies with the species. Contrary to the situation with cattle, sheep rely mainly on respiratory evaporative heat loss, according to Brockway, McDonald, and Pullar.[5] Other devices or activities for regulating body heat are changes in the amount of thermal insulation, such as subcutaneous fat, hair, and feathers; changes in the exposed body surface by huddling in a cold environment and the reverse in hot weather; and locating a more favorable environment, such as shade or sun, and shelter from or exposure to wind.

Chemical regulation is an important means of combating the effects of low environmental temperature. Under hormonal stimulus body oxidations are speeded up and more heat is produced accordingly. Shivering is an involuntary form of muscular activity, the function of which appears to be to increase heat production when physical regulation proves insufficient. At the environmental temperature at which the heat loss is equal to the minimum heat production the body temperature remains normal with little regulation. Actually there is a small range of environmental temperatures for each species under specific conditions where the regulation required is slight and is limited to the physical variety. This range is known as the *zone of thermal neutrality* or *comfort*. Here are some ranges reported for different species under basal conditions:

	°C		°C
Rat	28–29	Human (nude)	28–32
Dog (long hair)	13–16	Sheep (shorn)	21–31
Dog (short hair)	20–26	Pig	20–26
Fat hen	16–26	Goat	20–28
Turkey	20–28		

Under practical conditions the range is modified, for a given species, by food intake, activity, degree of fatness, hair, feather or wool coat, exposure to wind, humidity, and other factors. Thus the limits recorded above are more of theoretical interest than of practical importance.

The low point of a range is referred to as the *critical temperature* (Fig. 4.1). Below it, chemical regulation must come into play. As the environmental temperature continues to drop, there comes a point where the homeothermic mechanisms prove insufficient and the tissues freeze. As the environmental temperature rises above the high point of the range, physical regulation operates without any increase in metabolism until this regulation becomes insufficient to cool the body. At this point a supernormal body temperature ensues as a result of increased metabolism, referred to as the *hyperthermal rise,* which leads to death. Hamada[6] has estimated the critical temperature for lactating dairy cows varies from 2 to -10°C depending on level of milk yield.

NUTRITIONAL BALANCES

The physiologists of four centuries ago, though they knew nothing about respiration, recognized that there must be some other loss from the body besides those in the feces and urine. They referred to this loss as the *insensible perspiration,* by which they meant the invisible exhalations which are known today as carbon dioxide and water. Sanctorius, a professor in the Medical School at Padua, who died in 1616, spent much of his life trying to measure this insensible loss by weighing himself, his food, and his excreta. An old print shows Sanctorius eating while seated in a chair balanced on a steelyard. He weighed himself before eating, added a weight corresponding to the amount of food he proposed to eat, and stopped eating when his chair dipped.

In making these various measurements, Sanctorius performed what may be termed the first balance experiment. Such an experiment, as we know it today, involves a quantitative accounting for the intake of a given nutrient in the food and for its outgo in the excreta, providing data for determining whether there is a gain or loss of this nutrient by the body. Such an experiment constitutes another method of measuring nutritive value and the state of nutrition of the body. It gives specific information comparable to that of a slaughter experiment, previously described, and has the obvious advantage that it can be carried out with the living animal. Balance measures are commonly divided into two classes: those which deal with substances that can be weighed or measured, the *balance of matter;* and those which include heat losses, the *balance of energy.* A distinction between matter and energy is untenable since they are inseparable according to the theory of relativity,[7] but it remains useful for the present discussion.

Boussingault[8] carried out the first real balance experiment in 1839. He measured the carbon, hydrogen, oxygen, nitrogen, and ash in the food of a dairy cow receiving a ration that maintained her weight, and the outgo of these nutrients in the feces, urine, and milk. He recognized that he had not accounted for gaseous forms of the elements, and he used his data to estimate the atmospheric oxygen that was required by the cow. Later he made similar studies with a horse and other species. In our nutrition studies of today frequent use is made of the nitrogen balance, of various mineral balances, and to a lesser extent, of the carbon balance.

THE BALANCE OF ENERGY

By far the largest purpose which food serves is the production of energy for body processes, including energy storage. Since all the organic nutrients can serve this purpose, energy value provides a common basis for expressing their nutritive value. The fact that all these nutrients, notably protein, may have specific and unique functions as well does not alter their common usefulness as sources of energy. This holds whether they are used for the purpose immediately upon absorption or are built into body tissue, because the glycogen and fat of the body constitute reserves which can be used as needed, and when these supplies are exhausted, the protein of structural tissues can be broken down to serve as energy. Thus a measure of the gain or loss of energy provides a useful measure of the state of nutrition of the body and of the relative value of various foods. Specifically, in addition to furnishing a measure of the overall energy needs, the energy balance provides a basis for the prediction of the gross chemical changes in body composition resulting from a given dietary or other treatment.

9.4 Energy Distribution in Body Processes

In Sec. 5.2 the gross energy of food, measured as kilocalories when the food is burned to its ultimate oxidation products, has been discussed. This is the starting point for arriving at the energy utilized in body processes, as measured by balance studies. Fig. 9.2 outlines how the gross energy intake (GE) is lost in various body processes in arriving at the useful portion, and provides a convenient basis for the following discussions. The abbreviations here used are those recommended by the Committee on Animal Nutrition of the U.S. National Academy of Sciences—National Research Council.[9]

9.5 Apparent Digestible Energy (DE)

The first loss of the gross energy occurs in digestion. By determining the heat of combustion of the feces and subtracting this value from the GE, one obtains the apparent digestible energy. This value is labeled "apparent" because the fecal energy (FE) includes that of metabolic products of the body as well as that of undigested feed. The metabolic portion (FE_m) consists of digestive fluids and abraded intestinal mucosa. Strictly speaking, this loss is a part of the maintenance requirement of the animal. The term *true digestible energy* is used to denote the value arrived at by subtracting the fecal energy of food origin only from the gross energy intake. The fecal losses represent a substantial part of the gross energy intake. In cattle and sheep the losses are of the order of 40 to 50 percent in the case of roughages and 20 to 30 percent in the case of concentrates. On commonly fed rations horses lose around 35 to 40 percent of the gross energy intake in the feces. In pigs on productive, well-balanced rations the average losses are approximately 20 percent. Fecal losses per unit of feed intake increase with level of feeding.

The measurement of DE takes account of digestible losses only, which TDN (total digestible nutrients) does not, as customarily calculated. It is preferred to TDN on this basis as well as in the interest of expressing all measures of food energy and its requirements for various purposes in terms of calories. DE can readily be determined by the use of a bomb calorimeter to measure the

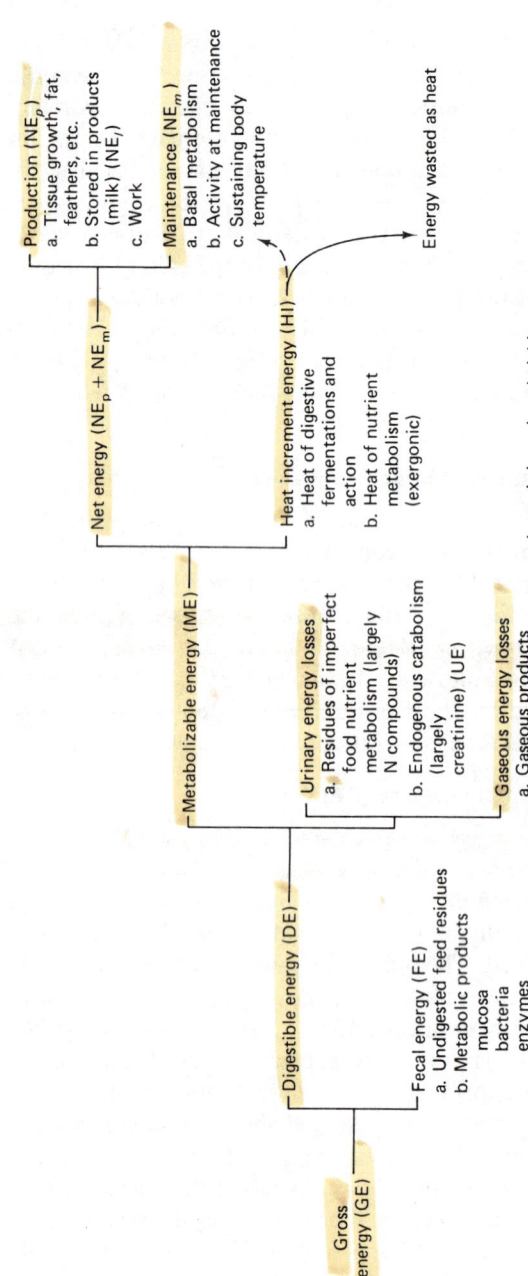

Fig. 9.2 Energy distribution in body processes.

Gross energy (GE)

Digestible energy (DE)

Fecal energy (FE)
a. Undigested feed residues
b. Metabolic products
   mucosa
   bacteria
   enzymes

Metabolizable energy (ME)

Urinary energy losses
a. Residues of imperfect food nutrient metabolism (largely N compounds)
b. Endogenous catabolism (largely creatinine) (UE)

Gaseous energy losses
a. Gaseous products of fermentation (CH_4) → Lost via bowels or belching

Net energy (NE_p + NE_m)

Heat increment energy (HI)
a. Heat of digestive fermentations and action
b. Heat of nutrient metabolism (exergonic)

Energy wasted as heat

Production (NE_p)
a. Tissue growth, fat, feathers, etc.
b. Stored in products (milk) (NE_l)
c. Work

Maintenance (NE_m)
a. Basal metabolism
b. Activity at maintenance
c. Sustaining body temperature

gross energy of the feed and feces. No chemical analyses are required. In 1894 Atwater and Woods determined the heats of combustion of feeds and feces to arrive at the "potential energy of digested nutrients." Since then, initial studies of apparent digestible energy values have been determined for only a limited number of feeds. Thus factors have been developed for computing this measure from TDN.

The first factor, developed some forty or more years ago, was based on the assumption that TDN had taken account of the nutrients on an equivalent carbohydrate basis and thus that TDN could be changed to calories by multiplying it by the carbohydrate factor 4:

$$4 \text{ kcal/g} \times 453.6 \text{ g/lb} = 1814 \text{ kcal/lb}$$

This figure, which erred by being too low, remained in texts and in use for many years. Schneider[10] from a worldwide study of nearly 500 experiments with feeds for which data on TDN and DE were presented, found that the approximate average factor for kilocalories per gram TDN was 4.38 in contrast to the factor 4 used in the equation listed above.

Crampton and associates[11] have carried out studies with swine, and Swift[12] with sheep and cattle, in which both DE and TDN were measured. In both series of experiments, it was concluded that the approximate factor 2,000 kcal per pound (4.41 Mcal per kilogram) was an appropriate one for calculating DE from TDN. As is to be expected, the data of these investigators show that the value 4.41 is somewhat variable according to species and type of ration, but this variation is small, indicating the reliability of the average figure. It corresponds fairly closely with the one Schneider arrived at from his statistical study.

DE is subject to the same variables which affect digestion and the same additional losses in metabolism as have been mentioned for TDN.

9.6   Total Digestible Nutrients (TDN)

In order to arrive at the actual useful portion of a ration, we must deduct losses in the feces, urine, heat increment, and in the case of ruminants, combustible gases. It is obvious that the determination of digestibility is only one step in this direction. The fecal loss is a significant one in all species, and particularly for ruminants in which, when certain roughages are fed as the sole ration, it may exceed the sum of all other losses.

Digestibility data and the TDN values calculated from them are available for all feed commonly fed to cattle and sheep. There is a considerable body of data for swine and a limited number for horses, as summarized by Schneider. The extent of the data available is an important consideration from the standpoint of their usefulness in practice. Published average digestion coefficients make possible the calculation of the digestible nutrients of a given supply of a feed from its specific nutrient content as determined by analysis, rather than relying on figures calculated from average analyses. This is frequently very important in experimental work as well as in the compounding of mixed feeds or rations on a large scale.

It must be borne in mind that digestibility data and TDN values calculated from them are not constants but are subject to many variables which affect the digestibility of individual feeds and rations, as has been discussed in Chap. 3. The principal limitation of the usefulness of TDN as a measure of food energy is that it does not take account of other important losses, such as the combustible gases in the case of herbivora, and, most important, the heat loss. These losses are relatively larger per unit TDN for roughages than for concentrates, thus a kilo of TDN in roughage has less value for productive purposes than a kilo in concentrates. This fact was shown experimentally by Wolff as early as 1888, but the recognition of the extent of the difference and of its practical importance has come primarily from modern experiments, such as the one by Smith and coworkers.[13] In a statistical study, Moore and associates[14] report the following approximate relationships between TDN and net energy:

> 1 lb TDN in corn ⇔ 1 Mcal net energy
> 1 lb TDN in better hays ⇔ 0.75 Mcal net energy
> 1 lb TDN in poor roughage ⇔ 0.5 Mcal net energy

Thus, as the roughage component of the ration, especially low-grade roughage, is substituted for grain, the productive value drops when the substitution is made on a TDN basis. This fact is of importance, both in practical feeding and also in experiments where shifts in the relative amounts of roughages and concentrates fed or in the quality of the roughage are involved. The recognition of this limitation of the TDN measure constitutes an important reason why active consideration is being given at the present time to other measures which account for additional losses.

The term *total digestible nutrients* implies that digestion losses only are taken into account. Actually this is not the case for reasons which, even today, remain unappreciated by some workers in the field because of a misconception which arose long ago. Wolff multiplied fat (ether extract) by the factor 2.4 to compute its carbohydrate equivalent on the basis of its relative calorie values. The factor was modified to 2.25 by American workers, and the latter came into general use by 1910. It was the correct factor for arriving at digestible carbohydrate equivalent from digestible fat according to Atwater fuel values (carbohydrate, 4 kcal per gram; fat, 9 kcal per gram), because the values took account of digestible losses only in the case of these nutrients.

When, however, digestible protein was added directly to the sum of digestible carbohydrate and fat ($\times$ 2.25), on the basis that its fuel value of 4 was the same as that for carbohydrate, it was overlooked that this value for protein took account of urine as well as digestion losses. Atwater's value for digestible protein was 5.2 kcal per gram. Thus to put protein on an equivalent carbohydrate basis, as was done for fat, digestible protein should have been multiplied by a factor also, namely, 1.3 ($5.20 \div 4 = 1.3$). Failure to do this resulted in a formula for calculated TDN which took account of urine as well as digestion losses. Actually, as calculated, it is a measure similar to ME for those species having no gaseous losses.

In 1952 Maynard[15] reviewed these early events in the development of the TDN values in use today and pointed out some of the complications or misuses which have resulted. Fortunately the usefulness of TDN values for feeds in today's standards is not seriously impaired by the error in their calculation as measures of digestible nutrients, because the requirements set forth in the standards have been arrived at on the same basis. Nevertheless, it is not very satisfactory scientifically to continue to use a measure which does not mean what it implies. Unfortunately the older literature contains formulas for calculating DE, ME, and NE from TDN, which are incorrect because they are based on the assumption that TDN takes account of digestion losses only.

9.7   Metabolizable Energy (ME)

When the energy lost in the gaseous products of digestion and that lost in the urine (UE) is subtracted from the apparent digestible energy, one arrives at the portion of the total energy ingested which is actually capable of transformation in the body. This value is called *metabolizable energy*. The gaseous products of digestion result from fermentations which take place in the digestive tract. The gases produced which contain energy, and thus result in an energy loss, consist very largely of methane, with traces of hydrogen and hydrogen sulfide. These gaseous losses are of largest significance in the case of ruminants. They are very small in the case of man, pigs, dogs, and chickens and are usually not taken account of in arriving at the metabolizable energy of a feed for these species.

The urine contains energy, which constitutes a further loss to be subtracted in arriving at metabolizable energy. This is of the order of 2 to 3 percent of the gross energy intake in the case of pigs and 4 to 5 percent in the case of cattle. The urine loss results from the excretion of incompletely oxidized nitrogenous products, primarily urea in mammals and uric acid in birds. This urinary loss arises in large part from the food, but there is always a portion of body origin, called endogenous urinary nitrogen. On a nitrogen-free diet or during fasting it represents the total excretion. The distinction between these two forms of urinary nitrogen is important for various purposes in connection with studies of both protein and energy nutrition. There are small amounts of energy which escape as perspiration, epidermal scales, and shed hair. If accounted for, they should be subtracted in arriving at metabolizable energy, but they are so small that no significant error is ordinarily involved in neglecting them.

An illustration of the data involved in the determination of metabolizable energy is given in Table 9.1. It is noted that by far the largest part of the losses is that which occurs in the feces as a result of lack of digestion. This illustrates the fact that metabolizable energy takes account of the same losses, for the most part, as does digestible energy. It does represent a more accurate measure, since further losses are accounted for. The much larger fecal loss for the

TABLE 9.1. Metabolizable Energy of Feeds for Sheep

Feed	Dry matter eaten per day per head, kg	Energy per kilogram of dry matter, Meal				
		Intake in food	Losses			Metab-olizable
			Feces	Urine	Methane	
Soybean hay	0.795	4.333	2.033	0.196	0.208	1.896
Soybean straw	0.674	4.345	2.676	0.042	0.229	1.398

Source: T. S. Hamilton, H. H. Mitchell, and W. G. Kammlade, The digestibility and metabolizable energy of soybean products for sheep, *Illinois Agr. Expt. Sta. Bull.* 303, 1928.

straw is the primary factor causing its lower metabolizable energy, corresponding to its lower digestibility. The straw has a smaller urinary loss than does the hay because of its lower content of protein, the nutrient which is less completely oxidized in metabolism than in the bomb. The urine data in this table contain the corrections to arrive at exact metabolizable energy when body protein is lost or gained, using correction factors suggested by Rubner. These factors call for subtraction from the urinary energy excreted of 7.45 kcal for each gram of nitrogen lost from the body (represented by a negative N balance) and the addition of 7.45 kcal for each gram of nitrogen stored (represented by a positive N balance). The resulting data from such a calculation are referred to as *nitrogen-corrected metabolic energy*. The Rubner factors were obtained in studies with dogs in 1885 but they have continued in use for other mammals, largely because of the absence of more specific data. Diggs and coworkers[16] made a study with pigs which resulted in a mean value of 6.77 kcal per gram of urinary nitrogen. In determinations of metabolizable energy in poultry, use has been made of the factor 8.7 kcal, representing approximately the average energy content of the urine per gram of urinary nitrogen, or 8.22 kcal based on the uric acid content of the urine.

The determination of the metabolizable energy of a feed for a nonruminant requires no accounting for methane and thus is similar to a nitrogen- or mineral-balance experiment. The determination of the metabolizable energy of feeds for chicks is illustrated by the study of Hill and Anderson,[17] in which the chromium oxide indicator method was used to avoid the need for quantitative measurements of feed intake and excretion.

Clearly, by taking account of additional losses, ME provides a more satisfactory measure of nutritive value than do TDN or DE. A special case where this is particularly true has been reported by Cook and coworkers.[18] Certain species of forage were found to have high gross energy and high TDN values due to essential oils, but low metabolizable energy values because the oils, though absorbed, were not metabolizable, and thus there was a large energy loss in the urine.

There are actually determined ME values for comparatively few feeds, but there have been a sufficient number of experiments in which the digestion losses and the additional losses have been measured to establish factors for the computation of ME values from TDN or DE. In the case of ruminants, if the urine losses are determined, the CH_4 losses can be calculated by one of the formulas given in Sec. 9.8. Alternatively ME can be calculated from DE by the use of an appropriate factor. The many years of experiments at the Institute of Animal Nutrition[19] provide data from which ME can be calculated as a percentage of DE. As an example, in two series of experiments in which corn and alfalfa were fed at varying levels in a total of 18 periods, the average percentage figure was 82.3.

Blaxter and coworkers,[20] from a study of data on cattle and sheep, found that the ratio of ME to digested energy varies little from mean values of 0.82 for roughage, 0.85 for cereals, and 0.79 for oil cakes and meals. The N.R.C. Committee on Animal Nutrition has adopted a figure of 0.82 for such calculations when determined data are not available.

Metabolizable energy is easier and cheaper to obtain than net energy value. It is the heat of combustion of a feed less that in the feces, the urine, and the methane produced. Losses of energy in the urine are proportional to the protein content of the feed. Energy losses as methane are proportional to the apparent digestibility of the diet. The efficiency of utilization of ME varies with the nature of the diet and the purpose for which the energy is used by the animal, as shown in Table 9.2.

The large differences in the efficiency of utilization of ME are primarily due to wide variations in the energy losses as heat increment. For combating cold the efficiency is 100 percent, because ME serves without loss for this purpose, particularly at maintenance levels of feeding. However, when cattle are heavily fed for rapid growth or high milk yields the dissipation of the excess heat increment may become a burden, especially at high environmental temperatures when heat stress depresses productivity.

A comparison of the energy value of selected feeds for cattle (Table 9.3) shows the magnitude of the various losses of energy. In the case of the roughages, alfalfa and timothy hay, the energy losses in digestion and as heat increment are considerably larger than for concentrates. For the forages 35 to 41 percent of the gross energy is lost in digestion compared with from 6 to 28

TABLE 9.2. Examples of the Efficiency of Utilization of ME in Ruminants

Type of diet: Mcal ME/kg diet DM	Efficiency of ME use, %		Relative value (Maintenance = 100)	
	Roughage 2.0	Concentrate 3.0	Roughage 2.0	Concentrate 3.0
Maintenance	68	75	100	100
Fattening	40	58	59	77
Lactation	66	68	97	90

Source: British A.R.C., Nutrient requirements of farm livestock: Ruminants, Agr. Research Council (London), 1965, p. 243.

TABLE 9.3. The Energy Value of Selected Feeds (dry basis)

Feed	GE	DE	ME	NE_m	NE_g	NE_l	TDN, %	Protein, %
			Mcal per kg					
Alfalfa hay	3.89	2.51	2.03	1.35	0.49	1.25	57	18.4
Barley grain	4.14	3.66	3.10	2.13	1.40	2.14	83	13.0
Corn grain	4.41	4.01	3.43	2.28	1.48	2.42	91	10.1
Cottonseed meal	4.84	3.47	2.56	1.81	1.20	1.12	79	44.7
Sorghum grain	3.73	3.53	2.96	1.86	1.24	2.05	82	11.7
Soybean meal	4.69	3.63	2.98	1.93	1.29	2.23	82	52.4
Timothy hay	4.44	2.63	2.16	1.26	0.62	1.41	60	7.6

Source: Atlas of nutritional data on United States and Canadian feeds, National Academy of Sciences, Washington, D. C., 1971.

percent for the concentrate feeds. The percentage of ME lost as heat increment for growing cattle (NE_g) was 71 to 76 percent versus 55 to 68 percent for the concentrates, but the heat increment losses were considerably less when the ME was utilized for maintenance or lactation (NE_l). The end products of rumen fermentation influence the value of ME (Rook[21]). High-concentrate, low-roughage rations increase the proportion of propionic and butyric acids and lowers the acetic acid in the rumen contents, which favors lipogenesis in growing ruminants, but which also lowers milk fat percentage in lactating cattle producing large amounts of milk.

A feeding standard based on ME should take account of the factors which determine the efficiency with which it is utilized. This had been done in the British Agricultural Research Council standards of ruminants.[22]

9.8   Methane Loss in Herbivora

The measurement of methane loss in herbivora requires an airtight chamber for housing the animal. This makes the determination a complicated and expensive one in terms of both equipment and operation. For this reason most of the metabolizable-energy values reported in the literature have involved the calculation of methane production rather than its actual measurement. There are several experiments, however, which provide a basis for establishing factors for this purpose in the case of ruminants. Forbes and associates analyzed the data from 12 experimental periods in the respiration calorimeter with cattle receiving corn and alfalfa, fed at levels ranging from one-half maintenance to twice maintenance, and found that the percentages of the gross energy lost as CH_4 energy varied from 6.42 to 9.83. The values tended to be lower as the level of feed intake increased. Many later studies with both cattle and sheep have shown a somewhat similar range and have also indicated that the values are lower at higher intakes. It would appear that in the case of ruminants on full rations, the energy lost as methane is of the order of 7 percent of the total intake. Methane contains 13.34 kcal per gram.

A substantial fermentation of cellulose occurs in the greatly enlarged cecum and colon of the horse. Other carbohydrates, as well as protein and fat, are largely digested before the feed reaches the cecum. Thus the production of methane is less in the horse than in ruminants.

Several studies have been made of the data of metabolism experiments in which the combustible gases were actually determined to develop formulas for the calculation of these methane losses from the total dry matter or from the digested nutrients of the ration. Swift and coworkers[23] developed a formula for sheep and compared it with the formula developed earlier at the Pennsylvania Institute of Animal Nutrition for cattle. The two formulas are as follows:

(sheep) $E = 2.41X + 9.80$
(cattle) $E = 4.012X + 17.68$

Blaxter and Clapperton[24] also reported on methods of estimating methane production. A number of studies have been reported on the effects of diet additives to reduce the energy loss by suppressing methane production in the rumen. Clapperton[25] reviewed these and reported that his own studies with trichloroethyl adipate caused a large initial reduction in methane and a shift from acetate to propionate production. The effect decreased greatly after 2 to 3 weeks due to adaption in rumen fermentation. Feed intakes and weight gains were reduced.

9.9 Physiological Fuel Values

These are calorific values for nutrients as originally set forth by Atwater for use in human nutrition to calculate the portion of the gross energy which is available for transformation in the body, a calculation which resulted in figures having a similar significance as metabolizable energy. Account was taken of losses in digestion on the basis of the following average figures for digestibility of a mixed diet: carbohydrates, 98 percent; fats, 95 percent; protein, 92 percent. The figures for gross energy were multiplied by these coefficients, and in the case of protein a subtraction of 1.25 kcal per gram was made for the energy lost in the urine. This factor was obtained by studies with human beings eating mixed diets. It varies somewhat with the makeup of the diet. By these calculations the average physiological fuel values became:

Carbohydrates	$4.15 \times 98\% = 4$ kcal per g
Fats	$9.4 \ \times 95\% = 9$ kcal per g
Protein	$(5.65 - 1.25) \times 92\% = 4$ kcal per g

These values are not applicable to feeds and rations of farm animals because the digestibility figures on which they are based are too high. The factor 1.25 kcal is too low to estimate the urine loss in the case of herbivora because of the relatively large amount of hippuric acid excreted. The average physiological fuel values have limitations, for use with certain human foods and diets, which can be overcome by making use of Atwater's basic data, as is done in the U.S. Department of Agriculture publication[26] on food composition. Physiological fuel values are similar to metabolizable energy.

9.10   Heat Increment (HI)

In addition to the losses which are subtracted in obtaining metabolizable energy, account must be taken of the energy lost as heat in arriving at the portion which is completely useful to the body, namely, the *net energy* (NE), the portion which actually appears as a product, viz., for tissue maintenance and growth, milk, eggs, wool, and work. In every cell of a living organism, chemical reactions are constantly occurring as an essential accompaniment and manifestation of life processes, and some of the chemical energy involved is lost as heat energy. There results a continual outgo of heat from the body, which is at a minimum in the postabsorptive state in a thermoneutral environment and increases with the amount of food consumed. This increase is called the *heat increment* (HI) and consists, as shown in Fig. 9.2, of the *heat of fermentation* (HF) and the *heat of nutrient metabolism* (HNM). The former is heat produced in the digestive tract as a result of microbial action and is, of course, much larger in ruminants than in other species.

Since HF is a loss in the digestive tract, from an exact point of view it should be subtracted in arriving at ME, as is done for combustible gases. It cannot be determined directly, however, and the results of indirect measures are questionable. Thus it seems preferable to include this loss as a part of the heat increment. The major component here is the loss in nutrient metabolism. Below the critical temperature (Sec. 4.6) heat produced in metabolism serves to keep the body warm and thus is a part of the net energy. The total heat loss from the body may constitute 25 to 40 percent or even more of the gross calorie intake. It is therefore of great importance in nutrition from the standpoint both of the economy of food utilization and of body-temperature relations.

The recognition that the additional heat loss produced by the ingestion of food arose from metabolic processes rather than from "work of digestion," with the exception of a small loss in mastication and in the propulsion of food through the digestive tract, resulted from experiments made early in the present century, notably by Rubner and Voit. Rubner coined the term *specific dynamic effect* to denote the increase in metabolic rate following food intake. The German word for "effect" was mistranslated into English as "action" and thus the abbreviation SDA arose, which has since been replaced by SDE. Rubner attributed specific stimulating effects to protein, carbohydrates, and fats, which were found largest in the case of protein. We now know the specific chemical changes in the metabolism of each of these nutrients which result in a part of the total chemical energy being lost as heat.

The term *heat of nutrient metabolism* corresponds to the SDE of Rubner. The term heat increment is a broader one, which includes also the heat of fermentation, an important loss in the case of ruminants. The heat increment varies according to the makeup of the ration, the level at which it is fed, and the body function being supported—maintenance, growth, fattening, or lactation. These variations must be taken into account in evaluation of different measures of food energy, i.e., apparent digestible, metabolizable, or net energy.

Heat increment is subject to many variables according to the nature of the ration, the purpose for which it is fed, and others. It is clear, therefore that a specific NE value of a feed or ration can have the exact meaning ascribed to it only for the specific purpose for which it was measured. A value measured at the maintenance level in a warm environment is not a suitable measure of maintaining an animal in a cold climate, because the heat produced under these conditions does serve a useful purpose in keeping the body temperature up to normal, a requirement which would otherwise call for chemical regulation through tissue breakdown. A value measured at maintenance does not apply to productive functions. Thus, specific NE values have been set up for maintenance (NE_m), growth (NE_g), and lactation (NE_l) in the N.R.C. reports.

"Foodstuffs express their characteristic net energy values only as they are of components of nutritionally complete rations, and the same must apply to metabolizable energy."

E. B. Forbes
J. Agr. Research, **46**:753–770, 1933.

9.11 Net Energy (NE)

The net-energy system, which originated with Kellner's studies of the fat-producing power of feeds, and with Armsby's respiration-calorimeter experiments, conceives of the measurement of that portion of the feed which is completely useful to the body for maintenance, growth, milk, or work. Armsby recognized that the assimilation of feed resulted in an energy cost to the organism in addition to those losses accounted for in arriving at metabolizable energy and that this could be measured as the heat lost from the body. He therefore measured the heat resulting from the ingestion of a feed at a given level of intake, increased the intake, and, by a second measurement, obtained by difference the *heat increment* (Sec. 9.10) corresponding to the amount by which the level of food intake was increased. He then subtracted the heat increment, expressed in terms of a given unit of intake, from the metabolizable energy of the same intake to obtain the net-energy value. In the case of concentrates, it was necessary to add them to a basal roughage ration in measuring their net-energy value with steers. Some of Armsby's values and the data on which they were based are given in Table 9.4. Later studies by Forbes and Kriss[27] resulted in improved methods of computing these values from Armsby's data and in the publication of a table of revised values.

Kellner's Starch Values. The *starch equivalent (SE) system* originated by Kellner[28] is similar in principle to the net-energy system of Armsby. Differing

TABLE 9.4. Net-energy Values of Feeds for Ruminants (Mcal per 100 kg of dry matter)

Feeds	Gross energy	Losses in excreta	Metabolizable energy	Heat increment	Net energy
Timothy	451.8	266.4	185.4	78.2	107.2
Red clover	446.2	246.1	200.1	97.3	102.8
Corn stover	433.2	238.0	195.2	106.5	88.7
Corn meal	444.2	111.5	332.7	128.6	204.1
Hominy feed	470.9	118.7	352.2	136.5	215.7
Wheat bran	453.2	202.1	251.1	117.7	133.4
Wheat straw	444.4	306.2	138.2	116.0	24.4

Source: Henry Prentiss Armsby and J. August Fries, Net energy values for ruminants, *Pennsylvania Agr. Expt. Sta. Bull.* 142, 1916.

from Armsby's values, which were determined near maintenance, Kellner's were measured above maintenance and thus were values for body-fat production.

Using the nitrogen-carbon-balance method, Kellner added pure carbohydrate, protein, and fat to a basal-maintenance ration and thus determined the relative amounts of these pure digestible nutrients required to produce a unit of body fat. When he tested feeding stuffs instead of pure nutrients, he found that the fat-producing power was less than calculated from their content of digestible nutrients and that the discrepancy was larger with those feeds high in fiber. He concluded, therefore, that some of the calculated fat-producing power was lost as a result of the "work of digestion" which increased with fiber content. He expressed this fat-producing power of the feed in terms of the number of kilograms of starch that would be required to produce the same amount of fat as 100 kg of the feed. Hence his values were called *starch equivalents,* or *starch values.* For example, the starch value of corn (maize) meal, 81.5 kg, was the amount of starch which would produce as much fat as 100 kg of the meal. Kellner published such values for approximately 300 feeds, a few of which were as follows: oats, 59.7; wheat, 71.3; linseed oil meal, 71.8; wheat bran, 45.0; timothy hay, 29.1; and oat straw, 17.0.

Nehring and Haenlein[29] at the University of Rostock, Germany, reviewed the early research of Kellner and the more recent research of others in Europe and compared the results with the Agricultural Research Council metabolizable energy and the National Research Council net-energy systems. Equations are presented for estimating the net-energy fat (NE_l) values of feeds and rations.

The *Scandinavian feed-unit system,* usually associated with the name of Hansson, is based on the results of practical group-feeding experiments, with 1 kg of barley as the standard unit. It is expressed in accordance with the net-energy concept by computations making use of Kellner's starch values. As so computed for cattle, 1 food unit is equivalent to 1.65 Mcal net energy for fattening or 2.1 Mcal milk energy. The determination of the value requires no elaborate apparatus, but the feeding experiments needed are laborious, requiring large groups of animals. For this reason many of the feed units in use have been calculated from digestible nutrients. Studies which illustrate how replacement equivalents of other feeds for barley are arrived at are presented in the report by Breirem and associates.[30]

Fraps[31] utilized data from published experiments and the results of his own feeding trials with sheep, chickens, and rats to arrive at the *productive energy* of feeds. In 1947 Morrison proposed a method of determining net-energy values by means of feeding experiments. Later he made a study of the previous work of Armsby, Kellner, and others, and particularly of data from feeding experiments. He constructed a table of *estimated net-energy* values for some 350 feeds designed to be applicable for productive purposes of nutritionally complete rations. This table is published as Appendix II in his book.[32] The applications and limitations of the values are discussed in the pages preceding the appendix. Breirem[33] has published a detailed review of various measures of energy value of feeds, including TDN, ME, starch equivalents, and feed units.

None of the measures discussed in this section has found any wide use in the United States.

The *nitrogen-carbon balance* has been used for many years to obtain data on the gain or loss of body protein or fat. Determinations are made of the nitrogen and carbon in the food, feces, and urine and of the carbon in the gaseous output. Prior to 1870 Henneberg, following the studies of Voit and associates with humans, began experiments with farm animals. Later a respiration chamber for large animals was built at the Möckern Experiment Station under the direction of Gustav Kühn, and studies were undertaken which were carried out for the most part by Kühn's successor, O. Kellner, which led to his starch values.

As an example of a nitrogen-carbon balance, the data from one of Kellner's experiments are presented in a condensed form in Table 9.5. It is noted that from the nitrogen balance the amount of carbon gained as protein is calculated and that this value subtracted from the total carbon gained gives a figure which represents that gained as fat, from which the amount of fat can be computed. The computation is based upon the fact that the carbon content of the body exists almost entirely as protein and fat. It disregards the small amount of glycogen which is normally present, since it is considered that any changes in this constituent are so small as to be of very minor importance under normal feeding conditions. This is less true when the diet is such as to cause a loss, rather than a storage, of fat. For experiments over an extended period, disregarding the glycogen is of no concern, but in experiments of only a few hours' duration, a considerable error may be introduced. The glycogen changes can be estimated by including determinations of hydrogen and oxygen in the balance data, a procedure which makes the experiment much more difficult and laborious. The most important use of the carbon- and nitrogen-balance method today is in connection with indirect calorimetry (Sec. 9.14).

TABLE 9.5. Example of a Nitrogen-Carbon Balance

Item	Nitrogen, g		Carbon, g	
	Intake	Outgo	Intake	Outgo
Feed	390.55	—	5668.2	—
Feces	—	105.69	—	1456.9
Urine	—	263.76	—	283.3
Gases	—	—	—	3247.9
Gain to body	—	21.10	—	680.1
Total	390.55	390.55	5668.2	5668.2

Note: Based upon a content of 52.54[a] percent carbon and 16.67[a] percent nitrogen in fat-free, ash-free flesh and of 76.5 percent carbon in fat, the following calculations gave the protein and fat gained: (1) 21.1 g nitrogen gain divided by 0.1667 equaled 126.6 g *protein gain.* (2) 126.6 g protein times 0.5254 equaled 66.5 g carbon in protein. (3) 680.1 g carbon gain minus 66.5 equaled 613.6 g carbon gained as fat. (4) 613.6 divided by 0.765 equaled 802.1 g *fat gain.*
[a]Figures used by later workers are slightly different.
Source: O. Kellner and A. Kohler, Untersuchungen uber den Stoff- und Energie-umsatz des erwachsenen Rindes der Erhaltungs- und Produktionsfütter, *Landw. Vers. Sta.,* 53:1–16, 1900.

The old systems of expressing the energy requirements of animals and the energy value of feeds including Kellner's starch units, the British starch equivalent system, the Scandinavian and Russian feed units, and the American TDN system all assume that a single value could be assigned to each feed whether the feed was given to a growing animal, a fattening or lactating animal, or was used for maintenance only, and that this single value is not altered by the level of feed given. None of these assumptions are entirely correct. While not expressed in the feeding systems, most scientists recognized that individual feeds had to be fed in balanced rations if their energetic potential was to be fully realized. While it appeared that net energy (NE) should overcome most of these difficulties, the heat increments and thus the final values varied among animals, the balance of the rations, the level of intakes, and the type of production. After years of study Forbes concluded that net energy values of feeds were useful for theoretical considerations but they were too sensitive to conditions under which they were determined to be useful in practical feeding systems. For some years TDN entirely replaced NE in feeding practice in the United States, although Morrison's estimated net-energy (ENE) values were used to some extent.

Later when the difficulties with the TDN system became more apparent, especially for rationing high-producing dairy cattle and fattening beef cattle interest was revived in using ME and NE. L. A. Moore and W. P. Flatt developed an energy metabolism laboratory, K. L. Blaxter began energy studies in Scotland, G. P. Lofgreen and W. N. Garrett in California devised a more practical method of determining net energy for beef cattle, and better types of equipment became available. While many problems still exist and much more research is needed it is now generally accepted that either the NE system as used in the United States or the ME system as used by the British A.R.C. system is an improvement over TDN. ME appears to be the preferred system for nonruminants.

The net-energy system used in the United States proposed by Lofgreen and Garrett[34] uses an expression to represent the net-energy requirement and the net-energy content of the feed when used for maintenance (NE_m) and a second expression for the net energy used for the production of weight gain (NE_g). Data from comparative slaughter trials indicate that the NE_m requirement for both steers and heifers is equal to 0.077 Mcal $W_{kg}^{0.75}$. Since heifers (Sec. 14.10) deposit more fat per unit of body gain than steers, separate equations were used to estimate the NE required for weight gain. The National Research Council has used this system in recent beef cattle reports.[35] Further studies are providing data on additional feeds and on the body composition of various types of beef cattle.[36]

Comparative studies were carried out with the same feeds for growing and lactating cattle.[37] When the total efficiency of feed conversion was considered, growing beef steers were 49 percent as efficient as lactating dairy cattle. The beef steers were 57 percent as efficient as dairy animals in converting feed above maintenance. These results along with more recent tests of Moe and associates[38] form the basis of the N.R.C. dairy cattle requirements in which different NE values of feeds are applied for maintenance, growth, and lactation. The NE values for maintenance and lactation are higher than for growth. Weight gain in lactating cows appears to be as efficient as for lactation alone or for maintenance. Many of the NE values of dairy cattle feeds have been calculated from TDN or DE values.[39]

In a series of experiments Evan and associates[40] have studied the energy metabolism and values of some feeds for young pigs by using feeding studies, balance experiments, and slaughter data. They reported the NE required for maintenance was 87.3 kcal/day/$W_{kg}^{0.75}$ and that young pigs utilized ME as efficiently for growth as for maintenance, namely, 69 percent. They reported the following energy values of corn and oats for young pigs:

	GE	DE	ME	ME_n	NE
		Mcal/kg, dry basis			
Corn	3.96	3.43	3.32	3.16	2.33
Oats	4.13	2.84	2.73	2.59	1.40

The differences in DE values reflect the indigestibility of the fiber in oats and the low NE value reflects the greater heat increment loss of the nitrogen-corrected metabolizable energy (ME_n) of oats. Other feeds for pigs are being studied.

9.12 Respiration Calorimeter

The heat produced by an animal can be measured directly in a respiration calorimeter. Indirect measurements can also be made by use of *respiration apparatus* which allows measurement of the oxygen consumed and the carbon dioxide expired. Many different types of units have been used.

Lavoisier was the first to recognize that animal heat is produced by oxidations in the body. He and Laplace[41] devised the first animal calorimeter to measure this heat, using a guinea pig as the subject. The animal was enclosed in a chamber surrounded by ice and the amount of ice melted in a given period of time was recorded as a measure of the amount of heat given off by the animal. More precise animal calorimeters were later developed in which water replaced the ice. The latest type is known as the gradient-layer calorimeter and is so constructed that the average temperature gradient between the inner and outer surfaces is proportionate to the total heat produced by the animal. The details are given in an article by Pullar.[42]

The direct measurement of heat loss can also be carried out in a respiration calorimeter which combines the features of a respiration chamber and a calorimeter. Such an apparatus makes possible an accounting for the income of feed, water, and oxygen and the outgo of the solid, liquid, and gaseous excreta and of the heat eliminated. The first accurate respiration calorimeter was constructed by Rubner in 1881 for use with dogs. Using this apparatus he made many basic findings on heat loss as affected by diet. His studies provided the first proof that the law of the conservation of energy holds in the animal body as well as in the inanimate world.

In 1892 Atwater, who had studied with Rubner, began the construction of a respiration calorimeter with the cooperation of Rosa, a physicist, for use with human subjects, which was completed in 1897. It was equipped with a bed, desk, chair, and a bicycle. The apparatus was used in pioneer studies of heat production, energy requirements for various body purposes, and the nutritive

TABLE 9.6. Daily Energy Balance of a Steer

Items	Intake kcal	Outgo kcal
6,988 g timothy hay	27,727	
400 g linseed meal	1,811	
16,619 g feces	. . . . .	14,243
4,357 g urine	. . . . .	1,210
37 g brushings	. . . . .	88
142 g methane	. . . . .	1,896
Heat	. . . . .	11,493
Gain by body	. . . . .	608
Total	29,538	29,538

Source: Henry Prentiss Armsby and J. August Fries, The available energy of timothy hay, *U.S. Dept. Agr. Bur. Animal Ind. Bull.* 51, 1903.

value of foods. Following Atwater, Armsby built a similar apparatus for experiments with cows at Pennsylvania State University, as described by Braman.[43] It has not been used for years, but has been preserved for its historical interest.

An illustration of the data obtained in an experiment carried out in a respiration calorimeter is given in Table 9.6. It is noted that an accounting for all energy losses left a balance of 608 kcal as the net gain to the animal from the feed ingested. The striking feature of these data is the large loss of energy as heat, representing approximately 40 percent of the total intake. The importance of giving attention to heat losses in measuring the usefulness of feeds is thus indicated.

9.13 Respiratory Quotient (RQ)

The relation between the oxygen consumed and the carbon dioxide given off in respiration is expressed as the respiratory quotient (RQ), computed as follows:

$$\frac{\text{Volume of } CO_2 \text{ produced}}{\text{Volume of } O_2 \text{ consumed}} = RQ$$

The numerical value of this quotient is dependent upon the chemical nature of the substance being oxidized within the body. The burning of a molecule of glucose, the form in which carbohydrates are catabolized, takes place according to the following equation:

$$C_6H_{12}O_6 + 6O_2 \longrightarrow 6CO_2 + 6H_2O - \Delta H \text{ (675 kcal/mole)}$$

Dividing 675 kcal by the molecular weight of glucose (180.16) gives it a value of 3.75 kcal per gram. Since the carbohydrate molecule contains hydrogen and oxygen in the proportion to form water, oxygen from the outside is required only for the oxidation of the carbon. One molecule of carbon dioxide is formed for each molecule of oxygen consumed, and therefore the RQ is 1.0.

The fat molecule, on the other hand, does not contain nearly enough oxygen to take care of the hydrogen present, and thus a part of the oxygen used in burning fats appears in water. More oxygen is consumed, therefore, than is represented by the carbon dioxide given off, and the RQ becomes less than 1.0. For most body and food fats, it is approximately 0.7. Differences in fatty-acid makeup of fats result in only minor variations. The figures from the breakdown of stearic acid, for example, are as follows:

$$CH_3(CH_2)_{16}COOH + 26O_2 \longrightarrow 18CO_2 + 18H_2O - \Delta H \text{ (2711 kcal/mole)}$$

Dividing 2711 by 284.5, the molecular weight of stearic acid, gives a value of 9.533 kcal per gram. The RQ figures as 0.692.

For protein the basis for computing the RQ is less certain because it is incompletely oxidized in catabolism in the body and because its amino acid makeup is a determining factor. In the complete oxidation of protein, the end products are carbon dioxide, water, and nitrogen (N_2). Taking the amino acid alanine as an example, its complete oxidation is represented as follows:

$$4CH_3CH(NH_2)COOH + 15O_2 \longrightarrow 12CO_2 + 14H_2O + 2N_2$$

The heat produced $(-\Delta H)$ is 388 kcal per mole or 4.35 kcal per gram, and the RQ figures out at 0.8. In the body, however, the nitrogen is excreted as urea and small amounts of other incompletely oxidized compounds, such as hippuric acid, creatinine, allantoin, ammonium salts, uric acid (the principal nitrogenous excretory product in birds), and others. Considering urea as the principal incompletely oxidized product, the following equation represents the breakdown in the mammalian body:

$$4CH_3CH(NH_2)COOH + 12O_2 \longrightarrow 10CO_2 + 10H_2O + 2CO(NH_2)_2$$

Here $-\Delta H = 312$ kcal per mole or 3.5 kcal per gram and the RQ is 0.83. Certain amino acids have RQ values quite different from those shown above. For example, glutamic acid, in which the relationships of carbon, hydrogen, and oxygen in the molecule are similar to those in a carbohydrate, has an RQ of 1. Obviously the RQ of a protein depends on its amino acid makeup, but for the "average" dietary the figure is about 0.83.

Though the RQ is an inadequate index of the nature of intermediate metabolism, its magnitude gives an approximate idea of the kind of nutrient which is being burned in the body. The closer the quotient approaches unity, the larger is the proportion of carbohydrates being used, while values lying close to 0.7 indicate that fat predominates as the body fuel. The meaning of an intermediate value is less clear, since the quotient for protein lies in between those for carbohydrate and fat, and since a determined quotient may represent the result of the burning of variable proportions of all three. Respiratory quotients considerably higher than unity may be obtained when carbohydrate is being converted into fat, because oxygen-poor fat is being formed from relatively oxygen-rich glucose. Hence the volume of expired carbon dioxide may be greater than the inspired oxygen. Wierzuchowski and Ling report quotients of 1.4 and higher in rapidly fattening hogs, and they cite a quotient of similar magnitude obtained by Benedict for the goose. On the other hand, RQs below 0.7 have been observed in fasting, particularly in hibernating animals, and they may be the result of the conversion of fat into carbohydrate.

The determination of gaseous exchange can be carried out either by placing the subject in a chamber, the atmosphere of which can be controlled and measured, or by the use of a facepiece which provides for the analysis of the inspired and expired air. The use of the chamber makes possible an accounting for the water lost as perspiration and for the intestinal gases produced, as well as the pulmonary exchange. These gaseous losses resulting from fermentations are of sufficient magnitude in herbivora, especially ruminants, to require that they be determined or calculated in arriving at a carbon or energy balance. The various devices which are used in either of these methods are referred to as *respiration apparatus*. The earliest forms consisted of closed chambers in which the subject was placed and in which the change in the composition of the air was determined. The limitation of this procedure, which failed to provide for any renewal of the air or removal of waste products during the course of the experiment, is obvious. Two types of apparatus were later devised to remedy this defect: the *closed-circuit type* designed by Regnault and Reiset and the *open-air current type* developed by Pettenkofer.

The closed-circuit type derived its name from the fact that the same air is continuously circulated, with provision for the removal of the waste products and the addition of oxygen. This apparatus is illustrated diagrammatically in Fig. 9.3. It is noted that the carbon dioxide and water are removed from the outgoing current by absorbents. Their output is determined by recording the increase in weight of the absorbing vessels. The oxygen of the circulating air is renewed through a meter by means of which the volume added is recorded. The residual air at the close of the experiment is analyzed to take account of any changes in composition from that at the start. In this apparatus, the intestinal carbon dioxide is absorbed along with that from the lungs. The other intestinal gases, chiefly methane, can be determined in the residual air. Methane is determined by drawing the air sample over platinized kaolin or a similar substance at red heat. The methane is thus oxidized and determined from the carbon dioxide produced. Methane and other oxidizable gases present are thus referred to as combustible gases, a term which has special significance in connection with the energy balance. Regnault and Reiset used their apparatus for studies with sheep, calves, pigs, and poultry. The same principle is employed in the apparatus designed for man by Atwater and Benedict,[44] but the larger the animal, the greater the difficulty and cost of constructing an airtight unit in which the temperature and humidity are well defined.

The open-air current type differs from the one just described in that the circulating air is drawn from the atmosphere, and the outgoing air or a measured fraction of it is passed through the absorbents. When it is desired to account for the intestinal gases other than carbon dioxide, provision for their

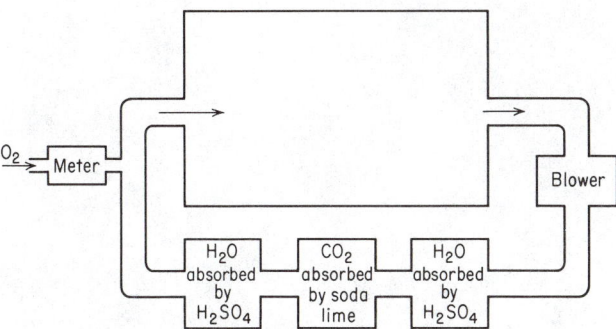

Fig. 9.3 Closed-circuit respiration apparatus.

determination in the outgoing air as well as in the residual air of the chamber must be made. The Pettenkofer apparatus, originally designed for studies with men, was adapted for use with farm animals by later German workers, a development with which the names of Henneberg and Stohmann, Kühn, and later Kellner are especially associated. The great advantage of the open-air current type over the closed system is that leaks are not important, for as long as suction is maintained, no air leaks out and it makes no difference whether the air enters through leaks or the designed inlet.

In either the closed- or open-air current type the chamber can be replaced by a facepiece or some other device for determining the pulmonary exchange only. Zuntz modified the Pettenkofer apparatus for use with the horse by eliminating the chamber and collecting the expired air by a tracheal canula, thus providing a portable device which was widely used by early German workers. Lavoisier devised a copper face mask to study gaseous exchange in human beings. Several later workers devised face masks for farm animals. The problem is to obtain an airtight fit which does not unduly disturb the animal. Fig. 9.4 shows the portable closed-circuit equipment used by Brody, including the spirometer for measuring the oxygen consumption and the tube of soda lime through which the expired air passes for absorption of carbon dioxide. Equipment using a much larger and somewhat modified spirometer, as described by Blaxter and Howells[45] and used with calves, provides for observations over considerably longer periods and for the measurement of carbon dioxide production as well as oxygen consumption. Today increasing use is being made of automatic equipment for the determination of respiratory exchange, such as those employed in the respiration chamber designed by Flatt and associates and shown in Fig. 9.6.

9.14 Indirect Calorimetry

The measurement of heat loss is referred to as *direct calorimetry* in contrast to *indirect calorimetry,* which is based on a calculation of the heat production

Fig. 9.4 Apparatus for measuring pulmonary exchange. (*Courtesy of S. Brody, Missouri Agricultural Experiment Station.*)

TABLE 9.7. Energy Values of Oxygen and Carbon Dioxide at Different Respiratory Quotients

RQ	kcal/liter O_2	kcal/liter CO_2	kcal/g CO_2
0.70	4.686	6.694	3.408
0.75	4.739	6.319	3.217
0.80	4.801	6.001	3.055
0.85	4.863	5.721	2.919
0.90	4.924	5.471	2.785
0.95	4.985	5.247	2.671
1.00	5.047	5.047	2.569

Source: Zuntz and Schumberg.

responsible for the loss measured directly. Such a calculation is possible if the chemical metabolism is known, since every chemical process is related to a definite transformation of energy.

The measurement of respiratory exchange and the calculation of heat production thus represented constitutes another procedure of indirect calorimetry. One method takes account of the oxygen consumption, carbon dioxide production, and the urinary nitrogen. These data are used to calculate the nonprotein RQ. Then from a table showing the calorie values of the O_2 used and CO_2 produced at the nonprotein RQ, the heat production from the carbohydrate and fat metabolism is obtained. Illustrative values from the original table of the German workers Zuntz and Schumburg, published in 1901 and containing data for all nonprotein RQs between 0.7 and 1.0, are presented in Table 9.7. The basic data obtained by this method are also used to compute the heat production of the protein metabolized.

Illustrative figures produced by the method are presented in Table 9.8 from figures obtained by Blaxter and coworkers,[46] for a young calf, using a respiratory chamber. No account was taken of methane production because none was to be expected in the young animal (40 kg). This publication also presents data showing, generally, a close correspondence of data obtained with sheep by the method here used and those obtained by the nitrogen-carbon balance procedure.

Flatt and coworkers have developed a method for measuring oxygen consumption, carbon dioxide production, and urinary nitrogen, for use with grazing animals, based on the open-air current procedure. The equipment used is shown in Fig. 9.5. It combines a tracheal canulae installation and lightweight portable gas meter with a continuous portable aliquoting device. Note that the cow is also equipped with a bag for collecting feces and urine. The animal breathes outdoor air and the expired air is measured, aliquoted, and analyzed for CO_2, O_2, and N_2. The urinary nitrogen excretion is also obtained. The equipment was later modified to provide for the collection of the ruminal as well as the respiratory gases.

TABLE 9.8. Heat Production from Oxygen Consumption, Carbon Dioxide Production, and Urinary Nitogen (24-hour period)

Basic data		
O_2 consumption	392.	liters
CO_2 production	310.7	liters
Urinary N	14.8	g
Protein metabolism data		
Protein oxidized (14.8 × 6.25)	92.5	g
Heat produced (92.5 × 4.3 kcal/g)	398	kcal
O_2 used (92.5 × 0.96[a] per g)	88.8	liters
CO_2 produced (92.5 × 0.77[a] per g)	71.2	liters
Carbohydrate and fat metabolism data		
O_2 used (392 − 88.8 liters used for protein)	303.2	liters
CO_2 produced (310.7 − 71.2 liters used for protein)	239.5	liters
Nonprotein RQ	0.78	liters
Heat produced from 303.2 liters O_2 at RQ 0.78	1452.	kcal
Total heat production (398 + 1452)	1850.	kcal

[a]In the combustion of one gram of protein 0.96 liter of O_2 is used and 0.77 liter of CO_2 is produced.
Source: Based on data from Blaxter, Graham, and Rook, *loc. cit.*

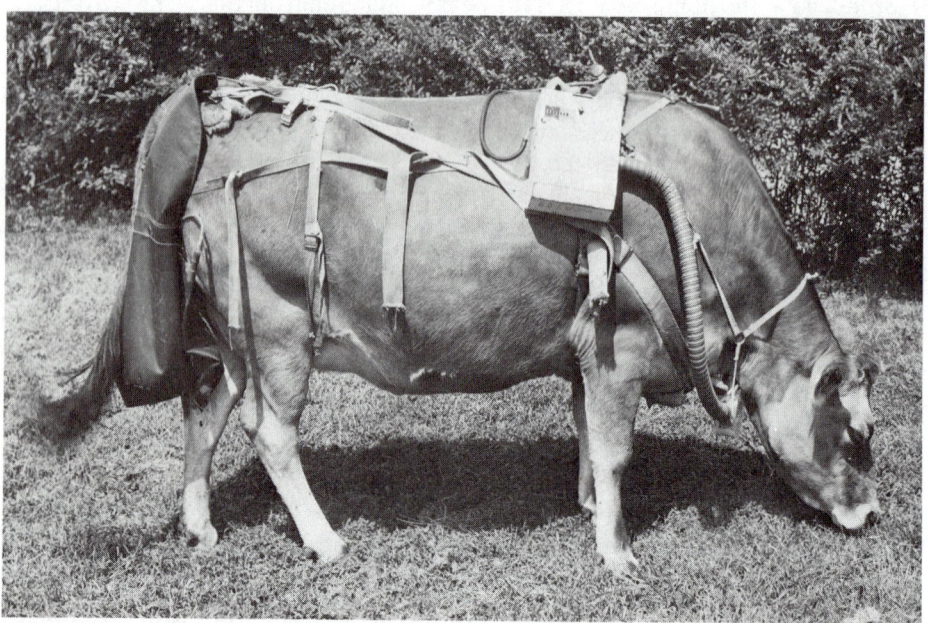

Fig. 9.5 Cow with tracheal canula and equipment for collecting respiratory gases, feces, and urine while grazing. (*Courtesy of W. P. Flatt, U.S. Department of Agriculture.*)

The method based on oxygen consumption, carbon dioxide production, and urinary nitrogen has certain limitations, notably for ruminants, where a substantial production of carbon dioxide in the intestinal tract remains unmeasured. In part because of these limitations and in part in the interest of employing a simpler procedure, a commonly used method disregards the protein catabolism entirely and uses the total RQ instead of the nonprotein RQ. A further simplification is to measure the oxygen consumption only and assume an intermediate RQ, 0.82. The differences in the results arrived at by these shorter methods and those by the longer one are small. In his extensive studies of the metabolism of farm animals—horses, cattle, sheep, swine, and goats—as described in his book, Brody[47] measured the oxygen consumption and multiplied it by the factor 4.852 kcal per liter, the average value at RQ 0.82, to arrive at the heat production. Other workers have used the factor 5.7 or 5.8. Chambers large enough for studies with cattle were constructed in this country by Ritzman and Benedict at New Hampshire, Mitchell and associates at Illinois, and Kleiber and associates at California.

A much more modern chamber has been constructed at the U.S. Department of Agriculture research center at Beltsville, Maryland, and is described by Flatt and associates.[48] Actually, there are six air-current chambers in the

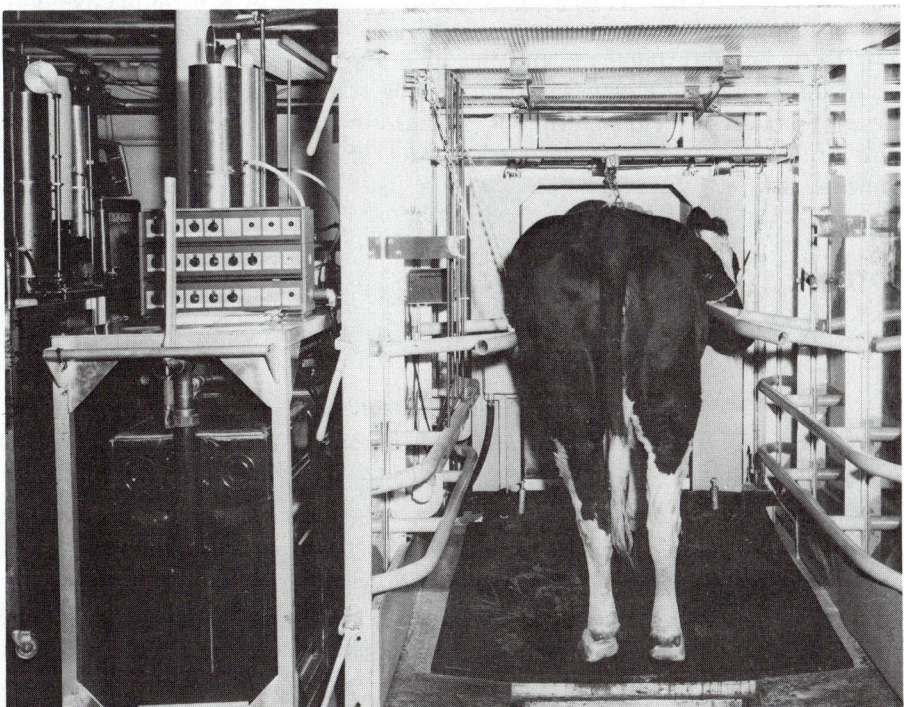

Fig. 9.6 Rear view of the respiration chamber unit for indirect calorimetry, U.S. Research Center, Beltsville, Md. (*Courtesy of W. P. Flatt, U.S. Department of Agriculture.*)

unit, constructed from transparent acrylic plastic, making it possible for the animals to see each other and the operator to observe the animals much more closely. A rear view of one of these chambers is shown in Fig. 9.6, including the ventilating and recording device. The special features of this unit include many automatic measuring and recording devices, equipment for recording and processing data, and others, which all contribute to the accuracy of the data obtained as well as result in large savings of time and labor. The chambers are adjustable for animals of different size. Respiration chambers have been found to produce highly reliable data. Accordingly, their use has replaced that of the respiration calorimeter, which was much more costly to construct and to operate.

9.15 Food Energy. A Field for Further Study

The foregoing discussion has indicated that our present bases for measuring and expressing the energy value of feeds and rations are subject to various limitations and uncertainties. Although no measure of useful food energy can be exact in its general application because of the many variables involved, the field is clearly an important one for further research. The superiority of a given method of evaluating energy cannot be established, however, merely from theoretical considerations or by emphasizing the limitations of other measures. Research must definitely show its greater usefulness for general application, both in experimental comparisons and in feeding practice. Such research is now being actively pursued, and the time and labor involved can be fully justified in view of the importance of arriving at a more useful measure of the largest function of feeds and rations. For use in a feeding standard, any measure arrived at must be accompanied by data for requirements set forth on the same basis.

NOTES

1. R. A. McChance, Famines of history and today, *Proc. Nutrition Soc.*, **34:**161–166, 1975.

2. S. Wortman et al., Food and agriculture, *Scientific American,* **235:**30–205, 1976.

3. W. F. Wedin, H. J. Hodgson, and N. L. Jacobson, Utilizing plant and animal resources in producing human food, *J. Animal Sci.*, **41:**667–686, 1975.

4. World Food and Nutrition Study, National Academy of Sciences, Washington, D.C., 1977.

5. J. M. Brockway, J. D. McDonald, and J. D. Pullar, Evaporative heat loss mechanisms in sheep, *J. Physiol.*, **179:**554–568, 1965.

6. T. Hamada, Estimation of lower critical temperature for dry and lactating dairy cows, *J. Dairy Sci.,* **54:**1704–1705, 1971.

7. According to this theory the relationship is represented by Einstein's equation $E = mC^2$, in which E is the amount of energy locked in the mass as ergs, m equals mass in grams, and C is the velocity of light (centimeters per second).

8. J. B. Boussingault, Analyses comparées des aliments consommés et des produits rendus par une vache laitière; recherches entreprises dans le but d'examiner si les animaux herbivores empruntent de l'azote à l'atmosphère, *Ann. Chim. et Phys.*, (2) **71**:113–127, 1839.

9. L. E. Harris, Biological energy interrelationships and glossary of energy terms, 1st rev. ed., *Natl. Acad. Sci. Natl. Research Council Publ.* 1411, 1966.

10. B. H. Schneider, Feeds of the World: Their Digestibility and Composition, West Virginia Univ., Morgantown, W. Va., 1947.

11. E. W. Crampton, L. E. Lloyd, and V. G. Mackay, The calorie value of TDN, *J. Animal Sci.*, **16**:541–545, 1957.

12. R. W. Swift, The caloric value of TDN, *J. Animal Sci.*, **16**:753–756, 1957.

13. V. R. Smith, I. R. Jones, and J. R. Haag, Alfalfa with and without concentrates for milk production, *J. Dairy Sci.*, **28**:343–354, 1945.

14. L. A. Moore, H. M. Irvin, and J. C. Shaw, Relationship between T.D.N. and energy value of feeds, *J. Dairy Sci.*, **36**:93–97, 1953.

15. L. A. Maynard, Total digestible nutrients as a measure of feed energy (ed. rev.), *J. Nutrition*, **51**:15–22, 1953.

16. R. G. Diggs and coworkers, Energy value of various feeds for the young pig, *J. Animal Sci.*, **24**:555–558, 1965.

17. F. W. Hill and D. L. Anderson, Comparison of the metabolizable energy and productive energy determinations for chicks, *J. Nutrition*, **64**:587–603, 1958.

18. C. Wayne Cook, L. A. Stoddart, and Lorin E. Harris, Determining the digestibility and metabolizable energy of winter range plants by sheep, *J. Animal Sci.*, **11**:578–590, 1952.

19. E. B. Forbes and associates, The energy metabolism of cattle in relation to the plane of nutrition, *J. Agr. Research*, **37**:253–300, 1928; Further studies of the energy metabolism of cattle in relation to the plane of nutrition, *ibid.*, **40**:37–78, 1930.

20. K. L. Blaxter, J. L. Clapperton, and A. K. Martin, The heat of combustion of the urine of sheep and cattle in relation to its chemical composition and to diet, *Brit. J. Nutrition*, **20**:449–460, 1966.

21. J. A. F. Rook, Ruminal volatile fatty acid production in relation to animal production from grass, *Proc. Nutrition Soc.*, **23**:71–80, 1964.

22. The nutrient requirements of farm livestock, No. 2: Ruminants, Agr. Research Council (London), 1965.

23. R. W. Swift and coworkers, The effect of dietary fat on utilization of the energy and proteins of rations by sheep, *J. Animal Sci.*, **7**:475–485, 1948. Raymond W. Swift (1895–1975) was trained at Massachusetts Agricultural College, Pennsylvania State University, and the University of Rochester. In 1922 he joined the staff of E. B. Forbes and devoted much of his career to studies on energy metabolism with the Armsby calorimeter. He promoted the use of digestible energy to replace TDN and contributed greatly to evaluating foods for humans and animals.

24. K. L. Blaxter and J. L. Clapperton, Prediction of the amount of methane produced by ruminants, *Brit. J. Nutrition*, **19**:511–522, 1965.

25. J. L. Clapperton, The effect of a methane suppressing compound, trichloroethyl adipate on rumen fermentation and growth of sheep, *Animal Prod.*, **24**:169–181, 1977.

26. B. K. Watt and A. L. Merrill, Composition of foods, *U.S. Dept. Agr. Handbook* 8, rev. 1963.

27. E. B. Forbes and Max Kriss, Revised net-energy values of feeding stuffs for cattle, *J. Agr. Research*, **31**:1083–1099, 1925. Ernest Browning Forbes (1876–1966) obtained his university training at Illinois and Missouri where he worked before becoming Chief of the Department of Nutrition at the Ohio Experiment Station in 1907. He succeeded Dr. H. P. Armsby as Director of the Institute of Animal Nutrition at Pennsylvania in 1922. His research included mineral balances with pigs and dairy cattle; numerous studies on utilization of feed energy for maintenance, growth and milk production; the associative effects of feeds; effects of level of nutrition, levels of protein and fat on energy metabolism, and the net energy value of feeds.

28. Oscar Kellner (1851–1911), following short periods of service in the agricultural experiment stations at Proskau and Hohenheim and an extended period as professor of agricultural chemistry at the University of Tokyo, became director of the experiment station at Möckern in 1893. Here he served until his death. His many accurately conducted respiration studies with farm animals made a large contribution to the fundamental knowledge of nutritional physiology and found practical application in his feeding standards. Kellner's textbook, Die Ernährung der landwirtschaftliche Nutztiere, the first edition of which was published in 1905, contains an extensive account of his respiration studies and describes his feeding standards.

29. K. Nehring and G. F. W. Haenlein, Feed evaluation and ration calculation based on net energy, *J. Animal Sci.*, **36**:949–964, 1973.

30. K. Breirem and associates, Adjustments of replacement equivalents by the aid of feeding experiments, Proc. Second Symposium on energy metabolism, *European Assoc. Animal Production Publ.* **10**:292–304, 1961.

31. G. S. Fraps, Composition and productive energy of poultry feeds and rations, *Texas Agr. Expt. Sta. Bull.* 678, 1946.

32. F. B. Morrison, Feeds and Feeding, The Morrison Publishing Co., Ithaca, N.Y., 1956.

33. K. Breirem, Neuere Gesichtspunkte zur Frage der Futterbewertung (summary in English), *Landwirtschaftliche Forschung*, **19**:8–32, 1965.

34. G. P. Lofgreen and W. N. Garrett, A system for expressing the net energy requirements and feed values for growing and finishing beef cattle, *J. Animal Sci.*, **27**:793–806, 1968.

35. The nutrient requirements of beef cattle, National Academy of Sciences, Washington, D.C., 1976.

36. R. J. Lipsey, M. E. Dikeman, and R. R. Schalies, Carcass composition of different cattle types related to energy efficiency, *J. Animal Sci.*, **46**:96–101, 1978.

37. D. L. Bath and coworkers, Comparative utilization of energy by cattle for growth and lactation, *J. Animal Sci.*, **25**:1138–1144, 1966.

38. P. W. Moe, W. P. Flatt, and H. F. Tyrrell, Net energy value of feeds for lactation, *J. Dairy Sci.,* **55:**945–958, 1972; P. W. Moe, The net energy approach to formulating dairy cattle rations, Proc. Cornell Nutrition Conf., 1977, pp. 72–76.

39. The nutrient requirements of dairy cattle, National Academy of Sciences, Washington, D.C., 1978.

40. L. W. DeGoey and R. C. Evan, Effect of level of intake and diet dilution on energy metabolism in the young pig, *J. Animal Sci.,* **40:**1045–1051, 1975; Energy value of corn and oats for young swine, *ibid.,* **40:**1052–1057, 1975.

41. A. L. Lavoisier and P. S. Laplace, Memoire sur la chaleur, *Mem. Acad. R. Sci.,* 1780, pp. 355–408.

42. J. D. Pullar, Direct calorimetry on animals by the gradient layer principle. Proc. First Symposium on energy metabolism, *European Assoc. Animal Production Publ.* 8, 1958, pp. 95–101.

43. W. W. Braman, The respiration calorimeter, *Pennsylvania Agr. Expt. Sta. Bull.* 302, 1933.

44. Wilbur Olin Atwater (1844–1907) served for thirty-four years as professor of chemistry at Wesleyan University, Middletown, Conn. The first agricultural experiment station in the United States was established at Middletown under his direction in 1875. It was later moved to New Haven. Atwater also served as the first chief of the Office of Experiment Stations of the U.S. Department of Agriculture. In 1892, with the assistance of E. B. Rosa, professor of physics at Wesleyan, Atwater began the construction of the first human-respiration calorimeter. In this work Francis Gano Benedict early became associated. Upon Atwater's death, the calorimetric studies were continued at Boston in the Nutrition Laboratory of the Carnegie Institute of Washington, where for forty years Benedict and his associates carried on outstanding studies of the energy metabolism in man and in various species of animals.

45. K. L. Blaxter and A. Howells, The nutrition of the Ayrshire calf. II: A spirometer for the determination of respiratory exchange in the calf, *Brit. J. Nutrition,* **5:**25–29, 1951.

46. K. L. Blaxter, N. McC. Graham, and J. A. F. Rook, Apparatus for the determination of energy exchange, *J. Agr. Sci.,* **45:**10–18, 1954.

47. S. Brody, Bioenergetics and Growth, Reinhold Book Corporation, New York, 1945. Samuel Brody (1890–1956) was born in Lithuania, came to the United States, and became a naturalized citizen in 1912. Trained in biochemistry and physiology, he joined the staff of the department of dairy husbandry of the University of Missouri in 1920. Here he served until he died, making outstanding contributions through his studies of growth, metabolism, and environmental physiology. His book has become a classic in its broad field.

48. W. P. Flatt and associates, A description of the energy metabolism laboratory at the U.S. Department of Agriculture, Agricultural Research Center in Beltsville, Maryland, Proc. First Symposium on Energy Metabolism, Copenhagen, *European Assoc. Animal Production Publ.* 8, 1958, pp. 53–64; 101–109.

SUPPLEMENTARY LITERATURE

Brody, S.: Bioenergetics and Growth, Reinhold Book Corp., New York, 1945.

Bull, L. S., B. R. Baumgardt, and M. Clancy: Influence of caloric density on energy intake by dairy cows, *J. Dairy Sci.*, **59**:1078–1086, 1976.

Close, W. H., and L. E. Mount: Critical temperature in relation to feeding level in the growing pig, in M. Vermorel (ed.), Energy Metabolism of Farm Animals, de Bussac, Clermont-Ferrand, 1976, pp. 343–350.

Close, W. H., L. E. Mount, and I. B. Start: The environmental temperature and plane of nutrition on heat losses from groups of growing pigs, *Animal Prod.*, **13**:285–294, 1971.

Denham, A. H.: Influence of energy and protein supplements on grazing and feedlot performance of steers, *J. Animal Sci.*, **45**:1–7, 1977.

Frisch, J. E., and J. E. Vercoe: Food intake, eating rate, weight gain, metabolic rate and efficiency of feed utilization in *Bos taurus* and *Bos indicus* crossbred cattle, *Animal Prod.*, **25**:343–358, 1977.

Holmes, C. W., and A. W. F. Davey: The energy metabolism of young Jersey and Friesian calves fed fresh milk, *Animal Prod.*, **23**:43–53, 1976.

Holter, J. B.: Fasting heat production in lactating versus dry cows, *J. Dairy Sci.*, **59**:755–759, 1976.

Irvin, H. M., and associates: Net energy vs. T.D.N. in evaluating the efficacy of an all-alfalfa hay ration for milk production, *J. Animal Sci.*, **10**:947–960, 1951.

Kleiber, M.: Dietary deficiencies and energy metabolism, *Nutrition Abst. & Revs.*, **15**:207–222, 1945.

Maxson, W. E., R. L. Shirley, J. E. Bertrand, and A. Z. Palmer: Energy values of corn, bird resistant and non-bird resistant sorghum grain in rations fed to steers, *J. Animal Sci.*, **37**:1451–1459, 1973.

Moe, P. W., H. F. Tyrrell, and W. P. Flatt: Partial efficiency of energy use for maintenance, lactation, body gain and gestation in the dairy cow, in A. Schurch and C. Wenk (eds.), Energy Metabolism of Farm Animals, Jurvis Druck & Verlag, Zurich, 1970, pp. 65–67.

Moir, K. W., and J. K. Comor: A comparison of three fibre methods for predicting the metabolizable energy content of sorghum grain for poultry, *Animal Feed Sci. Tech.*, **2**:197–203, 1977.

Mount, L. E.: The use of heat transfer coefficients in estimating sensible heat loss from the pig, *Animal Prod.*, **25**:271–279, 1977.

Phillips, B. C., and R. C. Evan: Utilization of energy of milo and soybean oil by young swine, *J. Animal Sci.*, **44**:990–997, 1977.

Protein Metabolism and Nutrition, Proc. Second International Symp., Centre Agr. Publ. and Doc., Wageningen, The Netherlands, 1977.

Rattray, P. V., et al.: Net energy requirements for growth of lambs age three to five months, *J. Animal Sci.*, **37**:1386–1389, 1973.

Ribero, J. M. de C. R., J. M. Brockway, and A. J. F. Webster: A note on the energy cost of walking cattle, *Animal Prod.,* **25:**107–110, 1977.

Schake, L. M., and J. K. Riggs: Calorie efficiency of beef production, *J. Animal Sci.,* **40:**561–566, 1975.

Schulz, A. R.: Simulation of energy metabolism in the simple-stomached animal, *Brit. J. Nutrition,* **39:**235–254, 1978.

Swift, R. W., and C. E. French: Energy Metabolism and Nutrition, The Scarecrow Press, Washington, D.C., 1954.

Trenkle, A.: Relation of hormonal variations to nutritional studies and metabolism of ruminants, *J. Dairy Sci.,* **61:**281–293, 1978.

Webster, A. J. F., H. Donnelly, J. M. Brockway, and J. S. Smith: Energy exchanges of veal calves fed a high-fat milk replacer diet containing different amounts of iron, *Animal Prod.,* **20:**69–75, 1975.

Webster, A. J. F., J. G. Gordon, and J. S. Smith: Energy exchanges of veal calves in relation to body weight, food intake and air temperature, *Animal Prod.,* **23:**35–42, 1976.

Webster, A. J. F., J. S. Smith, and G. S. Mollison: Prediction of the energy requirements for growth in beef cattle. III: Body weight and heat production in Hereford X British Friesian bulls and steers, *Animal Prod.,* **24:**237–244, 1977.

Chapter 10

The Inorganic Elements and Their Metabolism

The following mineral elements are recognized to perform essential functions in the body and thus must be present in the food: calcium, phosphorus, sodium, potassium, selenium, molybdenum, chlorine, magnesium, iron, sulfur, iodine, manganese, copper, cobalt, zinc, fluorine, nickel, vanadium, silicon, chromium, and tin. The proof that each of these elements is essential rests upon experiments with one or more species. In these experiments, symptoms produced by diets adequate in all nutrients except the mineral in question have been prevented or overcome by adding that mineral to the diets. All the elements mentioned have not been tested with all species, but it is highly probable that there are few exceptions to the need for all of them by all higher animals.

In recent years, improved technology such as ultraclean, more or less, trace-element-sterile isolators[1] and highly purified amino acid diets has resulted in the demonstration of the dietary essentiality of several minerals such as silicon and vanadium. Furthermore, more precise and accurate methods of determining minute quantities of trace elements have been developed. Accurate analysis is important because certain minerals may be needed only in traces to function either as constituents or as activators of enzymes. The body contains two dozen or more other minerals that are not yet considered essential. Some of the minerals may be there simply because they were present in the ingested food but others may have an essential role in the body. It was recently suggested that the improved techniques may lead to the demonstration of the essentiality of some of these minerals. It is not unreasonable to expect that in the future minerals such as lead, tungsten, mercury, and even silver and gold may be shown to be essential.[2]

In some cases it may be necessary to feed experimental animals the deficient diets for more than one generation. Anke and coworkers[3] fed diets containing low levels of arsenic to goats and pigs for three generations and reported reduced fertility and growth rate.

10.1 Factors Affecting Requirements

In many respects, the dietary requirements for minerals are more difficult to accurately define than those for the organic nutrients because many factors determine the utilization of minerals. For example, interrelationships among minerals or relationships between minerals and organic fractions may result in enhanced or decreased mineral utilization. Because of the many interrelationships among minerals, almost any mineral—essential or nonessential[4]—may influence directly or indirectly the utilization of any other mineral. Many of these interrelationships are discussed in the following sections, but further information can be readily obtained in more comprehensive books by Underwood[5] and Hoekstra et al.[6]

The actual amount of mineral in the diet may also influence utilization. For example, if the diet contains more calcium than required, the efficiency of absorption is usually decreased. The mineral status of the animal may also influence absorption. An iron-deficient animal is more efficient in the absorption of iron than an animal with adequate iron stores. The form of the mineral is also an important determinant of utilization. Iron oxide is not available but ferrous sulfate can be readily utilized. Many genetic-nutrition relationships have also been demonstrated. Examples of the extreme genetic effects are shown by one strain of mice that requires a high dietary level of copper[7] and by another strain of mice that requires a very high level of manganese.[8] A strain of cattle that has a genetic defect in zinc metabolism has been identified.[9] On the other hand, a strain of mice that is resistant to zinc depletion has been reported.[10]

Many more subtle differences exist. The copper requirement of Merino sheep is reported to be 1 to 2 ppm higher than that for the British breeds.

Changes in management practices may also influence the mineral requirement. Thus perhaps it is not so surprising that there is still a considerable amount of discussion as to the exact requirements of minerals. For example, the calcium nutrition of growing pigs has been studied at many experiment stations for over 70 years but there is still disagreement as to the optimum dietary level. Therefore, because of the new techniques, many interactions, and potential breakthroughs in the understanding of mineral metabolism, the study of minerals appears to be one of the most exciting areas of animal nutrition.

10.2 Area Deficiencies and Excesses of Mineral Elements

In the case of several of the essential minerals, the initial knowledge of their need and of the symptoms resulting from their deficiency in the ration was gained by observations with grazing animals. Thus phosphorus deficiency was first established as a result of the correlation of the symptoms with the low-phosphorus content in the herbage resulting from a corresponding deficiency in the soil. In a pioneer study Theiler and associates[11] showed that large losses which occurred in both growing and adult cattle on the range were due to a very low content of phosphorus in the herbage as a result of a deficiency in the soil, a situation that has since been noted in other parts of the world including various areas in the United States. This and similar observations have led to surveys of the occurrence of animal troubles that might be correlated with specific soil deficiencies and to the prevention of the troubles with mineral supplementation.

10.3 Ash

The mineral elements as a group are determined in a feed or animal tissue by burning off the organic matter and weighing the residue, which is called ash. Such a determination tells nothing about the specific elements present, and the ash may include carbon from organic matter as carbonate when base-forming minerals are in excess.

The specific elements present in ash can be determined, but it must be remembered that an analysis of the ash tells us nothing as to the combination in which a given mineral occurs either in a body tissue or in a feed. When the organic matter is oxidized, the minerals present in organic combination are

Fig. 10.1 A cow chewing bone in the "Llanos" region. Bone chewing is often associated with a phorphorus deficiency. (*Courtesy of David Morillo, Centro de Investigaciones, Maracay, Venezuela.*)

changed to an inorganic form. Many of the minerals in the body function primarily as specific organic and inorganic combinations, and in the case of the food also, the combination is important insofar as the usefulness of certain elements is concerned. For example, the primary need for sulfur in the food is as a constituent of the amino acids cystine and methionine. No information as to the amount of the element so combined is furnished by determining the sulfur content of the ash of the ration. Thus the nutrition chemist must resort to special methods to not only determine the amounts of minerals but also the forms in which the mineral elements occur in the body tissues and in foods. As mentioned earlier, many recent developments have brought about increased capabilities for mineral analysis.

10.4 General Functions of Mineral Elements

The essential elements serve the body in many different ways. As constituents of the bones and teeth, they give rigidity and strength to the skeletal structures. They are also constituents of the organic compounds, such as protein and lipids, which make up the muscles, organs, blood cells, and other soft tissues of the body. They are important in the activation of many enzymes. Further, they serve a variety of functions as soluble salts in the blood and other body fluids. Here they are concerned in the maintenance of osmotic relations and acid-base equilibrium and exert characteristic effects on the irritability of muscles and nerves. Many of their vital functions are due to an ionic interrelationship which finds expression in the terms "antagonistic action" and "balanced solution." For example, a certain balance between calcium, sodium, and potassium in the fluid which bathes the heart muscle is essential for the normal relaxation and contraction which constitute its beating. In addition to these general functions in which several minerals may take part, each essential one has various specific roles.

CALCIUM AND PHOSPHORUS

Over 70 percent of the ash of the body consists of calcium and phosphorus. These two elements are discussed together because they are closely associated with each other in metabolism. They occur in the body combined with each other for the most part, and an inadequate supply of either in the diet limits the nutritive value of both. As early as 1842 it became recognized through the work of Chossat[12] with pigeons that poor bone developed on a diet low in calcium. When fed wheat alone, the birds died after 10 months, and on autopsy, the bones were found very much depleted. Calcium carbonate prevented the trouble. Chossat used chickens, rabbits, frogs, eels, lizards, and turtles in later studies. During the next twenty years, studies in both France and Germany showed that skeletal development in various species of farm animals was dependent upon the supply of calcium and phosphorus in the ration and that the deficiencies could be corrected by feeding bone meal and other calcium and phosphorus sources.

Fig. 10.2 Cattle consuming soil. Excessive soil consumption may be associated with pronounced mineral deficiencies. (*Courtesy of Eliecer Alberto Velasco, Hato El Frio, Apure State, Venezula.*)

10.5 Interrelation of Calcium, Phosphorus, and Vitamin D

Adequate calcium and phosphorus nutrition is dependent upon three factors: a sufficient supply of each element, a suitable ratio between them, and the presence of vitamin D. These factors are interrelated. While an adequate supply of the elements is the first essential, they are more effectively utilized when they are present in a certain ratio to each other. A ration containing 10 parts of calcium to 1 of phosphorus will not provide for efficient assimilation of phosphorus even though the phosphorus is present in what is normally a sufficient amount. The same is true when this relation between the elements is reversed. While the desirable calcium-phosphorus ratio is often defined as one lying between 2:1 and 1:1, adequate nutrition is possible outside these limits. In calves ratios of 1:1 to 7:1 were entirely satisfactory, but wider or narrower ratios gave poor results. The effective ratio varies somewhat according to the levels of the elements. With plenty of vitamin D in the ration, the ratio becomes of less importance, and more efficient utilization is made of the amounts of the elements present. In the entire absence of the vitamin, assimilation is poor even though the other factors are optimum. The relative importance of these various factors differs considerably in different species and according to the physiological function in question. In the present discussion only incidental mention is made of vitamin D, since the role of this nutrient is taken up in Chap. 11.

10.6 The Composition of Bone

Approximately 99 percent of the calcium and 80 percent of the phosphorus of the body are present in the bones and teeth. Though somewhat variable according to age, state of nutrition, and species, normal adult bone may be considered to have the following approximate composition: water, 45 percent; ash, 25 percent; protein, 20 percent; and fat, 10 percent. The organic matrix of bone in which the mineral salts are deposited consists of a mixture of proteins, of which the principal one is ossein. The water content of bone decreases with age, and the fat is variable according to the nutritive state, since the bone marrow serves as a fat depot; thus ash content is expressed most frequently on the basis of the moisture-free, fat-free bone. In mammals the ash is made up approximately as follows: calcium, 36 percent; phosphorus, 17 percent; and magnesium, 0.8 percent. There are trace amounts of many other minerals.

Bone mineral has an amorphous or noncrystalline phase that is hydrated tricalcium phosphate and a crystallized phase that resembles hydroxyapatite, $Ca_{10}(PO_4)_6(OH)_2$. The young bone has relatively more amorphous phase than crystallized, whereas the reverse is true for mature bone.[13]

There is little variation in the elementary composition of bone ash. The calcium and phosphorus usually occur in approximately a 2:1 ratio. Bone contains considerable amounts of carbonate and citrate and small amounts of magnesium, sodium, potassium, chlorine, fluorine, and traces of other elements. Just how these constituents occur in the basic structure remains in debate.

The nature of the diet can affect somewhat the mineral relationships in bone, even though the ash content is not appreciably changed. Magnesium is somewhat higher than normal in rachitic bone and lower in magnesium deficiency (Sec. 10.19). These various changes need further study and their effects on bone quality need to be measured.

Teeth are similar to bone in chemical composition and in mineral relationships, but characteristic differences exist between the enamel, dentine, and pulp. The enamel is the hardest substance in the body and has the lowest water content, approximately 5 percent. It contains only 3.5 percent of organic matter.

10.7 Calcium and Phosphorus in Soft Tissues

The 1 percent of body calcium which occurs outside the bones is widely distributed throughout the organs and tissues and has many important functions. Calcium is required for normal blood clotting as it must be present for the formation of thrombin from prothrombin. Calcium is necessary for muscle contraction, myocardial function, normal neuromuscular excitability, activation of several enzymes, and secretion of several hormones and hormone-releasing factors. Pathological deposits of calcium in certain soft tissues sometimes occur as a result of upset mineral relations, such as a low magnesium intake relative to calcium and phosphorus (Secs. 10.19 and 10.21). The large amounts of phosphorus which are found other than in the bones are present mostly in organic combinations such as phosphoprotein, nucleoprotein, phospholipids, phosphocreatine, and hexose phosphate. Phosphate is a component of many enzyme systems. The discussions of phosphorus compounds in earlier chapters have indicated their distribution and functions and serve to show the many roles which the element plays in the organism other than as a structural element in bone. Phosphorus makes up 0.15 to 0.2 percent of the soft tissues of the body.

10.8   Calcium and Phosphorus in Blood

The blood cells are almost or entirely devoid of calcium but the plasma, in health, contains from 9 to 12 mg per 100 ml in most species. In the laying hen, levels three or four times higher may occur during egg production. The plasma calcium occurs in two forms. The soluble, ionized form makes up about 60 percent of the total. The other fraction is bound with protein, primarily albumin and plasma proteins.

The level of blood calcium is not readily influenced by the dietary intake though there are species differences in this respect.[14] Various physiological factors tend to maintain a constant level in the blood despite high intakes or marked body losses. The most important factor is a hormone secreted by the parathyroid glands which mobilizes calcium from the bones (Sec. 10.10). If the parathyroids are removed or fail to function, the blood level drops (hypocalcemia) and tetany occurs. There is a hyperirritability of the neuromuscular system which in severe cases results in convulsions. If the glands are abnormally active, as occurs in certain diseases, an excessive mobilization of calcium takes place with a consequent demineralization of the bones. There is an excessive loss of calcium from the body, principally in the urine. Blood calcium is low in "milk fever," but a low dietary intake is not the cause. It seems probable that the parathyroid glands fail to mobilize blood calcium rapidly enough to meet the drain at parturition which results from the onset of active milk secretion.[15] Thus the cattle develop hypocalcemia and tetany. (Milk fever is a misnomer. The cattle do not develop fever; on the contrary, body temperature may be depressed.) High dietary calcium prior to calving is not helpful, but rather harmful. It increases the incidence of milk fever because the parathyroid becomes less active during the prolonged period of high dietary supply (Sec. 17.29).

Another hormone that controls blood calcium is calcitonin, which decreases the rate of calcium mobilization from bone and, therefore, decreases blood calcium.

Whole blood contains from 35 to 45 mg of phosphorus per 100 ml, most of which is in the cells. The element occurs in a variety of forms, principally organic combinations. From the standpoint of mineral nutrition, our main interest lies in the inorganic phosphorus which occurs in the plasma, although it is evident that an interchange of phosphate between organic and inorganic forms continually occurs. In health its level generally lies between 4 and 9 mg per 100 ml, depending upon the age and species. The level is higher at birth than at maturity, the most rapid decline occurring early in life. The plasma phosphorus level is more easily changed by dietary means than the calcium levels. A dietary deficiency of phosphorus could produce hypophosphatemia (low blood levels) and high dietary intakes produce hyperphosphatemia (high blood levels).

10.9   Absorption and Excretion of Calcium and Phosphorus

One function of vitamin D is the promotion of phosphorus and calcium absorption as is detailed later (Sec. 11.10) through the formation of the calcium-binding protein.

Irrespective of the forms in which calcium and phosphorus are ingested, their absorption is dependent upon their solubility at the point of contact with the absorbing membranes. This applies both to the soluble compounds in the feed and also to the insoluble ones which are rendered soluble in passing down the digestive tract. The absorption of both calcium and phosphorus is thus favored by factors which operate to hold them in solution.

Lactose may promote absorption by interacting with the absorptive cells of the intestine to increase their permeability to calcium ions.[16] Large intakes of iron, aluminum, and magnesium interfere with the absorption of phosphorus by forming insoluble phosphates. An experimentally produced "beryllium rickets" is due to the effect of the beryllium in rendering phosphorus insoluble. Oxalates and phytates decrease the absorption of calcium as is discussed later (Sec. 10.15). Fatty acids may form insoluble calcium soaps which are assimilated with difficulty, yet a certain amount of fat seems to favor the absorption of this element.

The level of dietary calcium influences calcium absorption as high dietary levels depress efficiency of absorption. This adaptation to intake is probably mediated by modulation of renal 25-hydroxycholecalciferol-1-hydroxylase activity[17] (Sec. 11.10).

A great excess of either calcium or phosphorus interferes with the absorption of the other, a fact which helps to explain why a certain ratio between them in the diet is desirable for their best absorption. For some species, such as the horse, the effect of high phosphorus on calcium absorption is much more dramatic than the effect of high calcium on phosphorus absorption. As mentioned earlier, the importance of the ratio varies with species. In general, the ratio is of greater importance for nonruminants than for ruminants.

Differing from the organic nutrients previously discussed, a determination of apparent digestibility as measured by the difference between the amounts in the feed and feces is of no value as an indicator of the useful calcium and phosphorus. The reason for this is that the feces are a path of excretion of the minerals which have been absorbed and metabolized and thus served the body, as well as those which have escaped absorption. This fact has long been recognized, but our knowledge of the quantitative relations involved has been made much more specific by the use of the modern isotope techniques. These techniques provide for a distinction between the fecal fraction which is a metabolic excretion (the endogenous fraction) and the fraction of the diet which was not absorbed. The method is illustrated by the experiment of Kleiber and coworkers.[18] A milking cow eating 11 kg of air-dry feed showed an apparent digestibility of phosphorus of 12 percent, based on an anlysis of the intake and fecal outgo. By injection of P^{32} it was shown that 43 percent of the total fecal output was endogenous and thus that the true digestibility was approximately

50 percent. Later studies with lambs have shown a considerably higher true digestibility. Other studies with ruminants have shown that the major portion of the calcium found in the feces is endogenous. The true digestibility which is arrived at by correcting for the fecal endogenous loss is commonly referred to as *availability*—hence the terms available calcium and available phosphorus. The same terminology is applied to other minerals.

The urine is also a path of calcium and phosphorus excretion. The distribution between the urine and feces varies with the species and is somewhat influenced by dietary and age factors. In all species, the feces is a primary path for calcium excretion. Some species, such as the equine and rabbit, may also excrete considerable amounts of calcium in the urine when high levels of calcium are fed. The feces is the primary path for phosphorus excretion in the case of herbivora, but the urine is the principal path for carnivora, and the output is about equally divided between the two channels in the case of humans.

10.10 Deposition and Mobilization of Calcium and Phosphorus

The bones serve not only as structural elements but also as storehouses of calcium and phosphorus which may be mobilized at times when the assimilation of these minerals is inadequate to meet body needs. Thus the mineral metabolism of bone involves not only the deposition of calcium and phosphorus during growth but also processes of storage and mobilization which occur throughout life. As an aid to an understanding of how these various processes take place, a diagram of a longitudinal section of bone is given in Fig. 10.3.

The growth of bone in length takes place at the junction of the epiphysis and diaphysis. The cartilage in between is a temporary formation which grows by the multiplication of its own cells and continues to be replaced by calcified bone. When the cartilage ceases to regenerate and is entirely replaced by bone itself, the epiphysis unites with the diaphysis and growth ceases. This is re-

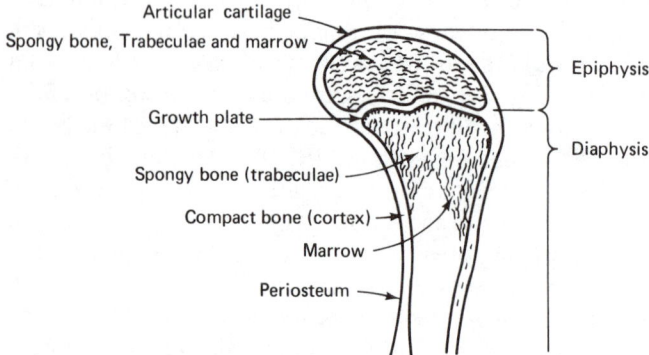

Fig. 10.3 Diagram of a longitudinal section of a growing bone. (*Courtesy of Katharine Hummel, Cornell University.*)

ferred to as the closing of the epiphysis. In the process of ossification, cartilage is replaced by osteoid, which is then calcified. The zone where this is taking place is referred to as the proliferation zone of cartilage, or the zone of provisional calcification. It is the area which is examined in the "line test" (Sec. 11.11) used as a measure of the stage of calcium and phosphorus nutrition. Bone formation also occurs under the periosteum. Internal reconstruction is responsible for the formation of the more complex structure called compact bone.

As distinguished from the shaft, the trabeculae are lacelike structures comprising the principal site in which a reserve of calcium and phosphorus is deposited for mobilization to meet needs not currently supplied by the diet.

These structures are located close to the epiphyseal ends of the bones where the blood supply is greatest. They provide the calcium mobilized by the parathyroid to maintain the level in the blood. During heavy lactation, they are drawn upon to meet a part of the requirements for the minerals secreted in the milk (Sec. 17.29) and they also may be drawn upon in pregnancy. This depletion of the reserves involves no physiological harm, since they can be readily restored with an adequate diet during periods when the body needs for calcium are less, e.g., during the dry period for the dairy cow.

The above discussion implies a more static condition of the bone minerals than actually exists. Isotope studies have shown that there is a continuous interchange of calcium and phosphorus between the blood and bone and between various parts of the bones. The rate of exchange varies from region to region but is most active in the case of the spongy bone. Thus the calcium and phosphorus in the body are in a dynamic state, similar to the situation for fat and protein previously described, and the net result of the interchange determines the nutritional status with respect to a given physiological need. In contrast to the skeleton, the calcium and phosphorus of the teeth are very little subject to mobilization and replacement. Once formed the teeth are comparatively little influenced by either the metabolic needs for these minerals or their supply in the ration.

A more detailed account of bone structure and development is presented by Wasserman.[19]

10.11 Rickets

It is evident from the previous discussion that a failure of normal calcium and phosphorus nutrition may occur at any time of life when the supply of the elements and the factors concerned in their assimilation, notably vitamin D, are not adequate to meet functional needs. In the adult the failure is reflected in a negative balance of the minerals, and in growth the balance data show inadequate retention. At both stages there is a decreased ash content of the bones. Their consequent weakening may eventually result in certain external symptoms, such as lameness and fractures, which are alike at all ages, though during the formative stage, abnormalities of growth which result in misshapen bones are the more common. There are, however, marked differences in the bone pathology, particularly histological, according to the stage of bone development and also according to the specific nutritional deficiency primarily concerned. Therefore various terms are used to designate different failures of calcium and phosphorus nutrition. Unfortunately there is a lack of uniformity in the use of these terms.

In its broadest sense, rickets represents a disturbance of the mineral metabolism in such a way that the calcification of the growing bones does not take place normally. This is the sense in which the term is used in this book. The formation of the organic matrix, osteoid, takes place, but calcium and phosphorus are not deposited in it. There is a lowering of the level of inorganic phosphorus or calcium or both and an increase in phosphatase, in the plasma. The blood picture varies according to the specific dietary deficiency involved. Some authorities limit the use of the term rickets, or *true rickets,* to the specific bone pathology found in very early growth, involving changes which are produced experimentally on a low-phosphorus, high-calcium diet deficient in vitamin D and which are accompanied by a low blood phosphorus. There is a widening of the epiphyseal-diaphyseal cartilage, an excessive production of osteoid tissue which accounts for the enlargement of the ends of the long bones, and other characteristic histological changes. But bone abnormalities can develop at any age. (The condition similar to rickets in mature animals is often called osteomalacia.) The abnormalities can occur at any time because of a lack of calcium as well as of phosphorus and of the vitamin, and the blood picture may vary as regards the mineral relations. Though the specific bone pathology may differ, the broad definition of rickets includes all of these nutritive failures, recognizing, however, that tetany may also occur if the blood calcium becomes very low. Frequently the terms *low-phosphorus rickets* and *low-calcium rickets* are used where a distinction is made.

The failure of bone nutrition during growth results not only in an arrest of its normal development but also in various structural abnormalities. There is an enlargement of the ends of the bones which shows itself in a beading of the ribs and certain other bones. In severe and prolonged failure of adequate nutrition, the tension of the muscles pulls the weakened bones out of shape and the weight of the body causes the leg bones to bend and even to fracture. Rapid growth accelerates the development of rickets, probably caused in part by the demands of muscle formation for phosphorus (Sec. 15.35).

Rickets were produced in calves and pigs. In calves its occurrence is revealed in decreased growth, stiffness of gait, enlarged and painful joints, arching of the back, and, as an extreme symptom, the birth of weak or deformed calves. A low blood calcium appears to be a less marked and regular finding than low phosphorus. In swine, retarded growth, enlarged joints, and bone deformities which may produce posterior paralysis are the most evident physical symptoms. In pigs low blood phosphorus is a characteristic finding, but on low-calcium rations the blood calcium may be low also. In lambs enlarged joints which are frequently painful, stiffness, and irregular gain are common. Rickets has also been reported in horses, chickens, dogs, and several

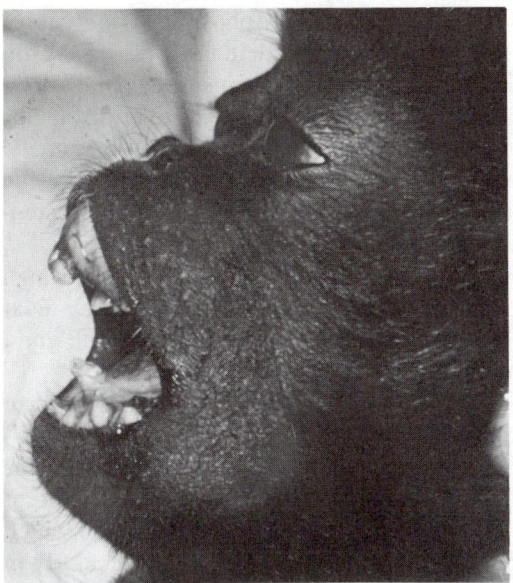

Fig. 10.4 Monkey with nutritional secondary hyperparathyroidism (NSH). The calcium is removed from the bone and the area invaded with fibrous connective tissue. Thus, the head actually increases in size but the bone does not have density or strength. A similar condition occurs in horses with NSH. Note the swelling of gums and loss of teeth. (*Courtesy of L. Krook, Cornell University.*)

other species. In all species, retardation of growth, inadequate calcification, and bone malformations result.

10.12 Nutritional Secondary Hyperparathyroidism

When animals are fed diets containing low or marginal levels of calcium but excessive amounts of phosphorus, a condition called nutritional secondary

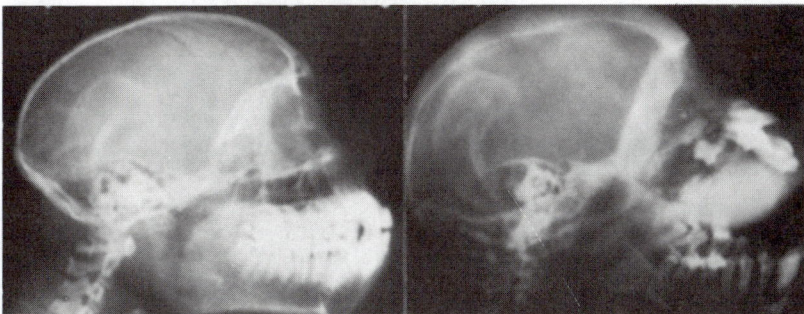

Fig. 10.5 Radiographs of control (left) and monkey with nutritional secondary hyperparathyroidism (right). Notice the widening of the bone but lack of density. There is loss of bone from around the teeth. The teeth become very loose and may fall out. (*Courtesy of L. Krook, Cornell University.*)

hyperparathyroidism (NSH) may develop. The excess phosphorus depresses calcium absorption and therefore the blood level of calcium drops. This drop triggers the parathyroid to release a hormone which causes calcium to be removed from the bone to replenish the blood level. A high level of dietary phosphorus also causes a greater turnover rate in the bone. In some animals, particularly the horse and the monkey, fibrous connective tissue invades the area and enlarged facial bones result. Hence the disease is also called osteo-dystrophia fibrosa and big-head disease.[20]

Field cases of NSH have been reported in horses and rabbits fed high intakes of grain, carnivores such as dogs, cats, lions, and tigers fed all-meat diets, reptiles such as turtles and lizards fed all-meat diets, birds fed all-grain diets, and primates fed only fruit and nuts.[21]

NSH also has been suggested as a cause of periodontal disease (loss of bone around the teeth) in dogs and humans.[22]

Osteoporosis is another term used to denote a failure of normal bone metabolism in the adult. It differs from ostemalacia in that the mineral content is normal but the absolute amount of bone is decreased. It is a condition which becomes prevalent in the human population after the age of 50, particularly in women. While there are various other causes, which include a defective syn-thesis of the protein matrix, there is evidence that a long-continued low intake of calcium plays a role in its development and that the resulting condition can be corrected, in some cases, by an intake of calcium double the commonly recommended adult allowance. In the literature one finds the term osteoporosis used in different ways. For example, those who consider the low-phosphorus variety as the only true rickets use it to denote the histology where low calcium is the factor involved. Osteomalacia and osteoporosis are sometimes used in-terchangeably in referring to bone troubles in the adult. Osteopenia is another term often used to describe bone pathology and simply means too little bone.

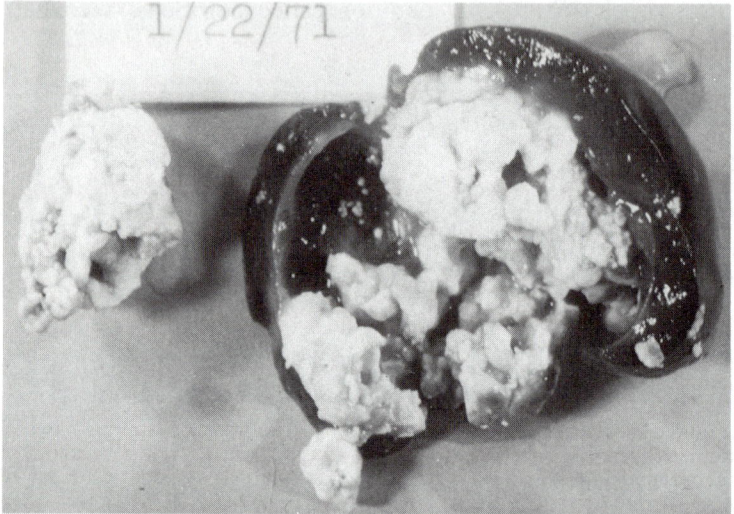

Fig. 10.6 Kidney stones in a rat kidney as a result of high calcium intake. (*Cornell University.*)

10.13 The Effects of High Intakes of Calcium

Deleterious effects on overall nutrition can result from excessive intakes of calcium. By "excessive" is meant levels markedly higher than body requirements. Within an appropriate calcium-phosphorus ratio the effects noted have been due to an unfavorable influence on the assimilation of other mineral nutrients rather than to an action of calcium per se. Particularly when the intake of magnesium, iron, iodine, manganese, zinc, or copper is borderline in terms of needs, deficiency symptoms of the borderline element have been produced experimentally. The extent of the practical importance of these findings for farm animals has not been established. Experimental results showing that increasing the calcium content of the diet well above requirements without concurrently increasing the zinc results in a severe parakeratosis in swine and that high levels of calcium and/or phosphorus in the rations of swine and poultry intensify the effect of manganese deficiency are examples. A further discussion of the effects of relatively excessive levels of calcium is presented later as the minerals in question are taken up. In general, it would appear that harmful effects may occur in practice from rations containing calcium markedly in excess of body needs unless attention is given to ensuring liberal intakes of those minerals which, otherwise, may become relatively deficient.

Chronic intakes of high dietary calcium may also cause a hypersecretion of calcitonin and bone abnormalities such as osteopetrosis (dense bone).[23] High calcium intakes have also been reported to cause the formation of kidney stones.

10.14 The Calcium and Phosphorus Content of Feeds

The different concentrates and roughages vary widely in their content of calcium and phosphorus. Certain combinations furnish a sufficient supply of these minerals, while others are deficient. Thus an important aid in providing for adequate calcium and phosphorus nutrition is a general knowledge of the composition of the common feeds. This knowledge enables the feeder to consider minerals, as well as protein and energy, in making up his rations and provides him with the information for determining when supplementary sources of these elements are needed.

The relative amounts of calcium and phosphorus in certain typical feeds are shown in Table 10.1. For comparative purposes the values from skim milk and corn silage are given on a dry basis. The cereal seeds are all low in calcium. Legume seeds, notably soybeans, are higher, and the same is true for the oil meals. All seeds and their products, however, must be classed as poor sources of calcium in terms of the requirement of the animal body. Grass hays, such as timothy, are also poor in contrast to legume hays, which are rich. On a dry basis, skim milk exceeds grass hays in calcium content. Much richer than any of the feeds shown in the chart, however, are the animal by-products containing bone, such as tankage, meat scrap, and fish meal. A 60 percent protein tankage will furnish four or five times as much calcium as will legume hay or skim milk and twenty times as much as will vegetable protein concentrates such as the oil meals.

TABLE 10.1. Calcium and Phosphorus Content of Some Common Feeds

Feed	Calcium, %	Phosphorus, %
Alfalfa hay, midbloom	1.20	0.20
Animal carcass residue, dehydrated, ground	5.94	3.17
Animal carcass residue, with bone, dehydrated, ground	7.34	3.73
Barley	0.08	0.42
Brewers' dried grains	0.27	0.50
Cattle, milk, skim, dehydrated	1.26	1.03
Citrus pulp	1.96	0.12
Clover hay	1.15	0.22
Corn, U.S. No 2	0.02	0.31
Corn silage (air dry)	0.35	0.21
Corn gluten feed	0.46	0.77
Cottonseed meal, solvent extracted	0.15	1.10
Flax seed, solvent extracted	0.40	0.83
Fish meal	5.49	2.81
Oat, grain	0.10	0.35
Soybean seeds	0.25	0.59
Soybean, seeds, solvent extracted, ground	0.32	0.67
Timothy hay, midbloom	0.36	0.16
Wheat bran	0.14	1.17
Wheat flour by-products	0.15	0.91
Wheat, grain	0.05	0.36

Source: Joint United States–Canadian Tables of Feed Composition, *Natl. Acad. Sci. Natl. Research Council Publ.*, 1684, 1969.

As is shown in Table 10.1, these same foods present a very different picture in phosphorus content. Here the seeds are uniformly higher than the roughages, and seed by-products, such as wheat bran and the oil meals, are especially rich in phosphorus. Skim milk is the only feed included in the charts which can be classed as rich in both calcium and phosphorus. The bone-carrying animal by-products are also very rich in both elements. Animal-carcass residue (tankage) supplies more than twice as much calcium and phosphorus as any feed shown. The milled flours are lower in both calcium and phosphorus than the whole seeds.

While the data presented in Table 10.1 are useful to show the differences between the various kinds of feeds, they must not be considered to be exact values, because the calcium and phosphorus contents of feeds, especially the roughages, are variable according to the nature of the soil on which they are grown, the fertilizer used, and the water relations. Timothy hay grown on fertile soil may contain two or three times as much calcium as that grown on a worn-out acid soil. On the other hand, legume hays require a soil rich in lime for satisfactory growth, and thus, while they show some variation according to soil, they can always be relied upon as rich sources of calcium. All hays are highly variable in phosphorus content according to the supply available in the soil and to other factors. Thus the feeder must know something about the mineral content of his roughage in order to tell the exact conditions under which he needs a mineral supplement. General statements that one type of hay always needs a supplement while another type never does cannot always hold. Pasture grass is subject to even larger variations than dry roughage because the species of the grasses is a large factor and because climatic conditions also play a role.

On the basis of the survey reported by Collins and associates, it may be calculated that, aside from a few limited areas, the water consumed provides an insignificant amount of calcium in terms of body needs (Sec. 4.8).

10.15   Availability of Calcium and Phosphorus of Feeds

The possibility that the availability of calcium and phosphorus may vary considerably according to their chemical combination or physical association with other compounds in feeds must be recognized. Much research has centered around the question of the availability of phosphorus present as *phytin,* or *phytic acid*. The latter compound is an acid hexaphosphoric acid ester of inositol (Sec. 11.60). It occurs as salts of calcium, magnesium, etc., the complex being referred to as phytin. Half or more of the phosphorus of most mature seeds and their products, notably wheat bran which is a rich source, is so combined. Thus the question of its availability is an important one in animal nutrition. There is no simple answer.

As a result of many studies, some of which have produced seemingly contradictory results, it is now clear that the answer is different for different species and also varies according to the level of vitamin D in the ration. There is clear evidence that ruminants can utilize phytin phosphorus satisfactorily. Horses utilize phytin phosphorus less efficiently than ruminants but more efficiently than pigs. Poultry are poor utilizers of phytin phosphorus.[24]

In those species which utilize phytin phosphorus less effectively than the inorganic form, phytin is incompletely broken down in the digestive tract. Both dietary *phytase* and microbiological activity appear to be concerned in the extent of this breakdown. Experiments with dogs and man have produced evidence that undigested phytate precipitates calcium, preventing its absorption and thus resulting in poorer bone calcification. On the other hand, in the reports of the experiments with rats and chickens a lessened supply of absorbable phosphorus is stressed. It seems reasonable to believe that species differences are here involved. While the end results can be definitely measured for all species in terms of bone calcification, the differences in intermediate metabolism which may be concerned are particularly difficult to study because of the interrelations of calcium, phosphorus, and vitamin D.

Both rat and human experiments have shown that the availability of the calcium of certain leafy materials is impaired by the presence of oxalic acid. The acid precipitates the calcium and prevents its absorption. In the case of spinach, for example, there is usually enough oxalic acid present to render all of its calcium unavailable. On the other hand, the calcium of kale, which contains practically no oxalic acid, is nearly as well assimilated as that of milk.

Certain tropical grasses contain high levels of oxalate and field cases of calcium deficiency (nutritional secondary hyperparathyroidism) have been reported in horses grazing such grasses.[25]

10.16 Phosphorus Deficiency and Appetite

Free-choice feeding of calcium and phosphorus is often recommended but several recent studies demonstrate that ruminants do not have the nutritional wisdom to select the amount of these nutrients needed. Therefore, the addition of proper amounts of calcium and phosphorus to the diet of domestic livestock appears to be a prudent practice.[26,27]

But deficiency of phosphorus has a specific effect in causing a loss of appetite and even a depraved appetite, frequently referred to as "pica," which is exhibited in the eating of bones, wood, clothing, and other materials to which the animal may have access (Fig. 10.3). The animal becomes very emaciated. This condition is most frequently met in grazing animals in areas where the soil, and thus the forage, is very low in phosphorus. It may also occur in barn feeding, and here a lack of vitamin D may also be involved. Very large losses have occurred among grazing animals in different parts of the world as a result of this severe phosphorus deficiency, even where the forage was abundant and nutritionally adequate in other respects. Many of the deaths have resulted from diseases to which the weakened animals become especially susceptible, notably from infections contracted by eating decaying bones of animals which had died.

10.17 Calcium and Phosphorus Supplements

In selecting rations for their calcium and phosphorus content, consideration should first be given to supplying the minerals, insofar as possible, by choosing those feeding stuffs which are rich in the elements needed. However, the farmer must build his rations around the available feeding stuffs on his own farm, and it may not be practicable for him to make up his rations so that they will be rich in the needed minerals. Wherever it is not practicable or possible

TABLE 10.2. Analyses of Calcium and Phosphorus Supplements

Supplement	Calcium, %	Phosphorus, %
Animal bone, steamed, dehydrated	29	14
Dicalcium phosphate	26	21
Defluorinated phosphates	29–36	12–18
Limestone, ground	34	
Calcium phosphate	17	21
Sodium phosphate	–	22
Diammonium phosphate	–	20
Oyster shell	35	

to provide for adequate mineral nutrition by an appropriate combination of the feeding stuffs available, mineral supplements should be used.

The qualities which determine the feeding value of a supplement are content of calcium and phosphorus, fineness of division, and freedom from harmful impurities. The analyses of commonly available ones are shown in Table 10.2.

The chemical composition of these supplements varies according to the purity of the raw material and the method of processing. Standard products should have a composition approximating that represented by the figures given in the table. The figure for ground limestone is representative of the high-calcium products available. Dolomitic limestones have much smaller and variable contents of calcium.

Many studies have shown that the minerals of these supplements are readily utilizable by both animals and humans. Assuming that no harmful substances are present in the supplements and that the latter are in a suitable physical condition, all may be considered to be of approximately equal value per unit of calcium and phosphorus present. Though insoluble in water, they are similar in value to soluble salts of the minerals because, as previously explained, it is the solubility at the point of absorption that counts. As regards chemically prepared salts, orthophosphates are readily available, but the meta- and pyrophosphates are of very limited availability, at least for chickens. The calcium in complex iron calcium silicates has also been found to be unavailable. Phosphoric acid has been found to be a highly available source of phosphorus for swine and cattle. Diammonium phosphate contains 18 percent of nitrogen along with 20 percent of phosphorus, both elements being well utilized by ruminants.

10.18 Rock Products

Defluorinated Phosphates The calcium and phosphorus of raw rock phosphate and of superphosphate are absorbable, but the feeding of these products is harmful because of the fluorine present (Sec. 10.52). Spurred on particularly by wartime needs, processes were developed for removing this fluorine, and thus the production of "defluorinated phosphates." The calcium and phosphorus contents of these phosphates are given as ranges, reflecting the variations in the various products according to source of raw material and method of manufacture. Defluorinated superphosphate contains more of both minerals than does defluorinated rock phosphate, as is to be expected. Dicalcium phosphate can be made from either bone or rock, but where the latter source is used, the fluorine must be removed.

Various defluorinated phosphates have been tested experimentally with rats, chickens, pigs, and calves. Several have been found satisfactory, having a calcium and phosphorus availability equal to that of steamed bone meal or dicalcium phosphate. Some products have given poorer results, however. The differences in availability are attributable in large part to the amounts of pyrophosphate and metaphosphate present. These forms are less available than the usual ortho form. The temperature at which the fluorine is eliminated is a large factor in determining the relative amounts of the different forms in the final product.

Colloidal phosphate, or soft rock phosphate, frequently offered for feeding purposes, is a mixture of fine rock and clay. *Phosphatic limestone* is a product having a composition similar to a mixture of limestone and rock phosphate. Both the colloidal phosphate and the limestone product have nearly as much fluorine per unit of phosphorus as does the rock and, thus, are not satisfactory phosphorus supplements unless defluorinated.

Curacao phosphate is a special rock phosphate naturally low in fluorine (0.4 percent). It contains approximately 33 percent of calcium and 14.5 percent of phosphorus, both of which are highly available to animals.

MAGNESIUM

Magnesium is closely associated with calcium and phosphorus, both in its distribution and in its metabolism. Approximately 70 percent of the body supply is in the skeleton, the remainder being found widely distributed in the various fluids and other soft tissues. Approximately one-third of the supply in the bones is subject to mobilization for soft-tissue use when the intake is inadequate. Depending on the species and individual, blood serum normally contains 2 to 5 mg of the element per 100 ml. The considerable percentage which is found elsewhere than in the bones indicates that its distribution in the body as a whole follows that of phosphorus rather than of calcium. In fact, though the calcium content of the entire body is many times that of magnesium, the soft tissues actually contain much more of the latter. Like calcium and phosphorus, magnesium is excreted in both the urine and feces. The major output is found in the feces but the kidney can be an important homeostatic mechanism.

As well as serving as an essential constituent of bones and teeth, magnesium is required for various body processes, notably as an activator of various enzymes. Specifically it activates all enzymes transferring phosphate from ATP to ADP and must therefore influence all life processes.

Magnesium is a cofactor for decarboxylation for certain peptidases and for alkaline and acid phosphatases. In the plant world, magnesium is a constituent of chlorophyll, a metalloprotein complex that is essential for photosynthesis and, thus, for all plant and animal life.

10.19 Symptoms of Magnesium Deficiency

In 1926 Leroy in France showed that magnesium is essential for the growth of rats. Later, Kruse, Orent, and McCollum described the specific symptoms. Lowering the magnesium content of the diet to 1.8 ppm resulted, in rats, in vasodilation, hyperirritability, convulsions, and death. In general the same symptoms occurred in dogs. A characteristic blood finding was a lowered magnesium content but normal calcium and phosphorus. This picture led the investigators to call the trouble "magnesium tetany," thus distinguishing it from the usual tetany in which a low blood calcium is characteristic. Later workers with rats found a decrease in the ash content of the bone, reflecting a large loss of magnesium and some loss of calcium and phosphorus as well. Clinical and neurological changes similar to those found in the rat have been reported for the chick. Pigs fed magnesium-deficient rations become irritable, show a reluctance to stand, lose equilibrium, go into tetany, and eventually die.

A tetany in calves, characterized by a low blood magnesium with normal calcium and phosphorus, was reported by Duncan, Huffman, and Robinson[28] of the Michigan Experiment Station. The calves were fed for extended periods either on milk alone, a food which is rather low in magnesium, or on milk plus special supplements which contained little or none of the element. In calves showing the typical symptoms, the magnesium in the blood serum is frequently as low as 0.1 mg per 100 ml compared with a normal of around 2.5 mg. This lowering of the blood level is preceded or accompanied by a large loss from the bones. On postmortem examination the calves which died revealed marked calcium deposits in the arteries, myocardium, and skeletal muscles. The finding of tetany in calves fed a magnesium-deficient diet suggested that "grass tetany" in cattle may be related to magnesium deficiency.

The signs of grass tetany include excessive nervousness, twitching of muscles, labored breathing, rapid pulse, convulsions, and death. The disease has been reported frequently in countries such as The Netherlands and New Zealand where the climate is cool and wet. It often occurs in fresh cows within a week or two after they are turned out to pasture. In the United States, lactating beef cattle have been the most common victims. The incidence varies greatly from year to year, probably because of different climatic conditions. Tetany usually occurs in the spring of the year when the cattle are on lush pasture. The disease has also been called lactation tetany, winter tetany, and oat tetany. It may be called wheat tetany in certain areas such as the Texas panhandle because ruminants pastured on wheat have been reported to develop tetany (Sec. 17.32).

The serum level of magnesium is usually, but not always, low (1.0 mg per 100 ml or less is common) but the level of magnesium in the feed may not be low. Factors such as transaconitate, citrate, potassium, and nitrogen in the grass influence the utilization of magnesium. The ratio of potassium to calcium and magnesium content might also be important. Potassium apparently depresses the absorption of magnesium from the rumen, a primary site of magnesium absorption.[29] Rumen pH, dietary phytic acid, vitamin D, fat, energy content, and type of carbohydrate also influence magnesium utilization.

Magnesium supplements are usually helpful in preventing the problem. Magnesium "bullets" which can be placed in the rumen and release magnesium slowly over a long period of time have also been used.[30]

10.20 Body Needs and Feed Supplies

Although there are some species differences, it appears that the magnesium requirement of growing animals is of the order of 0.06 percent of the dry ration, assuming that the calcium and phosphorus intakes are adequate but not excessive. Most of the commonly fed roughages and concentrates contain at least 0.1 percent of magnesium on a dry-matter basis, and many have three or four times this figure. Thus, assuming that the needs of other species are similar to those of calves and chicks, it would seem likely that the commonly fed rations of farm animals can be relied upon to supply an adequate amount without supplementation except in special cases as discussed above.

10.21   Interrelations of Magnesium with Calcium and Phosphorus

It has been shown with laboratory animals, notably the guinea pig and cotton rat, that, with diets adequate in magnesium and other nutrients, increasing the calcium or phosphorus or both results in symptoms of magnesium deficiency. The effects are much more marked on the production of calcification in the soft tissues and also on growth with borderline intakes of magnesium. The physiological explanations for these effects are not clear. There is evidence of decreased absorption or increased excretion of magnesium, or both. Quite evidently, high calcium and phosphorus intakes have the practical effect of increasing the minimum magnesium requirement for the species studied. Whether the findings discussed above hold for farm animals and have practical importance for them accordingly remains unknown. Magnesium may influence calcium and phosphorus utilization but when these two elements are present in liberal amounts, the harmful effect of the magnesium is slight or nil. Palmer, Eckles, and Schutte,[31] for example, showed that the ingestion of magnesium sulfate by cattle on a low-phosphorus diet results in serious and continuous losses of calcium which are overcome by increasing the phosphorus content of the ration.

Experiments with pigs, chickens, and rats, using rations liberal in phosphorus, have shown that dolomitic limestone is a satisfactory source of calcium for bone formation despite its magnesium content. Thus it appears that, provided both calcium and phosphorus are plentifully supplied, the ingestion of at least a moderate excess of magnesium, either in a mineral supplement or in water or other foods, will not markedly disturb calcium retention though it may tend to increase the requirements for calcium and phosphorus in the ration.

SODIUM, POTASSIUM, AND CHLORINE

Differing from the minerals previously discussed, these minerals occur largely in the fluids and soft tissues. They function in maintaining osmotic pressure and acid-base equilibrium, in controlling the passage of nutrients into cells, and in water metabolism in general. Nutritionally they are often considered of minor importance because meeting body needs usually presents no practical problem and dangers of excessive intakes exist only in special situations. There is a regular dietary need, however, because of limited storage, the current excess being rapidly excreted. A deficiency of any of these elements results in lack of appetite, a decline of growth, loss of weight and production in the adult, and decreased blood levels—general symptoms which reflect more specific physiological and pathological changes for each.

10.22 Sodium

The body contains approximately 0.2 percent of sodium. Some of this amount is localized in the skeleton in an insoluble, rather inert form, but by far the larger proportion is found in the extracellular fluids where it undergoes a very active metabolism. The element makes up 93 percent of the bases of the blood serum, and thus it is the predominant basic element concerned in neutrality regulation. Sodium seems to be absent from blood cells, but it does occur in considerable amounts in the muscles, where it is associated with their contraction. A lack of the element also lowers the utilization of digested protein and energy and prevents reproduction. In laying hens, a deficiency results in lowered production, loss of weight, and cannibalism.

Sodium salts are readily absorbed and circulate throughout the entire body. Excretion takes place through the kidneys as chlorides and phosphates. There is some loss in the perspiration, which, in human beings at hard work, particularly in warm weather, may represent by far the major portion of the total excretion. Sodium requirements for growth ranging between approximately 0.1 percent and 0.2 percent of the ration have been reported from studies with rats, chicks, pigs, and calves. The exact figure depends on the species and certain mineral relations in the diet. The commonly used feeds do not contain sufficient sodium to meet body needs, but this deficiency is usually covered by the practice of including common salt in the ration.

10.23 Potassium

The potassium content of the body is similar to the sodium content, but it exists primarily as a cellular constituent. Human blood cells, for example, contain over twenty times as much of the element as does the plasma. Potassium plays a vital role in muscle, where its content is six times that of sodium. While blood plasma contains many times as much sodium as potassium, in milk the reverse is true.

Potassium deficiency has been experimentally produced in several species. In addition to nonspecific gross symptoms, there is a lowered content of the element in the heart and other organs, heart lesions, tubular degeneration of the kidneys, muscular dystrophy, and other pathological changes. As studied with rats, pigs, and chickens, the potassium requirement markedly exceeds that of sodium. Reported figures range from approximately 0.2 percent to approximately 0.6 percent of the dry ration. The commonly fed rations of farm animals can be counted on to meet these requirements. But changes in management practices, such as decreased use of hay and increased use of grain, may result in less than optimum intakes of potassium because grains contain a much lower level of potassium than hay. The replacement of protein supplements such as soybean meal by urea may also decrease dietary potassium intake.

Potassium, like sodium, is readily absorbed, and the excess over body needs is immediately excreted. This excretion normally takes place in the urine to the extent of 90 percent, but profuse sweating diverts a large portion through this channel.

Adrenal hormones cause the kidney to conserve sodium but increase the excretion of potassium.[32] The kidney has only limited capacity to conserve potassium even when the diet is deficient.

In 1873 Bunge[33] evolved the theory that an excessive intake of potassium impoverished the organism of sodium and chlorine. Recent studies with rats and poultry have shown that there are indeed certain interrelationships between sodium and potassium in metabolism; for example, at inadequate levels of either, the deficiency symptoms are aggravated by a large excess of the other. But it seems apparent that sodium-potassium antagonism is not a matter of practical importance in the feeding of farm animals.

10.24 Chlorine

Differing from sodium and potassium, chlorine is found in large concentrations both within and without the cells of the body tissues. Blood cells contain about one-half as much as the plasma. Approximately 15 to 20 percent of the chlorine of the body appears to be in organic combination. The chlorides of the blood make up two-thirds of its acidic ions. This indicates their large role in acid-base relations. The gastric secretion contains chlorine as free acid and in the form of salts. The body has a certain capacity to store chlorine in the skin and subcutaneous tissues. Its excretion follows that of sodium and potassium. The body's requirement is approximately half of that for sodium.

The role of chloride in dairy cattle nutrition has been recently reviewed by Coppock and Fettman.[34]

10.25 Common Salt

Since salt serves as a condiment as well as a nutrient, the intake tends to be highly variable and frequently in excess of needs. Its use as a condiment has physiological support in evidence that it stimulates salivary secretion and promotes the action of diastatic enzymes. When the intake is at a minimum, the body makes an adjustment whereby the output of sodium and chlorine in the urine nearly ceases. In contrast, large intakes involve a correspondingly large excretion, the water requirement being increased accordingly. The kidney is the regulating organ which, through its secretory activity, controls the concentration of electrolytes in the blood.

In 1905 Babcock[35] reported a long-time study of the role of salt in the dairy ration. He found that cows receiving no salt exhibited an abnormal appetite for it after two or three weeks, but that a much longer time elapsed, even a year, before any ill effect on health was noted. Eventually there was a loss of appetite, an unthrifty condition, and a marked decline in weight and milk yield. These symptoms appeared first in the higher producers, and the breakdown most frequently occurred at calving or shortly after the height of milk flow. The feeding of salt produced rapid recoveries in animals showing acute symptoms. The long period which elapsed before health was affected by salt deprivation illustrates the ability of the body to husband its supply of sodium and chlorine, reducing their excretion to a minimum, when the intakes are very small.

This early study was confirmed and extended over fifty years later by Smith and Aines[36] in a much more comprehensive experiment with higher-producing cows. Sodium was found to be the primary deficiency. Its level in the blood plasma remained unchanged in cows receiving no salt, but the urine content dropped almost to zero within a month. The sodium, chloride, and potassium contents of milk remained unchanged. The supplemental salt (NaCl) required for lactating cows producing about 5000 kg of milk per year was estimated to be about 30 g per cow per day in addition to that in the feeds used, involving a total sodium requirement of about 21.3 g daily. The condition of one of the salt-deficient animals, reflecting a large loss of body weight and gaunt appearance, is shown in Fig. 10.7.

The human kidney may excrete as little as 1 g or as much as 40 g of sodium chloride per day, depending on the intake. Given normal kidneys and an appropriate water intake, large amounts taken for short periods can be excreted without harm. Excessive intakes can, however, result in water retention in the body, causing edema. These results have been reported in chicks on rations containing little more than 3 percent of salt. Ruminants have a higher tolerance, but with a low water intake toxicity has been produced with a level of 2.2 percent. It appears that the primary mode of action of salt poisoning is through a disturbance of water balance.

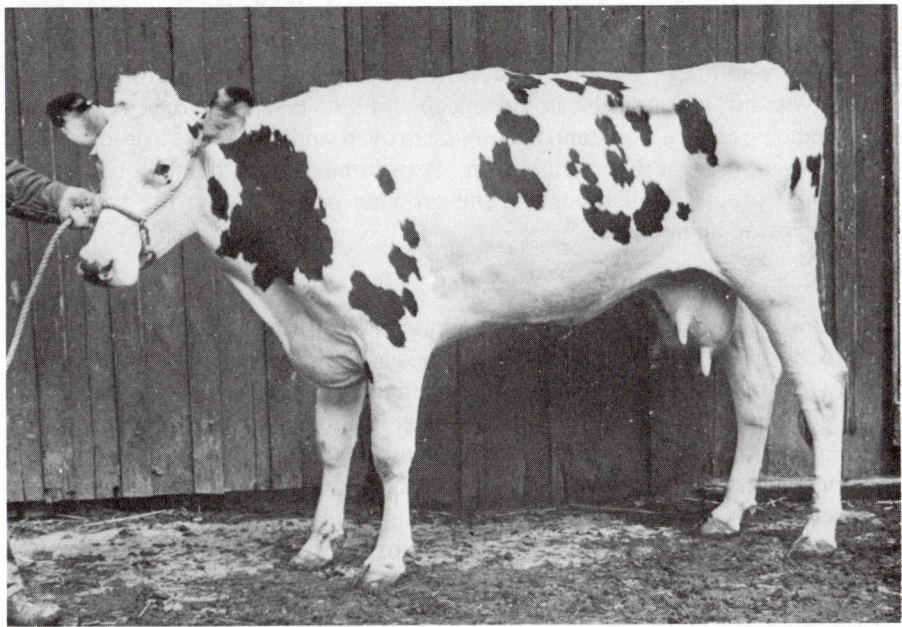

Fig. 10.7 Extreme salt deficiency in lactating cow. (*Courtesy of S. E. Smith, Cornell University.*)

Chronic intakes of high levels of salt have been reported to raise blood pressure in some people and low-sodium, high-potassium diets are often recommended. For persons with hypertension and those prone to have hypertension, the role of sodium in human nutrition has recently been reviewed.[37] Few studies have been conducted on effects of chronic excessive salt intake by livestock.

The salt requirement is greatly increased under conditions which cause heavy sweating because of the large loss in this secretion. Miners have been noted to lose 2 1/2 kg of sweat per hour, containing 2 g of sodium chloride. If large amounts of water are drunk under these conditions, cramps result. The cramps disappear on drinking water containing salt. On a low-salt diet, however, the body gradually makes an adjustment whereby the concentration in the sweat, as well as in the urine, is gradually decreased. Equilibrium between intake and outgo is thus established at a much lower level than is possible initially.

IRON

Although the body contains only about 0.004 percent of iron, this element plays a central role in life processes. As a constituent of the respiratory pigment hemoglobin, iron is essential for the functioning of every organ and tissue of the body. It occurs as an iron-porphyrin nucleus, known as heme, not only in hemoglobin but also in proteins that are components of cytochrome c, peroxidase, catalase, and other enzymes. Thus, iron is a constituent of oxygen carriers and of oxidizing catalysts or enzymes. Over half of the iron present in the body is in the form of hemoglobin. Some is present in another protein, myoglobulin. In addition, a variable store as ferritin and hemosiderin is located in the liver and, secondarily, in the spleen and kidneys. Since the red cells and their hemoglobin are constantly being destroyed and replaced, iron obviously undergoes a very active metabolism. Its synthesis into hemoglobin occurs throughout life, as well as during the growing period, when the total blood supply is being augmented.

10.26 Hemoglobin Formation; Anemia

The blood cells which contain the hemoglobin are formed in the bone marrow, the process being referred to as hematopoiesis. These red corpuscles are continuously undergoing destruction and replacement. Their average life span is 127 days, according to isotope studies with rats. In the course of their destruction, the hematin of the hemoglobin is split into an iron compound, bilirubin, and other pigments which are carried to the liver and secreted in the bile. Iron released by the normal blood-cell destruction can be used again to form hemoglobin, practically without loss. In certain diseases, however, this destruction may be accelerated, and iron formed by toxic destruction cannot be reutilized. If the cells are not renewed as rapidly as they are destroyed, or if the increase in the number of cells which are required to enlarge the blood supply with growth does not occur, *anemia* results. The condition of the blood in this respect is commonly determined by measuring its hemoglobin content. The normal content for most mammals lies within the range 10 to 18 g per 100 ml of blood, depending on species, sex, and age. In severe anemia the value may drop to one-half or even one-third of the normal.

While the reduced hemoglobin content of the blood in anemia results primarily from a reduction in the number of red cells, there are also certain changes in the size and hemoglobin content of the cells. Thus, anemias are referred to as *microcytic, normocytic,* or *macrocytic* in accordance with cell size, and as *hypochromic, normochromic,* or *hyperchromic* in accordance with their color index. These various terms are useful in characterizing the different anemias that result from malnutrition or specific diseases. Deficiencies of protein, iron, copper, and certain vitamins can result in anemias, most of which differ in morphological type. There are also species differences for a given deficiency. Anemia may result from an interference with or cessation of the production of hemoglobin, from a block in cell maturation, from increased destruction, or from blood loss. Various specific nutrients play a role in hemoglobin production and cell maturation, and thus the term *nutritional anemia* has a broader significance than earlier attributed to it. In addition to the anemias which result from dietary or pathological causes, there are hereditary anemias such as "sickle-cell anemia" in humans and other types identified in animals.

10.27 Iron-deficiency Anemia

The anemia resulting from iron deficiency, like that produced on a low-protein intake, is obviously due to a lack of a building stone necessary for hemoglobin formation. In pigs and chickens iron deficiency results in hypochromic, microcytic anemia; in calves the microcytic, normochromic type has been reported. Anemia may occur at any time of life when the available supply of the mineral becomes deficient relative to the needs for hemoglobin formation. It is particularly likely to develop in certain species during the suckling period, since milk is very low in iron. Our knowledge of iron nutrition during this period is due particularly to the early work of Bunge and Abderhalden. A comparison of the mineral constituents of milk and of the newborn showed that, while the other constituents were in similar concentrations, the percentage of iron in the milk ash was only one-sixth of the figure for the iron in the ash of the newborn. It was also found that there was a much larger percentage of iron present at birth than later in life. The explanation was therefore made that nature provides for the iron requirements of the suckling largely by means of a store in its body at birth which may be drawn upon for blood formation and other essential functions during the period when milk is normally the principal or sole food. Bunge found that the guinea pig, which normally commences to eat leafy material within a day after birth, is born without any special store of iron.

This provision of nature for iron nutrition in the suckling does not always prove adequate. The store in the newborn is influenced by the diet of the mother during gestation. If the birth occurs prematurely, there is a smaller store because most of the storage occurs late in gestation. If the number of young born is larger than usual for the species, for example, twins in humans and extra-large litters in hogs, the individual's supply tends to be smaller. Even if the store is normal, a long nursing period without supplementary iron-rich food may exhaust it. The reserve of the human young is usually exhausted before the end of the sixth month. These factors are responsible for many cases of nutritional anemia in babies. In farm animals, the trouble occurs as a practical problem only in the case of suckling pigs. In both species the anemia can be prevented by the feeding of iron and a trace of copper to the sucklings. It does no good to feed the mineral to the lactating mother, for the iron content of milk cannot be increased in this way. A detailed description of this anemia in suckling pigs and its treatment is given later (Sec. 15.39). While iron-deficiency anemia has been produced experimentally in lambs and calves by restricting them to a milk diet, in practice the disease is not met in these species because they early begin to eat supplementary food which supplies the needed iron.

Other effects of iron deficiency may include reduced growth rate, elevated serum triglyceride levels, and depressed folic acid levels.

Baby pigs with low hemoglobin levels have labored and spasmodic breathing (thumps). Iron-deficient mink may develop "cotton fur."[38]

10.28 Iron Absorption and Conservation

Iron is absorbed primarily from the small intestine, but some may be taken up from the stomach. Many studies have shown that, once iron is absorbed, it is tenaciously held by the body and not excreted to any appreciable extent. This means that, after the stores are filled, the adult animal, apart from blood loss or other pathological condition, needs little iron in its ration except for productive purposes. The chief storage site is the liver, but significant amounts are also found in the bone marrow and spleen. Both gestation and egg production call for substantial intakes, as is discussed later (Chap. 16). Given an adequate supply in its food, the animal normally regulates iron absorption in accordance with its needs. Studies using Fe^{59}, for example, have shown that there is an increased absorption during gestation to meet the needs for fetal growth.

The mechanism of iron absorption has recently been reviewed by Linder and Munro.[39] They concluded that ionic iron is first adsorbed to receptors in the brush border of the mucosal cells and then passes into the cell by an energy-dependent process. The iron is then probably transported to the serosal surface of the cells as a small molecular weight compound and transferred across the serosal cell membrane by a nonenergy-dependent system. The remaining iron is returned to the gut when the mucosal cells are shed from the villus. They further concluded that all the compounds in the regulatory process have not been identified but suggested serum ferritin was a possible candidate.

Many factors in addition to the level in the diet influence iron absorption. Ferrous salts are more efficiently absorbed than ferric salts. High dietary intakes of clay, phytate, phosphorus, zinc, cadmium, cobalt, copper, and manganese may reduce iron availability. Organic acids such as ascorbate, citrate, lactate, pyruvate, and succinate may enhance iron absorption perhaps by reducing and chelating the mineral. The type of dietary carbohydrate may also influence absorption.

Although the body is very effective in the conservation of iron, some is lost in urine, feces, and sweat. For example, endogenous iron losses in humans may amount to 0.2 to 0.5 mg per day for each route.

10.29 Iron Requirement and Content in Feeds

The N.R.C. requirements[40] for iron are for chickens 40 mg per kilogram diet, and for pigs 80 mg per kilogram. For ruminants the requirements are of the order of 25 to 40 mg per kilogram. Aside from milk, most of the feeds for animals contain very liberal amounts of iron relative to the need by the body. Leafy materials are rich sources, and so are many seeds. Since most of the iron in cereal seeds is in the outer coatings and germ, milling results in increasing the supply to animals but in decreasing it in so far as humans are concerned. Insofar as farm animals, including chickens, are concerned, their usual rations are so rich in iron that they can normally be counted upon to supply enough even though its availability is low. On the other hand, benefits have been reported from adding iron to swine rations based on corn and soybean oil meal, without access to soil.

The largest metabolic demand for iron comes during high egg production, because eggs contain 1.1 mg each. Probably the usual rations contain more than enough, but anemia has been reported in laying hens. The roughages which make up two-thirds of the ration of cattle and sheep, species which have lower requirements than pigs, commonly contain from 100 to 200 mg per kilogram of iron. This should certainly ensure enough except in iron-low soils.

There have been reports, notably from Florida, of areas where the soil is so low in iron as to result in a deficiency in the forage and, thus, anemia in grazing animals. Later studies suggest, however, that other mineral deficiencies, such as cobalt or copper, may be primarily involved. From soil surveys one would not expect iron deficiency on any herbage producing a reasonably good yield. In general, there seems to be no need to supply additional iron to farm animals except for suckling pigs on concrete. In fact, unneeded additions may be deleterious. Too much iron in the diet interferes with phosphorus absorption by forming an insoluble phosphate, and rickets may thus result on a diet otherwise adequate.

Iron toxicosis is not a common problem in farm animals, but it has been reported in children as a result of accidental excessive intake of iron pills, in Bantus of South Africa due to cooking in iron pots and drinking Kaffir beer which contains a high level of iron, and in certain genetic conditions.

COPPER

As a result of a series of studies beginning in 1925, Hart and associates at the University of Wisconsin discovered that a small amount of copper is necessary, along with iron, for hemoglobin formation. It is not a constituent of hemoglobin, but it does occur as hemocuperin in blood cells. In certain invertebrates copper is present in hemocyanin that functions as an oxygen carrier.

10.30 Specific Functions of Copper

The role of copper in iron metabolism is incompletely understood. When copper is deficient in the diet, there is a decreased absorption of iron, a lowering of its total content in the body, a decrease in its mobilization from the tissues, and the development of a severe microcytic hypochromic anemia. The failure of injected iron to correct the anemia indicates that copper is essential for the utilization of iron in hemoglobin synthesis. It has been found to play a role both in hemoglobin synthesis and in red-cell maturation. It seems clear that copper exerts an influence on iron metabolism at the cellular level. It has been suggested that iron is absorbed as Fe^{++} and transported as Fe^{3+} and that the conversion requires a copper-dependent enzyme such as ferroxidase II.

Copper has many basic functions besides its role in iron metabolism. It is an integral part of many metalloenzymes such as cytochrome C oxidase, uricase, tyrosinase, lysyl oxidase, benzylamine oxidase, and diamine oxidase. Copper is also important for normal bone formation as it is essential for osteoblastic activity and for normal collagen and elastin formation.

Approximately half of the total body supply of copper is found in the muscle mass. Stores are also present in the bone marrow, liver, and to a lesser extent elsewhere. The total supply is only a very small fraction of the amount of iron found in the body. The body supply of copper is greatly lowered when the diet is deficient. Like iron, a large store of copper at birth serves the special purpose of providing for growth needs during the suckling period, for milk is very low in both elements.

10.31 Copper Deficiencies as Area Problems

More than a decade before the role of copper in hemoglobin formation was discovered, investigators in Northern Europe were studying possible mineral deficiencies in the forage in areas where a wasting disease of cattle and sheep, called *lechsucht,* characterized by diarrhea, loss of appetite, and anemia, was common. In 1933 Sjollema established its cause as a copper deficiency by finding marked differences in the copper content of the forage in "healthy" and "sick" areas and by curing the trouble with copper therapy. Later Sjollema showed that in affected cows and goats the copper content of the blood falls to one-third of the normal concentration of 100 μg percent. The copper content of the liver, spleen, and hair was also decreased. It has also been established that a disease of lambs and calves, referred to as *enzootic ataxia,* or "sway-back," and by other names and characterized by ataxia and nervous symptoms, is caused by a deficiency of copper. The disease has been reported from several countries, and its prevention has been accomplished by feeding copper to the pregnant female. A chronic copper deficiency resulting from a low content in the herbage has also been reported responsible for a trouble in grazing cattle in

Australia, known as "falling disease." This disease is characterized by staggering, falling, and instantaneous death. Troubles in cattle and sheep ascribed to copper deficiency have also been reported from South Africa, New Zealand, Scotland, and elsewhere. In the United States a wasting disease, called "salt sick," was reported from Florida by Becker in 1931. It seems clear that a copper deficiency was involved, at least in some cases. In parts of the Southeastern United States use of copper supplements is necessary for successful livestock production. Some forms of copper, such as in lush pasture, may not be well utilized by animals. Excess molybdenum may cause a conditioned copper deficiency (Sec. 10.50). Excess copper can cause toxic effects.

10.32 Symptoms of Copper Deficiency

These various reports of troubles ascribed to copper deficiency record a wide variety of symptoms differing somewhat from area to area and from species to species. The lowering of the copper content of the blood and liver is a rather constant finding, but anemia and scouring are not.

Depigmentation of colored hair and black wool is a common finding in copper deficiency. The probable cause is a defect in melanin synthesis because of a reduced tyrosinase activity.

Copper deficiency in sheep results in a marked decrease in the rate of wool growth and a change in its character. The depigmentation of black wool and the development of "stringy" or "steely" wool, characterized by limp, glossy fibers lacking the normal crimp, are reported to be early symptoms. There is good evidence that these hair and wool changes reflect a failure of certain enzyme activities for which copper is essential.

The nervous symptoms in lambs and calves referred to as *ataxia* have been widely reported and have been found to be associated with a low copper content of the dam during pregnancy. Postmortem studies have revealed nerve lesions in these lambs. The reported occurrences in cattle of abnormal gait, staggering, and falling have been referred to as nervous symptoms, but these physical symptoms vary in their appearance and severity from area to area, and it remains uncertain whether they all have the same basic cause.

Bone troubles have been described in lambs, cattle, pigs, chickens, and dogs, including symptoms such as lameness and swelling of the joints. The bone cortices may be thin and spontaneous fractures may occur. The bone lesions of copper-deficient animals may also vary among species and among ages within species. A copper deficiency can affect bone formation in at least two ways: (1) decreased osteoblastic activity which means a lack of bone matrix formation, or (2) a reduction in lysyl oxidase activity which leads to diminished stability and strength of bone collagen because cross linkage is impaired. The impairment of cross linkage also affects elastin formation. Aneurysms and ruptures of the major vessels may result when the elastin content of the blood vessels is reduced.[41]

Copper deficiency may also result in impaired reproduction.

It has now become evident that other minerals are also involved in many

Fig. 10.8 A: rabbit reared on diet adequate in copper. B: rabbit showing graying of hair resulting from copper deficiency. (*Courtesy of S. E. Smith, Cornell University.*)

of the area problems reported. In some cases a combined copper and cobalt deficiency is concerned (Sec. 10.35). In others an excess of molybdenum, as is discussed later (Sec. 10.51), is an important factor in the occurrence of the symptoms observed. Sulfate may also be indirectly involved. These relationships help explain why a copper content in the forage which prevents troubles in one area proves inadequate in another. It has also been suggested that in some forages copper occurs in complexes of reduced availability.

Copper antagonists such as molybdenum and zinc may also be obtained by ingestion of soils rich in these minerals.[42]

Studies seeking to clarify this complicated picture of copper deficiency and related troubles and thus to identify more clearly the specific effects of a deficiency of a given mineral are now in progress in several laboratories.

10.33 Requirements and Feed Supply

On the basis of studies with men, dogs, and rats, the copper requirement for the prevention of anemia is considered to be about one-tenth that of iron. Such an intake provides the amount of copper that is needed along with iron to prevent anemia in suckling pigs.

The copper requirement for cattle is probably higher than of sheep[43] but the level needed in the diet depends on the dietary intake of several minerals as discussed above. In some areas, herbage with 5 to 8 ppm produced healthy animals, in other areas, considerably higher levels of dietary copper are needed.

Although the copper requirement of swine is less than 10 mg per kilogram, a growth response is sometimes obtained when 125 ppm or more of copper is added to the diet. The response to copper varies with the content of the other minerals and type and level of protein. In some cases, 125 ppm may even be toxic. Furthermore, using the manure from pigs fed high levels of copper as fertilizer may increase the copper content of pasture to levels that are toxic to sheep.[44]

TABLE 10.3. Requirements and Toxicities of Trace Mineral Elements

Element	Species	Requirement, mg/kg ration	Toxic level, mg/kg ration
Copper	Cattle	5–8	115
	Sheep	5	?
	Swine	6	250
	Chickens	4	?
Cobalt	Cattle	0.05–0.07	60
	Sheep	0.08	120
	Horses	?	?
Iodine	Livestock	0.1	?
	Chickens	0.35	?
	Swine	0.2	?
Manganese	Cattle	16–40	2000
	Swine	20	4000
	Chickens	55	?
Zinc	Cattle	20–30	900->1200
	Sheep	20–30	1000
	Swine	50	2000
	Chickens	35	?
Selenium	Cattle	0.1	3–4
	Sheep	0.1	10
	Horses	0.1	5–40
	Swine	0.1	5–8

Sheep are more susceptible to copper toxicity than any other farm animal. Thus, at present, it is not recommended to routinely use high levels of copper. Table 10.3 data are given for the approximate requirements and toxicity levels of copper and other trace elements for different species. These data are based primarily on growth studies. For most of the elements listed, requirement figures are subject to variation according to the levels of other mineral elements in the ration, as is discussed later for specific ones.

COBALT

The knowledge of cobalt's essential role developed as a result of long-time studies of certain peculiar wasting diseases of grazing animals, known by dif-

Areas of critical cobalt deficiency. Both grasses and legumes, especially those growing on sandy soils, are very low in cobalt.

Areas of marginal cobalt supply. In these areas, grasses are often low in cobalt, while legumes generally contain adequate amounts.

Fig. 10.9 Distribution of cobalt deficiency in Eastern United States. There are no large cobalt-deficient areas in the states not shown. (*Source: U.S.D.A. Agr. Inf. Bull. 299, 1965.*)

ferent names in different areas throughout the world but having similar symptoms that suggested a common cause. Each disease was early recognized to be limited to certain areas of the country in question, and prevention and cure by transferring the animals from "sick" to "healthy" areas were successfully practiced long before anything definite was known about the cause. The discovery of a lack of cobalt, resulting from its deficiency in the soil and thus in the herbage grazed, was reported independently in 1935 by Filmer and Underwood and by Marston and Lines, working in Australia.

Following the reports from Australia, cobalt-deficient areas were discovered elsewhere. They have been noted in the Eastern and Midwestern states of the United States in areas as outlined in Fig. 10.9 and also in western Canada. In the United States they have been noted in the Eastern Seaboard states, the Northeast, and certain of the Midwestern states. Other areas are suspected. According to the reports from Canada and New Hampshire, cobalt deficiency can occur on dry, winter rations as well as on pasture.

10.34 Symptoms of Cobalt Deficiency

The symptoms in cattle and sheep are similar to those of general malnutrition. The animals become listless, lose appetite and weight, become weak and anemic, and finally die. Uncomplicated cobalt deficiency has been experimentally produced in sheep (Fig. 10.10) and the lesions studied. The anemia is of the normocytic, normochromic type, thus differing from that occurring in iron or copper deficiency. Careful studies of organs and tissues at autopsy have provided little basis for the accurate diagnosis of cobalt deficiency. General inanition, a fatty degeneration of the liver, and deposits of hemosiderin (a breakdown product of hemoglobin) in the spleen are commonly found changes. Wool growth is retarded, and the fibers are weak. As is explained later, a lowering of the vitamin B_{12} content of the blood appears to be a specific symptom. The only certain diagnosis of cobalt deficiency rests upon the response of the animal to cobalt feeding. This response is rapid. Appetite picks up in about a week, and weight gains follow, but the remission of the anemia occurs more slowly, indicating that it may be a secondary effect.

10.35 Physiological Role of Cobalt

The mystery as to the role of cobalt was solved as a result of the discovery of vitamin B_{12}, an antianemic and growth factor, which was found to contain cobalt in its molecule. This vitamin is a metabolic essential for all species studied, but it is not a dietary essential for cattle and sheep because it is synthesized adequately by rumen organisms, provided an appropriate supply of cobalt is present. Thus the long observed cobalt-deficiency trouble in ruminants is actually a vitamin B_{12} deficiency, for this vitamin is not present in the plant products which comprise their feed. This conclusion has been reached as a result of several experiments. It has been shown that cobalt-deficient sheep have a very low vitamin B_{12} content in the blood compared with normal animals, that injecting the vitamin alleviates the cobalt-deficiency symptoms

Fig. 10.10 Sheep No. 8 reared on a diet adequate in cobalt. Sheep No. 12 reared on cobalt-deficient diet. (*Courtesy of S. E. Smith, Cornell University.*)

whereas injecting the mineral itself does not, and that there is little B_{12} in the rumen contents of animals showing cobalt deficiency compared with the situation in normal animals.

Nonruminant herbivores such as rabbits and horses may thrive on pasture containing levels of cobalt that would be deficient for ruminants. But studies with dogs, pigs, and rats suggest that cobalt may have an essential function in nonruminants beyond that of a component of vitamin B_{12}.[45]

10.36 Cobalt Requirements and Content in Feeds

While cobalt deficiency in ruminants is essentially a vitamin B_{12} deficiency, the most practical way of preventing or curing the trouble in situations where it otherwise may occur is to add cobalt to the ration. The evidence of the quantitative needs for cobalt rests on studies of its content in the herbage where troubles occur. It has been reported that the dry matter of grass of healthy areas contains around 0.1 ppm of cobalt on the average compared with figures from 0.004 to 0.07 ppm for ''sick'' areas. As little as 0.1 ppm has restored sick animals to health. The approximate requirements for cattle and sheep have been presented in Table 10.3.

Aside from the pasture grass and roughages produced in the areas where cobalt deficiency has been definitely identified, feeds in general appear to contain more than the level of cobalt that is considered essential. This is not true, however, in the case of milk and certain samples of corn and possibly other grains. Analytical data obtained by reliable methods are far too few to provide any adequate picture of the cobalt content of our feed supplies. Certainly, in those areas where cobalt deficiency has been identified, the simplest way of ensuring its prevention in the future is to feed a cobalt supplement to sheep and cattle. For this purpose 150 g of cobalt sulfate per 100 kg of salt fed free-choice should suffice. For milking cows cobalt sulfate can be added to the grain mixture at the rate of 2 g (1/15 oz.) per metric ton. Cobalt "bullets" which can be

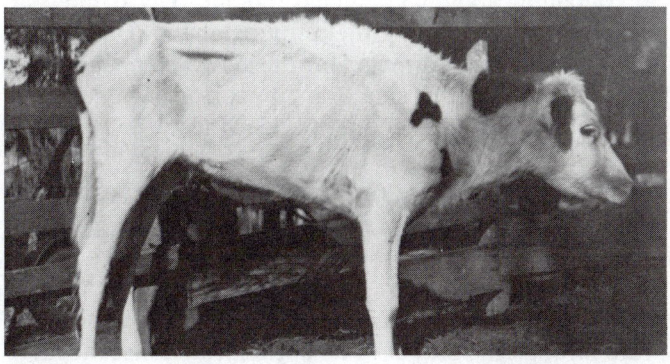

Fig. 10.11 Above picture shows a cobalt-deficient heifer that had access to an iron-copper-salt supplement. Note the severe emaciation. Below is the same heifer fully recovered with an iron-copper-cobalt salt supplement while on the same pasture. (*Florida Agr. Expt. Station Bull. 699, 1965. R. B. Becker, J. R. Henderson and R. B. Leighty, University of Florida, Gainesville.*)

placed in the rumen and slowly release cobalt over a long period of time have also been used.

Large intakes of cobalt are toxic but there is a very wide margin between the harmful and the essential levels.[46]

IODINE

The mature animal body is estimated to contain less than 0.00004 percent of iodine, but if this minute amount is not maintained through the food, disaster results. More than half of this iodine is in the thyroid gland, and it is in connection with the functioning of this gland that the body's need for iodine occurs.

10.37   The Thyroid Gland

This gland consists of two parts lying on each side of the trachea at its upper end. In the case of an adult man it weighs about 30 g. It produces an internal secretion which contains the hormone thyroxine, isolated by Kendall in 1914 as a crystalline product containing about 65 percent of iodine. Thyroxine is an iodine-containing amino acid present as the protein thyroglobulin. Diiodotyrosine is also present in this protein. The formulas for these two compounds are shown in Sec. 8.2. If the diet contains tyrosine or its precursor phenylalanine, these compounds can be synthesized in the body, provided the needed iodine is present.

The removal of the thyroid early in life results in all species in a stunting of physical, mental, and sexual development. After removal of the gland, in adult animals the hair and skin show premature aging, and mental and physical sluggishness may develop. In all cases there is a lowered basal metabolism (Sec. 14.1). It is probable that primary function of the thyroid gland is to control the metabolic rate through the output of its hormone and that the more evident effects of thyroid deficiency are a result of a failure of this control. Most of the iodine that occurs in the tissues and fluids of the body other than the thyroid is probably in thyroxine, serving its function in the control of metabolism.

The administration of thyroxine and of iodine-containing proteins stimulates body processes, notably milk and egg production. On the other hand, certain specific compounds such as thiourea and thiouracil suppress the gland's action and metabolic processes, thus promoting increased fattening. The application of these physiological effects to feeding practice is discussed later (Secs. 15.7 and 17.3).

10.38   Symptoms of Deficiency

Goiter is an enlargement of the thyroid gland.[47] Medical men recognize two types: simple or endemic goiter, caused primarily by a lack of iodine, and exophthalmic goiter, which is due to a hyperactive gland. Simple goiter is the most common type, and it is the one with which we are concerned in nutrition. Hyperplastic changes begin as a result of a failure of the thyroid tissue to supply enough secretion, owing either to a reduced supply of iodine for its manufacture or to an increased demand for the secretion by the body. The demand for thyroxine varies in accordance with the activity of the body functions it controls, and thus, given a fairly constant supply of iodine in the diet, simple goiter is most likely to develop during periods of greatly increased need. In the human these critical periods are pregnancy and puberty. In farm animals, however, goiter usually shows itself in the young at birth as a result of a deficiency of iodine in the rations of the mother during gestation. The young thus affected are born weak or dead. On a deficient diet the pregnant mother is not able to supply the fetus with enough iodine. The danger is thus increased in the case of multiple births.

In calves, lambs, and kids the enlargement of the gland is very evident in the newborn. In pigs the most outstanding symptom of the deficiency is hairlessness. They are bloated and have thick skins and puffy necks. In foals the only symptom may be extreme weakness at birth, resulting in an inability to stand and suck. A limited amount of data indicates that ''navel ill'' in foals may be lessened by feeding iodine to brood mares, but further evidence is required. Birds and fish as well as mammals have enlarged thyroids as a result of iodine deficiency. Animals born alive with a well-developed goitrous condition usually fail to survive or remain weaklings. No treatment has been found particularly effective. Studies of goiter troubles in humans have clearly established that, while iodine is effective as a preventive it may be harmful rather than beneficial as a treatment after the goiter has developed. While the nutrition scientist is concerned with prevention, treatment belongs entirely to the field of medicine.

Though a lack of iodine is the primary cause of simple goiter, it is recognized that other factors may contribute, notably the high-calcium content of the water in many goitrous regions. In addition, there are specific goitrogenic substances in certain foods, notably various members of the Brassica family,

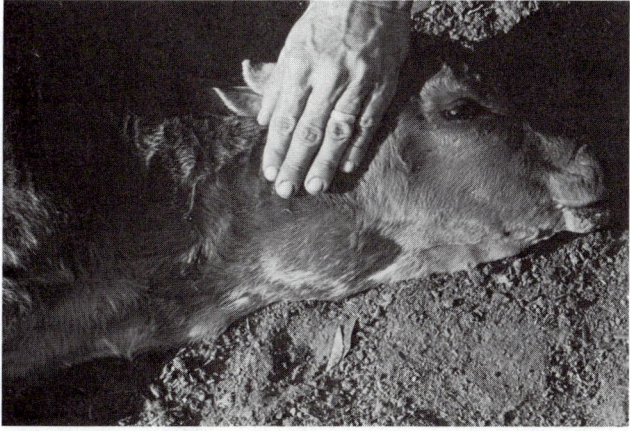

Fig. 10.12 Goiter caused by iodine deficiency in calves in Mato Grosso, Brazil. (*Courtesy of Jürgen Döbereiner and Carlos H. Tokarnia, EMBRAPA, Rio de Janeiro, Brazil.*)

peanuts, and soybeans. These foods contain specific substances which slow down the thyroxine-secreting activity of the gland. Srinivasan and coworkers[48] have reviewed earlier studies and reported experiments of their own showing that peanuts are goitrogenic for the rat. They later found the active principle to be the glucoside, arachidoside. Mild goitrogenic efforts have been produced experimentally in both rats and chickens by feeding soybeans and counteracted by small amounts of additional iodine. The antithyroid principle is partially removed or destroyed in processing the beans to produce the low-fat meal.

10.39   Iodine Deficiency an Area Problem

The need for additional iodine in the rations of farm animals, as well as of humans, exists primarily in certain areas where the soil and thus the water and food crops are low in this element. There are various regions throughout the world where goiter troubles of varying degrees are very common in all species unless additional iodine is fed, and there are others where the trouble is entirely unknown. In the United States, the goiter areas are primarily in the Northwest and in the Great Lakes region. It is estimated that, before iodine feeding was practiced in Montana, goiter caused an annual loss of many thousand pigs. Records from other areas show that serious losses in the sheep and cattle industries occurred which were largely prevented following the discovery of the lack of iodine as the causative factor. There are borderline regions in which goiter occurs only occasionally. When the usual iodine intake is little above the minimum requirement, an enlarged physiological demand by an individual may be responsible for the occasional troubles. A barely sufficient intake may be changed to an inadequate one by a change in the makeup of the ration.

Hundreds of years before iodine was discovered, people living in goitrous areas learned the usefulness of certain products, now known to be rich in iodine, as a preventive of goiter. The value of sea salt in comparison with certain inland deposits was early recognized. Our real understanding of the problem is comparatively recent, as indicated by the fact that the discovery by Baumann of iodine in the thyroid gland was not made until 1896.

10.40   Iodine Requirements

We do not have specific knowledge of the minimum iodine requirements of the various species, but we do know the levels that are at least high enough to prevent goiter. Approximate requirements for livestock and chickens have been set forth in Table 10.3. Since goiter troubles in cattle, sheep, and hogs are associated primarily with reproduction, their specific needs for iodine during this function are discussed in Chap. 16. Although it appears that these needs must be greatest during reproduction, iodine is obviously required throughout life to keep the thyroid gland functioning normally. The appearance of goiter may result as an advanced stage of deficiency. On this basis the continued feeding of an additional supply might be considered desirable in areas of severe iodine deficiency. There is some evidence from certain areas that additional iodine improves growth in various species, but this evidence is not entirely convincing. It has also been reported that, with sows fed in dry lot in an area borderline with respect to the occurrence of goiter, additional iodine lessens the number of pigs dead at birth or which die in the first week, although there was little physical evidence of iodine deficiency. Recommendations for year-round feeding are usually put on an insurance basis or are suggested as the simplest way of making certain that the needs of breeding stock are taken care of in areas where goiter is very common. The amounts fed should be limited to recommended levels, for excesses can cause harm. For example, the feeding of large doses of iodine to the pregnant mare may result in goiter in the foal. Excessive iodine supplementation has also been shown to produce goiter in rats, humans, and birds.

Iodine is primarily absorbed as iodide. Iodide is efficiently absorbed and very little appears in the feces. Urine is the primary pathway of iodine excretion.[49]

10.41 Iodine Supplements

Where the occurrence of goiter shows that attention to iodine nutrition is needed, the most practical method is the use of some special source such as iodized salt or sodium or potassium iodide. The addition of 0.0076 precent iodine to the salt represents the customary and effective level.

None of the common feedstuffs, with the exception of fish meal made from salt-water fish, can be relied upon to be rich in iodine. Dried kelp, a sea plant, is rich in the element, and so is cod-liver oil. While special advantages are sometimes claimed for organic sources of iodine, the preponderance of the evidence indicates that the cheaper inorganic iodides are equally satisfactory. When massive doses of iodine are given as a therapeutic agent, an organic source has the advantage of slower absorption and less risk of harm from overdosage, but in the amounts needed to prevent goiter, this is not a factor.

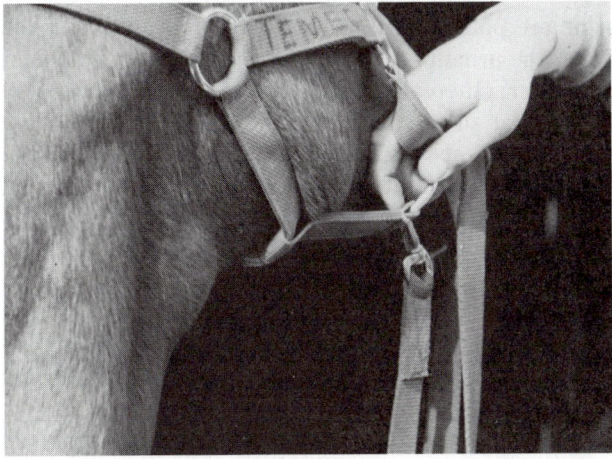

Fig. 10.13 Goiter in a foal caused by feeding excess iodine to its dam during pregnancy. (*Cornell University.*)

Iodized salt loses its iodine rather readily under certain conditions because of the catalytic action of impurities present. Thus, after storage it may be an unreliable source of iodine unless stabilized. Some processes have been developed for accomplishing this stabilization. Iodized salt in mixed feeds is less subject to deterioration because the presence of proteins and unsaturated fats tend to stabilize the iodine. Both human and animal experiments have tested the usefulness of salt iodized with potassium iodate as a product of higher stability than where iodide is used. Salt blocks containing KIO_3 retained a much higher percentage of iodine when fed in the manger or subjected to outside exposure than did a product containing a commercial iodide complex.

Any required addition of iodine need not be supplied every day, because the thyroid has a considerable capacity to store the element. This is illustrated by the fact that goiter has been prevented in children in goitrous areas by feeding sodium iodide periodically for a month and repeating twice yearly.

SULFUR

The body contains approximately 0.15 percent of sulfur. This element occurs almost entirely in organic compounds, notably in proteins in which it is present as the sulfur-containing amino acids cystine and methionine. Wool contains approximately 4 percent of sulfur present as these acids. The body also utilizes these acids in the manufacture of the two sulfur-containing regulators of its metabolism, glutathione and insulin. It is clear that the sulfur needs of the body are primarily a matter of amino acid nutrition. Two vitamins, thiamin and biotin, contain sulfur, but these vitamins are not made inside the body proper of any species. Sulfur is present in inorganic form in chondroitin sulfate, a constituent of cartilage. The blood contains small amounts of sulfate. Thiocyanate ions are also present in blood, as well as in the saliva and other secretions.

Both the feces and urine are paths of sulfur excretion. In the urine three forms occur: inorganic sulfate, the principal fraction, which represents the final stage of oxidation of organic sulfur; ethereal sulfur, which is present in complex detoxication products; and neutral sulfur, which occurs as cystine, taurine, thiosulfates, and other compounds. Since excreted sulfur arises primarily from protein catabolism, there is a rather constant ratio between it and the nitrogen in the urine. There is evidence that the excretion of neutral sulfur is proportional to the basal metabolism.

It is evident from the above discussion that the sulfur needs of the body call for it in the form of organic complexes, notably amino acids, rather than in inorganic form. Mention has been made, however, of the fact that rumen bacteria can utilize the latter to build the sulfur-containing amino acids. Benefits have been demonstrated in ruminants from adding either elemental sulfur or sulfate to purified rations low in sulfur and having a part of the protein replaced by urea.[50,51]

Studies using S^{35} have shown that the sulfate form can be used by the body to synthesize chondroitin sulfate. It has been reported that S^{35} injected into laying hens was found in the albumin of the egg in organic form, in part as cystine. A cystine-sparing action for growing chicks has also been reported. These effects in poultry appear to be small quantitatively. The various findings mentioned in this paragraph indicate that inorganic sulfur does play a role in nutrition, but, except when there is a substantial need for the rumen synthesis of protein, the effect is a small one compared with the role of that supplied as amino acids. A special case, however, in which the sulfate ion is of importance is its influence on molybdenum toxicity (Sec. 10.51). Sulfur deficiency in ruminant diets may result in reduced feed intake and reduced cellulose digestion. It is frequently recommended that the nitrogen-sulfur ratio of ruminant diets should be approximately 10:1.

MANGANESE

Manganese occurs in the body principally in the liver, but it is also present in appreciable amounts in various other organs and in the skin, muscle, and bones. Despite the very small total supply in the body, this element has several essential functions and its consideration is of practical importance in animal nutrition.

10.42   Functions in Growth and Reproduction

Manganese deficiency may result in a wide variety of signs, including reduced rate of growth, delayed sexual maturity, degeneration of germinal epithelium, irregular ovulation, resorption of fetus or birth of small weak offspring, impaired glucose tolerance, impaired egg shell formation, defective blood clotting, ataxia, and skeletal deformities.

Mice from dams fed a manganese-deficient diet may have a specific congenital ataxia because of defective development of the otoliths.[52] Such animals do not have the ability to orient themselves when submerged in water. A similar condition exists in mice having the *pallid* mutant gene.[8] Apparently, the *pallid* mice have impaired manganese metabolism because the feeding of large amounts of manganese to the pregnant mouse will prevent the ataxia in the offspring.

Leach[53] suggested that perhaps most of the defects, including skeletal abnormalities observed in manganese-deficient animals, might be accounted for by the need for manganese in glucosyltransferase activity. This enzyme is needed for the formation of mucopolysaccharides and glycoproteins. A deficiency of these products leads to decreased chondrogenesis (cartilage formation).

10.43   Manganese and Bone Development

In 1936 Wilgus and associates reported evidence that manganese is markedly effective in reducing the incidence of *perosis,* or "slipped tendon," a malformation of the leg bones of growing chicks, thus providing the first specific information regarding the prevention of a long-known and serious trouble in practice. In this trouble the hock joints become swollen and the Achilles tendon slips from its condile. The initial findings were confirmed by others, and further studies showed that a shortening of the leg bones was involved. Definite changes in the physical and chemical structure of the bones were noted. Manganese did not always prove 100 percent effective, however, and support was thus furnished for an earlier view that some organic factor was also involved. In 1940 clear evidence was produced that choline, in addition to manganese, is necessary to prevent perosis in birds. Just how choline acts is not clear, but its action has no apparent connection with its role in fat metabolism (Sec. 7.14). It appears from present studies that other B-vitamins may also be concerned. The high-calcium and -phosphorus rations normally fed to poultry are a contributing cause of the occurrence of perosis, probably by interfering with manganese absorption, for they also aggravate the adverse effect of a low-manganese intake on growth. The gross and histological lesions of perosis are described in detail by Wolbach and Hegsted.[54]

Bone malformations, grossly most evident as crooked front legs, have also been produced in rabbits by feeding a diet low in manganese, as are illustrated in Fig. 10.14. The humeri are shorter than normal and lower in ash, density, and breaking strength. There are also marked histological changes. Crooked legs and enlarged hocks have been produced in pigs on manganese-deficient rations. In the studies where the crooked legs were noted, the ration was high in calcium and phosphorus. In rats fed a manganese-low diet, bone changes appear in the second generation. Calves from cows fed diets containing less than 16 ppm manganese may have skeletal abnormalities.[55]

10.44   Requirements and Feed Supplies

The manganese requirements for most species of livestock have been estimated to be 20 to 40 mg per kilogram of feed. The chick is estimated to need 55 mg per

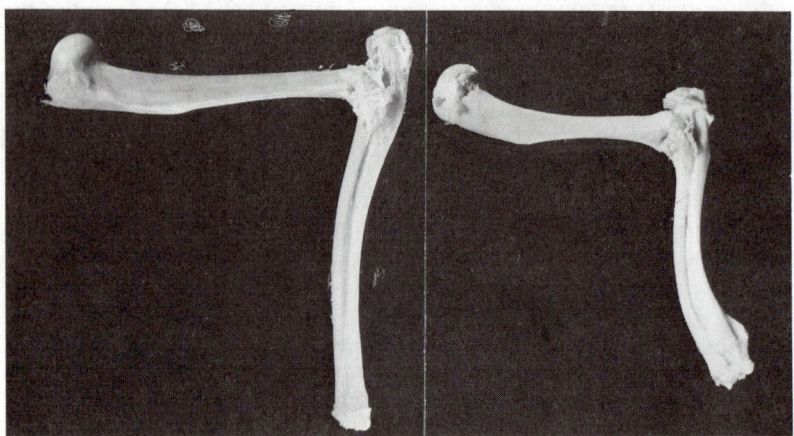

Fig. 10.14   Front leg bones from a control (left) and a manganese-deficient rabbit. (*Courtesy of S. E. Smith, Cornell University.*)

kilogram. It has become common practice to add a manganese supplement to poultry rations, primarily as an insurance measure.

Most roughages contain 40 to 140 mg per kilogram, most grains contain 15 to 45 mg per kilogram, but corn may contain 5 mg per kilogram.

There is no problem of toxicity from moderate excesses of manganese, but 125 ppm have been found to depress hemoglobin synthesis in baby pigs. This apparently represented a manganese-iron antagonism since additional iron overcame the depression. Growth was depressed by 1250 ppm.

ZINC

The zinc content of the body is only approximately 3 mg percent. The highest concentrations are found in the epidermal tissues, such as skin, hair, and wool, but traces also occur in the bones, muscles, blood, and various organs. The mineral is present in milk and, in a much higher concentration, in the colostrum. The initial studies of Bertrand and Berzon, in 1922, indicating that this element played an essential role, have been confirmed and extended by others in experiments with both rats and mice. On a nearly zinc-free diet, growth was retarded and hair development was interfered with. These early findings, plus the identification of zinc in certain enzyme systems, led to the conclusion that the element is an essential one for all species, but several years elapsed before specific information on farm animals was obtained.

10.45 Zinc Deficiency in Farm Animals

After a detailed description by Kerncamp and Ferrin in 1953 of a dermatitis in swine, designated as parakeratosis, Tucker and Salmon[56] in 1955 reported that a zinc deficiency was concerned. This report was rapidly confirmed by other workers. The disease is characterized by specific skin lesions, retarded growth, lowered feed utilization, and other symptoms. The trouble is aggravated by excess calcium intakes.

In 1958 at least three different laboratories described the symptoms of zinc deficiency in the chick. The prominent symptoms reported were slow growth, shortened and thickened long bones, and poor feathering, with keratosis resulting when the deficiency was severe. It has also been noted that the deficiency results in lower hatchability and embryonic anomalies. Excess calcium is apparently an aggravating factor in this species as well.

Zinc deficiency has been produced experimentally in dairy calves on purified diets containing 3.6 ppm of zinc.[57] The deficiency symptoms included inflamed skin around the nose and mouth; stiffness of the joints; alopecia; breaks in the skin around the hoofs; rough scaly skin on the rear legs, ears, and neck involving parakeratosis; and retarded growth. Blood and tissue values for zinc and blood carbonic anhydrase were depressed.

Similar signs of zinc deficiency have also been produced in lambs, goats, and horses.

Until recent years, it was thought that field cases of zinc deficiency in ruminants were unlikely, but recent studies suggest the borderline deficiencies

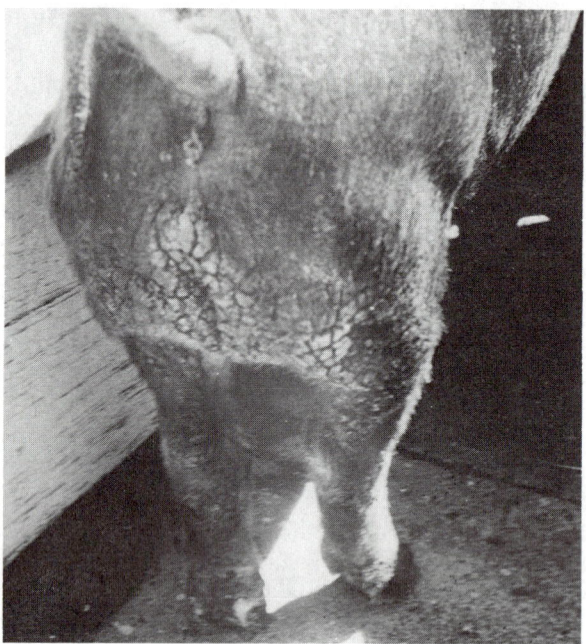

Fig. 10.15   Pig with parakeratosis because of zinc deficiency. (*Cornell University.*)

may be much more common than realized.[58] Zinc deficiency has also been reported in humans. Dwarfism and absence of sexual maturation have been reported in severe deficiencies. Poor growth, poor appetite, and impaired taste acuity in children have been reported to be caused by a less severe deficiency of zinc.

The specific biochemical lesions responsible for the symptoms of zinc deficiency remain unknown. There is a decrease in the trace amounts of the mineral normally found in the body. Zinc functions in several enzyme systems, notably the respiratory enzyme carbonic anhydrase, which is found in the red blood cells and elsewhere in the body, which plays an essential role in eliminating carbon dioxide, and which contains 0.3 percent of zinc. The element can serve as an activator of alkaline phosphatase. There is some evidence that it plays a role in keratinization and calcification. It is evident that zinc functions at the enzyme level, and this appears to be true also for the antagonizing effect of calcium.

10.46   Zinc Requiremens and Feed Supplies

Parakeratosis, or dermatosis, sometimes occurs on practical swine rations and supplemental zinc will prevent or cure the syndrome. The zinc requirement of swine varies greatly depending upon level of calcium, source and level of protein, source of corn, and other factors. Phytic acid may depress zinc absorption from the intestinal tract. On a diet containing isolated soybean protein and recommended calcium levels, the zinc requirement is approximately 50 mg per kilogram of diet. On corn diets and high calcium diets, 100 mg per kilogram will not always prevent growth depression and lowered feed efficiency associated with parakeratosis. A level of 2000 mg per kilogram of diet produced toxicity in pigs, but 1000 mg per kilogram (1000 ppm) of zinc was not harmful.

SELENIUM

In 1935 Franke and Potter, of the South Dakota Experiment Station, identified selenium as the toxic factor in forage which causes a long-recognized peculiar disease in livestock locally known as "alkali disease," or blind staggers. Beginning in 1950, evidence accumulated that selenium in traces is an essential nutrient despite its toxicity in larger intakes; that animals grazing on certain soils suffered from retarded growth and reproduction troubles which could be overcome by the feeding of traces of the element; and that a specific disease, referred to as white muscle disease, could be prevented.

10.47 Selenium as an Essential Nutrient

Studies beginning about 1950 established that brewer's yeast contained an unidentified factor which would prevent dietary liver necrosis in rats and exudative diathesis, a hemorrhagic disease in chicks, and that the factor was different from vitamin E, which would also prevent these troubles. Continuing research established the fact that the unidentified factor, designated as *factor 3* by Schwarz who was the pioneer in the rat studies, is a selenium compound. Only minute traces of the element are required, the amount in the case of exudative diathesis being approximately 0.1 ppm. Thus selenium became recognized as a useful dietary ingredient, as well as a harmful one.

In 1958 both Oregon and Cornell workers reported the complete prevention of white muscle disease in newborn lambs by the additon of trace amounts of selenium to the diet of the ewes. New Zealand investigators reported that the periodic feeding of selenium in certain grazing areas lessened the mortality, improved the growth rate, and prevented white muscle disease in the newborn lambs and calves. This disease, earlier found to be prevented by vitamin E (Sec. 11.18), is so named because of the appearance of the muscles on autopsy. It is a dystrophic condition which occurs in several species. In chicks and mink, as well as lambs and calves, the disease is prevented by either selenium or the vitamin. The relationship here represented is discussed in Sec. 10.20, where the pathology of muscle dystrophy is outlined and diseases preventable by the vitamin but not by the mineral are mentioned.

Because liver necrosis, white muscle disease, and exudative diathesis in chicks could also be prevented by vitamin E, some nutritionists were reluctant to list selenium as an essential mineral but recent studies have greatly clarified the situation. Cornell workers formulated a purified diet containing 0.005 ppm selenium.[59] Chicks fed the diet grew poorly, had poor feathering, and developed atrophy of the pancreas even though high levels of vitamin E were added

Fig. 10.16 Control lamb and lamb with white muscle disease (nutritional muscular dystrophy) caused by selenium-vitamin E deficiency. (*Cornell University.*)

to the diet. The pancreatic changes were dramatic and rapid. The degeneration of the pancreas could be detected histologically within 6 days and, after 14 days of selenium supplementation, the pancreas returned to normal.

Wisconsin workers reported that selenium was a component of glutathione peroxidase.[60] This enzyme catalyzes the removal of peroxides which may explain the antioxidant role of selenium and why vitamin E, which is also a biological antioxidant, may have similar effects to selenium in certain situations. Other studies have suggested that selenium has essential functions in addition to that in glutathione peroxidase but further studies are necessary to elucidate the mechanisms. Selenium supplementation has also been reported to improve the reproductive efficiency of sheep and reduce the incidence of retained placentas in dairy cattle.[61] The requirement for sheep and cattle is of the order of 0.1 mg per kilogram of ration.

10.48 Selenium Deficiency an Area Problem

There are wide areas in the United States where most grains and forages are low in selenium in terms of animal needs and where white muscle disease of calves and lambs frequently occurs. These areas are located in the Pacific Northwest, Northeast, and along the Southeastern Seaboard, as shown in Fig. 10.17. This map, supplied by W. H. Allaway, is based on a publication by J. Kubota and associates[62] of the U.S. Plant, Soil and Nutrition Laboratory at Ithaca, New York. Deficiency symptoms result where the animals are fed very largely or entirely on feeds grown in the area in question. In the Northeast, where the rations of dairy cows and poultry contain feeds shipped in from selenium-adequate areas, their needs may thus be met.

10.49 Selenium Toxicity

The physical symptoms of selenium poisoning have been noted in many parts of the world in animals grazing on or consuming feeds from certain areas. In the United States the areas are in the western part of the country, as shown in Fig. 10.17. In chronic cases of the disease there is a loss of hair from the mane and tail in horses, from the tail in cattle, and a general loss of hair in swine. The hoofs slough off, lameness occurs, food consumption decreases, and death may occur by starvation. These external symptoms are accompanied by marked pathological changes, notably liver injury, which are revealed on autopsy. An animal which exhibits the external symptoms is shown in Fig. 10.18. In some areas the soil may contain as much as 40 ppm, but any soil which contains more than 0.5 ppm is potentially dangerous. Both the forage and the grains contain toxic levels. Different plants vary greatly in the amounts they take up, but the concentration in the plant is generally much greater than in the soil. On a soil containing 9 ppm of selenium certain crops have been found to contain as much as 1200 ppm. Chronic toxicity is caused by rations containing as little as 8.5 ppm of selenium. Acute cases of poisoning have been reported from levels of 500 to 1000 ppm. Young animals are especially susceptible, and growth is retarded with levels too low to cause other evident symptoms. In swine, levels as low as 10 ppm have been found to lower the conception rate and result in a higher percentage of pigs dead at birth or weak and smaller in size.

Reproduction troubles have also been noted in sheep and poultry. This selenium injury is not limited to animals, for human cases have also been

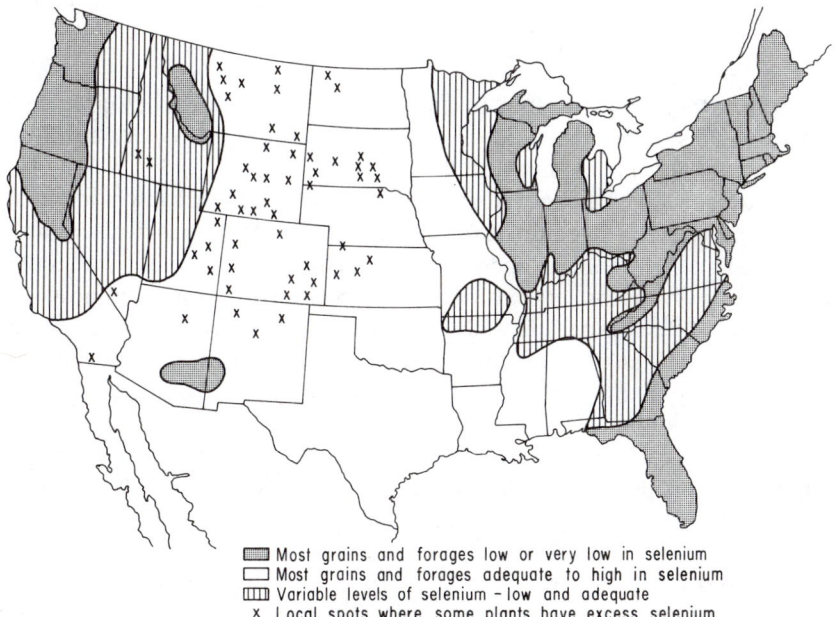

Most grains and forages low or very low in selenium
Most grains and forages adequate to high in selenium
Variable levels of selenium – low and adequate
x Local spots where some plants have excess selenium

Fig. 10.17 Selenium in crops in relation to animal needs. (*Courtesy of W. H. Allaway, U.S.D.A.*)

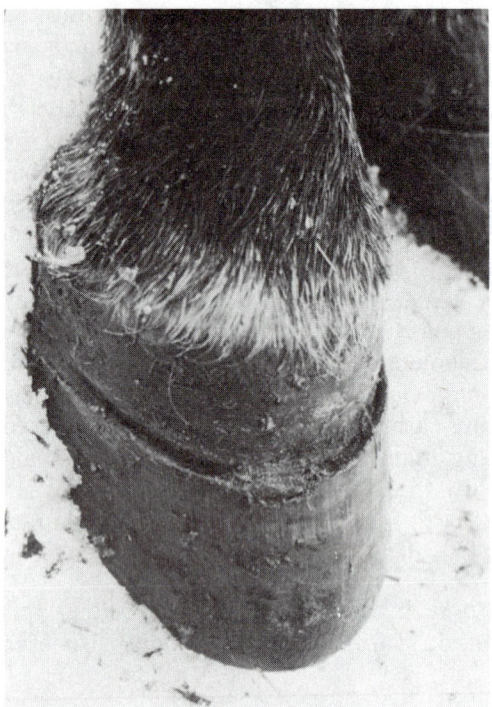

Fig. 10.18 Horse with selenium toxicity. The hoof may be sloughed off. (*Colorado State University*)

reported. White flour milled from wheats grown in a seleniferous area may contain toxic levels. Selenium occurs in the milk and eggs from cows and hens fed rations containing the element.

Many studies with both laboratory and farm animals have been made in an endeavor to counteract selenium toxicity. Certain protein sources, notably linseed oil meal, exert some protective action. The oil meal is less protective for cattle than for swine or rats. Studies have shown that various compounds exert a favorable effect experimentally, but, in general, they have been found to have only a very limited practical application. The incorporation of arsenicals of the organic type into swine rations has proved helpful. Sulfate alleviates the growth depression in rats and chicks, but does not lessen the liver damage which is the critical pathological effect. Sulfate, used as a fertilizer, also lessens selenium uptake by plants, but this does not appear to be of any practical importance as a control measure. No satisfactory practical solution has been found.

Mapping the areas with seleniferous soils is essential. Some of the areas can be utilized by pasture rotation and adding supplemented feeds. Pasturing during the last part of the growing season may also be helpful because the selenium content of the plant declines with maturity.[63]

MOLYBDENUM

Molybdenum is another element recently found essential in traces, though long known to be the apparent cause of toxic symptoms exhibited by grazing animals in certain areas of the world.

10.50   Molybdenum as an Essential Element

The conclusion that molybdenum is an essential nutrient is based on the finding that it is a constituent of the enzyme xanthine oxidase, a metalloflavo protein. This enzyme, which plays an essential role in purine metabolism, is found in liver and intestinal tissue and probably others, and also in milk. Molybdenum has been reported to be essential for the growth of chicks and poults hatched from eggs from molybdenum-depleted hens or in chicks fed diets containing tungsten, which is an antagonist of molybdenum. But negative results have been obtained with rats. A species difference might be expected on the basis that the chick has a large need for xanthine oxidase for uric acid formation in contrast to the rat, in which, like other mammals, the principal end product of nitrogen metabolism is urea. A nutritional role of molybdenum in the growth of lambs has been reported and explained on the basis of a stimulating action on rumen organisms. While present evidence indicates that molybdenum should be classed as an essential nutrient, it seems highly probable that any need by farm animals is always met by their usual rations.

10.51   Molybdenum Toxicity

A trouble in cattle referred to as *teartness* and known for over a hundred years to be definitely associated with certain pasture areas in England and not with others was identified in 1938 by Ferguson and associates as a molybdenum toxicity. The trouble was found to affect ruminants, particularly calves and cows in milk. The prominent physical symptoms were extreme diarrhea with consequent loss in weight and production. Molybdenum was discovered to be the cause of the trouble by analyses of the forage from unhealthy and normal areas and by its experimental production with sodium molybdate. The trouble was observed where the forage contained 0.002 percent or more of the element. Later, potentially toxic levels in the soils and herbage and the actual occurrence of molybdenosis were found in various areas in the world, including Florida, California, and Manitoba in North America. The symptoms reported generally included diarrhea, emaciation, anemia, and stiffness. Molybdenum toxicity has been found to be a practical problem only in grazing animals and apparently only in limited areas, but its toxicity to other species, notably rats, rabbits, chicks, and pigs, has been demonstrated by feeding experiments.

The early English workers noted that the symptoms they observed were similar to those reported for cattle in copper-deficient areas in Holland. They gave copper sulfate to affected animals and found that it cured the diarrhea. This result was obtained despite the fact that the forage contained levels of copper considered adequate. It became apparent that a copper-molybdenum interrelationship was involved. A new area of study was opened accordingly. It was shown with both laboratory animals and ruminants that feeding excess molybdenum brought on the physical symptoms of copper deficiency and interfered with copper metabolism. The addition of copper restored the animals to normal. In ruminants a molybdenum content ranging from 2 to 25 ppm in the growing forage was found to increase the copper requirements. The bone changes noted in copper deficiency were aggravated where the molybdenum content of the forage was high. In laboratory animal experiments the symptoms of chronic molybdenum poisoning and of copper deficiency were found similar. These various studies led to the tentative conclusion that molybdenosis was primarily the result of a relative copper deficiency or, expressed another way, that copper deficiency as seen in the field resulted from an excess intake of molybdenum. Later studies have shown that the interrelationship is much more complicated than this.

The sulfate content of the ration is a determining factor, as established by a series of studies by Dick of Australia.[64] In experiments with sheep it was found that increasing the sulfate intake increases molybdenum excretion. But in some cases, sulfate may enhance the effect of molybdenum on copper status. Thus copper-sulfate-molybdenum interrelationships are complex and further research is necessary. The natures of the interrelationships among the minerals may vary with differing levels of the minerals and may vary among species, not only between ruminants and nonruminants, but also between sheep and cattle.[65] Species differences in tolerance to molybdenum are large. Underwood[5] concluded that cattle and sheep are the least tolerant and horses and pigs are the most tolerant. Rats, rabbits, guinea pigs, and poultry were less tolerant than pigs but more tolerant than cattle.

FLUORINE

Fluorine occurs in an apatite form as an integral part of the structure of bones and teeth (Sec. 10.6) to the extent of 0.04 to 0.06 percent or more in the adult. Intakes of rather minute amounts have been shown definitely harmful cumulatively. This is an aspect of real practical importance in the feeding of animals.

10.52   Harmful Effects of Fluorine

These effects result from the continuous ingestion of waters high in fluorides, of rations supplemented with mineral phosphates high in the element, or of forage contaminated with fluorine from fumes released by smelting plants. There is a gradual excessive accumulation in the bones and teeth. The bones lose their normal color and luster, become thickened and softened, and the breaking strength is decreased. Bony outgrowths from the surface, called exostoses, occur. The fluorine content of the bone increases many times, and its magnesium content also increases, but there is a decrease in carbonate. The total ash content is lowered by high levels of intake. There are also characteristic histological changes. The effects upon the teeth are similar, though they manifest themselves somewhat differently, particularly in certain species. In the rat the enamel loses its glistening yellow color and becomes chalky and brittle. The permanently growing incisors do not wear away normally, and either the upper or lower incisors become elongated. These changes are illustrated in Fig. 10.19, which also shows the thickening of the skull bones resulting from fluorine feeding. In hogs and cattle, defects in the enamel are produced and the teeth become soft and worn down until in some cases the pulp cavities are exposed. The teeth become sensitive to cold water, and food consumption is interfered with.

In children an excessive fluorine intake is responsible for the development of mottled enamel. This condition is characterized by the presence of chalky-white patches on the surface of the teeth. Frequently the entire tooth surface is dull white in color, and the enamel becomes pitted and may chip off. Secondarily, the teeth may become stained, showing a coloration which varies from yellow to black. Mottled teeth are structurally weak owing to an interference with the normal development of the enamel. Mottling is chiefly a defect of the permanent teeth which results during their formation. Normally formed teeth do not become mottled later. In the permanently growing teeth of the rat, however, bleaching may be produced at any time during life. The fact that mottling will occur in children who regularly drink water containing as little as 2 to 5 ppm of fluorine illustrates how small an amount may cause this change. Mottled enamel occurs in cattle and sheep in areas where the water is high in fluorine, and erosion of the teeth can result.

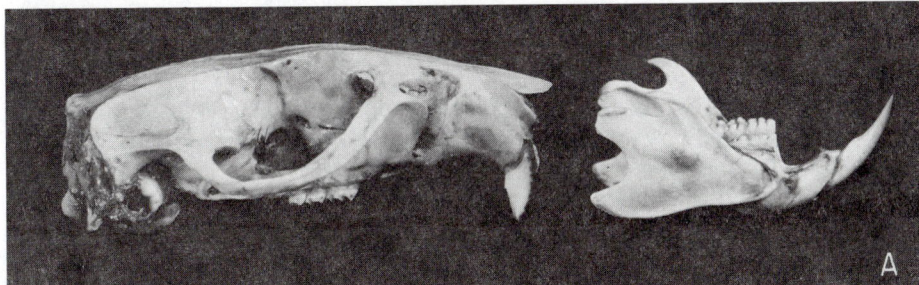

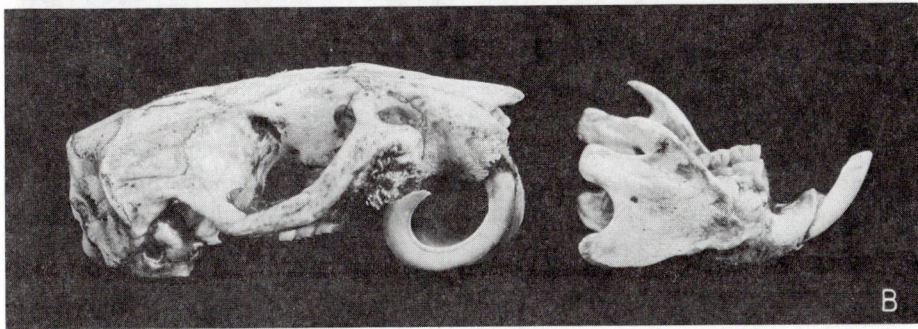

Fig. 10.19 Skulls of rats fed on diets with and without fluorine. Top, normal skull. Bottom, thickened skull with elongated and chalky incisors caused by excessive fluorine intake. (*Tolle and Maynard, Cornell Agr. Expt. Sta. Bull. 530, 1931.*)

While injuries to the bones and teeth are the initial or the most evident effects of fluorine, higher intakes or long-continued feeding interfere with food consumption, growth, reproduction, and lactation. Of course, the wearing down of the teeth interferes with food consumption, and this, in turn, with growth and production. There are generalized toxic effects, however, reflected in degenerative changes in various organs and soft tissues. The form in which fluorine is fed has an influence on its toxicity. Sodium fluoride is more toxic than calcium fluoride and certain other insoluble products. There are species differences in susceptibility. The chicken will tolerate a considerably higher level than other farm animals.

Fluorine is a cumulative poison. Short-time experiments in which no harmful effect is noted are not reliable measures of the safety of a given level in feeding practice. At first the fluorine merely accumulates in the bones and teeth without evident harm, and considerable time elapses before structural injury becomes evident. The avidity of the bones and teeth for fluorine tends to protect the soft tissues against excessive concentrations. As the bones become saturated, however, the greater part of the absorbed fluorine is free to produce its general toxic effects on the organs and soft tissues. Wisconsin workers found that while the deleterious effects of higher levels were evident much earlier, an intake of rock phosphate which provided approximately 0.008 percent of the total dry matter fed to dairy cows had a marked effect on production only after 3 years of feeding. Reproduction was affected in these animals.

An excellent summary of the effects of fluorides has been published by the National Academy of Sciences.[66]

The safe level varies with species, age of animal, duration of feeding period, source of fluorine, and general level of nutrition. However, general guidelines as shown in Table 10.4 have been assembled. Poultry appears to have a greater tolerance than other livestock.

10.53   Is Fluorine an Essential Element?

Evidence that, despite its harmful effects at higher intakes, the ingestion of minimum amounts of fluorine may be distinctly beneficial has come from observations with respect to tooth decay in man and experimental animals. It has been definitely shown by both epidemiological and experimental studies that, where the water supply contains 1 to 2 ppm, in contrast to lower levels, the incidence of dental caries is decreased during the period of tooth development. The records have indicated that decay can be reduced 50 percent or more in young children and that the effect is permanent so long as the individual continues to ingest an adequate amount of the element. Thus, as a public health measure, many municipalities are adding 0.7 to 1 ppm of fluorine as fluoride to water supplies which otherwise contain much smaller traces. This added level does not result in mottled enamel or any other deleterious effect observable in longtime studies. There is some epidemiological evidence that fluoride helps retard osteoporosis in adults.

Clearly it has been proven that fluorine in traces is a useful mineral for the

TABLE 10.4. Dietary Fluoride Tolerances for Domestic Animals[a]

Animal	Performance[b] (ppm F)	Pathology[c] (ppm F)
Beef or dairy heifers	40	30
Mature beef or dairy cattle[d]	50	40
Finishing cattle	100	NA[e]
Feeder lambs	150	ID[f]
Breeding ewes	60	ID
Horses	60	40
Finishing pigs	150	NA
Breeding sows	150	100
Growing or broiler chickens	300	ID
Laying or breeding hens	400	ID
Turkeys[g]	400	ID
Growing dogs	100	50

[a] The values are presented as ppm F in dietary dry matter and assume the ingestion of a soluble fluoride, such as NaF. When the fluoride in the ration is present as some form of defluorinated rock phosphate, these tolerances may be increased by 50 percent.
[b] Levels that, on the basis of published data for that species, could be fed without clinical interference with normal performance.
[c] At this level of fluoride intake pathologic changes occur. The effects of these changes on performance are not fully known.
[d] Cattle first exposed to this level at 3 years of age or older.
[e] NA = not applicable.
[f] ID = insufficient data.
[g] This level has been shown to be safe for growing turkey females. Very limited data suggest that the tolerance for growing male turkeys may be lower (Shupe et al.[66]).

prevention of dental caries in children, and some workers have considered it an essential dietary constituent on this basis.

Early attempts to demonstrate that fluorine was essential were unsuccessful, but in recent studies rats raised in isolation units and fed diets containing less than 0.04 ppm fluorine responded with improved growth when fluorine was added to the diet.[67] Mice fed diets containing low levels of fluorine developed anemia and reproduction was impaired.[68]

10.54 Chromium

Rats fed chromium-deficient diets have an impaired glucose tolerance. The addition of trivalent chromium to the diet restores the ability to metabolize glucose.[69] Apparently, the chromium deficiency leads to a decreased sensitivity of peripheral tissues to insulin. Corneal lesions, reduced growth rate, impaired protein metabolism, and reduced longevity have also been attributed to chromium deficiency. It has also been reported that glucose intolerance in humans in certain cases may be treated with chromium supplementation.[70] The most active form of chromium is called the glucose tolerance factor in an organic complex that has not been identified.

Few studies have been conducted on the chromium nutrition of livestock and nothing is known of their requirements or toxicity.

10.55 Newly Recognized Essential Trace Minerals

Several trace minerals, silicon, tin, vanadium, and nickel, have recently been reported to be essential for animals. There is little evidence at present that farm animals are likely to have a deficiency of any of these four minerals. The deficiencies were produced in laboratory animals under very stringent conditions and most feedstuffs contain ample amounts of these minerals. However, Miller[71] pointed out, "For more than 20 years after the essentiality of zinc was established, no one appreciated its tremendous practical importance. We trust that scientists will be open-minded concerning the possible practical importance of the more recently established essential trace elements."

10.56 Silicon

The first reports of a physiological role of silicon were by Carlisle[72] who demonstrated that silicon was essential for normal calcification of chick bone. Bone joints were small and the bone contained 34 to 35 percent less water. Silicon may have a role in mucopolysaccharide synthesis and may function as a biological cross-linking agent contributing to the structural integrity of connective tissue. Growth depression has also been reported in rats[73] and chicks[74] fed silicon-deficient diets. The rats also demonstrated impaired incisor pigmentation and the deficient chicks had no wattles and their combs were very small.

High levels of silicon in the diet of farm animals may be detrimental. In a manner not yet fully understood, silicon of the urine may be deposited in the kidney, bladder, or urethra to form calculi. Such calculi may become large enough to block passage of urine. However, the condition is not simply due to high dietary intakes of silicon as attempts to produce calculi by adding silicon to the diets of ruminants have not been successful.[75]

The digestibility of plants containing high levels of silicon as silicates is greatly decreased.[76]

10.57 Tin

Tin is claimed to be an essential mineral because of a report by Schwarz et al.[77] that the average daily gain of rats raised in plastic isolation chambers and fed a purified diet was increased from 1.1 g per day for controls to 1.75 g per day for rats supplemented with 2 ppm of tin in the form of stannic sulfate. The biological value of different tin compounds differs greatly. The biological function of tin is unknown but it has been suggested that tin could be an oxidation-reduction catalyst and function as the active site of metaloenzymes.

10.58 Vanadium

Vanadium deficiency has been reported to decrease feather growth and rate of gain in chicks and depress growth rate and impair reproduction in rats.[78] It has been suggested that vanadium probably functions as an oxidation-reduction catalyst but the dietary requirement is less than 500 ppb.

High intakes of vanadium have been reported to inhibit cholesterol synthesis in human and animal tissues, perhaps by inhibiting squalene synthetase. Vanadium has also been reported to decrease the incidence of caries in rats, hamsters, and guinea pigs, but other investigators have reported no benefit.

10.59 Nickel

Chicks fed a diet containing less than 14 ppb nickel exhibited several abnormalities corrected by the addition of nickel.[79] The abnormalities included impaired liver metabolism such as reduced ability to oxidize α-glucerophosphate, increased lipid content and decreased phospholipid content, and ultrastructural degeneration of the liver cells. Gross changes included a dermatitis and change in pigmentation of the shank skin and a decrease in friability of the liver.

Other studies have suggested that nickel is essential for the rat[80] and the pig and goat.[81] It has been suggested that nickel has a role in the metabolism or structure of membranes and may have a structural role in nucleic acids.

Nickel has also been reported to be essential for urease activity of rumen microbes.[82]

10.60 Lead

Lead should be of interest to nutritionists because it is a mineral but one of the most common causes of accidental poisoning in man and domestic animals. Sources of lead include ingestion of lead-base paints, used motor oil, storage batteries, and linoleum.[83] Lead poisoning may also occur when cattle, sheep, and horses eat forage contaminated by lead from fumes and dusts emitted from industrial lead operations. Aronson[84] concluded that 1.7 mg of lead per kilogram of body weight can produce clinical signs of lead poisoning in horses but 6 to 7 mg per kilogram of body weight is required to produce clinical signs of lead poisoning in cattle. Therefore, lead concentrations of 80 ppm in forage could be toxic to horses but cattle could tolerate 200 ppm or more.

Lead interacts with several other minerals. High levels of calcium may decrease lead toxicity. Studies on zinc and lead interactions have been contradictory. Some reports suggest zinc may enhance lead tolerance in animals whereas others suggest a decreased lead tolerance in animals fed high levels of zinc.

Lead is found in trace amounts in almost all foods and animal tissues. It has also been suggested that lead may even have an essential function because rats grown in plastic isolation units and fed purified diets had improved growth when lead was added to the diets. However, further studies are needed before lead can be listed as an essential nutrient.

10.61 Nitrates

As with lead, the greatest concern is because of toxic effects. Nitrate per se is not noxious but nitrate can be converted by microorganisms of the gastrointestinal tract to nitrite. Nitrite oxidizes the ferrous iron of hemoglobin to the ferric iron of methemoglobin which does not transport oxygen. In severe cases, the blood becomes almost chocolate brown and there is a brownish discoloration of nonpigmented areas of the skin and mucous membranes. The pulse is rapid and breathing is labored. Death may result because of anoxia. Nonruminants can tolerate nitrate but ruminants do not because the bacteria in the rumen convert the nitrate to nitrite. Nitrate/nitrite sources include water contaminated with animal wastes or industrial wastes and feeds containing high levels of nitrate. Cornstalks and oat hay were two of the feeds first reported to occasionally contain high levels of nitrate. The environmental factors that cause nitrate accumulation have not been defined.

Reviews of the role of nitrates in animal nutrition have been published by Wright and Davidson[85] and by Deeb and Sloan.[86]

NOTES

1. J. C. Smith and K. Schwarz, A controlled environment system for new trace element deficiencies, *J. Nutrition,* **93**:182–188, 1967.

2. K. Schwarz, Recent dietary trace element research, *Fed. Proc.,* **33**:1748–1757, 1974.

3. M. Anke, A. Hennig, M. Grün, M. Partschelfeld, B. Groppel, and H. Ludke. Arsenic, a new essential trace element, *Archiv Tierernährung,* **26**:742–743, 1976.

4. H. A. Schroder and A. P. Nason, Interactions of trace metals in mouse and rat tissues: Zinc, chromium, copper, and manganese with 13 other elements, *J. Nutrition,* **106**:198–203, 1976.

5. E. J. Underwood, Trace Elements in Human and Animal Nutrition, 4th ed., Academic Press, New York, 1977.

6. W. G. Hoekstra et al. (ed)., Trace Element Metabolism, Univ. Park Press, Baltimore, Md., 1974.

7. L. S. Hurley and L. Bell, Amelioration by copper supplementation of mutant gene effect in the crinkled mouse, *Proc. Soc. Expt. Biol. Med.,* **149**:830–834, 1975.

8. L. Erway, L. S. Hurley, and A. Frazer, Neurological defect: Manganese in phenocopy and prevention of a genetic abnormality of inner ear, *Science,* **152**:1766–1768, 1966.

9. T. Flagstad, Lethal trait A46 in cattle. Intestinal zinc absorption, *Nord. Vet-Med.,* **28**:160–169, 1976.

10. B. L. Reis and G. W. Evans, Genetic influence on zinc metabolism in mice, *J. Nutrition,* **107**:1683–1686, 1977.

11. A. Theiler, H. H. Green, and P. J. duToit, Phosphorus in the livestock industry, *Union S. Africa J. Dept. Agr.,* **8**:460–504, 1924.

12. M. Chossat, Note sur le systeme osseux, *Compt. Rend. Acad. Sci.,* **14**:451–454, 1842.

13. A. S. Posner, Bone mineral on the molecular level, *Fed. Proc.,* **32**:1933–1937, 1973.

14. R. E. Chapin and S. E. Smith, Calcium requirements of growing rabbits, *J. Animal Sci.,* **26:**67–71, 1967.

15. J. T. Yarrington and coworkers, Effects of a low calcium prepartal diet on calcium homeostatic mechanisms in the cow, *J. Nutrition,* **107:**2244–2256, 1977.

16. H. J. Armbrecht and R. H. Wasserman, Enhancement of Ca^{++} uptake by lactose in the rat small intestine, *J. Nutrition,* **106:**1265–1271, 1976.

17. J. L. Omdahl and H. F. DeLuca, Moderation of calcium absorption of 1,25-dihydroxycholecalciferol, *J. Nutrition,* **107:**1975–1980, 1977.

18. M. Kleiber and coworkers, Radio-phosphorus (P^{32}) as tracer for measuring endogenous phosphorus in cow's feces, *J. Nutrition,* **45:**253–263, 1951.

19. R. H. Wasserman, Bones, in M. J. Swenson (ed.), Duke's Physiology of Domestic Animals, 9th ed., Cornell Univ. Press, Ithaca, N.Y., 1977.

20. L. Krook and J. E. Lowe, Nutritional secondary hyperparathyroidism in the horse, *Pathol. Vet. I. Suppl.,* 1964.

21. Normally, wild animals will eat a variety of foods and avoid NSH. For example, carnivores such as tigers and lions may eat small rodents, including the skeleton or the bones of the larger animals. But one report suggested that some cheetahs in a game reserve park developed NSH because the impala population became too large. The impala became very easy prey for the cheetah. The cats' jaws were not powerful enough to crack the large bones of the impala so they ate only the meat and developed NSH.

22. L. Krook, Periodontal disease in dogs and man, *Adv. Vet. Sci. Comp. Med.,* **20:**171–190, 1977.

23. L. Krook, L. Lutwak, and K. McEntee, Dietary calcium, ultimobranchial tumors and osteopetrosis in the bull, *Am. J. Clinical Nutrition,* **22:**115–118, 1969.

24. T. S. Nelson, The hydrolysis of phytate phosphorus by chicks and laying hens, *Poultry Sci.,* **55:**2262–2264, 1976.

25. S. Groenendyk and A. A. Seawright, Osteodystrophia fibrosa in horses grazing *Setaria sphacelata, Aust. Vet. J.,* **50:**131–135, 1974.

26. C. E. Coppock, R. W. Everett, and W. G. Merrill, Effect of ration on free choice consumption of calcium-phosphorus supplements by dairy cattle, *J. Dairy Sci.,* **55:**245–256, 1972.

27. D. E. Pamp, R. D. Goodrich, and J. C. Meiske, A review of the practice of feeding minerals free choice, *World Rev. Animal Prod.,* **12:**13–31, 1976.

28. C. W. Duncan, C. F. Huffman, and C. S. Robinson, Magnesium studies in calves. I: Tetany produced by a ration of milk, or milk with various supplements, *J. Biol. Chem.,* **108:**35–44, 1935.

29. F. M. Tomas and B. J. Potter, The effect and site of action of potassium upon magnesium absorption in sheep, *Aust. J. Agr. Research,* **27:**873–880, 1976.

30. W. A. House and H. F. Mayland, Magnesium and calcium utilization in sheep treated with magnesium alloy rumen bullets or fed magnesium sulfate, *J. Animal Sci.,* **42:**506–514, 1976.

31. L. S. Palmer, C. H. Eckles, and D. J. Schutte, Magnesium sulfate as a factor in

the retention of calcium and phosphorus in cattle, *Proc. Soc. Expt. Biol. Med.,* **26:**58–62, 1928.

32. Several cases of potassium deficiency have been reported in persons eating large amounts of licorice. Licorice contains a compound that has corticoid-like activity and causes a loss of potassium. M. T. Epstein et al., Effect of eating liquorice on the renin-angiotensin aldosterone axis in normal subjects, *Brit. Med. J.,* **1**(6059):488–490, 1977.

33. G. von Bunge (1844–1920), trained both in chemistry and in medicine, had a long and outstanding career as a teacher and investigator, serving for many years as professor of physiological chemistry at Basel. He made many contributions to the knowledge of the nutrition of minerals, notably iron, and wrote a textbook, Physiologie des Menschen, which contains a wealth of information for the modern student.

34. C. E. Coppock and M. J. Fettman, Chloride as a required nutrient for lactating dairy cows, Proc. Cornell Nutrition Conf., 1977, pp. 43–51.

35. S. M. Babcock, The addition of salt to the ration of dairy cows, *Wisconsin Agr. Expt. Sta. Ann. Rept.* 22, 1905, pp. 129–156.

36. S. E. Smith and P. D. Aines, Salt requirements of dairy cows, *Cornell Agr. Expt. Sta. Bull.* 938, 1959.

37. G. R. Meneely and H. D. Battaree, Sodium and potassium, in Present Knowledge in Nutrition, 4th ed., Nutrition Foundation, New York, 1976.

38. F. M. Stout, J. E. Oldfield, and J. Adair, Aberrant iron metabolism and the "cotton fur" abnormality in mink, *J. Nutrition,* **72:**46–52, 1960.

39. M. C. Linder and H. N. Munro, The mechanism of iron absorption and its regulation, *Fed. Proc.,* **36:**2015–2023, 1976.

40. This term, as used here and later, refers to the requirements set up by the Committee on Animal Nutrition of the National Research Council (Sec. 13.13).

41. B. C. Starcher, C. H. Hill, and G. Matrone, Importance of dietary copper in the formation of aortic elastin, *J. Nutrition,* **82:**318–322, 1964.

42. N. F. Suttle, B. J. Alloway, and I. Thornton, An effect of soil ingestion on the utilization of dietary copper by sheep, *J. Agr. Sci.,* **84:**249–254, 1975.

43. N. F. Suttle, The role of organic sulphur in the copper-molybdenum-S interrelationship in ruminant nutrition, *Brit. J. Nutrition,* **43:**411–420, 1975.

44. N. F. Suttle and J. Price, The potential toxicity of copper-rich animal excreta to sheep, *Animal Prod.,* **23:**233–241, 1976.

45. E. E. Roginski and W. Mertz, A biphasic response of rats to cobalt, *J. Nutrition,* **107:**1537–1542, 1977.

46. Cobalt toxicity has been reported as a result of excessive beer drinking; cobalt salts were added to the beer to make the foam last longer. The practice has been discontinued but apparently the cobalt was the cause of several outbreaks of severe cardiac failure. Anon, Epidemic cardiac failure in beer drinkers, *Nutrition Rev.,* **26:**173–175, 1968.

47. Goiter has long been a problem of humans and animals. It was reported in ancient

times. Some authorities have suggested that even Cleopatra had an enlarged thyroid.

48. V. Srinivasan, N. R. Moudgal, and P. S. Sarma, Studies of goitrogenic agents in foods. I: Goitrogenic action of groundnut, *J. Nutrition,* **61**:87–95, 1957.

49. J. K. Miller, E. W. Swanson, and G. E. Spalding, Iodine absorption, excretion, recycling, and tissue distribution in the dairy cow, *J. Dairy Sci.,* **58**:1578–1593, 1975.

50. R. Bouchard and H. R. Conrad, Sulfur requirements of lactating dairy cows, *J. Dairy Sci.,* **56**:1429–1438, 1973.

51. T. S. Kahlon, J. C. Meiske, and R. D. Goodrich, Sulfur metabolism in ruminants, *J. Animal Sci.,* **41**:1154–1160, 1975.

52. L. Erway, L. S. Hurley, and A. S. Frazer, Congenital ataxia and otolith defects due to manganese deficiency in mice. *J. Nutrition,* **100**:643–654, 1970.

53. R. M. Leach, Jr., Biochemical role of manganese, in W. G. Hoekstra et al. (ed.), Trace Element Metabolism, Univ. Park Press, Baltimore, Md., 1974.

54. S. Burt Wolbach and D. Mark Hegsted, Perosis: Epiphyseal cartilage in choline and manganese deficiencies in the chick, *Arch. Pathol.,* **56**:437–453, 1953.

55. M. A. Rojas, I. A. Dyer, and W. A. Cassatt, Manganese deficiency in the bovine, *J. Animal Sci.,* **24**:654–670, 1965.

56. H. F. Tucker and W. D. Salmon, Parakeratosis or zinc deficiency disease in the pig, *Proc. Soc. Expt. Biol. Med.,* **88**:613–616, 1955.

57. J. K. Miller and W. J. Miller, Development of zinc deficiency in Holstein calves fed a purified diet, *J. Dairy Sci.,* **43**:1854–1856, 1960.

58. H. F. Mayland, Zinc increases range cattle weight gain, *J. Animal Sci.,* **41**:337 (Abstr.) 1975.

59. J. N. Thompson and M. L. Scott, Impaired lipid and vitamin E absorption related to atrophy of the pancreas in selenium-deficient chicks, *J. Nutrition,* **100**:797–809, 1970.

60. J. T. Rotruck et al., Biochemical role of selenium as a component of glutathione peroxidase, *Science,* **179**:588–590, 1973.

61. W. E. Julien et al., Selenium and vitamin E and incidence of retained placenta in parturient dairy cows, *J. Dairy Sci.,* **59**:1954–1959, 1976.

62. J. Kubota and associates, Selenium in crops in the United States in relation to selenium-responsive diseases in animals, *J. Agr. Food Chem.,* **15**:448–453, 1967.

63. O. E. Olson, Selenium as a toxic factor in animal nutrition, Proc. Georgia Nutrition Conf., 1969, pp. 68–72.

64. A. T. Dick, D. W. Dewey, and J. M. Gawthorne, Thiomolybdates and the copper-molybdenum-sulphur interaction in ruminant nutrition, *J. Agr. Sci.,* **85**:567–568, 1975.

65. C. F. Mills, A. C. Dalgarmo, and G. Wenham, Biochemical and pathological changes in tissues of Friesian cattle during the experimental induction of copper deficiency, *Brit. J. Nutrition,* **35**:309–331, 1976.

66. J. L. Shupe and coworkers, Effects of Fluorides in Animals, National Academy of Science, Washington, D.C., 1974.

67. K. Schwarz and D. B. Milne, Fluorine requirement for growth in the rat, *Bioinorganic Chem.*, **1**:355–362, 1972.

68. H. H. Messer, W. D. Armstrong, and L. Singer, Influence of fluoride intake on reproduction in mice, *J. Nutrition*, **103**:1319–1326, 1973.

69. K. Schwarz and W. Mertz, Chromium (III) and the glucose tolerance factor, *Arch. Biochem. Biophys.*, **85**:292–294, 1959.

70. C. T. Gurson and G. Saner, Effect of chromium on glucose utilization in marasmic protein-caloric malnutrition, *Am. J. Clinical Nutrition*, **24**:1313–1319, 1971.

71. W. J. Miller, Newer candidates for essential trace elements, *Fed. Proc.*, **33**:1747, 1974.

72. E. M. Carlisle, Silicon: A possible factor in bone calcification, *Science*, **167**:279–280, 1970.

73. K. Schwarz and D. B. Milne, Growth-promoting effects of silicon in rats, *Nature*, **239**:333–334, 1972.

74. E. M. Carlisle, In vivo requirement for silicon in articular cartilage and connective tissue formation in the chick, *J. Nutrition*, **106**:478–484, 1976.

75. M. D. Nottle and J. M. Armstrong, Urinary excretion of silica by grazing sheep, *Aust. J. Agr. Research*, **17**:165–173, 1966.

76. G. S. Smith and N. Scott Urquhart, Effect of sodium silicate added to rumen cultures on digestion of siliceous forages, *J. Animal Sci.*, **41**:882–890, 1975.

77. K. Schwarz, D. B. Milne, and E. Vinyard, Growth effects of tin compounds in rats maintained in a trace element-controlled environment, *Biochem. Biophys. Res. Commun.*, **40**:22–29, 1970.

78. L. L. Hopkins, Jr., and H. E. Mohr, Vanadium as an essential nutrient, *Fed. Proc.*, **33**:1773–1775, 1975.

79. F. H. Nielsen and D. A. Ollerich, Nickel: A new essential trace element, *Fed. Proc.*, **33**:1767–1772, 1974.

80. F. H. Nielsen and coworkers. Nickel deficiency in rats, *J. Nutrition*, **105**:1620–1630, 1975.

81. M. Anke and coworkers. Nickel, ein essentielles Spurenelement, *Arch. Tierenährung*, **27**:25–38, 1977.

82. J. W. Spears, C. J. Smith, and E. E. Hatfield, Rumen bacterial urease requirement for nickel, *J. Dairy Sci.*, **60**:1073–1076, 1977.

83. Lead poisoning caused by ingestion of spent lead shot has been reported to be an important cause of mortality of waterfowl; L. Silco, R. N. Jones, and R. C. Hatch, The effect of ingested lead shot in the electrocardiogram of Canada Geese, *Avian Disease*, **17**:308–313, 1973.

84. A. L. Aronson, Sources and pathways of lead in domestic animals, *J. Air Pollution Control*, **75**:2–12, 1975.

85. M. J. Wright and K. L. Davidson, Nitrate accumulation in crops and nitrate poisoning in animals, *Adv. Agron.*, **16:**197–247, 1964.

86. B. S. Deeb and K. W. Sloan, Nitrates, nitrites, and health, *Univ. Illinois Agr. Expt. Sta. Bull.* 750, 1975.

SUPPLEMENTARY LITERATURE

Adams, R. S.: Variability in mineral and trace element content of dairy cattle feeds, *J. Dairy Sci.*, **58:**1538–1544, 1975.

Aikawa, J. K.: The Relationship of Magnesium to Disease in Domestic Animals and in Humans, C C Thomas, Springfield, Ill., 1971.

Amini, E. K., and D. M. Hagsted: Effect of diet on iron absorption in iron-deficient rats, *J. Nutrition*, **101:**927–936, 1971.

Ammerman, C. B., and S. M. Miller: Selenium in ruminant nutrition: A review, *J. Dairy Sci.*, **58:**1561–1577, 1975.

Apgar, J.: Effects of zinc deficiency and zinc repletion during pregnancy on parturition in two strains of rats, *J. Nutrition*, **107:**1399–1403, 1977.

Capen, C. C., and S. L. Martin: Calcium metabolism and disorders of parathyroid glands, *Vet. Clin. N. America*, **7:**513–548, 1977.

Cunningham, H. M., J. M. Brown, and A. E. Edie: Molybdenum poisoning of cattle in the Swan River Valley of Manitoba, *Can. J. Agr. Sci.*, **33:**254–260, 1953.

Ellis, W. C., and coworkers: Molybdenum as a dietary essential for lambs, *J. Animal Sci.*, **17:**180–188, 1958.

Hurley, L. S., G. Cosmo, and L. Theriant: Teratogenic effects of magnesium deficiency in rats, *J. Nutrition*, **106:**1254–1260, 1976.

Jacobs, A.: Iron balance and absorption, *Bibl. Nutrition Diet*, **22:**61–73, 1975.

Kennedy, P. M., E. R. Williams, and B. D. Siebert: Sulfate recycling and metabolism in sheep and cattle, *Aust. J. Biol. Sci.*, **28:**31–42, 1975.

Krehl, W. A.: Water, sodium, potassium, and electrolyte balance, *Prog. Food Nutrition Sci.*, **1:**669–692, 1975.

Lofgreen, G. P.: The availability of the phosphorus in dicalcium phosphate, bone-meal, soft phosphate, and calcium phytate for wethers, *J. Nutrition*, **70:**58–62, 1969.

Malm, O. J.: Calcium and magnesium, *Prog. Food Nutrition Sci.*, **1:**173–182, 1975.

Mason, R. W.: Milk iodine content as an estimate of the dietary iodine status of cattle, *Brit. Vet. J.*, **132:**374–379, 1976.

Mertz, W.: The newer essential trace elements, chromium, tin, vanadium, nickel, and silicon, *Proc. Nutrition Soc.*, **33:**307–313, 1974.

Meyer, H., and F. W. Busse: Storage and mobilization of body magnesium, *Fentralblatt Veterinärmedizin*, **22:**864–876, 1975.

Miller, W. J.: New concepts and developments in metabolism and homeostasis of inorganic elements in dairy cattle. A review, *J. Dairy Sci.*, **58:**1549–1560, 1975.

Muth, O. H., et al.: White muscle disease (myopathy) in lambs and calves. VI: Effects of selenium and vitamin E on lambs, *Am. J. Vet. Research,* **20:**231–235, 1959.

Poitevint, A. L., and J. D. Nelson: Molybdenum in animal nutrition, Proc. Georgia Nutrition Conf., 1978, pp. 36–58.

Pond, W. G.: Trace elements in the nutrition of the pig, *Festskrift til Hjalmar Clausen,* pp. 183–198, 1975, Copenhagen.

Pope, A. L.: Mineral interrelationships in ovine nutrition, *J. Am. Vet. Med. Assoc.,* **166:**264–268, 1975.

Ramberg, C. F., Jr., and coworkers: Dietary calcium, calcium kinetics, and plasma parathyroid hormone concentration in cows, *J. Nutrition,* **106:**671–679, 1976.

Sandstead, H. H.: Some trace elements which are essential for human nutrition: zinc, copper, manganese, and chromium, *Prog. Food Nutrition Sci.,* **1:**371–392, 1975.

Stadtman, T. C.: Biological function of selenium, *Nutrition Rev.,* **35:**161–166, 1977.

Standish, J. F., C. B. Ammerman, A. Z. Palmer, and C. F. Simpson: Influence of dietary iron and phosphorus on performance, tissue mineral composition, and mineral absorption in steers, *J. Animal Sci.,* **33:**171–178, 1971.

Ward, G. M.: Molybdenum toxicity and hypocuprosis in ruminants. A review, *J. Animal Sci.,* **46:**1078–1085, 1978.

Ward, G. M.: Potassium metabolism of domestic ruminants. A review, *J. Dairy Sci.,* **49:**268–276, 1966.

Whanger, P. D., N. D. Pederson, and P. H. Weswig: Selenium proteins in ovine tissues. II: Spectral properties of a 10,000 molecular weight selenium protein, *Biochem. Biophys. Res. Comm.,* **53:**1031–1035, 1973.

Chapter 11

The Vitamins

Over a century and a half ago Prout[1] stated that there were three great staminal or proximate principles—a saccharine principle, an oily principle, and an albuminous principle—which provided the essential nutritive constituents of all organized bodies. For many years these principles, which later became known as the carbohydrates, fats, and proteins, were considered to be adequate to meet all the nutritive needs of the body other than its mineral requirements. Then came the discovery that there were other organic dietary essentials, previously unrecognized because needed in only minute amounts, which were not supplied by the early known principles. These are the nutrients which we class as vitamins. Knowledge as to the chemical nature of these dietary essentials lagged far behind the discovery of their nutritional importance, and thus, in the absence of any chemical basis for classifying them, they were grouped together and the term *vitamine,* coined by Funk in 1912 to designate a single one, was taken over to cover the group. From a physiological and nutritional standpoint, there are many advantages in considering the vitamins as a group, but it should be borne in mind that most of them are unrelated chemically and that the group name has no chemical significance.

Over the years since the first vitamin was discovered, our knowledge in this field has greatly advanced. New ones have been reported from time to time, their physiological functions have been worked out, and their chemical nature has been established. There are some 15 vitamins for which the information is sufficiently complete and definite that their existence is generally accepted, but much more needs to be learned about them. There are several others which have been proposed as the result of various experiments. It is unlikely that all of them are distinct essentials. On the other hand, the probability that there are still undiscovered vitamins must be recognized. A textbook on nutrition can deal only with those which are well established by a substantial amount of generally accepted evidence. For a knowledge of the status of others which have been proposed, the student must consult the voluminous literature in this field, and as is true for an up-to-date knowledge of the subject of vitamins in general, he must follow the new contributions as they appear in the various journals. He will find this no easy task, and he will also come to realize that complete knowledge of the subject of the vitamins lies far in the future.

A generally recognized vitamin is one that has been proved an essential dietary constituent for one or more species. Some vitamins are metabolic essentials, but not dietary essentials, for certain species, because they can be synthesized readily from other food or metabolic constituents.

Various B-vitamins are essential for normal ruminant metabolism and yet are not usually needed in the food because of bacterial synthesis in the rumen. While metabolic needs are similar, dietary needs for the vitamins differ widely among species. No generalizations can, therefore, be made regarding the nutritive requirements of farm animals for the vitamins as a group. Neither can generalizations be made with respect to feed sources. Each vitamin has a somewhat different distribution from the others in terms of the materials which make up the food supply. Our knowledge here must be specific, as is the case for the requirements of the different animals.

The vitamins now recognized as distinct dietary essentials are differentiated chemically and on the basis of their physiological functions, particularly as indicated by the metabolic and other symptoms of their deficiency. Chemical evidence is required to make certain that a distinct essential is being dealt with, but in the case of several of the vitamins there are different, though closely related, chemical compounds that have the same physiological effects.

11.1 Development of the Vitamin Concept

There is still an unknown substance in milk, which even in very small quantities, is of paramount importance to nutrition. If this substance is absent the appetite is lost and with apparent abundance the animals will die of want.

C. A. Pekelharing (1905)

Though the incidence of some of them has doubtless increased in modern times owing to changes in dietary habits, the specific diseases which we now know to be due to the absence of the recently discovered vitamins date far back in history. Scurvy was described as early as 1500 B.C. and beriberi was probably recognized even earlier. Though the specific evidence is comparatively recent, it is clear that, by the trial-and-error method, various individuals and peoples gradually learned that certain of these troubles were associated in some way with the nature of the diet and that specific foods were helpful in their treatment. The very early use by the Chinese of substances rich in vitamin A as remedies for night blindness, now known to be caused by a lack of this vitamin, is evident from the recent studies of these old remedies. Around 400 B.C. Hippocrates prescribed raw ox liver dipped in honey. In 1747, Lind, a British naval surgeon, showed that the juice of citrus fruits was a cure for scurvy. Cod-liver oil was used as a specific treatment for rickets long before anything was known about the cause of this disease.

During the nineteenth century, many isolated observations were made which gradually led up to the discovery of vitamins as the causes of these disorders now called deficiency diseases. Prior to 1816, Magendie[2] observed in a dog what was undoubtedly xerophthalmia, in an experiment constituting a forerunner of the purified-diet method which was responsible one hundred years later for the discovery of vitamin A. In the latter part of the century, several men made observations which led them to suggest that there were other dietary essentials besides the early recognized proximate principles, but these unorthodox suggestions at first received little attention. In 1881, Lunin reported studies made in Bunge's laboratory showing that mice would not grow on an artificial mixture made up of the proximate principles of milk. He expressed the view that there might be "unknown substances" essential for life in addition to proteins, fats, carbohydrates, and salts.

During the last decade of the century, Eijkman, a physician working in the Dutch East Indies, was led to study polyneuritis in birds in view of its similarity to beriberi in man. He found that the disease was caused by an exclusive diet of polished rice and cured by adding the polishings. He also noted that beriberi in prisoners eating polished rice tended to disappear when a less highly milled product was fed. In these various studies, published in 1897, he was clearly dealing with the factor which later became known as vitamin B. The studies of Eijkman were extended by Grijns, another Dutch scientist who made important contributions to the early knowledge of vitamins. Prior to the work of Eijkman, Takaki, director-general of the Japanese Navy, sent two ships in 1887 on a nine-month voyage to test the effect of dietary additions on the incidence of beriberi. Of the crew which received mostly polished rice and dried fish, 60 percent developed the disease, while in the other ship, where this diet was supplemented with more meat, vegetables, and milk, only 14 cases occurred among the 276 men. At that time the beneficial effect was erroneously ascribed to the larger amount of protein in the diet.

With the opening of the twentieth century, the earlier work of Lunin with artificial diets was repeated by others, notably by Pekelharing of the University of Utrecht, who was familiar with the course of the studies in the Dutch East

Indies and by Hopkins in England. Again it was concluded that the proximate principles would not suffice. As stated by Hopkins in 1906: "No animal can live on a mixture of pure protein, fat, and carbohydrate, and, even when the necessary inorganic material is carefully supplied, the animal still cannot flourish. The animal body is adjusted to live either upon plant tissues or other animals, and these contain countless substances other than protein, carbohydrate and fats." He coined the term *accessory food factors* for these substances.

In 1902 Holst,[3] a Norwegian physician, was commissioned by his government to study "ship beriberi" (actually scurvy), a disease of Norwegian sailors. Studies with poultry following the procedures used by Eijkman for "tropical beriberi" were unsuccessful. He decided to use a mammal instead and fortunately chose the guinea pig. For this study he was joined by Frølich,[3] a pediatrician. On a cereal diet they produced a disease characterized by hemorrhages, bone fragility, loose teeth, and bleeding gums. They prevented the disease with fresh foods and concluded that the disease produced was identical with human scurvy. Their first publication was made in 1907, followed by others, thus laying the foundation for the later specific discovery of vitamin C.

Following work published in 1909 by Stepp showing the necessity of some constituent contained in the lipid fraction of certain natural foods, definite proof of the existence and specific physiological function of vitamin A was furnished in 1913 by the independent investigation of McCollum and Davis[4] and of Osborne and Mendel.[5] This specific knowledge resulted from carefully controlled experiments by the purified-diet method (Sec. 13.3). Dating from 1913, the extension of the knowledge of vitamins proceeded very rapidly.

In this brief statement of the historical background of the vitamin concept, only a few of the men concerned in its development have been mentioned. The discussion suffices to show that many scientists throughout the world contributed and that it is impossible to name any one person or group as the discoverer of this far-reaching concept. Such is the often-repeated story in scientific investigations. Links in the chain of facts are supplied and gradually put together by various workers over a period of years. Finally some one man may complete the chain, and a discovery is announced; but he may deserve no more, and sometimes even less credit than others who made the previous observations which he used and extended in making the final contribution.

11.2 Fat-soluble and Water-soluble Vitamins

McCollum proposed the names *fat-soluble A* for the factor found in butter and *water-soluble B* for the one concerned with beriberi as descriptive terms, since the first was extractable from foods with fat solvents and the second with water. On a similar basis, the antiscorbutic vitamin was later called water-soluble C. Though these descriptive adjectives were eventually given up, they are still frequently used as general terms in classifying the vitamins. Thus the fat-soluble vitamins include A, D, E, and K, while the members of the B complex, C, and others are classed as water soluble.

The fat-soluble vitamins are primarily excreted in the feces via the bile, whereas water-soluble vitamins are excreted in the urine. Water-soluble vitamins are relatively nontoxic but excesses of the fat-soluble vitamins A and D can cause serious problems. The fat-soluble vitamins consist only of carbon, hydrogen, and oxygen, whereas some of the water-soluble vitamins also contain nitrogen, sulfur, or cobalt.

The signs of fat-soluble vitamin deficiency can sometimes be related to the function of the vitamin. For example, vitamin D is required for calcium metabolism and a deficiency results in bone abnormalities. On the other hand, the signs of B-vitamin deficiency are much less specific and the signs of deficiency are difficult to relate to function in most cases. Most B-vitamin deficiencies result in dermatitis, rough hair coats, poor growth, and reduced feed efficiency. Several vitamin deficiencies may produce achromotricia (loss of pigment in hair) and several can produce anemia which is a deficiency of hemoglobin. The type of anemia may vary among deficiencies or among species of animals with the same deficiency. Of course, there are also some specific signs of deficiency such as curled toed paralysis caused by riboflavin deficiency, but even in such cases it is difficult to relate the signs to vitamin function.

Fat-soluble vitamins are absorbed by passive diffusion through the lipid phase of the mucosal cell membrane. Most B-vitamins are also absorbed by passive diffusion but some may be absorbed by active process when the diet contains low levels of the vitamin. Most of the studies on active transport of B-vitamins have been conducted with thiamin. The method of absorption of B_{12} is unique and is discussed in more detail in the section on B_{12}.

In the present chapter, these various vitamins are discussed as regards their physiological effects, chemical nature and properties, and distribution in feeds. The specific requirements for growth, lactation, and other body functions are discussed in later chapters. Principal attention is given to those vitamins which have been demonstrated to be of practical importance in the feeding of farm animals.

Methods of vitamin analysis can be divided into several categories such as biological, microbiological, chemical, and radioisotope. In general, the results of the biological tests have the greatest practical application but such tests are usually expensive and time consuming, may require a large sample, and may have application only for the species tested. Chemical tests are adequate for some vitamins but the results can be very misleading for others. For example, Ullrey[6] concluded "chemical analysis of the diet for the concentration of fat-soluble vitamins does not establish biopotency. This is true because certain widely used chemical assays tend to lack specificity while biopotency is dependent on precise and specific structural and stereochemistry." Analysis can also be misleading for B-vitamins such as vitamin B_{12}. For example, some microbiological assays are specific whereas others include the pseudovitamins. For further details, the reader is referred to sources on vitamin analysis such as the publications of the Association of Vitamin Chemists and the Association of Official Agricultural Chemists.

VITAMIN A

All animals require a dietary source of vitamin A. The vitamin does not occur as such in plant products but rather as its precursor, carotene. This compound is commonly spoken of as provitamin A because the body can transform it into the active vitamin. This is the way in which the vitamin A needs of farm animals are met, for the most part, because their rations consist mainly or entirely of foods of plant origin. The combined potency of a feed, represented by its vitamin A and carotene content, is referred to as its vitamin A value. Retinol is the alcohol, retinal is the aldehyde, and retionic acid is the acid of vitamin A.

11.3 Physiological Function and Symptoms of Deficiency

Vitamin A is combined with a protein in visual purple, a compound that breaks down in the physiological process of sight, as a result of a photochemical reaction. A deficiency of the vitamin, in terms of the needs for the resynthesis of visual purple, results in night blindness (nyctalopia), which is a symptom in all animals. The deficiency first manifests itself as a slow, dark adaptation and progresses to total night blindness. In man, the measurement of this physiological response after exposure of the eye to a calibrated source of light is used in estimating quantitative needs for the vitamin. The reactions which take place in vision and the role played by vitamin A are shown in Fig. 11.1.

There are various other eye symptoms that vary markedly among species, some of which represent secondary infections. Xerophthalmia (from the Latin words for dry and eye), an advanced stage of vitamin A deficiency noted particularly in children, dogs, fox, and rats, is characterized by a dry condition of the cornea and conjunctiva, cloudiness, and ulceration. It is not a common

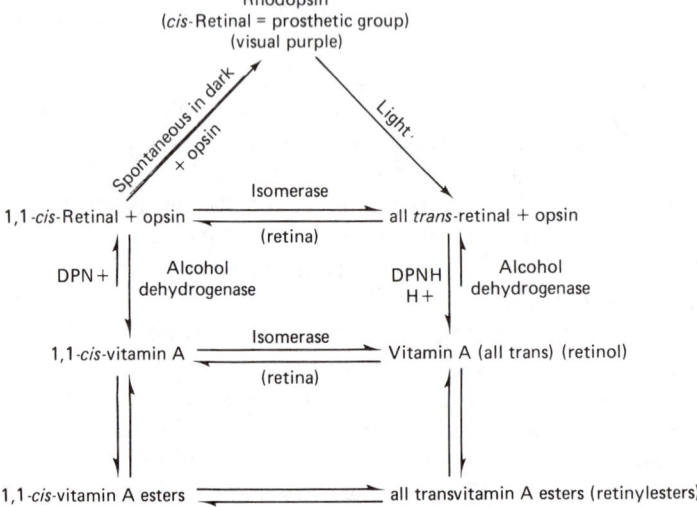

Fig. 11.1 Vitamin A and vision. (*Adapted from George Wald, Science* 162:230–239, 1968.

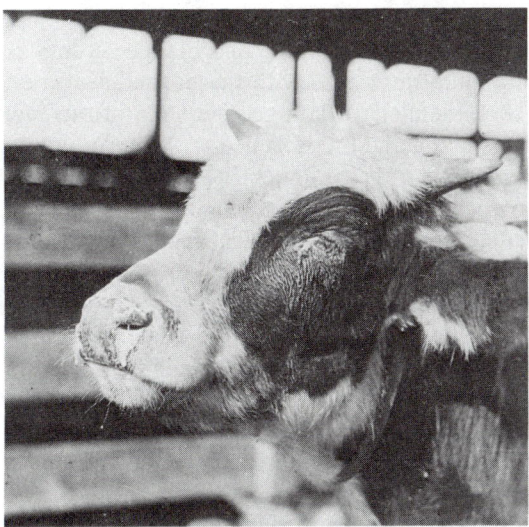

Fig. 11.2 Advanced stage of eye lesion in viatmin A deficiency. (*Courtesy of S. W. Mead, University of California.*)

symptom in other species, although corneal changes occur. Copious lacrimination is a more prominent eye symptom in cows and horses (Fig. 11.2). In the case of chickens, on the other hand, the secretions of the tear glands dry up, and an infection may then occur, resulting in a discharge that causes the lids to stick together. At least some of these symptoms develop as a result of basic epithelial changes caused by a deficiency of the vitamin.

Wolbach and Howe[7] were responsible for discovering a generalized pathological effect of the absence of the vitamin. They found that the normal epithelium in various locations throughout the body became replaced by a stratified, keratinized epithelium. Subsequent studies showed that the epithelial cells from deficient animals fail to differentiate beyond the squamous type to the mucus-secreting cells and mesenchymal cells fail to differentiate beyond the blast stage.[8] This effect has been noted in the respiratory, alimentary, reproductive, and genitourinary tracts, as well as in the eye. This keratinization lowers the resistance of the epithelial tissues to the entrance of infective organisms. Thus respiratory troubles, such as colds and sinus infections, tend to be more severe in vitamin A deficiency. A diet adequate in the vitamin is necessary to help maintain the normal powers of resistance, but additional intakes will not increase resistance to infections which enter through the epithelium. Aside from its curative effect on xerophthalmia, which may be secondary to a bacterial invasion, there is no evidence that the administration of the vitamin after an infection has become established will shorten its course or lessen its severity.

These observations do not mean that this keratinization of the epithelium is of little importance. There are many troubles, noninfective in character, which increase following its occurrence. This is true for certain gastrointestinal disturbances such as diarrhea. The formation of kidney and bladder stones is favored because the damaged epithelium interferes with the normal secretion and elimination of the urine and the sloughed keratinized cells may form foci for the formation of stones. There is a specific interference with reproduction caused by this altered epithelium which is of great importance, as is discussed later (Sec. 16.13). Jungherr and coworkers[9] found that squamous metaplasia in the parotid gland was an early change in vitamin A-deficient calves and that it proved useful in diagnosing a deficiency.

Even the elevated cerebrospinal fluid pressure observed in vitamin A-deficient animals and which is a most sensitive measure of the onset of vitamin A deficiency is the result of cell changes. The increased ground substance in the dura mater surrounding the arachnoid villus and the altered epithelial cells cause a decreased absorption of the fluid.[10]

Vitamin A is concerned in the normal development of bone through a control exercised over the activity of the osteoclasts and osteoblasts of the epithelial cartilage. The bones are altered in shape during growth, as is clearly illustrated by the studies of Mellanby.[11] The teeth are also affected. A failure of the spinal and some other bones to develop normally results in turn in pressure on the nerves and in their degeneration. For example, a blindness in calves results from a constriction of the optic nerve caused by a narrowing of the bone canal through which it passes. Avitaminosis A can result in deafness in dogs, owing to an injury to the auditory nerve. Bone changes may also be responsible for the muscle incoordination and other nervous symptoms shown by A-deficient cattle, sheep, and swine. They may be also concerned in the increase in cerebrospinal fluid pressure shown by Moore and coworkers[12] to be characteristic of the deficiency. While the pathological basis is unknown, several studies have shown that a lack of vitamin A causes congenital malformation in certain soft tissues. Examples are the birth of pigs without eyeballs, as studied by Hale,[13] and hydrocephalus in rabbits, reported by Lamming.[14] There is biochemical evidence that the vitamin plays a role in the synthesis of mucopolysaccharides which are prosthetic groups in glyco- or mucoproteins present in cartilage and mucous epithelium, a function that could explain some of the deficiencies previously mentioned.

While the basic physiological functions which have been found for vitamin A are, in general, common to all animals, it should be understood that the physical deficiency symptoms, or their prominence, vary from one species to another. This is true for other vitamins.

11.4 Chemical Nature of Vitamin A and Carotene

Vitamin A is a nearly colorless substance. It does not occur as such in plant materials, but rather as its precursors, certain carotenoid pigments which are converted into the vitamin in the animal body. The development of the knowledge as to the relationship between vitamin A and carotene and as to their chemical nature represents a highly interesting chapter of vitamin research, which illustrates the course of such studies.

In 1914 McCollum and Davis found that the vitamin was contained in the unsaponifiable fraction of milk fat, and later studies confirmed its identity as an unsaponifiable constituent extractable by lipid solvents. In 1919 Steenbock called attention to the fact that among vegetable foods vitamin A potency was associated in a rather remarkable way with yellow color, and shortly thereafter, he and his associates published many data demonstrating this association. They went so far as to suggest that carotene was the source of the vitamin. This view was not accepted by other workers, and Steenbock came to recognize that the vitamin was not carotene itself because certain potent sources of the vitamin were colorless. It was ten years before the riddle was solved. Von Euler and associates in Stockholm obtained a definite growth response when carotene was added to a vitamin A-deficient diet. In 1929, Moore produced proof that the animal body transformed carotene into vitamin A. He fed rats on a diet which resulted in the symptoms of deficiency. Some of the animals were then killed, and their livers found devoid of A. The rest of the animals were fed carotene with a resulting disappearance of the deficiency symptoms, and on autopsy, their livers were found to be rich in the vitamin.

Meanwhile halibut-liver oil had been found an especially concentrated source of the vitamin, and fractionation methods were applied to it to isolate the active substance. These methods brought success by 1932 in the isolation, by Karrer and his associates in Switzerland and by Drummond and his coworkers in England, of a very active fraction which was identified as an unsaturated alcohol having the formula $C_{20}H_{30}O$. Karrer proposed the structural formula given below which shows the compound to have a β-ionone ring and an unsaturated side chain:

Vitamin A, $C_{20}H_{30}O$ (Retinol)

The formula shown for vitamin A is the alcohol *retinol*. Replacement of the alcohol group by an aldehyde group gives *retinal,* and by an acid group, *retinoic acid*. Except for vision and reproduction, this acid carried out all the functions of vitamin A.

Brilliant researches by Karrer, by Kuhn in Germany, and by others established the chemical structure of the highly complex hydrocarbon carotene, which was previously known to have the empirical formula $C_{40}H_{56}$. Several isomeric forms of carotene were isolated, and their structural formulas worked out. All were reddish-yellow crystalline compounds, which differed, however, as regards optical activity and the wavelength at which maximum color absorption occurred. It is now recognized that there are many different carotenes which have vitamin A activity. Some of the carotenes are shown in Figs. 11.3 and 11.4.

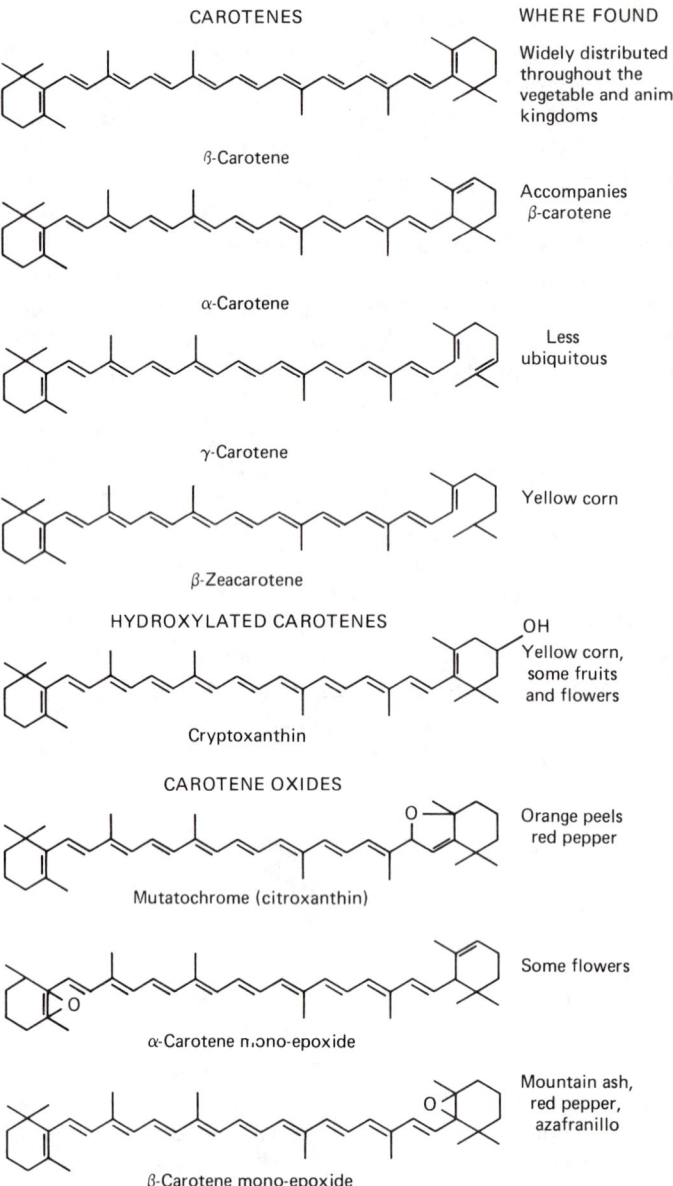

Fig. 11.3 Vitamin A active carotenes, hydroxylated carotenes and carotene oxides. (*Adapted from Ullrey[6].*)

The vitamin A activity of β-carotene is substantially greater than that of the other carotenoids. The conversion of β-carotene into vitamin A involves two enzymes.

β-carotene-15,15'-dioxygenase catalyzes the cleavage of β-carotene at the central double bond to yield two molecules of retinaldehyde. The second en-

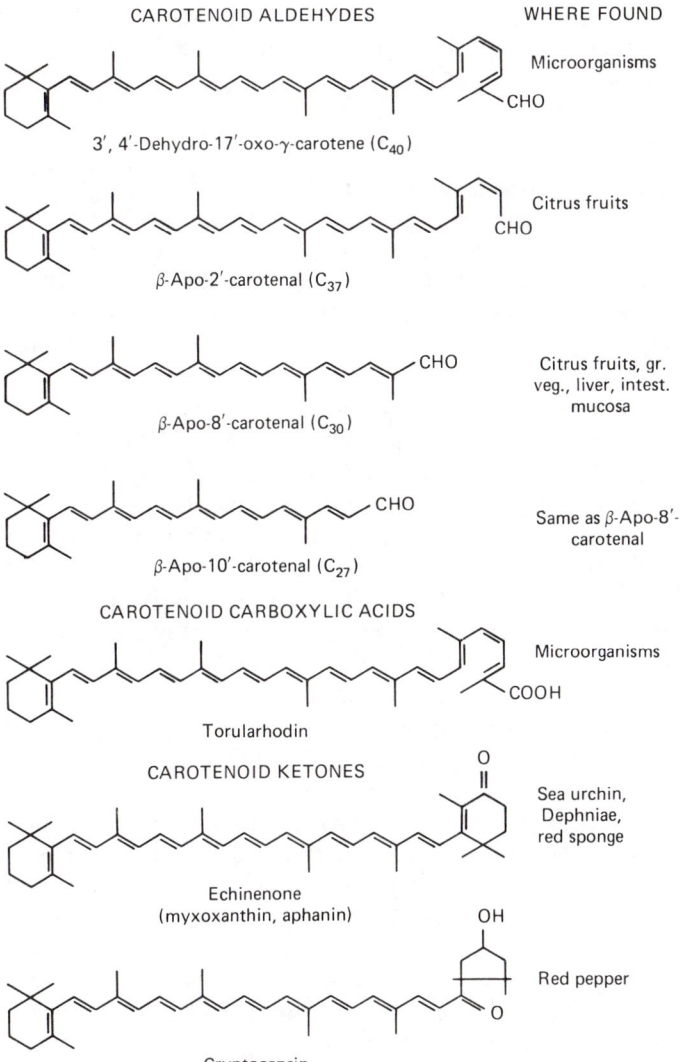

CAROTENOID ALDEHYDES   WHERE FOUND

Microorganisms

3', 4'-Dehydro-17'-oxo-γ-carotene (C$_{40}$)

Citrus fruits

β-Apo-2'-carotenal (C$_{37}$)

Citrus fruits, gr.
veg., liver, intest.
mucosa

β-Apo-8'-carotenal (C$_{30}$)

Same as β-Apo-8'-
carotenal

β-Apo-10'-carotenal (C$_{27}$)

CAROTENOID CARBOXYLIC ACIDS

Microorganisms

Torularhodin

CAROTENOID KETONES

Sea urchin,
Dephniae,
red sponge

Echinenone
(myxoxanthin, aphanin)

Red pepper

Cryptocapsin

Fig. 11.4 Vitamin A active carotenoid aldehydes, carotenoid carboxylic acids and carotenoid ketones. (*Adapted from Ullrey*[6].)

zyme retinaldehyde reductase reduces the retinaldehyde to retinol. The cleavage enzyme has been found in many vertebrates but it is not present in the cat or mink.

Vitamin A exists only in the animal kingdom where it is found free and as esters of higher fatty acids. In addition to retinol, the vitamin portrayed by the preceding formula, there is another form later isolated from freshwater fish. It was originally distinguished on the basis of a different maximum spectral absorption and named A$_2$ to differentiate it from the one earlier isolated from halibut-liver oil. Later, it was found to have the formula, $C_{19}H_{25}OH$, and to be structurally related to retinol, and was named 3,4-didehydroretinol. It has much less activity than retinol for farm animals.

11.5   Stability of Vitamin A and Carotene

The vitamins as a class are susceptible to destruction in varying degrees by certain physical and chemical agents which may become operative in the course of some of the processes to which feeds are subjected. The same is true during storage under certain conditions. The different vitamins vary greatly in their susceptibility to the action of these agents. Both carotene and vitamin A are destroyed by oxidation, and this is the most common cause of any depreciation which may occur in the potency of sources of them. The process is accelerated at high temperatures, but heat without oxygen has a minor effect. Butter exposed in thin layers in air at 50°C loses all its vitamin A potency in 6 hours but in the absence of air there is little destruction at 120°C over the same period. Cod-liver oil in a tightly corked bottle has shown activity after 31 years, but it may lose all of its potency in a few weeks when incorporated in a feed mixture stored under the usual conditions. The distribution of the oil over the feed particles provides a large surface for oxidative action, and this process is accelerated by the presence of prooxidants in any rancid fats present. Certain metals also, especially the trace-mineral elements, catalyze the destructive action. The effective prevention of rancidity and of destruction of vitamin A activity presents a practical problem in mixed rations, as is shown by the studies of Kamstra and associates.[15] The carotene in alfalfa meal is much more stable than carotene in oil under these conditions.

Large losses of carotene take place in the curing of roughages and in their later storage. Bernstein and Thompson[16] have shown that the destruction of carotene in leaves is partly enzymic and partly photochemical. Enzymatic destruction requires oxygen, is greatest at high temperatures, and ceases after complete dehydration. The photochemical action increases with decreasing moisture content in curing but slows down as this content reaches 20 to 30 percent. Walsh and Hauge[17] have shown that the same type of reactions occur in alfalfa. In curing and preserving forages both the enzymatic and the photochemical destruction must be controlled to produce roughages high in carotene (Sec.11.10). Corn has been reported to lose as much as 60 percent of its carotene on seven months' storage. With both carotene and vitamin A the nature of the associated substances and the temperature and moisture conditions have a marked influence on the rate of destruction.

The cooking processes commonly used in human food preparation do not cause much destruction to the vitamin potency. Hydrogenation of fats lessens their vitamin A value, and thus the commercial products prepared for culinary purposes commonly contain none of the vitamin, though it may have been present in the original material. Saponification does not destroy the vitamin if oxidation is avoided.

11.6 Metabolism of Vitamin A and Carotene

The first step in the utilization of carotene and vitamin A is absorption. This absorption varies, particularly for carotene, according to the nature of the diet and the species. Differences in digestibility of food sources are reflected in variations in the amounts of the provitamin available for absorption. Fats promote the absorption of both A and carotene, and emulsifying agents may have an additional effect. Some of the ingested provitamin is destroyed in the intestine. The presence of vitamin E, an antioxidant, in the ration lessens this destruction. The effects of these variable factors have been studied with various feed sources and species. One cannot generalize from the results, however, as applied to farm animal nutrition, other than to note that the utilization of the vitamin A value of the ration may be low in many situations because of limited absorption. For therapeutic use, certain vitamin A preparations in aqueous solution have been found more absorbable than in the usual oily medium.

Dietary vitamin A is usually in the form of long-chain retinyl esters. The esters are hydrolyzed in the intestinal lumen by pancreatic retinyl ester hydrolase or by a hydrolase bound to the brush border of the intestinal mucosa. The retinol is absorbed and reesterified in the intestinal mucosal cell. The esters are then transported via the lymphatic route in association with chylomicrons which are removed from circulation by the liver.

The absorption of carotene varies with species. In some species, such as the rat, pig, goat, sheep, rabbit, zebra, donkey, buffalo, and dog, almost all of the carotene is cleaved in the intestine. In man, cattle, horses, and carp, significant amounts of carotene can be absorbed. The absorbed carotene can be stored in the liver and fatty tissues. Hence, these animals have yellow body and milk fat, whereas animals that do not absorb carotene have white fat. It appears that carotene can be converted by various extra-intestinal body tissues to vitamin A in some species. For example, water-soluble carotene injected into the bloodstream of rats was changed into vitamin A.[18] But the efficiency of converting absorbed carotenoids to vitamin A in ruminants is considered to be low.

The vitamin A level of the blood may reflect the nutritional status with respect to this vitamin. This level is governed, however, by the extent of the liver stores, discussed later, as well as by the current intake. In fact, in some species the blood level tends to be maintained until the liver stores are exhausted when the diet is devoid of the vitamin. For this reason a normal blood level cannot be interpreted to ensure that the current intake is adequate, but a low level indicates a deficiency.

Eaton and associates[19] have shown that in dairy calves approximately four months of age, fed a depletion ration devoid of vitamin A activity, the plasma vitamin A level decreased until values of 4.0 μg per 100 ml are reached. Calves exhibiting such low plasma values were shown to be depleted of liver stores of vitamin A.

Total vitamin A body pools can be measured with radioactive vitamin A or deuterated vitamin A. Such measurements could reflect vitamin A status much more accurately than plasma levels.[20]

Highly chlorinated naphthalene interferes with vitamin A metabolism. For example, in the 1950s, it was discovered that cattle developed signs of vitamin A deficiency when exposed to lubricating oil that contained naphthalenes but now the compounds have been largely eliminated from oils.[21,22,23]

11.7   Vitamin A Storage

Some vitamins are stored in the body in large amounts, others to only a very limited extent. The liver can store large amounts of vitamin A in the parenchymal cells. The vitamin is usually stored as the retinyl ester, primarily palmitate. Several studies have shown that the liver can store enough vitamin A to protect the animal from long periods of dietary scarcity. Steers fed a vitamin-deficient diet required four months to exhaust initial liver reserves of 20 to 40 mg per kilogram.

A quantity sufficient to protect a rat for several months can be given in a single dose. Animals on good pasture can store extensive reserves to help meet their needs during the winter feeding period when their rations may be deficient. This large capacity to store the vitamin must be taken account of in studies of requirements in order to make sure that intakes which appear adequate for a given function are not being supplemented by reserves stored up prior to the period of observation. The measurement of the liver store of vitamin A at slaughter or in samples obtained on biopsy is a useful technique in studies of vitamin A status and requirements.

11.8   Vitamin A Value of Feeds

Many factors influence the utilization of carotene. The activity of β-carotene decreases with increasing intake. Carotene from late-cut grass is not utilized as efficiently as that from early-cut grass. The carotene of silage may not be as effective as the carotene of hay. Species of plant and dry-matter content of plant may also influence utilization. The presence of antioxidants in the digestive tract may help protect carotene from oxidation. In humans an I.U. of carotene is considered to be only one-half to one-third as valuable, depending on its food source, as an I.U. of vitamin A. Guilbert found that it took approximately twice as many I.U. of carotene as of vitamin A to prevent night blindness in cattle and that the ratio was wider at the higher levels needed for storage and reproduction. The pig and sheep also utilize carotene less efficiently than does the rat. In the case of poultry, however, both forms of the vitamin have been found equal per I.U.

Therefore, it is becoming an increasing practice to state animal requirements as vitamin A and to express the carotene and vitamin A content of feeds separately, both in micrograms. This makes it possible to calculate how the needs of a given species can be met in terms of a specific feed supply.

The richest sources of vitamin A are the fish oils. Some swordfish-liver oils contain as many as 250,000 units of vitamin A per gram. Halibut-liver oil may run even higher. Thus both are many times more potent than cod-liver oils. Products from the same species, however, may be highly variable in potency, and thus, in their manufacture for use as vitamin A supplement, they are subjected to a biological assay in order that the user may be assured of a certain minimum potency. Among the common foods of animal origin, milk fat, egg yolk, and liver are rated as rich sources, but this is not the case if the animal from which they came has been receiving an A-deficient diet for an extended period. Since the vitamin is present in the fat, skim milk contains very little.

In the nutrition of herbivorous animals, we are primarily interested in the potency of plant products. Though the yellow color is masked by chlorophyll, all green parts of growing plants are rich in carotene and thus have a high vitamin A value. Good pasture always provides a liberal supply, and the kind of pasture plant, whether grass or legume, appears to be of minor importance. At maturity, however, leaves contain much more than stems, and thus legume hay is much richer in vitamin content than timothy or other grasses. With all hays and other forage, the vitamin value decreases after the bloom stage, and much of the carotene is destroyed by oxidation in the process of field curing. Russell[24] found that there may be a loss of more than 80 percent of the carotene of alfalfa during the first 24 hours of the curing process. It occurs chiefly during the hours of daylight, owing in part to photochemical activation of the destructive process. Hays which are cut in the bloom stage or earlier and cured without exposure to rain or to too much sun retain a considerable proportion of their carotene content, while those which are cut in the seed stage and

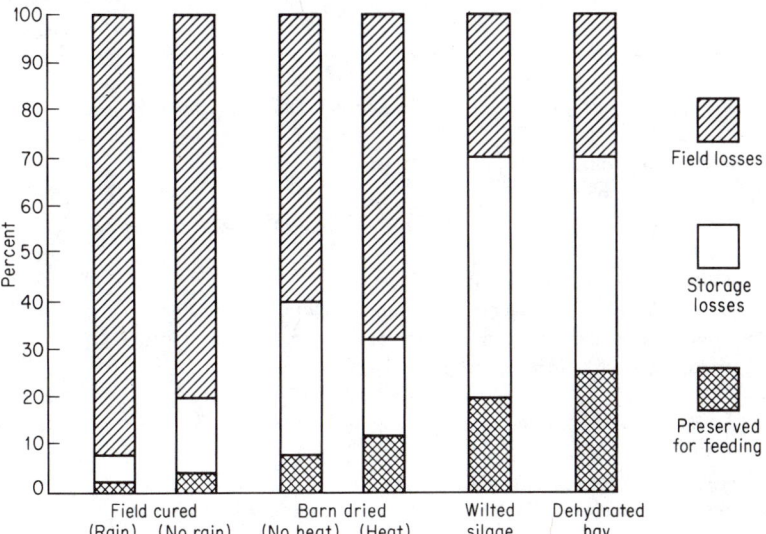

Fig. 11.5 Percentage of carotene lost in the field and during storage of alfalfa hay and silage. (*Courtesy of J. B. Shepherd, U.S.*)

exposed to rain and to the sun for extended periods lose it almost entirely. Under similar conditions of curing, alfalfa and other legume hays are much richer than grass hays because of their leafy nature, but a poor grade of alfalfa may have less than a good grade of timothy.

In the artificial curing of hay with a "hay drier," there is only a slight loss of carotene because of the rapidity of the process. Russell and coworkers found the machine-dried product to have two to ten times the value of field-cured alfalfa. Severe heating of hay in the mow or stack reduces the vitamin content, and there is a gradual loss in storage so that old hay is poorer than new. Owing to their higher initial content machine-dried hays are subject to larger percentage losses than field-cured. Losses as high as 60 percent have been reported from July to November, but they were found much smaller during the colder months. Temperature is the major factor causing variations in loss during storage. The extent of carotene losses which occur under practical conditions is shown by the report of Shepherd and associates,[25] as illustrated in Fig. 11.5.

The degree of greenness in a roughage is a good index of its carotene content. The data in Table 11.1 are useful to indicate the order of the differences found among various roughages differing as to color, kind, and other factors. Published average values as to carotene content can serve only as approximate guides in feeding practice because of the many factors affecting the actual potency of individual samples as fed.

Aside from yellow corn and its by-products practically all of the concen-

TABLE 11.1. Estimated Carotene Content of Feeds in Relation to Appearance and Methods of Conservation

Feedstuff	Carotene, mg/lb
Fresh green legumes and grasses, immature	15–40
Dehydrated alfalfa meal, fresh, dehydrated without field curing, very bright green color	110–135
Dehydrated alfalfa meal after considerable time in storage, bright green color	50–70
Alfalfa leaf meal, bright green color	60–80
Legume hays, including alfalfa, very quickly cured with minimum sun exposure, bright green color, leafy	35–40
Legume hays, including alfalfa, good green color, leafy	18–27
Legume hays, including alfalfa, partly bleached, moderate amount of green color	9–14
Legume hays, including alfalfa, badly bleached or discolored, traces of green color	4–8
Nonlegume hays, including timothy, cereal, and prairie hays, well cured, good green color	9–14
Nonlegume hays, average quality, bleached, some green color	4–8
Legume silage	5–20
Corn and sorghum silages, medium to good green color	2–10
Grains, mill feeds, protein concentrates, and by-product concentrates, except yellow corn and its by-products	0.01–0.2

Source: Compiled by H. R. Guilbert, University of California, and reproduced with his permission.

trates used in feeding animals are devoid of the vitamin or nearly so. The potency of yellow corn is only about one-eighth that of good roughage. Roots and tubers as a class supply practically no vitamin A, but carrots are a very rich source and so are sweet potatoes, as might be expected from their yellow color. Pumpkins and squash also supply considerable amounts. The green leafy vegetables used in human nutrition are rich.

Tankage, meat scraps, and similar animal by-products have little if any vitamin A potency. Certain fish meals are fair sources, but variation in the raw material and in the methods of processing which may entirely destroy any potency originally present make generalizations here of no value.

11.9 Toxicity of Vitamin A

Large doses can be toxic and many cases of overdoses have been reported. Arctic and Anarctic explorers have suffered from acute toxicity after eating polar bear or seal liver. Even sled dog liver may be toxic. Hypervitaminosis A has occurred in children taking therapeutic doses prescribed for various skin problems.[26] Domestic cats fed exclusively liver may suffer from chronic hypervitaminous A.[27] Bone abnormalities appear to be the most common development of chronic toxicity but other signs such as degeneration of organs, poor growth, loss of weight, skin lesions, and congenital malformations also may be exhibited. The bone abnormalities may include extensive bone resorption, erosion, and narrowing of the bone shaft, bone fragility, cervical spondylosis, short bones because of retarded growth, hyperostoses, exostoses, and kyphoscoliosis. The cartilage matrix may be destroyed. For example, high levels of vitamin A injected into rabbits may cause the ears to curl because of the destruction of cartilage. Retinoic acid is more toxic than retinol.

VITAMIN D

Only a few years after vitamin A was discovered, it became clear, through the work of Mellanby in England with dogs, that a dietary deficiency was concerned in rickets. It was first thought that a lack of vitamin A was the causative factor, but proof that the disease was due to a distinct vitamin was furnished by McCollum and associates[28] in 1922. This proof was obtained by oxidizing cod-liver oil until vitamin A was destroyed, as shown by the inability of the oil to cure xerophthalmia, and then by demonstrating that the oxidized oil was still effective in curing rickets. Following the initial use of the term *antirachitic factor,* in 1925 McCollum proposed the name vitamin D, which was adopted.

There are two forms of vitamin D: ergocalciferol (vitamin D_2) and cholecalciferol (vitamin D_3). Their structures which are shown in Fig. 11.6 differ only on the side chains. Both have the same value for rat, dog, pig, calf, and man, but D_3 is much more effective for the chick, turkey, and New World primate. D_2 and D_3 are formed by ultraviolet irradiation of the steroids, ergosterol, and 7-dehydrocholesterol, respectively. Ergosterol is produced in plants and 7-dehydrocholesterol in animals.

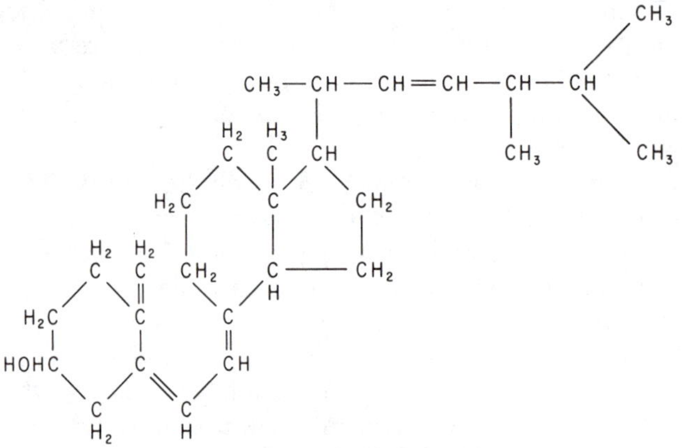

$$CH_3-CH-CH=CH-CH-CH\diagdown\begin{matrix}CH_3\\CH_3\end{matrix}$$

Ergocalciferol (vitamin D_2), $C_{28}H_{44}O$

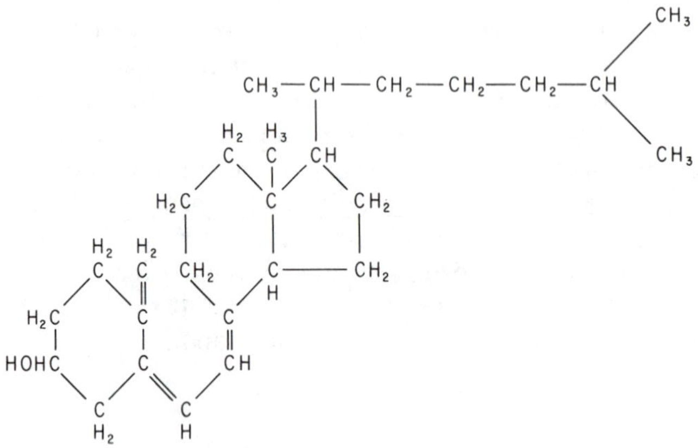

$$CH_3-CH-CH_2-CH_2-CH_2-CH\diagdown\begin{matrix}CH_3\\CH_3\end{matrix}$$

Cholecalciferol (vitamin D_3), $C_{27}H_{44}O$
Irradiated 7 – dehydrocholesterol

Fig. 11.6 Vitamin D_2 and vitamin D_3.

11.10 Physiological Functions and Symptoms of Deficiency

It is evident from previous discussions (Sec. 10.11) that, however the term is defined, rickets is a disturbance of calcium and phosphorus metabolism and that the mineral relations in the diet as well as the vitamin are involved. There can be no calcification without calcium and phosphorus. On the other hand, the vitamin has a nutritional significance in addition to its relation to this disturbance of bone growth.

Undoubtedly vitamin D is always required for the normal calcification of the bone, but the amount needed varies with the mineral relations in the diet and also with the species. More is required when the amount of either element or the ratio between them is suboptimum. But no amount will compensate for severe deficiencies of either mineral. Theiler,[29] for example, has shown that rickets develops in calves on a low-phosphorus ration despite a very large supply of D in the form of radiant energy (Sec. 11.12). The species differences are illustrated by the fact that with adequate intakes of calcium and phosphorus a ration which contains only enough vitamin D to produce normal bone in the rat or pig will quickly cause the development of rickets in chicks. Surprisingly, the human baby is more like the bird in this respect than the other mammals mentioned. Turkeys and pheasants have higher requirements than chicks. Radiographs of the width of the ulnar epiphyseal cartilage have been effectively used to detect the presence of rickets in calves and its severity. The method is described in the report by Thomas and associates.[30]

The necessity of vitamin D for normal calcification during growth has been demonstrated for many different species. The physical symptoms and bone and blood changes characteristic of inadequate calcification have been described (Sec. 10.11). The role of vitamin D in the adult animal appears much less important except during reproduction and lactation. Congenital malformations in the newborn result from extreme deficiencies in the diet of the mother during gestation, and the mother's skeleton is injured as well.

A lack of the antirachitic factor decreases egg production and hatchability, and the few eggs laid have thin shells and are easily broken. The vitamin content of the eggs produced is influenced by the amount present in the diet of the hen. In view of the intense calcium and phosphorus metabolism that takes place in lactation, one would expect vitamin D to play a large role in milk secretion. This is not true in the case of the cow, as is discussed later (Sec. 17.36). Other species have been little studied. Apparently the vitamin is not readily secreted into milk, for massive doses in the feed are required to influence its concentration in this secretion. Ordinary levels of cod-liver oil are ineffective for this purpose.

A primary function of the vitamin is the promotion of calcium absorption. D_3 is converted in the liver to 25-hydroxycholecalciferol (25-OH-D_3) which in turn is converted in the kidney to 1α,25-dihydroxycholecalciferol (1α,25(OH)$_2$D$_3$). The latter is the physiologically active form of vitamin D. It stimulates the synthesis of calcium-binding protein. The binding protein is necessary for efficient calcium absorption. It has been found in many species of mammals and birds. It is also involved in phosphorus absorption.

Because 1α,25(OH)$_2$D$_3$ is produced in the body tissues and because of its mode of action, many scientists feel that it is more proper to call it a hormone than a vitamin. Its mode of action is probably through mRNA synthesis. Haussler[31] proposed that 1α,25(OH)$_2$D$_3$ enters the cell and is transported to the chromatin as a hormone-receptor complex. The complex and RNA polymerase II combine to result in the synthesis of new RNA which is probably mRNA or a precursor form. This compound is transported to the cytosal and translated into calcium-binding protein. Haussler concluded that the proposed mechanism resembles that of steroid hormones.

11.11 Metabolism of Vitamin D

Many theories have been proposed as to vitamin D metabolism and regulation. Some early reports indicated that the 25-hydroxylation could be suppressed by feedback inhibition but other studies suggest that 25-hydroxylase is not inhibited by 25-OH-D$_3$ or D$_3$. On the other hand, the 1α hydroxylation has been clearly demonstrated to be an important control point. The induction of hypocalcemia or vitamin D deficiency increases the activity of 1α hydroxylase in the kidney. The mechanism(s) for the stimulation of the hydroxylase activity are not well known but it has been proposed that the parathyroid hormone which is released in response to hypocalcemia stimulates the hydroxylase. Other studies indicate that hypocalcemia or even hypophosphatemia can directly affect the activity of the enzyme. Other metabolites such as 24,25-(OH)$_2$D$_3$ and 25,26-(OH)$_2$D$_3$ are formed from 25-OH-D$_3$ but their role is not known. Perhaps they are also converted to active forms such as 1,24,25-trihydroxycholecalciferol but more probably they are inactive forms that are more easily excreted. Hence this is a form of regulation because when 1α,25(OH)$_2$-D$_3$ is not needed, 24,25-(OH)$_2$D$_3$ is preferentially produced. Ergocalciferol probably has analogous derivatives and similar regulation to that of cholecalciferol.

The body has some ability to store the vitamin, although to a much lesser extent than is the case for vitamin A. The principal stores of vitamin D occur in the liver, but it is also found in the lungs, kidneys, and elsewhere. While there is no large transfer to the fetus, a liberal intake during gestation does provide a sufficient store in the newborn to help prevent early rickets. For example, newborn lambs can be provided enough in this way to meet their needs for 6 weeks. Excess vitamin D is excreted in the bile.

11.12   Vitamin D and Radiant Energy

While the value of sunlight in the treatment of rickets had been known for many years prior to the discovery of vitamin D, it was not until x-ray methods of diagnosis became available that positive proof was obtained of a specific effect on bone calcification. Using these methods, Huldschinsky demonstrated in 1919 that ultraviolet light caused the deposition of calcium salts in the bones of rachitic children and thereby cured the disease. Later work showed that sunlight was also effective.

When it became evident that both ultraviolet light and a factor present in cod-liver oil produced an identical effect in the healing of rickets, the question naturally arose as to why two such apparently unrelated factors could produce the same specific results. The answer was not long in coming. Following the studies of Goldblatt and Soames showing that livers of irradiated rats possessed antirachitic properties, Hess of Columbia University and Steenbock of Wisconsin, independently and almost simultaneously, announced in 1924 that food materials which were ineffective in preventing rickets could be made antirachitic by exposing them to ultraviolet light. The original announcements were published in detail by Hess and Weinstock,[32] by Steenbock and Black,[33] and by Steenbock and Nelson.[34] Several other papers quickly followed showing that a great variety of edible materials could be activated, that the same short wavelengths were here concerned as were effective in irradiating the body, and that the active substance was in the unsaponifiable fraction. These observations provided the working hypothesis that radiant energy cured rickets because it activated some precursor in the body to provide the active agent, that by similar action it produced in certain foods a similar agent which became effective upon ingestion, and that certain substances such as cod-liver oil naturally possessed this agent. Leads were thus provided for an attack on the problem of the chemical nature of the antirachitic factor.

11.13 Vitamin D in Foods

Both provitamins D are widely distributed in nature, but the active forms have the most limited distribution in natural foods of any of the vitamins. Although ergosterol occurs commonly in vegetation, ergocalciferol is never present in the living plant. In the animal world 7-dehydrocholesterol is the predominant provitamin which, through the action of sunlight on the body surface (Sec. 11.15), is changed to the active form. The tissues and products of animals contain vitamin D by this means, and also through consuming foods containing one of the provitamins which has been irradiated, or foods which have been commercially enriched (Sec. 11.14). Among animal products, eggs, especially the yolks, are a very good source, particularly where the diet of the hen is rich. Milk contains a variable amount in its fat fraction (5 to 40 U.S.P. units in cow's milk per quart), but neither cow's nor human milk contains enough to protect the baby against rickets. Other animal products are poor, as is to be expected from the fact that the storage of the vitamin in animal tissues is very limited. Certain fish meals, depending upon the nature of the raw material and its processing, contain fair amounts of the antirachitic factor.

Seeds and their by-products are practically devoid of the vitamin. The same is true for the living tissues of pasture grass and other growing forage crops. During the sun-curing of roughages, however, vitamin D is formed under the action of radiant energy upon ergosterol or some other provitamin, and the principal source of the antirachitic factor in the rations of farm animals is thus provided. Legume hay which is cured in such a way as to preserve most of its leaves and green color contains considerable amounts. Alfalfa, for example, will range from 650 to 2200 I.U. per kg. Timothy and other grass hay contain less. Stemmy hay, lacking in leaves and color, which has been exposed to a minimum of sunlight may contain none, whether legume or nonlegume.

Machine-dried and barn cured hay generally contains less than that which is properly sun-cured. Even hay dried in the dark immediately after cutting, however, has some of the vitamin present. This results because the dead or injured leaves on the growing plant are responsive to irradiation even though the living tissues are not. This fact is also largely responsible for the vitamin D found in corn silage. An extensive study of this general subject was reported by Moore and coworkers which showed that either barn-cured hay or wilted silage could supply sufficient D to prevent rickets in calves reared out of sunlight. Thomas and Moore[35] have shown that the antirachitic value of the alfalfa crop increases with state of maturity because of the increase in dead leaves which were found very high in the vitamin. A cooperative study carried out at ten agricultural experiment stations, involving the determination of vitamin D in 65 roughage samples, has been reported by Wallis and coworkers.[36] The results, which showed wide and unpredictable variations for a given type of forage, are summarized as follows in I.U. per kilogram on an air-dry basis:

Forage	I.U./kg
Sun-cured hay	150–3120
Mow-cured hay	350–1740
Winter range grass	200–590
Silage (wet basis)	150–240
Artificially dried roughages	170–620

Cod- and certain other fish-liver oils, as well as certain fish-body oils, are rich sources of vitamin D and thus are used in both human and animal nutrition to supplement the common foods which are deficient. More commonly, however, the much more concentrated products mentioned in the following section are employed.

11.14   Enrichment of Foods in Vitamin D

Steenbock patented his discovery that certain foods could be enriched in the antirachitic factor by irradiation with ultraviolet light, assigning the patent to the University of Wisconsin. The process has found wide application and has proved an outstanding contribution to better nutrition. Activation is dependent upon the presence of a provitamin in the substance in question, and thus certain materials develop a high potency on irradiation, while others acquire little or none. The most potent products are obtained by irradiating the sterols that are subject to activation. Thus irradiated ergosterol is produced and sold for human use in a variety of forms. Irradiated animal sterol, activated 7-dehydrocholesterol, is most frequently used in poultry feeds in view of the superior value of the D_3 form of the vitamin for this species. Yeast is rich in ergosterol and thus its irradiation results in a potent source that has been used for other farm animals. Milk can be irradiated to contain 400 U.S.P. units per quart, a level also obtainable by feeding irradiated yeast (Sec. 17.37). However, most foods are now enriched by the addition of vitamin D rather than by irradiation.

11.15   Sunlight and Vitamin D Nutrition

In the previous section we have seen how sunlight activates ergosterol in the curing of roughages. The action of sunlight on the body is even more important in its effect on vitamin D nutrition. The skin and sebaceous secretions contain the precursor, 7-dehydrocholesterol, and thus the activated substance produced on and in the skin is absorbed, or licked off, in the case of animals. That absorption can take place is clear from the fact that rickets can be successfully treated by rubbing cod-liver oil on the skin. Irradiation is less effective on dark-pigmented skin. This has been shown to be true for white and black breeds of hogs as well as for people. Irradiation is more effective on exposed skin than through a heavy coat of hair.

The effectiveness of the sunlight is dependent upon the length and intensity of the ultraviolet rays which reach the body. It is ineffective through ordinary window glass because the latter does not allow sufficiently short wavelengths to pass through. The radiations which reach the earth contain only a small part of the ultraviolet range which has an antirachitic effect. The shortest wavelength which ever reaches the earth is 290 mμ, shorter ones being absorbed by the atmosphere. This shortest available wavelength reaches the earth only in summer and only in the tropics. The greater the distance the rays have to travel, the longer is the minimum wavelength reaching the earth and the lesser the intensity of the effective radiations. Thus sunlight is more potent in the tropics than in the Temperate or Arctic zone, more potent in summer than in winter, more potent at noon than in the morning or evening, and more potent at high altitudes. These variations are of large importance in vitamin D nutrition. Animals which are on pasture during the summer never suffer from the lack of the antirachitic factor even though their diet is practically devoid of it. In the wintertime the story is different. At best, the animals are outside only a part of the time, there are generally fewer sunny days, and the sunlight which actually reaches the animal is much less effective than in summer. Under most conditions of practice in the latitude of the northern United States, it is unsafe to rely on exposure to sunlight to provide the antirachitic factor during the winter months, as has been definitely proved for pigs and calves.

Fortunately, especially for city dwellers, it is not necessary for the body to be in the direct sunlight in order that activation may take place. It can occur in the shadow on sunny days. "Skyshine" from the northern sky on bright days may be one-half to two-thirds as potent as direct sunlight. Rays reflected from snow and water are more potent than when direct. Clouds, smoke, and dust cut down the effectiveness of the light.

11.16   Overdosage with Vitamin D

Experiments with massive doses of irradiated ergosterol have shown that a condition of "hypervitaminosis" can be produced, characterized by hypercalcemia, the widespread deposition of calcium salts in the arteries and various organs and tissues (calcinosis) and bone changes.

Chineme and coworkers[37] reported that an intake of 625 times the recommended requirement of vitamin D_3 resulted in calcinosis in young pigs. Pigs fed 25 or 125 times the normal level did not show calcinosis but negative effects on osteoblasts and osteocytes were observed. They concluded that osteopenia (lack of bone) as a result of D_3 toxicosis is not due to increased resorption of bone but rather to decreased formation of bone because of the destruction of bone forming cells (osteocytes and osteoblasts).

The events in vitamin D toxicosis as described by L. Krook and coworkers are shown in Fig. 11.7.

Detailed human studies have shown low tolerances by both infants and pregnant women. The Committee on Nutrition of the American Society of Pediatrics found that the toxicity levels for infants may be as low as 3000 to 4000 I.U. per day compared with a recommended daily intake of 400 I.U. The

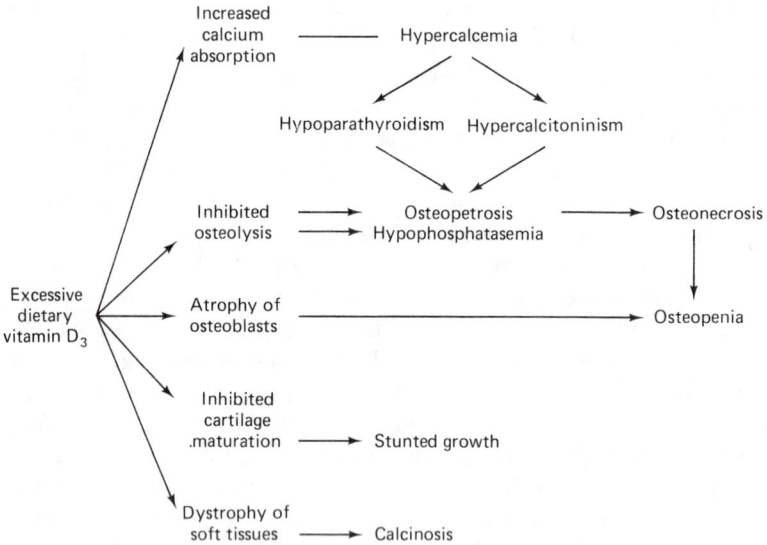

Fig. 11.7 Summary of events in vitamin D₃ toxicosis (*Chineme et al.*[37])

Committee has recommended that the enrichment of foods other than milk be discontinued and that products containing more than 400 I.U. in recommended daily doses be restricted to sale by prescription only.

It is recognized that there are situations, notably limited access to sunlight, where vitamin D supplements to the rations of farm animals are called for, but overdosage with the pure vitamins or concentrates of them should be avoided as a safety measure.

Although it is usually assumed that living plants do not contain D_2, some plants contain compounds which have $1\alpha,25(OH)_2D_3$ activity. Furthermore, these plants are extremely toxic to animals and have caused extensive economic losses. The ingestion of plants such as *Cestrum diurnum* and *Solanum malacoxylon* has caused lameness, hypercalcemia, osteopetrosis, and dystrophis calcinosis of soft tissues of horses, cattle, and pigs.[38,39] The amount ingested of $1\alpha,25(OH)_2D_3$ to result in toxicosis is much lower than for D_3 because when $1\alpha,25(OH)_2D_3$ is fed, the feedback regulation of conversion of $25(OH)D_3$ to $1\alpha,25(OH)_2D_3$ by the kidney enzyme is bypassed. *Cestrum diurnum* (common names are day-blooming, jessamine, wild jasmine, King of the Day) is a subtropical to tropical plant that is a native of the West Indies that has been introduced and cultivated as an ornamental in many parts of the United States such as California, Texas, Florida, and Hawaii. *Solanum malacoxylon* is a South American plant. Both *C. diurnum* and *S. malacoxylan* are members of the solanaceae family. *Trisetum flavescens* (yellow oat-grass) is a plant in the Alpine region of Europe that also produces calcinosis, but $1\alpha,25(OH)_2D_3$-like activity was not observed in one study[40] although the plant was shown to have vitamin D activity.

11.17 Clinical Uses of Vitamin D Metabolite

The discovery of the active form of vitamin D may permit treatment of several disorders that were previously only ineffectively treated with vitamin D_3.[41] Patients with chronic renal failure may develop hypocalcemia, bone abnormalities and negative calcium balance. Vitamin D_3 is not an effective treatment because the damaged kidney cannot effectively produce $1\alpha,25(OH)_2D_3$ but humans and dogs with chronic renal failure have been successfully treated with $1\alpha,25(OH)_2D_3$. Patients with familial vitamin D-dependent rickets respond to low levels of $1\alpha,25(OH)_2D_3$ but not D_3 or $25(OH)_2D_3$.[42] This indicates that the genetic condition is due to a lack of 1-hydroxylase activity.

The effectiveness of milk fever prevention in cattle may also be improved with vitamin D metabolites (Sec. 17.36).

VITAMIN E (TOCOPHEROLS)

As a result of the stimulus to experimentation with purified diets which followed the discovery of the first vitamin, it was frequently observed that on certain diets, which were satisfactory for growth and health, rats failed to reproduce. Studies of the cause of this failure resulted in the discovery in the early twenties, by Evans of the University of California, Mattill of the University of Rochester, Sure of the University of Arkansas, and their associates, that there is a specific dietary factor essential for reproduction in the rat. Sure coined the name vitamin E, which is now known to consist chemically of tocopherols. Much has been learned about the chemical nature of vitamin E and its distribution in foods, but despite many studies the knowledge of its physiological functions and need by various species remains in an unsatisfactory state. Specific information regarding its significance in farm-animal nutrition is especially limited.

11.18 Physiological Functions and Signs of Deficiency

Vitamin E has been demonstrated to be required by a large number of animal species but the deficiency signs may differ greatly among species and even within the same species. Vitamin E seems to be involved in a large number of apparently unrelated body functions. Vitamin E is essential for reproduction in several species such as rotifers, paramecia, snails, and crickets. Testicular degeneration has been reported in vitamin E-deficient dogs, hamsters, pigs, rabbits, rats, roosters, and fish. The changes are irreversible in most of the species but can be corrected in the rooster. Vitamin E deficiency may cause fetal death or weak offspring in rats, mice, guinea pigs, and hamsters. Vitamin E deficiency does not appear to cause decreased fertility in the male mouse nor have any experiments shown a connection between vitamin E deficiency with reproductive failure in cattle, sheep, or goats. The studies with horses on vitamin E and reproduction have been contradictory.

Muscular dystrophy has been experimentally produced in various animals on E-deficient rations. The dystrophy is found primarily in the skeletal muscles, but sometimes also in the heart. The external symptoms of muscle weakness and paralysis are reflected in degenerative histological changes. Creatine excretion is increased. Vitamin E does not appear to be concerned in human muscular dystrophy, however. Cornell workers[43] showed that the vitamin can prevent and cure "stiff-lamb" disease, which is characterized by stiffness and dystrophic lesions. Mention has been made of the findings that selenium also prevents this disease (Sec. 10.47). Muscle dystrophy has been produced in chicks on a diet low in both vitamin E and sulfur-containing amino acids. The addition of one of these acids alone will alleviate the trouble.

Chronic vitamin E deficiency has been noted to produce, in rabbits, lambs, poultry, and cattle, electrocardiogram changes which are considered to reflect heart muscle injury. Sudden deaths have resulted in some cases, and histological examinations have showed atrophy and scarring of the muscle fibers. Several studies have shown that a deficiency of the vitamin in the diet of the chick results in nutritional encephalomalacia, characterized by an uncoordinated gait, prostration, and brain lesions. While experimental muscular dystrophy in rabbits and guinea pigs, produced by an E-deficient diet, is not accompanied by changes in the central nervous system, there is evidence that a chronic deficiency in rats affects both the muscular and the nervous systems. Exudative diathesis, a hemorrhagic disease of chicks, and dietary liver necrosis in rats and pigs are other troubles which are prevented by vitamin E and also by selenium, as previously mentioned. A basic effect of E deficiency is the hemolysis of red blood cells. Premature infants fed formulas with inadequate vitamin E develop hemolytic anemia resulting in shortened life span of the red cell. Steatitis, "yellow fat disease," has been reported to be a vitamin E responsive disease in cats, mink, and reptiles. Several cases were reported in cats fed tuna fish containing rancid oils. The many diverse symptoms and pathological changes in vitamin E deficiency, many of them apparently unrelated, may have some common physiological link.

Tappel[44] concluded that vitamin E reacts or functions as a chain-breaking antioxidant, thereby neutralizing free radicals and preventing peroxidation of lipids within the membranes. Peroxidation of the lipids can destroy the structural integrity of the cells and cause metabolic derangements. These changes can account for many of the observable symptoms of vitamin E injury.

The findings with respect to the antioxidant properties of the vitamin explain earlier observations (going back to 1926) that large intakes of cod-liver oil, which is high in unsaturated acids but low in vitamin E, caused muscular dystrophy in various herbivora, which did not occur when hydrogenated oil was fed. It is clear that highly unsaturated acids in the diet increase the vitamin E requirement. This has been found true for various species. Mention has been made (Sec. 7.18) that adding the vitamin to the diet corrected the dystrophy in young calves which resulted from the substitution of vegetable oils for butterfat in their diet.

The related effects of vitamin E, selenium, and antioxidants are summarized in Table 11.2.

TABLE 11.2. Summary of Effects of Vitamin E, Selenium, and Antioxidants on Vitamin E-Deficiency Diseases

Species	Symptom	Prevented by		
		E	Se	DPPD
Chick	Muscular dystrophy[a]	+	+ (Partial)	+
	Exudative diathesis	+	+	±(Low Se)[c]
	Encephalomalacia	+	–	+
Rat	Resorption-gestation	+	–	+
	Liver necrosis[b]	+	+	+
Lamb	Muscular dystrophy	+	+	±[c]
Calf	Muscular dystrophy	+	+	+
Rabbit	Muscular dystrophy	+	–	+
Swine	Liver necrosis	+	+	
Mouse	Multiple necrotic degeneration (heart, liver, muscle)	+	+	
Monkey	Anemia	+	–	(Partial)
Mink	Skeletal and cardiac myopathy	+	+	

[a] Need diet deficient in vitamin E and sulfur amino acids. Cystine (or methionine) will prevent it.
[b] Sulfur amino acids reduce need for tocopherol.
[c] ± indicates reports in literature conflict as to effectiveness of antioxidants; may depend on fat in the diet.
Source: Adapted from table prepared by Louise Daniel, Cornell University.

Bieri[45] concluded that it remains to be demonstrated that any enzyme reaction has a specific requirement for α-tocopherol although many of the observed enhancements of inhibitions due to vitamin E status have not been adequately explained.

The study of the functions of vitamin E is complicated by the fact that the body has a large ability to store it. This storage is largest in the liver but also occurs in various organs and tissues. The extent of the store is shown by the fact that females born of mothers whose diets contained a liberal supply frequently have enough in their bodies at birth to carry them through a first pregnancy. Rats reared on natural foods rich in the factor and then placed on a deficient diet may produce three or four litters before exhausting their reserves. The vitamin E content of the body tissues can also be demonstrated by feeding these tissues to females which have failed as a result of a deficiency and noting the recovery that occurs. Storage doubtless complicates a study of questions now at issue as to the functions of the vitamin and its need by various species. Tocopherols pass through the placental membranes and also the mammary gland; thus the diet of the female influences the store of the young at birth and the amount it gets from its mother's milk.

11.19   Chemistry of Vitamin E

The group consists of four tocopherols (α, β, γ, and δ) and four tocotrienols (α, β, γ, and δ). The structure of α-tocopherol is shown below. The differences between α, β, γ, and δ are due to the placement of methyl groups on the ring. The difference between tocopherols and tocotrienols is due to the unsaturation of the side chain in the latter.

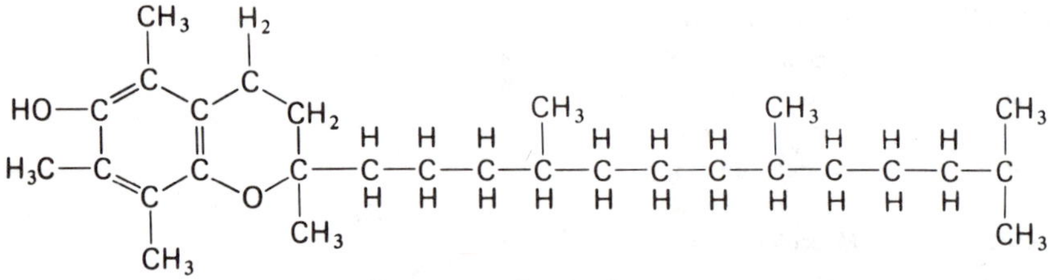

α-Tocopherol, $C_{29}H_{50}O_2$

Several active synthetic products have also been produced. Thus, vitamin E, like several others, has a multiple nature. The α form is much more active than the others, at least for the cure of resorption and sterility in the rat and for the prevention of dystrophy. The tocopherols are extremely resistant to heat but readily oxidized. They keep well in ordinary feeds and mixtures but are destroyed by rancid fats.

11.20   Vitamin E Assay and Feed Supply

Synthetic D,L-α-tocopheryl acetate is the international standard. One mg equals 1 I.U., defined as the amount needed to prevent resorption in 50 percent of the rats on an E-deficient ration.

Vitamin E is widely distributed in livestock feeds. In nature the synthesis of vitamin E is a function of plants and thus their products are by far the principal sources. It is abundant in whole cereal grains, particularly in the germ, and thus in the by-products containing the germ. Green forage and other leafy materials, including good-quality hay, are very good sources. Alfalfa is especially rich. Animal by-products supply only limited amounts, and milk and dairy products are poor sources. Eggs, particularly the yolk, make a significant contribution, depending, however, on the feed of the hen. Wheat-germ oil is the most concentrated natural source. Various other oils such as soybean, peanut, and particularly cottonseed are also rich. Unfortunately, most of the oil meals now marketed are devoid of these oils because of solvent extraction. Animal products contain mainly α-tocopherol. Cereals contain about equal amounts of α and non-α forms. Legumes supply mostly the non-α forms. Since vitamin E is readily oxidized, its supply deteriorates in ground feeds. Compilations of data on the vitamin E content of human and animal foods have been published by Dicks[46] and Bauernfeind.[47]

The wide distribution of vitamin E bears out the findings of experimental work that any need of farm animals should usually be met by the commonly fed rations.

Megadoses of Vitamin E. Many persons routinely consume large doses of vitamin E but the value of such a practice remains to be determined. Fortunately, in contrast to vitamins A and D, vitamin E is relatively nontoxic. Bieri[45] concluded that, for most humans, doses below 300 I.U. are innocuous but doses above 300 I.U. may result in minor complaints of nausea and intestinal distress.

VITAMIN K

In 1929, Henrik Dam of Denmark fed chickens a purified diet in an attempt to determine whether they were able to synthesize cholesterol. He noted that the chickens that were kept on an ether-extracted diet became anemic and developed subcutaneous and intermuscular hemorrhages. This was the first observation of the symptoms that we now know to be attributable to vitamin K deficiency. Since the factor was found to be concerned with blood coagulation, the Danish workers (Dam and Schønheyder) proposed the name *vitamin K* from the Danish word for coagulation.

11.21 Physiological Functions and Symptoms of Deficiency

The coagulation time of the blood is increased when vitamin K is deficient because the vitamin is required for the synthesis of prothrombin (factor II) and the other clotting factors VII, IX, and X. It has been proposed that the mode of action of vitamin K is to convert the precursor to the factor by the carboxylation of the glutamic acid residues. Vitamin K-dependent carboxylase enzyme systems may be widely distributed and may have other roles in addition to blood clotting such as in bone formation but further studies are necessary to identify such activities.

Though originally discovered in chicks, the symptoms of K deficiency have been reported in other birds (geese, pigeons, and ducks) and also in mice, rats, rabbits, baby pigs, and humans, under special conditions. Although the vitamin is needed for the physiological functioning of all species, any dietary need may be unimportant in species other than birds because of bacterial systhesis in the digestive tract (Sec. 11.23). Laying hens fed a diet deficient in vitamin K produce eggs low in the vitamin. When the eggs are incubated, the chicks produced have low reserves of the vitamin and a prolonged blood-clotting time. Vitamin K has been found to have a definite value in human therapy: (1) as a preoperative and postoperative measure to prevent risk of bleeding; (2) in cases where absorption is impaired, as in obstructive jaundice, because bile is necessary for the absorption of K; and (3) in hemorrhagic diseases of the newborn. The blood of infants at birth contains less than the usual amounts of prothrombin, and this deficiency increases for a time because the intestine of the newborn is sterile.

Vitamin K bears an interesting relation to "hemorrhagic sweet clover disease." This disease has been responsible for large animal losses. Studies have also shown that spoiled sweet clover contains dicumarol that decreases the blood prothrombin, resulting in the hemorrhages that are characteristic of the disease. More vitamin K will overcome this action by dicumarol. These findings have resulted from a series of brilliant studies by Link and associates of the University of Wisconsin. Dicumarol serves as an anticoagulant in medicine in certain situations, just as vitamin K under other conditions steps up the coagulation time. The action of dicumarol against the vitamin is similar to that of an antimetabolite.

11.22 Chemistry of Vitamin K

It is now recognized that there are several different compounds, similar in structure, that have vitamin K activity. In 1939, as a result of the activities of several different workers, two such compounds were isolated as fat-soluble substances. The more active one, designated as K_1, was isolated from alfalfa and found to have the following structure:

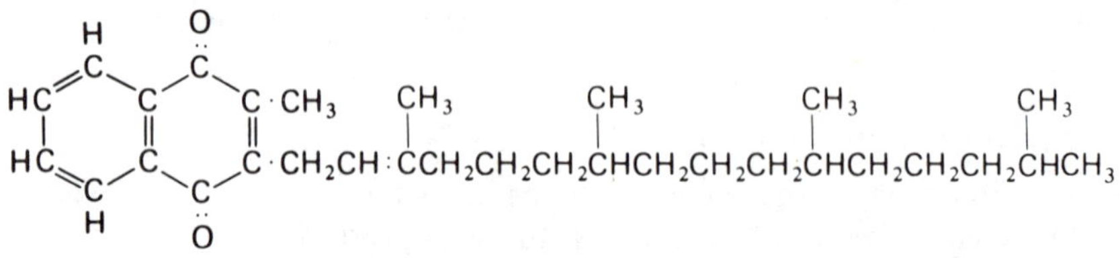

Vitamin K_1 (2-methyl-3-phytyl-1,4-naphthoquinone, phylloquinone)

The second compound, designated as K_2, was isolated from fish meal. It contains the same quinone nucleus, but there are some differences in the side chain. The preferred name for K_1 is phylloquinone and menaquinone is for K_2. Later other active compounds were isolated which were found to differ as regards the length and structure of the side chain. In addition, several synthetic naphthoquinones have been prepared that have vitamin K activity. 2-methyl-1,4-naphthoquinone, called menadione, is a synthetic product which is much more active than K_1. Some of the synthetic products are water soluble, in contrast to the natural products that caused the vitamin to be classed originally as fat soluble. The formula for dicumarol is shown below.

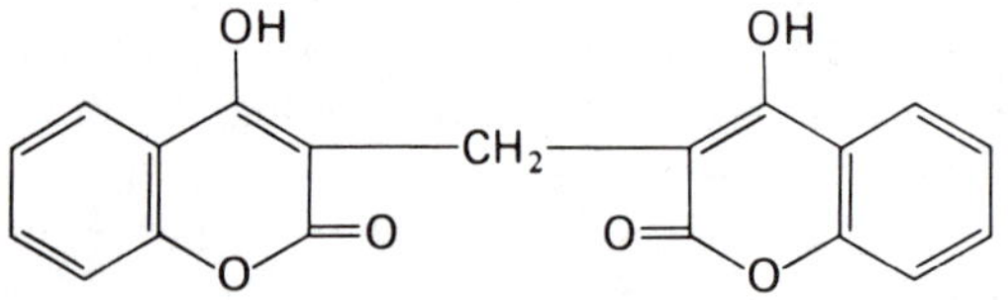

Dicumarol,3,3′—methylenebis(4—hydroxycoumarin)

11.23 Bacterial Synthesis of Vitamin K

Like the case for amino acids (Sec. 8.14), microorganisms in the digestive tract play important roles in the synthesis of certain vitamins and in modifying dietary needs accordingly. Vitamin K is the first example met in this chapter.

Like most of the B-vitamins discussed later, vitamin K is synthesized by bacteria in the rumen. In cattle and sheep the rumen is the principal seat of the synthetic activity. In nonruminants, however, synthesis takes place in the large intestine, as was demonstrated by the following rat experiment. Animals in which the blood-clotting time was normal were fed sulfonamides which check bacterial action in the intestine. As a result the clotting time was slowed down and hemorrhages developed. These defects were then corrected by feeding or injecting the vitamin. While this experiment proved that intestinal synthesis took place, its practical significance has been greatly lessened by a later finding indicating that *coprophagy* (feces eating) may have occurred despite the fact that the rats were kept on wire screens. Many years ago it was proved that rats caged on litter obtained B-vitamins by feces eating. The vitamins were apparently synthesized in the tract below the zone of absorption, excreted in the feces, and then absorbed when the feces were eaten. Thus the practice was established of keeping experimental animals on wire screens to prevent this complication in vitamin feeding experiments. The practice was assumed to be successful until in 1957 Barnes and coworkers[48] showed, by the use of a new technique, that rats may eat 50 to 65 percent of their feces even when maintained on wire screens. In a later experiment Barnes and Fiala[49] showed that, when rats were prevented by this new technique from feces eating, a vitamin K deficiency uniformly developed on a diet free of the vitamin. This did not occur when a dietary source of the vitamin was provided. These and other findings by Barnes and associates have called for a reexamination of previous experiments with feces-eating animals from which it has been assumed that the techniques used prevented coprophagy and thus that the vitamins synthesized in the intestine from diets devoid of them were directly absorbed and thus served the body. This fact must be borne in mind in reading the later discussions in which the intestinal synthesis of vitamins is mentioned.

There are physiological reasons why one would not expect any large absorption of vitamins synthesized in the large intestine. The bacterial cells containing them are little subject to digestive action, in contrast to what happens in the small intestine where rumen organisms containing vitamins are broken down. Furthermore, absorption of nutrients in general from the large intestine appears to be limited because of the nature of the epithelial lining.

These considerations do not affect the practical significance of conclusions previously drawn regarding the synthesis of vitamin K and various B-vitamins in the rumen. In the case of poultry little intestinal synthesis apparently occurs because of the short digestive tract.

11.24　Food Sources of Vitamin K

As far as is known there is no need for giving any special consideration to the vitamin K content of the rations of farm animals except in the case of poultry. The N.R.C. Committee has set a requirement of 0.53 mg per kilogram feed for starting chicks. Rumen synthesis should meet any needs of cattle and sheep not supplied by their rations. Specific deficiency symptoms have been produced in baby pigs housed in wire-bottom cages and fed a synthetic diet. The need for the vitamin was met by feeding 5 μg per kilogram body weight. It would appear that under practical conditions the needs of pigs would be met by intestinal synthesis, but this is a matter that requires further study. All green, leafy materials, fresh or dry, are rich sources of the vitamin, and some other plant products contain substantial amounts. Liver, egg, and fish meal are good animal sources.

11.25 Toxicity of Vitamin K

The natural forms of vitamin K, phylloquinone and menaquinones, are relatively nontoxic but large amounts of menadione may produce hemolysis, methemoglobinuria, and porphyrinuria.

THIAMIN (VITAMIN B₁)

The work of Eijkman and others which established the fact that there is a specific dietary factor essential for the prevention of beriberi in man and polyneuritis in pigeons has been referred to. Further studies of this factor, which became known as water-soluble B and later simply as vitamin B, caused it to be recognized as essential for growth and for certain other physiological functions besides its antineuritic properties. As more critical investigations were made of its distribution in foods, chemical nature, and properties, the realization gradually developed, between 1925 and 1930, that vitamin B actually consisted of at least two factors differing as regards chemical nature and physiological effects. Thus what was formerly spoken of as vitamin B came to be called "the vitamin B complex" or "group." The term *vitamin B* (or B₁) was reserved for the antineuritic factor. The name "thiamine" was introduced when its chemical nature was established. The spelling "thiamin" is now recommended. The vitamin is required in the metabolism of all species of animals and in plant metabolism as well. Higher plants synthesize it, and so do many of the lower forms. All animals, however, must have a dietary source, unless it is synthesized for them by microorganisms in the digestive tract, as is the case with ruminants.

11.26 Physiological Functions and Symptoms of Deficiency

Thiamin functions in all cells as the coenzyme *cocarboxylase* (thiamin pyrophosphate). It is the coenzyme for all enzymic decarboxylations of α-keto acids. Thus it functions in the oxidative decarboxylation of pyruvate to acetate, which in turn is combined with coenzyme A for entrance into the tricarboxylic cycle. This is an essential reaction for the utilization of carbohydrates to provide energy for body processes. When thiamin is deficient, pyruvic and lactic acids accumulate in the blood and tissues and give rise to deficiency symptoms. In the cycle itself there are two steps at which CO_2 is released under the action of the coenzyme. There are several other metabolic processes in which thiamin plays an essential role. Because its principal functions are concerned in energy metabolism, its requirement by the body bears a direct relation to the energy intake.

The classic diseases, beriberi in man and polyneuritis in birds, represent a late stage of the deficiency, resulting from a peripheral neuritis, perhaps caused by the accumulation of intermediates of carbohydrate metabolism, as was first indicated by the studies of Peters and associates at Oxford. It has also been proposed that the neurological disease of thiamin deficiency may be due to lack of membrane permeability and that thiamin is localized in the membranous structure in the nerve and plays a role in the sodium transport system. Other symptoms include a slowing of the heart beat (bradycardia), enlargement of the heart, edema, gastrointestinal troubles, and lack of appetite (anorexia). Muscle weakness, easy fatigue, and hyperirritability are less specific symptoms. There is a very marked effect on growth that is the result, at least in part, of loss of appetite. In swine the deficiency reveals itself particularly in a decrease of appetite and body weight, vomiting, a slow pulse, subnormal body temperature, nervous symptoms, and postmortem heart changes. In chickens and turkeys, there is a loss of appetite, emaciation, impairment of digestion, a general weakness, opisthotonos or star gazing, and frequent convulsions with polyneuritis as an extreme symptom. This characteristic trouble will develop in 9 to 12 days with day-old chickens on a thiamin-deficient diet. Recovery is amazingly prompt when the vitamin is given. In foxes, the deficiency causes a characteristic disease, Chastek paralysis. Horses fed experimental diets low in B_1 and other B-vitamins have shown incoordination and other nervous symptoms which were alleviated by feeding thiamin, indicating that this species requires a dietary source of this vitamin. A lack of vitamin B_1 causes reproductive failure in both sexes. There is a larger requirement for lactation than for growth because of the increased metabolism involved.

While the symptoms listed above have been proved experimentally to result from B_1 deficiency, most of them are not specific for this deficiency alone. In pigs, for example, nervous symptoms result from deficiencies of vitamin B_6 and pantothenic acid as well as of thiamin. Symptoms noted in practice may represent multiple deficiencies. There is no question about the specificity of such symptoms as polyneuritis and Chastek paralysis, but these are terminal stages. The problem of identifying the early stages by any specific sign is an unsolved one for thiamin and for several other of the B-vitamins. It is recognized that the first lesions are biochemical and that tissue changes and physical symptoms are later effects. Thus current research is endeavoring to find biochemical changes which may be considered diagnostic. With respect to thiamin, the urinary excretion is linearly related to the intake except at low levels, and use is made of this fact in the diagnosis of deficiency in man. A low level of excretion following a test dose is indicative of tissue depletion.

11.27 Chemistry of Thiamin

In 1926 the Dutch workers Jansen and Donath isolated vitamin B_1 in crystalline form and reported an empirical formula containing carbon, hydrogen, oxygen, and nitrogen. Later Windaus and coworkers prepared a crystalline product which was found to contain sulfur as well as the elements reported by Jansen and Donath. Other workers confirmed the presence of this additional element, and it was later found also in the crystals of the Dutch workers. In 1936, synthesis of the vitamin was accomplished and the following structure established by R. R. Williams:

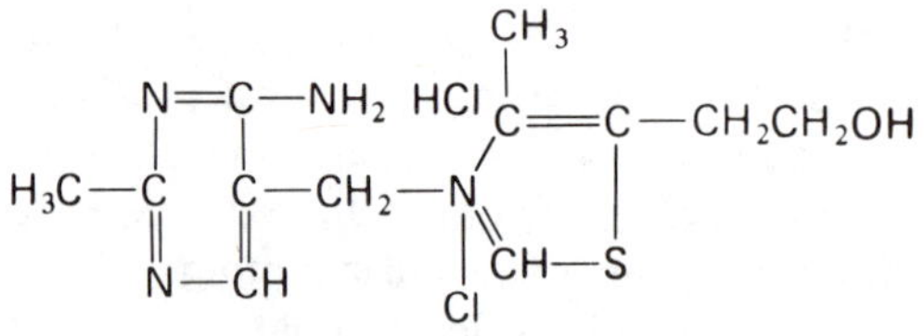

Thiamin hydrochloride

The vitamin consists of a molecule of pyrimidine and a molecule of thiazole linked by a methylene bridge. Thiamin is soluble in 70 percent alcohol as well as water and is readily destroyed by heat, especially in the presence of alkali. In a dry state it is stable at 100°C for several hours, but moisture greatly accelerates the destruction, and thus it is much less stable to heat in fresh than in dry foods. Autoclaving destroys vitamin B_1, an observation which played an important role in the discovery that what was originally considered to be a single vitamin contains more than one factor.

11.28 Metabolism of Thiamin

The vitamin is primarily absorbed from the small intestine. It is carried to the liver where it is phosphorylated under the action of ATP to form cocarboxylase. While a high level of intake may be reflected in a somewhat higher blood level, there is little storage in the body. Intakes in excess of current needs are rapidly excreted as such in the urine. This means that the body needs a regular supply and also that unneeded intakes are wasted. The pig is somewhat of an exception, however. For some reason which is not understood, its tissues contain several times as much thiamin as is the case with other species studied, and there is thus a store that can meet body needs on a thiamin deficient diet for as long as two months.

Though thiamin is a metabolic essential for all species studied, it is not usually needed in the diets of ruminants because of microbial synthesis, a fact established by the pioneer studies of Bechdel and coworkers.[50,51] However, thiamin deficiency can be produced in lambs and calves that do not have a functional rumen.[52]

Even the adult ruminant may require dietary thiamin under certain conditions. Loew[53] reported that cattle fed diets high in soluble carbohydrate and low in fiber may develop a lesion of the nervous system called polioencephalomalacia. Loew concluded that the condition was due to a thiamin deficiency because affected animals responded to thiamin therapy, biochemical findings were consistent with TPP inadequacy, and the condition could be produced with large doses of antithiamin compounds such as amprolium. The condition has caused significant economic losses in tropical countries where high levels of molasses are fed and in feed lots in the United States where high grain diets are fed.

Bacterial synthesis of thiamin in the lower intestinal tract has also been demonstrated. In the case of the horse, considerable quantities of B_1 and other B-vitamins are produced in the cecum and in the large intestine, according to the studies of Carroll and coworkers.[54] The production of nervous signs which were alleviated by feeding thiamin indicated that the synthesis of this vitamin was not sufficient to meet metabolic needs. The bacterial synthesis of thiamin in the intestine of rats and humans has also been demonstrated, but it is doubtful whether the amount thus produced and absorbed is large enough to make a significant contribution to body needs, except when the dietary intake is very low. Apparently, the same is true for the pig and the chick.

11.29 Antimetabolites of Thiamin

The term *metabolite* is used to denote any substance which is essential to the chemical process by which living cells are produced and maintained. Substances which block or inhibit the normal function of a metabolite are referred to as *antimetabolites* and also as *antivitamins* where a vitamin is the metabolite involved. Thiamin presents a case in point. In a diet which is adequate in thiamin for mice, the addition of pyrithiamine, which has the same structural formula as thiamin, except that the sulfur atom of the vitamin is replaced by the grouping —CH=CH— will produce deficiency symptoms of the vitamin which can in turn be overcome by adding more thiamin. Other structurally related compounds have been noted to produce similar effects. It is considered that the antimetabolite competes with thiamin in the enzyme systems in which the latter functions and thus antagonizes its action. In such a case the antivitamin is referred to as the *analog* of the vitamin. Later discussions refer to substances which act similarly in the case of various B-vitamins.

An antagonistic action on thiamin of a different sort is represented by the occurrence in practice of Chastek paralysis in foxes and other animals fed raw fish. Here the causative agent is an enzyme, *thiaminase,* which splits the thiamin molecule into two components and thus renders it inactive. The case seems to be similar in the disease in horses referred to as "fern poisoning" or "bracken poisoning." Following earlier evidence obtained with rats by Oregon workers that a factor antagonistic to thiamin was concerned, Evans and coworkers[55] experimentally produced the disease in horses by feeding a bracken hay ration and cured the symptoms, which were typical of thiamin deficiency, by the administration of the vitamin. It has been shown, however, that the causative agent in fern poisoning is not thiaminase but an antithiamin substance of nonenzymic nature.

Many different kinds of fish contain thiaminase and thiamin deficiency has been reported in penguin, seals, and dolphins fed diets primarily of fish in zoos. Wild aquatic animals apparently do not suffer from the thiamin deficiency even though they eat a diet primarily of fish, because the fish must undergo some putrefaction to release the enzyme.[56]

Other naturally occurring antithiamin factors include 3,5-dimethysalicylic acid found in cottonseed[57] and caffeic acid.[58]

11.30 Thiamin in Foods

The official unit (I.U. and U.S.P.) is the biological activity of 3 μg of pure thiamin hydrochloride. Thus, 1 mg of thiamin equals 333 I.U. Food content is expressed in milligrams.

Brewer's yeast is the richest known source of vitamin B_1. Whole cereal grains are rich sources. Since the vitamin is present primarily in the germ and seed coats, by-products containing the latter are richer than the whole kernel, while highly milled flour is very deficient. Wheat germ ranks next to yeast. Lean pork, liver, kidney, and egg yolk are rich animal products. The content in lean pork can be doubled by increasing the thiamin intake of the pig. The content in hays decreases as the plant matures and is less in the cured than in the fresh product. The content is correlated with leafiness, greenness, and protein content. In general, good-quality hay is a substantial source, and in a dry climate there is practically no loss in storage. Milk is not a rich source, and pasteurization for 30 min at 145°F destroys 25 percent of its content.

The fact that thiamin is water-soluble as well as unstable to heat results in large losses in certain cooking operations to which foods are subjected. There should be little concern about any lack of vitamin B_1 in the rations of farm animals in view of the generous supply in most of the feeds used. Further, rumen synthesis makes the question of feed supply of little or no importance in the case of cattle and sheep except in the conditions described earlier.

RIBOFLAVIN

Following the recognition that the original water-soluble B consisted of more than one factor, it was thought for a time that the effects not attributable to a lack of the antineuritic vitamin were owing to a deficiency of another single factor, designated G by some and B_2 by others. But soon it was learned that at least two factors were involved here also. Some confusing use of the letters followed. The situation became more complicated as evidence developed that there were more than two, and possibly several, factors concerned. Fortunately, one by one the different vitamins were identified chemically and given specific chemical names that have come into standard use.

The first factor identified was riboflavin. It is sometimes still referred to as G or B_2. The vitamin is required in the metabolism of all animals but not required in the rations of cattle and sheep because of bacterial synthesis in the rumen.

11.31 Physiological Functions and Symptoms of Deficiency

The vitamin functions as two coenzymes in a large number of enzyme systems. These coenzymes are flavine mononucleotide (FMN), also named riboflavin 5-phosphate and flavine adenine dinucleotide (FAD), mentioned in the previous discussion of nucleic acids (Sec. 8.3). Enzymes containing flavin are commonly referred to as flavoproteins. FMN is a constituent of L-amino acid oxidase, which participates in enzyme systems, oxidizing L-α-amino acids and L-α-hydroxyacids to α-keto acids; of lactic acid dehydrogenase; of Warburg's yellow enzyme; and of others. FAD is contained in succinate dehydrogenase, which serves as a carrier in the electron-transport chain of two of the hydrogen atoms released in the tricarboxylic acid cycle (Sec. 5.6). The coenzyme is also a constituent of several other enzymes such as D-amino acid oxidase, glycine oxidase, histiminase (diamine oxidase), and xanthine oxidase. It is thus apparent that riboflavin plays many essential roles in the release of food energy and the assimilation of nutrients. It is understandable therefore why a deficiency is reflected in a wide variety of symptoms which are variable with the species. It is not possible to relate the symptoms, however, to the specific biochemical roles which have been established.

In 1929 Norris and associates described a peculiar type of leg paralysis in chicks which was later shown to be the most characteristic symptom of riboflavin deficiency in this species. The chicks are first noted to be walking on their hocks with their toes curled inward ("curled-toe paralysis"). The legs become paralyzed, but the birds may otherwise appear normal. Diarrhea is another common symptom in chicks. In laying birds, a deficiency of riboflavin results in low egg production and poor hatchability. A deficiency in swine causes crooked and stiff legs, thickened skin, skin eruptions, and exudates over the back and sides, lens opacities, and cataracts.

In humans, there are both skin and eye symptoms, such as roughened skin, furrows around the mouth (cheilosis), dermatitis, "corneal vascularization," photophobia, blurred vision, conjunctivitis, and lacrimation. The corneal vascularization may be a result of the oxygen requirement of the corneal epithelium. The cells of the cornea have no hemoglobin supply but maintain integrity by intracellular oxidative processes which depend on riboflavin activity. Tears are a rich source of riboflavin. Foy and Mbaya[59] stated that in vitamin A deficiency the tear ducts are blocked with keratin and vascularization of the cornea follows from lack of tears. Thus, such patients should be given both vitamin A and riboflavin.

Foy and Mbaya concluded from their studies in the baboon that riboflavin deficiency is an important factor in the derangement of the squamos epithelium of the upper gut and may potentiate esophageal malignancy.

While studies with newborn calves and lambs have shown that riboflavin is a metabolic essential for these species, bacterial synthesis in the digestive tract is adequate to meet body needs after the rumen has developed. Riboflavin synthesis occurs in the cecum of horses, but it is not sufficient to meet metabolic needs.

A decreased rate of growth and a lowered feed efficiency are common symptoms in all species affected. Neither of these symptoms nor the eye, skin, and nervous symptoms noted above can be considered specific signs of riboflavin deficiency. Various ones have been produced experimentally by deficiencies of other B-vitamins and even of certain amino acids. The only certain way at present of identifying a riboflavin deficiency seems to be to cause a remission of the symptoms in question by feeding riboflavin alone. The level of the riboflavin excretion in the urine and riboflavin content of red blood cells are of some value in assessing the adequacy of the nutrition with respect to this vitamin.

The activity of glutathionine reductase in the blood appears to be a better measure, on the basis of human studies.[60] Requirements are determined by adding graded levels of the vitamin to purified diets devoid of it using growth as the measure. The method is illustrated by the report of Miller and associates[61] which also records the physical symptoms and the gross and microscopic pathology resulting from a deficiency of the vitamin. It was concluded that the optimum requirement for baby pigs approximates 3.0 mg per kilogram feed solids, although external, gross, and microscopic lesions were present only in animals receiving less than 2.0 mg.

11.32 Chemistry of Riboflavin

The biological importance of certain yellow pigments became apparent in 1932 when Warburg and Christian in Germany isolated a respiratory enzyme, called *yellow enzyme,* which could be split into a protein and a yellow pigment. Isolation of the pigment from other sources rapidly followed. Because of growth effects attributed to this vitamin in milk and whey, these products came under special study. Observations that the growth effect exhibited by whey seemed to be associated with its greenish-yellow fluorescent pigment led to the isolation of this pigment, first called lactochrome and later *lactoflavin,* as the biologically active material. A similar compound was also isolated as hepatoflavin from liver and as ovoflavin from eggs. Later riboflavin was synthesized, and evidence was obtained for its identity with the naturally occurring products, both chemically and biologically. The advances here resulted particularly from the brilliant work of György, Kuhn and associates in Germany, Karrer and coworkers in Switzerland, and von Euler and associates in Sweden. It is now accepted that the biologically active substance has the following structural formula:

Riboflavin, $C_{17}H_{20}N_4O_6$

The compound consists of a dimethyl-isoalloxazine nucleus combined with the alcohol of ribose as a side chain. It may be obtained in the form of orange-yellow crystals which, in solution, have a greenish-yellow fluorescence. Riboflavin is only slightly soluble in water but readily soluble in dilute basic or strong acidic solutions. It is heat-stable in acid solution but not in an alkaline medium. Visible light, particularly the blue and violet rays, quickly destroys it. Loss in milk during pasteurization and exposure to light is 10 to 20 percent. Much larger losses can occur if the bottled milk is left standing in bright sunlight for more than two hours. Poultry mashes left exposed to direct sunlight for several days and frequently stirred are subject to some loss accordingly.

11.33 Metabolism of Riboflavin

In addition to the rumen and cecal synthesis previously mentioned, intestinal synthesis has been demonstrated in rats and humans. Its extent is markedly influenced by the nature of the diet. High levels of protein, sucrose, cellulose, or lactose may inhibit synthesis. Dextrin or starch may increase synthesis. Under certain conditions, enough riboflavin is produced and absorbed to make a significant contribution to body needs. The vitamin is phosphorylated in the intestinal wall and carried by the blood to the cells of the tissues where it occurs as the phosphate or as a flavoprotein. Following a period on a riboflavin-deficient diet, a liberal intake of the vitamin results in some increase in the blood level and in the tissues. The total amount thus stored is small, however, and intakes above current needs are rapidly excreted in the urine, primarily in the form of free riboflavin.

Several catabolites of riboflavin are also excreted by the mammal but no degradation of the vitamin was detected in tissue homogenates and the excretion of the catabolites was decreased to less than half when microfloral activity of the intestine was decreased with the use of sulfa drugs. Therefore, it is suggested that the catabolites are of microfloral origin.

According to studies with pigs, the riboflavin requirement is substantially higher at a low than at a high environmental temperature. There is no adequate physiological explanation for this finding.

11.34 Riboflavin in Foods

The vitamin is synthesized by higher plants, yeasts, fungi, and some bacteria, but not by animal tissues. Thus animals must depend on foods of plant origin for their supply, with the exception of ruminants, which are supplied by rumen synthesis. In higher plants the vitamin is richest in the leaves, and thus leafy forages, particularly alfalfa, are very good sources. Cereals and their by-products have a rather low content, in contrast to their supply of thiamin. Oil meals are fair sources. Yeast is the richest natural source, containing up to 125 μg per gram. Milk, eggs, liver, heart, kidney, and muscle meat are rich sources. Riboflavin concentrates obtained from whey and distiller's solubles are important commercial sources, particularly for animal feeds.

NIACIN

Our knowledge regarding nicotinamide as a dietary essential grew out of the long-time search for a cure for pellagra, a disease that as late as 1935 was estimated to be taking several thousands of lives annually in Southern states. Outstanding pioneer studies covering several years were made by Goldberger of the U.S. Public Health Service, resulting in the conclusion in 1920 that the disease was caused by an ill-balanced diet. In 1925 he established significant resemblances between pellagra and black tongue in dogs, thus ushering in an era of research on the latter. In 1937 Elvehjem and coworkers[62] at the University of Wisconsin made the dramatic discovery that nicotinic acid, a compound that had remained idle on the chemist's shelf for many years, would cure black tongue. Proof that it would also cure uncomplicated human pellagra quickly followed. It became recognized as a dietary essential for pigs, chickens, monkeys, and other species. Nicotinamide was found equally useful and, in fact, the physiologically active compound.

11.35 Physiological Functions and Symptoms of Deficiency

In the body nicotinamide functions as a component of two coenzymes: I, nicotinamide adenine dinucleotide (NAD, formerly called DPN) and II, nicotinamide adenine dinucleotide phosphate (NADP, formerly TPN). Enzymes containing NAD and NADP are important links in a series of reactions associated with carbohydrate, protein, and lipid metabolism. They function in biological oxidation-reduction systems by virtue of their ability to serve as hydrogen-transfer agents. Although structurally nearly identical they are not interchangeable. NAD is specific for hydrogenases concerned in passing electrons on to O_2 via the electron-transport system in the tricarboxylic acid cycle. Here NAD serves as the electron acceptor in three of the four dehydrogenation steps. There are other important NAD-containing dehydrogenases. NADP is specific for dehydrogenases concerned with biosynthetic reductions. It is contained in the alcohol-dehydrogenase system, the lactic acid-dehydrogenase system, and others.

Nicotinamide was isolated from coenzyme II in 1934 and its function in the hydrogen-transport system demonstrated. The vitamin was isolated from coenzyme I in 1935. Thus, it is interesting to note that through these isolations, contrary to the history of other vitamins, a biochemical function for niacin was demonstrated before its nutritional significance was established.

Pellagra in humans is characterized by a fiery red tongue, ulcers of the mouth, dermatitis, loss of appetite, nausea, and other symptoms. Somewhat similar troubles are found in pigs and chickens. The need of pigs for nicotinic acid and the symptoms resulting from its deficiency were first established by the studies of Chick and associates in England and by Hughes of California. Loss of weight, diarrhea, vomiting, dermatitis, and normocytic anemia are commonly occurring symptoms. In chicks its deficiency is characterized by poor growth, mouth symptoms somewhat similar to those of black tongue in dogs, poor feathering, and occasionally a scaly dermatitis. On a niacin-low diet turkey poults develop an enlarged hock disorder. It is obvious that most of these symptoms are not specific.

The synthesis of the vitamin in the rumen of cattle and sheep has been demonstrated. A deficiency has been produced by Hopper and Johnson[63] in young calves on a low-tryptophan diet. The significance of the tryptophan level, which explains why niacin may not be a dietary requirement for various species on certain rations although it is always a metabolic need, is discussed later.

11.36 Chemistry and Metabolism of Nicotinic Acid and Nicotinamide

The structural formulas of nicotinic acid and the physiologically active nicotinamide are shown below:

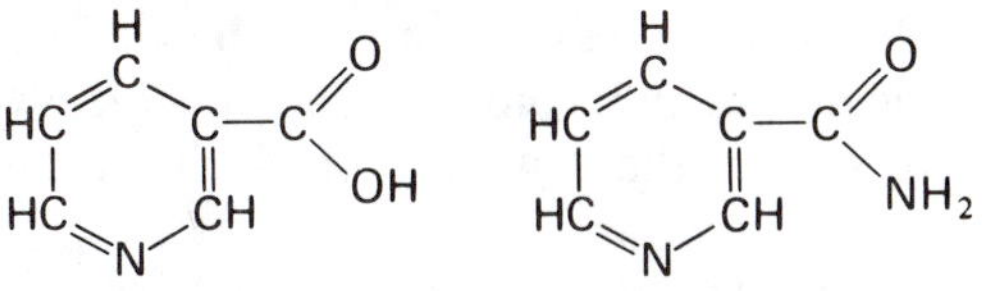

Nicotinic acid   Nicotinamide

The formulas reveal that the compounds are derivatives of pyridine, explaining the previously mentioned names of the compounds in which the vitamin functions. Both the acid and amide are colorless crystalline substances readily soluble in water and alcohol. They are very resistant to heat, air, light, and alkali and, thus, are stable in foods. There are four antivitamins for niacin, all of which have the basic pyridine structure, as indicated for two of them below:

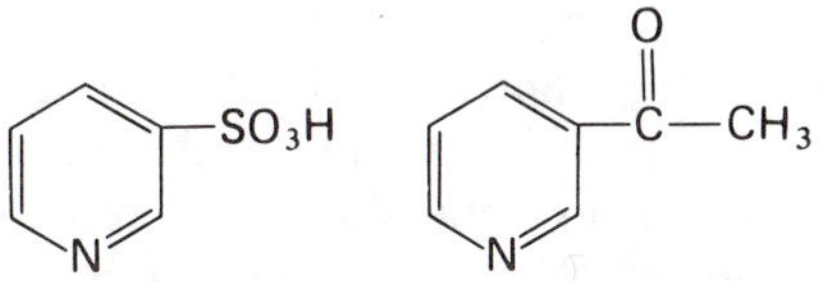

Pyridine-3-sulfonic   3-Acetylpyridine
acid

Niacin is readily absorbed from the intestine. The liver is the site of greatest niacin concentration in the body but the amount stored is not great. The urine is the primary pathway of excretion.

The principal excretory products in humans, dogs, rats, and pigs are the methylated metabolites N'-methylnicotinamide and N'-methyl-2-pyridone-5 carboxylamide. On the other hand, in herbivora niacin does not seem to be metabolized by methylation, but large amounts are excreted unchanged. In poultry the excretory product is dinicotinylornithine. The measurement of the excretion of these metabolities is carried out in studies of niacin requirements and of niacin metabolism. Such studies are complicated by the fact that the kinds and relative amounts of these products vary with the species and level of niacin intake, as shown by Perlzweig and coworkers.[64]

11.37   Nicotinic Acid and Tryptophan Interrelationships

The consumption of corn was early associated with the occurrence of pellagra. Even after the vitamin was discovered and its low content in corn established, evidence persisted that corn had some positive role in connection with the disease. This idea was furthered by studies first published in 1945 showing that a high-corn diet increased the niacin need for dogs and chicks and caused a deficiency in rats not normally requiring a dietary source of the vitamin. The deficiency in rats was overcome by adding nicotinic acid or casein or trypto-phan. Milk also helped correct the deficiency, an observation which was in line with the evidence that milk is effective in curing pellagra, though rather low in nicotinic acid. Other protein sources were also found effective. An interrela-tionship between protein and nicotinic acid needs was also established for pigs and chicks. As a result of further studies it became definitely established by 1947 that the effective proteins were those high in tryptophan and specifically that this amino acid served as a precursor for the synthesis of niacin in the body.

Many studies with rats have dealt with the question as to the site of this synthesis. There is considerable evidence, such as that by Ellinger and Kader,[65] that the synthesis can take place in the intestine. These investigators found that there was a much larger urinary excretion of the metabolite N'-methylnicotinamide where tryptophan was administered by stomach tube than when injected parenterally. When a sulfa drug was fed, the excretion of the metabolite decreased markedly and the feeding of tryptophan caused no in-crease in the excretion. There is also evidence that the synthesis can take place inside the body. This evidence is based on the observation that the parenteral administration of the amino acid causes increased output of the metabolites of niacin in the urine and more specifically on an experiment by Henderson and Hankes[66] showing that, despite the removal of the major portion of the intes-tinal tract, marked increases in the excretion of niacin metabolites occurred on the feeding of tryptophan. Synthesis occurs in the developing chick embryo.

Isotope studies have been used to demonstrate the synthesis of nicotina-mide from tryptophan and to explain the chemical pathway by which it occurs. Several steps are involved, and thiamine, riboflavin, and vitamin B_6 are essen-tial, each acting at a given step in the pathway. It has been shown that rats, pigs, and chicks meet their entire niacin needs by synthesis, assuming an ade-quate supply of tryptophan in the diet for the purpose and also for protein synthesis. The latter function takes preference, however, at least in rats, and a deficiency of the vitamin can be induced by a diet relatively deficient in tryp-tophan for protein synthesis. Approximately 35 mg of tryptophan is required for the synthesis of 1 mg of niacin in the body, as determined by rat studies.

Nicotinamide is clearly a metabolic need for all species. Ruminants have no dietary need, and the requirement for other species depends upon the extent of synthesis from tryptophan. It is evident that pigs and chickens on a liberal protein diet rich in tryptophan have a very low, if any, need for the vitamin. Human needs can also be met by tryptophan.

The establishment of this tryptophan-niacin interrelationship, together with the fact that corn protein is low in this amino acid, provided an explanation for the "pellagragenic" effect of corn. A further explanation is that much of niacin in corn is in a "bound form," unavailable to the rat, pig, or poultry without alkali treatment. The same is true for wheat, rice, and barley grains. Boiling releases niacin in sweet corn but not in mature corn.[67]

Pellagra may also be mediated through an amino acid imbalance.[68] Black tongue disease was induced in pups by the addition of leucine. Pellagra may be present in populations that eat large amounts of jowar, a millet. Jowar is not low in tryptophan and the niacin present is not bound but jowar has a relatively high content of leucine.

Some species, such as the cat and the mink, apparently lack the ability to convert tryptophan to niacin.

Pharmacologic Effects. Nicotinamide and nicotinic acid are relatively nontoxic but massive doses result in vascular dilation or "flushing" with a burning sensation. Large, chronic doses of nicotinic acid have also been used in attempts to lower serum cholesterol, but there may be also decreased mobilization of fatty acids from adipose tissue, hepatic injury, increased incidence of diabetes, and peptic ulcers.

11.38   Niacin in the Food Supply

The vitamin has widespread distribution among feeds, but that present in cereals, which may be the principal ingredients in the ration, cannot be counted on because of its low availability. Animal and fish by-products, distiller's grains and yeast, various distillation and fermentation solubles, and certain oil meals are good sources. Leafy materials, especially pasture grass, are fair sources. Milk, dairy products, fruits and eggs are poor sources. As has been mentioned, if the diet contains tryptophan in addition to that required to meet specific needs, then the food supply of niacin becomes less important accordingly.

VITAMIN B$_6$

In 1934 György separated the nonthiamin part of the B complex into riboflavin and a "complementary factor" which he named vitamin B$_6$ and defined as the factor "responsible for the cure of a specific dermatitis developed by young rats on the vitamin-free diet supplemented with B$_1$ and riboflavin." In 1938 Lepkovsky isolated the vitamin as crystals of his previously reported "factor I." Other workers announced the isolation during the same year. The structure of the vitamin was first explained by Kuhn and coworkers who gave it the name *adermin*. These same workers, along with Harris and Folkers of Merck and Company, deserve credit for the synthesis of the vitamin.

In view of the chemical structure of the compound here concerned, György proposed *pyridoxine* as the name for the vitamin, and this proposal was widely adopted. Later, however, two other compounds, *pyridoxal* and *pyridoxamine,* were identified. Thus, by official action of the Society of Biological Chemists and the American Institute of Nutrition, the original term B_6 is now the approved name for this vitamin, which, like several others, has a multiple nature. The vitamin is a dietary essential for the rat, pig, chick, dog, human, and other species including microorganisms. Vitamin B_6 is a metabolic essential for ruminants as well as other animals, as indicated by studies with newborn calves, but after the rumen is developed, cattle and sheep have no need for a dietary source. In horses the vitamin is synthesized in the cecum.

11.39 Physiological Functions and Symptoms of Deficiency

Vitamin B_6 functions in several enzyme systems concerned in protein metabolism. In the form of phosphorylated pyridoxal it serves as a coenzyme (codecarboxylase) for enzymes which decarboxylate several amino acids. As phosphorylated pyridoxal, it also is a coenzyme for transaminases, which catalyze the transfer of the amino group of glutamic acid and certain other amino acids to keto acids. The vitamin also functions in a variety of other amino acid enzymes, such as racemases, dehydrases, desulfhydrases, and hydroxylases. Pyridoxal phosphate is required for the synthesis of δ-aminolevulinic acid, a precursor of heme. It is essential for the complete metabolism of tryptophan. Otherwise, the abnormal metabolite xanthurenic acid is formed and excreted. The level of xanthurenic acid excretion has been used as an indicator of the status of B_6 nutrition in the pig. B_6 is also concerned in some way with fat metabolism. A deficient animal is shown in Fig. 11.8.

The specific dermatitis (acrodynia) that characterizes vitamin B_6 deficiency in rats is not found in other species, but convulsions are a common symptom in all. In pigs the symptoms include a microcytic, hypochromic anemia; epileptic-like fits or convulsions; and slow growth. Nerve degeneration and hemosiderosis, the deposition of a dark-yellow iron pigment, are regularly found on autopsy. In chicks, abnormal excitability, jerky, aimless movements, and later, convulsions occur, followed by complete exhaustion. Slow growth and suppressed appetite are accompanying symptoms. In laying hens egg production and hatchability are markedly decreased. A report by Miller and co-workers[69] details the physical symptoms of the deficiency in the baby pig, and also blood changes, organ weights, and levels of xanthurenic acid excretion. They found the requirement to lie between 0.75 and 1 mg per kilogram feed. As shown by rat studies, normal reproduction requires an adequate supply of the vitamin both prior to mating and during gestation.

The symptoms in animals have been produced experimentally with purified diets designed to be lacking in the vitamin and also with diets of natural feeds to which the antimetabolite desoxypyridoxine has been added. There is no evidence that deficiencies occur in farm animals on commonly fed rations, and their occurrence is not to be expected in view of the widespread distribu-

Fig. 11.8 Vitamin B_6 deficiency. (*Courtesy of R. W. Luecke, Michigan Agriculture Experiment Station.*)

tion of the vitamin in feedstuffs. The symptoms cannot be considered specific for B_6 deficiency only. For example, lack of pantothenic acid produces similar skin and nervous manifestations. Bacterial synthesis occurs in the rumen. There is also evidence that the vitamin is synthesized by intestinal bacteria.

The requirement for B_6 is increased when high-protein diets are fed. Oral contraceptives may also affect B_6 status and nutritionists have suggested that women using oral contraceptives may consider increasing their B_6 intake. Some individuals have inborn errors of metabolism that greatly increase their B_6 requirement. Fortunately, the number of such individuals appears to be small but unless they take doses of B_6, starting at birth, they develop convulsions, brain damage, and die.[70]

11.40 Chemistry of Vitamin B_6

The formulas for the three naturally occurring free forms of B_6 are as follows:

$$
\begin{array}{ccc}
\text{Pyridoxine} & \text{Pyridoxamine} & \text{Pyridoxal}
\end{array}
$$

As is to be expected in view of their common role in metabolism, pyridoxine can be converted into the amine and the aldehyde. The latter two are reversibly converted into each other. The three compounds are all water-soluble and fairly stable to heat, but the sterilization of milk products results in a substantial loss.

11.41 Food Supply

Vitamin B_6 is widely distributed in feeds. Yeast, liver, muscle meat, milk, cereal grains and their by-products, and vegetables are all excellent sources. Many other products contain substantial amounts. The usual food supply is so rich in the vitamin that deficiencies in practical rations seem very improbable.

PANTOTHENIC ACID

In 1933, R. J. Williams and associates fractionated bios, a growth factor for yeast, and obtained a very potent acid fraction which they named *pantothenic acid*. The name, meaning "from everywhere," was selected because of the widespread distribution of the factor. When the chemical nature of this substance was definitely established, it became clear that it was the same factor concerned in a specific dermatitis condition in chicks, first described by Ringrose in 1930. Now it is known that pantothenic acid is a dietary essential also in rats, dogs, pigs, turkeys, and other species. It is synthesized in the rumen of the cow and sheep to the e extent that no need for a supply in their rations has been shown. Intestinal synthesis of pantothenic acid has been found to occur in all species studied. In the case of the rabbit and the horse it appears to be sufficiently extensive to meet body needs, at least in large part.

11.42 Physiological Functions and Symptoms of Deficiency

In 1947 Lipmann reported that the coenzyme required for the acetylation of sulfanilamide contained 10 percent of pantothenic acid. Thus arose the name *coenzyme A,* meaning coenzyme for acetylation. The discovery of its role in other acetylations and many other biochemical reactions essential in metabolism followed. Its combination with two-carbon fragments from fats, carbohydrates, and certain amino acids to form acetyl coenzyme A is an essential step in their complete metabolism, because the coenzyme enables these fragments to enter the citric acid cycle. Its essential roles in the synthesis and catabolism of fats and the synthesis of steroids have been described in Chaps. 6 and 7. These various functions of the vitamin serve to illustrate the fact that it plays an essential role in many cellular reactions. Many of the manifold functions of CoA, only partially mentioned here, have been detailed by Lipmann.[71]

In view of the diverse functions of the vitamin, a variety of deficiency symptoms are to be expected. They include growth and reproductive failure, skin and hair lesions, gastrointestinal symptoms, and lesions of the nervous system. The metabolic roles of the vitamin are interrelated with those of several other B-factors. In chickens there is first a retardation of growth and feather development. Next, the dermatitis appears. The eyelids become granular and stick together, and scabs appear around the mouth, and vent, and on the feet. Liver damage, changes in the spinal cord, and several other postmor-

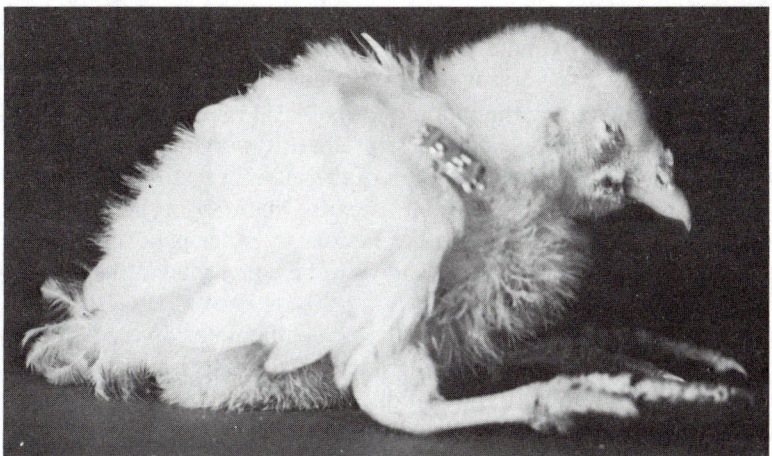

Fig. 11.9 Pantothenic acid deficiency. (*Courtesy of L. C. Norris*, Cornell University.)

tem findings are to be noted. Hatchability is decreased in adult birds. The condition produced in chickens is shown in Fig. 11.9. Pigs suffering from pantothenic acid deficiency have a scurvy skin and thin hair, a brownish secretion around the eyes, gastrointestinal troubles, slow growth, and a characteristic goose-stepping (Fig. 11.10). Nerve degeneration and organ changes are found on autopsy. None of these physical symptoms can be considered specific for pantothenic acid deficiency alone. The various ones described have been noted on experimental diets lacking in the vitamin. There is also evidence that deficiencies may occur on practical rations. McKigney and coworkers[72] have reported such evidence using a low-protein (about 14 percent) corn-soybean

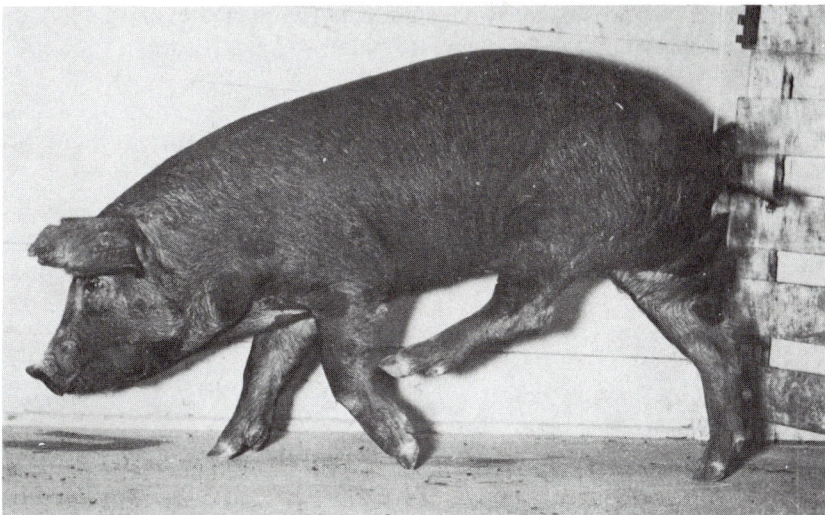

Fig. 11.10 Goose stepping in pantothenic acid deficiency. (*Courtesy of R. W. Luecke, Michigan Agriculture Experiment Station.*)

ration, fortified with minerals and vitamins except pantothenic acid. The ration contained, on analysis, approximately 6.7 mg of pantothenic acid per kilogram. Growth was not affected, but various other deficiency symptoms were noted, which were prevented by adding the vitamin to the basal ration.

A lack of pantothenic acid results in a premature graying of the hair in piebald rats, foxes, and dogs. Administration of the vitamin restores the hair to its natural color. The effect is not specific because such a graying has been observed as a result of a deficiency of other factors. Neither pantothenic acid nor any other nutritional factor has been shown to be concerned in the graying of hair in humans.

Porphyrin from the Harderian gland of the rat accumulates on the whiskers in pantothenate deficiency giving the appearance of "bloody whiskers." Deficiency signs in man include depression, reduced glucose tolerance, intestinal disorders, dermatitis, and burning sensations of the feet.

11.43 Chemistry of Pantothenic Acid

This vitamin is a peptide of β-alanine and 2,4-dihydroxy-3,3-dimethyl butyric acid:

Pantothenic acid

It is obtained as its calcium salt, which is a white, relatively insoluble powder. There are various analogs of pantothenic acid which act as antimetabolites in antagonizing its growth effect on microorganisms. Coenzyme A contains the vitamin combined with adenosine 3'-phosphate, pyrophosphate, and β-mercaptoethylamine, as shown by the following formula:

Another metabolically active form of pantothenic acid is acyl carrier protein (ACP).

11.44 Pantothenic Acid in Foods

The vitamin is widely distributed in foods of both animal and plant origin. A deficiency in the commonly fed rations of farm animals seems improbable. Alfalfa hay, peanut meal, cane molasses, yeast, rice bran, and wheat bran are especially rich in the factor. Cereal seeds and their by-products and many other feeds are good sources. In both plant and animal tissues, the vitamin occurs very largely as coenzyme A, which is referred to as the "bound form." Royal jelly of bees and the ovaries of codfish are exceptionally rich sources of pantothenic acid.

BIOTIN

The history of the discovery of the physiological significance of biotin is the history of the merging of three lines of investigation, for a long time apparently unrelated. It well illustrates how scientific information develops. In 1936, Kögl and Tonnis in Germany announced the crystallization of a factor called "biotin" necessary for the growth of yeast. Later it was discovered to be identical with a growth factor, "coenzyme R," found essential in 1933 for the growth of nodule bacteria. In 1927 Boas in England had reported that the feeding of Chinese egg white to rats produced a typical dermatitis. Parsons of Wisconsin studied this "egg-white injury" in detail for several years, finding that there was a "protective factor" in certain foods, notably liver and kidney. In the meantime, György studied the chemistry of this protective factor and in 1937 applied the term "factor H" to it.

In 1938 György and du Vigneaud teamed up on the problem and in 1940 announced that vitamin H and biotin were identical, and they gave further proof later. Next it was announced by Eakin and by György and Eakin that egg-white injury resulted from its rendering dietary biotin unavailable, owing to a specific constituent, "avid-albumin," or "avidin." At the same time R. J. Williams announced the isolation from egg white of a substance rendering biotin inactive for yeast growth. Du Vigneaud isolated the vitamin in 1941 and determined its structure in 1942. Harris and coworkers of Merck and Company synthesized it in 1943. Thus a new metabolic essential for both animals and lower plants was discovered. In animal nutrition, it presents a unique case in that its discovery rested on the identification of another previously unknown substance which, as a dietary constituent, combines with biotin in the intestine, preventing its absorption. Avidin is a protein which is a secretory product of the mucosa of the oviduct and thus found in the albuminous part of eggs. An extensive review of biotin has been written by Murthy and Mistry.[73]

11.45 Physiological Functions and Symptoms of Deficiency

Most of our knowledge of the basic metabolic functions of biotin has been obtained in studies with microorganisms. It is a constituent of various enzyme systems, some of which have been shown to function in animals as well as in bacteria. The vitamin is concerned in both carbon-dioxide fixation and decarboxylation. For example, it functions in the addition of carbon dioxide to pyruvate, adenine, and guanine, and in the decarboxylation of oxalacetate and succinate. It participates in the addition of carbon dioxide to acetyl CoA to form malonyl CoA and thus in fat synthesis. A decrease in propionyl-CoA carboxylase activity has been suggested as an indicator of biotin deficiency. The vitamin plays a role, at least in bacteria, in deaminating systems of certain amino acids.

Symptoms of biotin deficiency in several species have been produced by feeding raw egg white or by the use of biotin-free diets plus a sulfa drug to prevent intestinal synthesis. Dermatitis, loss of hair or feathers, and poor growth are found in all species.

Rats with biotin deficiency develop a peculiar alopecia called spectacle eye because the skin around the eye loses its hair. Black rats and yellow rats with biotin deficiency may demonstrate achromotrichia.

In chicks, symptoms have been produced on a biotin-low diet without the use of egg white or sulfa drugs and cured by supplying the vitamin. Biotin has also been found useful along with manganese, choline, and folic acid in preventing perosis (Sec. 10.43). In the mature fowl, hatchability is decreased by its deficiency. Cunha and coworkers[74] have reported the experimental production of biotin deficiency in pigs by feeding egg white. The symptoms found were as follows: spasticity of the hind legs, cracks in the feet, and a dermatitis characterized by dryness, roughness, and a brownish exudate. None of these symptoms in chicks or pigs can be considered specific for differentiating biotin deficiency from those of certain other of the B-vitamins.

In cattle and sheep, rumen synthesis apparently takes care of body needs adequately. There is a substantial intestinal synthesis of biotin in all other species studied, including man. The nature of the diet markedly influences the extent of this synthesis, as illustrated by the studies of Couch and coworkers[75] with hens.

11.46 Chemistry of Biotin

As determined by isolation and by synthesis, biotin is 2-keto-3,4-imadazilido-2-tetrahydrothiophenevaleric acid. It is a monocarboxylic acid. It is a crystalline compound, very stable chemically but soluble in alcohol and water. Structurally related compounds have been found to have biotin activity for the growth of microorganisms, and thus there may be more than one compound that has activity in animal nutrition. For example, oxybiotin, in which sulfur is replaced by oxygen, can substitute for biotin. There are also compounds which have antibiotin activity.

11.47 Biotin in Foods

Biotin is widely distributed in plant and animal tissues and also in feeds and foods generally, occurring in both the free and the combined form. Yeast, liver, milk, molasses, peanut meal, and safflower meal are rich sources of biotin. A tabulation of the biotin content of many feedstuffs has been recently published.[76]

$$
\begin{array}{c}
\text{O} \\
\parallel \\
\text{C} \\
\text{HN}\diagup \quad \diagdown \text{NH} \\
| \qquad\qquad | \\
\text{HC} ---- \text{CH} \\
| \qquad\qquad | \\
\text{H}_2\text{C} \diagdown_{\text{S}} \diagup \text{CH} - \text{CH}_2 - \text{CH}_2 - \text{CH}_2 - \text{CH}_2 - \text{COOH}
\end{array}
$$

Biotin

CHOLINE

Choline[77] has been previously discussed as a constituent of the phospholipid lecithin, and its formula has been given (Chap. 7). The importance of phospholipids as constituents of cells and tissues throughout the body has long been recognized, but the knowledge that the choline component is a specific dietary essential under certain conditions is a comparatively recent development. While the fatty acids and glycerol of lecithin can arise from either fat or carbohydrate metabolism, choline must be supplied in the diet as such except of the extent that it can be synthesized within the body through the special process of transmethylation (Sec. 11.49).

11.48 Functions and Symptoms of Deficiency

Choline is a metabolic essential for building and maintaining cell structure. It also plays an essential role in fat metabolism in the liver, preventing the abnormal accumulation of fat (fatty livers) by promoting its transport as lecithin or by increasing the utilization of the fatty acids in the liver itself. Because of this role in preventing fat accumulation, the compound is referred to as a "lipotropic" factor. Choline is essential for the formation of acetylcholine, a substance which makes possible the transmission of nerve impulses. It plays a nonspecific role as a source of "biologically labile methyl groups" (Sec. 11.53) and a specific one in the prevention of perosis.

Choline deficiency has been produced in rats, dogs, chickens, pigs, and other species. Slow growth is a nonspecific symptom. Fatty livers are produced for reasons mentioned above, although other factors are also concerned. In young rats the kidneys become hemorrhagic, owing presumably to a deficiency of choline for the phospholipid required to build cell structure at this critical growth period. Best and Hartroft[78] made the important observation that, when young rats which have developed hemorrhagic kidneys as a result of a deficiency of choline and its precursors are placed on a normal diet, they later develop moderate to severe hypertension. As the authors suggest, these results are of general significance because they indicate that extensive and often fatal pathological changes during adult life may have their origin in a very short period of faulty nutrition at a very young stage.

The need for choline by chicks was first noted in connection with perosis. Several experiments have shown its essentiality for the growth of this species. Neumann and coworkers[79] have produced choline deficiency in the pig, using a purified diet. The pigs made only slow growth, were very unthrifty, lacked coordination in movements, and showed fatty infiltration of the liver on autopsy.

In a cooperative study[80] at nine experiment stations, it was found that the addition of choline to a corn-soybean meal type diet increased the number of live pigs farrowed per litter an average of 0.7 pigs.

Choline deficiency causes growth depression and fatty livers in trout.[81] A need by baby calves has also been shown.

11.49 Labile Methyl Groups

In 1935 it was shown that casein had a lipotropic effect, preventing or curing fatty livers in rats, and two years later this effect was found due to the amino acid methionine. These observations provided the background for a brilliant series of investigations by du Vigneaud and associates showing that methionine can furnish methyl groups which combine with ethanolamine to form choline, and that, in reverse, methyl groups from choline can unite with homocysteine to form methionine. Thus it was recognized that both choline and methionine contain "biologically labile methyl groups" which can be transferred within the body, the phenomenon being called *transmethylation*. In this way methionine can partially replace choline as a dietary essential, and choline plus homocysteine can replace the essential amino acid methionine. In addition to the exogenous sources of methyl groups described above, they can arise de novo in the body from a formate carbon under the action of folic acid. Vitamin B_{12} is also involved in this process. (See Sec. 8.20.)

In view of these findings it is now recognized that the metabolic needs for choline can be supplied in two ways: either by dietary choline as such or by choline synthesis in the body which makes use of labile methyl groups. The synthesis in the body cannot take place fast enough to meet the choline needs for rapid growth and thus the symptoms of deficiency previously described result on choline-deficient diets. The dietary supply of choline, of methionine and cystine to furnish labile methyl groups, and of folic acid and vitamin B_{12} for the synthesis of the groups from formate carbon are all concerned in meeting the physiological requirements for choline. An excellent review of the role of methyl groups in choline nutrition is to be found in the paper by Griffith and Dyer.[82]

The interrelations of methionine and choline in chick nutrition, which result from transmethylation, are illustrated by the studies of McKittrick,[83] who concluded that the choline required by the chick may be divided into two parts, an essential part and another part replaceable by methionine.

Other interrelationships of choline are indicated by the findings that deficiencies of biotin and folic acid are also concerned in the production of perosis and that vitamin B_{12} is involved in methyl synthesis.

11.50 Food Supply

All naturally occurring fats contain some choline, and thus it is supplied by all feeds which contain fat. Since the metabolic need for choline can be met in part by body synthesis, it seems unlikely that dietary deficiency is apt to occur in commonly fed rations. This is a matter that may need further study, however, particularly in view of the increasing removal of fat from the feed supply.

Obviously, the dietary requirement for choline depends on the level of methionine in the ration. In the case of baby pigs it has been reported that, with the methionine level at about 0.8 percent, 0.1 percent of choline is required, but that no requirement could be demonstrated when the methionine level was 1.6 percent.

FOLACIN AND RELATED FACTORS

Folic acid is an antianemia factor which has several physiologically active derivatives. These various compounds first came into the literature as unidentified factors on the basis of their essentiality for the prevention of deficiency states in various animal species or for the growth of certain microorganisms. Thus, such names as "factor U," "vitamin M," "vitamin B_c," "B_c conjugate," "*Lactobacillus casei* factor," "SLR factor," folic acid," and others originated. There was much confusion regarding these various unidentified factors until in 1943 the Parke-Davis group and the Lederle group isolated a crystalline compound from liver which proved to be the *L. casei* factor. Next the structure and synthesis of the compound was accomplished by the Lederle group and named pteroylglutamic acid. This is the compound now called "folic acid." It also represents the factors previously referred to as "vitamin M," "vitamin B_c," and "factor U." Other unidentified ones were found to be derivatives of it. A later discovered active derivative is the *citrovorum* factor. The same compound is also represented by the names "folinic acid" and "leucovorum."

11.51 Chemistry of Folic Acid

The *L. casei* factor, identified as mentioned above, has the following formula:

Liver *L. casei* factor (folic acid)

Reading from left to right it is seen that this compound consists of glutamic acid, paraaminobenzoic acid, and a pteridine nucleus, the last two making up pteroic acid. Thus the name pteroylglutamic acid was suggested. The shorter term, folic acid, is now commonly used. A large number of folates occur in nature because there can be reduction of the pyrozine ring, the addition of glutamyl residues and the addition of a one-carbon unit at N^5 and/or N^{10}. Thus, there are folic acid glutamates, tetrahydrofolic acid, and 5 methyl-tetrahydrofolic acid plus many others. In fact, it has been stated that there are more biologically active forms of folacin than of any other known vitamin. Tetrahydrofolic acid is the coenzyme form. The main storage form is 5 methyl-tetrahydrofolic acid.

Another active pterin is xanthopterin, which was isolated from butterfly wings by Hopkins in 1889. At least five different pterins have been isolated from eye pigments of *Drosophila*. Twenty years ago, German workers reported that this compound caused red cell formation in rats made anemic by drinking goat's milk. Its effectiveness in curing an anemia in fish was later reported. These reports did not become explainable until later discoveries regarding folic acid were made.

Folic acid is a yellow crystalline solid, slightly soluble in water but unstable in acid solution. As much as 95 percent of folates in foods can be lost by oxidative heating processes. At pH 1, autoclaving destroys most of its activity. Its action is antagonized by many compounds structurally related to it, one of the most effective being 4-amino-pteroylglutamic acid, commonly called "aminopterin."

11.52 Functions and Symptoms of Deficiency

Folic acid plays a basic biochemical role in the transfer of single-carbon units in various reactions, a role analogous to that of pantothenic acid in the transfer of two-carbon units. The one-carbon units can be formyl, forminino, methylene, or methyl groups. Thus, it functions in the interconversion of serine and glycine, in the synthesis of purines, in histidine degradation, and in the synthesis of certain methyl groups. Deficient purine synthesis results in a deficiency of nucleoprotein formation for blood-cell maturation, and the characteristic anemia develops accordingly.

Some of the enzymes requiring folate are dehydrofolic reductase, thymidylate synthetase, glycine cleavage enzyme, plus many other reductases and transferases. Folic acid has been found to be a dietary essential for the monkey, chick, turkey, fox, mink, rabbit, mouse, and guinea pig. In the rat and pig a deficiency has not been produced except by the simultaneous feeding of sulfa drugs, indicating that intestinal synthesis is adequate to meet needs. Synthesis occurs in the rumen, but newborn lambs require a dietary supply. In species in which deficiency symptoms are produced, there is a characteristic macrocytic, hyperchromic anemia called megloblastic anemia. The red cells are large and immature and there are related changes in the bone marrow. *Leucopenia* (a reduced number of white cells) also occurs. In the chick, growth is retarded, poor feathering results, and depigmentation occurs in colored feathers.

Folate deficiency is considered by some to be one of the most common hypovitaminoses of human beings. Infants, adolescents, and pregnant women seem particularly vulnerable. Tropical sprue, parasitic infection, and cancer all increase the folate requirement. Chronic alcoholism is an important cause of folate deficiency. Alcohol interferes with folate absorption and transportation from storage sites in the body. Oral contraceptives may also interfere with folate absorption. Patients with chronic renal failure may develop folate deficiency.

Folic acid is also effective in treating the anemia but not the nervous symptoms of pernicious anemia, which is discussed in the section on vitamin B_{12}.

11.53   Interrelations of Folic Acid and Other Vitamins

Several studies have shown the actions of folic acid and other vitamins to be interrelated in ways incompletely understood. Both folic acid and vitamin B_{12} appear to function in the synthesis or metabolism of the various compounds making up nucleic acids. For some years it has been known, as shown initially by Schaefer and coworkers,[84] that folic acid has a sparing effect on the requirement of rats and chicks for choline. The effect is probably due to increased synthesis of methyl groups. An interrelationship with vitamin C is indicated by the finding that rats showing folic acid deficiency as a result of a diet containing sulfa drugs are benefited by feeding ascorbic acid. A relationship between folic acid and riboflavin has also been reported.

11.54   Feed Supplies and Requirements

Folic acid is widely distributed in plant and animal products. Green leafy materials and organ meats are rich sources. Cereals, soybeans, other beans, and various animal by-products are good sources. Milk contains the vitamin in limited amounts. From the previous discussion it is evident that the only farm animals found to require a dietary source are poultry. The requirements per kilogram of feed, as stated by the N.R.C., are 0.55 mg for the chick, 0.25 for the laying hen, and 0.35 for breeding hens. These needs are readily met by good practical rations. The chick is the preferred animal for assaying feeds for the vitamin. Microbiological methods are also used.

VITAMIN B_{12} (CYANOCOBALAMIN)

For many years following the discovery that liver contained a substance which could cause a remission of pernicious anemia, scientists strove unsuccessfully to isolate from liver the *antipernicious anemia factor* (APA). When hope that folic acid was the factor proved unfounded despite its action on the specific type of anemia involved, activities were redoubled in other directions. Within two years success was achieved through the discovery of vitamin B_{12}, which proved to be both APA and also a long-sought growth factor for animals.

11.55 The Discovery of Vitamin B$_{12}$

The story of the discovery of this vitamin is a dramatic one made possible by the combined efforts of microbiologists, biochemists, nutrition scientists, and physicians working in various laboratories. In 1947 Mary Shorb of the Department of Poultry Husbandry of the University of Maryland reported the finding in liver extract of a factor (LLD factor) required by *Lactobacillus lactis* Dorner in concentrations bearing an almost linear relationship to the APA activity of the extract. Making use of this organism, Rickes and coworkers of Merck and Company culminated several years of experiments by announcing, in 1948, the isolation of a red crystalline compound from liver which West found highly active hematopoietically in pernicious anemia. The dose required by intramuscular injection was only a few micrograms. The Merck group proposed the name vitamin B$_{12}$. The compound was found to have a high LLD activity and thus was apparently identical with the factor of Shorb. Almost simultaneously with the report by the Merck group came the announcement by E. L. Smith of England of the isolation from liver of two red pigments, later crystallized, which were found clinically active in treating both the anemia and the nervous symptoms of pernicious anemia. One of these was later crystallized and found to be identical with vitamin B$_{12}$. Within a year clinical trials by four other groups of workers showed that crystalline B$_{12}$ caused both hematological and neurological improvement in the disease. The above is a very incomplete story of the many contributions which resulted in the discovery of the long-sought APA factor.

11.56 Vitamin B$_{12}$ and the Animal Protein Factor

As a result of studies by various workers it came to be recognized by 1945 that there was an unidentified factor (or factors) essential for the growth of chicks fed diets entirely of plant origin and that the missing nutrient could be supplied by fish meal, fish solubles, liver, meat scrap, and other animal products. The association of the unknown factor with animal protein sources was responsible for its designation as the *animal protein factor* (APF). It was also referred to as the "chick growth factor." The same factor was also found essential for hatchability. Cow manure was also found to contain the factor, and an organism was isolated from hen's feces that could synthesize a factor effective in promoting chick growth. Cary of the U.S. Department of Agriculture found that milk, commercial casein, and liver extract contained a "factor X" necessary for rat growth on purified diets containing all the known essentials, but having, as its protein component, casein which had been highly purified. With the isolation of vitamin B$_{12}$ it was quickly found that this factor was required for chick growth and for hatchability and that under appropriate experimental conditions it could entirely replace sources of the APF.

11.57 Nomenclature and Chemistry

Vitamin B$_{12}$ is now considered by nutritionists as the generic name for a group of compounds having B$_{12}$ activity. The compounds have very complex structures. The structure of one B_{12} compound, cyanocobalamin, is shown below.

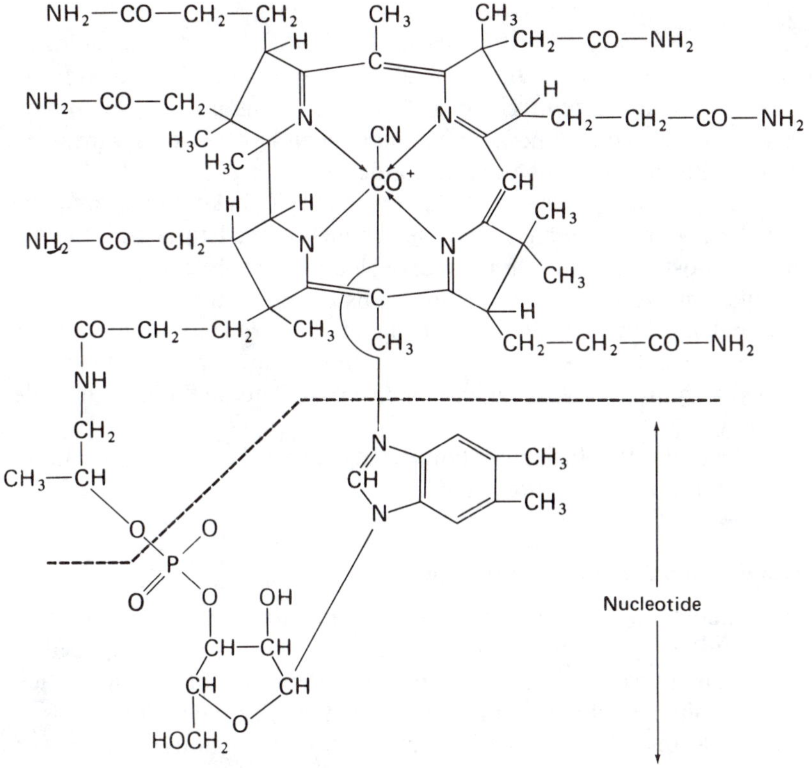

Vitamin B_{12}

The tremendous task of working out this structure was completed in 1955 through the concerted efforts of scientists at Cambridge and Oxford Universities in England and at Princeton University and the University of California at Los Angeles in the United States. Such diverse techniques as chemical and crystallographic analyses, electron density measurements, and those of electronic computers were used to accomplish the task. The reports of the different groups were published briefly in *Nature* (London) as two articles in 1955 and one in 1956. The articles bear the names of thirteen different authors, reflecting the large number of scientists involved. The size and international distribution of the overall group, and the diversity of techniques used, reflect the effectiveness of present-day team research.

The formula shows that the main part of this complex molecule has a cobalt atom in the center of a tetra-ring porphyrin structure.

The large ring formed by the four reduced rings is called "corrin" because it is the core of the vitamin. Any compound with a corrin is called a corrinoid. A cyanide group is attached to the cobalt atom, which is responsible for the name cyanocobalamin. It is noted that there is a nucleotide portion which is coordinated with the cobalt atom and joined to the ring structure through its phosphate group and amino-propanol. The cyanide group can be replaced by an hydroxyl group to form hydroxocobalamin.

Similarly, replacement with a nitrite group produces nitrocobalamin. All three compounds, referred to as the "cobalamins," are all active. In addition, several other compounds, referred to as "pseudo" vitamins B_{12} or vitamin B_{12}-like factors which have some activity, have been isolated or synthesized. Their structure differs as regards the nucleotide moiety.

Adenosylcobalamin, hydroxocobalamin, methylcobalamin, cyanocobalamin, and sulphitocobalamin have been determined in feedstuffs with the first three being the most common. Deoxyadenosylcobalamin, hydroxocobalamin, and methylcobalamin are the predominant forms in human tissue.[85]

Cyanocobalamin has a molecular weight of 1354, reflecting the complex structure of its molecule. It is a dark-red crystalline solid, very slightly soluble in water. Oxidizing and reducing agents and exposure to sunlight tend to destroy its activity.

Losses of vitamin B_{12} during cooking are usually not excessive because it is stable at temperatures lower than 250°C.

11.58   Functions and Symptoms of Deficiency

Vitamin B_{12} functions as a coenzyme in several metabolic reactions. It is required for methyl-group synthesis from one-carbon precursors. Of special interest in ruminant nutrition is its role in the metabolism of propionic acid produced in carbohydrate fermentation in the rumen. The coenzyme is essential in the step that involves the transformation of methylmalonyl CoA to succinyl CoA.

In B_{12}-deficient animals, there is a decreased ability to metabolize propionic acid which is one of the major products of rumen fermentation. This impairment is considered to be the basic metabolic lesion of vitamin B_{12} deficiency in the ruminant.

Propionic acid is converted to propionyl CoA which is converted to methylmalonyl CoA which in turn is converted to succinyl CoA. Succinyl CoA is changed to succinate and enters the tricarboxylic acid cycle. B_{12} is needed for methylmalonyl CoA isomerase. Thus a B_{12} deficiency impairs the utilization of propionate and causes an increase of methylmalonic acid in the urine.

A B_{12} deficiency may induce a folic acid deficiency by blocking the utilization of folate derivatives. Vitamin B_{12} is needed in the methyltransferase reaction in which methyltetrahydrofolic acid (methyl—THFA) and homocysteine form THFA and methionine. A deficiency of B_{12} results in an increase of methyl—THFA. The vitamin also plays a role in protein synthesis and purine metabolism.

Vitamin B_{12} is a metabolic essential for all animal species studied and vitamin B_{12} deficiency can be induced with the addition of high dietary levels of propionic acid.

Pernicious anemia in humans is a B_{12}-deficiency state in the tissues caused by the failure of the absorption of the vitamin, usually due to a lack of intrinsic factor. Intrinsic factor is a mucoprotein produced in the stomach and it facilitates the absorption of B_{12}. Intrinsic factor has been isolated from rats, pigs, guinea pigs, and hamsters but was not isolated from ruminant or dog intestine. The lack of intrinsic factor in man is hereditary.

Malabsorption of B_{12} in man can also result from partial gastrectomy, fish tapeworm infestation, sprue, gluten-induced interopathy, or other gastric abnormalities.

Injection of the vitamin alleviates the megaloblastic anemia and the nervous symptoms. Injections must be continued because the gastric abnormality is not corrected. This abnormality prevents any benefit from feeding the vitamin. Pernicious anemia does not occur in B_{12}-deficient animals but a normocytic or microcytic anemia may develop in some cases.

Decreased growth is the most evident symptom, as shown by studies with rats, mice, chicks, pigs, and others. In addition to retarded growth, posterior incoordination and unsteadiness of gait resulted. In hens, body weight and egg production are maintained despite a deficiency, but poor hatchability results. The newly hatched chicks show bone abnormalities similar to perosis. In rats, a deficiency in the diet of the mother can result in hydrocephalus, eye defects, and bone defects in the newborn and the mothers may eat the offspring. The vitamin is also essential for normal reproduction in pigs.

B_{12} is a metabolic essential for cattle and sheep and is a dietary requirement in the young before the rumen becomes functional. The deficiency symptoms which have been noted in young calves are cessation of growth, poor appetite, and, in some cases, incoordination. Later the vitamin is synthesized adequately in the rumen provided that sufficient cobalt, a constituent of the vitamin, is available. In Sec. 10.35 it was pointed out that the major, if not the entire, role of cobalt is as a constituent of B_{12}. An excellent review of the various physiological studies dealing with this cobalt-vitamin B_{12} relationship in ruminants is presented by Smith and Loosli.[86] The feces of ruminants, assuming the intake of cobalt is adequate, contain large amounts of the vitamin, explaining the discovery of the "cow manure factor."

The cobalt content of the diet is the primary limiting factor for the synthesis of vitamin B_{12} by ruminal microflora. However, recent studies indicate that the synthesis of vitamin B_{12} could be restricted even when the diet is adequate in cobalt as several factors can influence synthesis and perhaps utilization.[87,88] Restriction of roughage may decrease the synthesis of B_{12} and result in a relative increase in synthesis of pseudovitamin B_{12}. The effect of pseudovitamin B_{12} on the utilization of B_{12} is unknown. Benzimidazole bases or alfalfa meal may stimulate vitamin B_{12} production. Further studies are needed to determine if a decreased synthesis or utilization of B_{12} is related to ketosis or decreased fat test in dairy cattle.

Intestinal synthesis occurs in other species, which probably explains frequent failures to produce a B_{12} deficiency in pigs and rats on diets designed to be free of it. The deficiency can be readily produced in rats, however, when coprophagy is completely prevented, as shown by Barnes and Fiala.[89]

Strict vegetarians may develop a B_{12} deficiency. Large doses of vitamin B_{12} have been advocated for various disorders in people. Herbert[90] concluded "the red color of the vitamin along with its almost total lack of known toxicity makes it an almost ideal placebo, and it is so used by many physicians. Unfortunately, it is also used by megahustlers to make megabucks selling oral tablets containing megaquantities of the vitamin for a wide range of claimed but nonexistent effects."

Vitamin B_{12} is stored in the liver in most animals that have been studied but it is stored in the kidney in the bat.

11.59   Food Supply

The origin of vitamin B_{12} in nature appears to be microbial synthesis. There is no convincing evidence that it is produced in the tissues of higher plants or animals. It is widely distributed in foods of animal origin, either because of ingestion of the vitamin in animal products or, in the case of herbivora, because of rumen or intestinal synthesis. Plant products are apparently devoid of the vitamin. Thus herbivorous animals must depend on synthesis in the digestive tract for their needed supply. The best feed sources are fermentation residues. The amount required by all species is very small compared to those of other vitamins—1 μg per day for man, 3 μg per kilogram feed for a 12- to 23-kg pig, and 9 μg per kilogram feed for a starting chick.

The occurrence in the natural food supply of a number of B_{12}-like compounds with widely different activities, and the fact that animals are very selective as to the form used, makes assay difficult as referred to earlier.

OTHER VITAMINS OF THE B GROUP

11.60   Inositol

This compound occurs in plant products in the organic phosphorus substance phytin (Sec. 10.15). In the animal body it is a constituent of certain cephalins. It has the following formula:

Myo-inositol, $C_6H_{12}O_6$

In 1940 Woolley showed that inositol could prevent and cure a characteristic alopecia (deficient and patchy hair) in mice. Evidence for its usefulness otherwise is very limited despite much study. It has a lipotropic action in certain rat diets in which other vitamins may be deficient. Most workers feel that it is highly doubtful whether inositol is essential as such. It may merely be able to perform some of the functions of other vitamins. The symptoms of alopecia are very similar to those produced by a deficiency of B_6 or pantothenic acid. Relationships with biotin have also been established. Intestinal synthesis has been demonstrated. There is no evidence that inositol is needed in otherwise adequate rations of farm animals. Its occurrence in animal feeds is widespread, and thus commonly fed rations should supply it in abundance to meet any need which may exist. There is some question whether inositol should be called a vitamin. Shark muscle has a high inositol content and it has been suggested that in the shark inositol may serve in a manner similar to glycogen in mammals.

11.61 P-Aminobenzoic Acid

This compound was originally discovered as a growth essential for microorganisms and later classed with the vitamins on the basis of its reported growth effects with chicks and lactation effects in rats. Clear evidence that it performs essential functions on otherwise complete rations is still lacking. It is an essential group in folic acid. Thus in a diet lacking in this vitamin p-aminobenzoic acid may provide intestinal bacteria with an essential building stone for folic acid synthesis. Because of its essentiality for the growth of certain microorganisms it may promote the synthesis of the other B-factors in the intestine. In this connection it is interesting to note that aminobenzoic acid has the ability to reverse the bacteriostatic effects of sulfonamides. Here we have another example of antimetabolite action, explainable on the basis of similarity of structure:

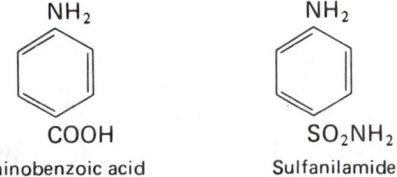

aminobenzoic acid   Sulfanilamide

ASCORBIC ACID (VITAMIN C)

Ascorbic acid is the vitamin that was designated as *water-soluble C,* or the *antiscorbutic factor,* about fifty years ago. The metabolic need appears to be a general one among species, but a dietary need is limited to humans, the guinea pig, subhuman primates, bats,[91] certain birds, certain fish[92] and perhaps certain reptiles. These species lack the enzyme L-gulonolactone oxidase which is necessary for vitamin C synthesis from 6 carbon sugars.

11.62   Physiological Functions and Symptoms of Deficiency

The most clearly established functional roles of vitamin C are exhibited in connection with the formation and maintenance of intercellular material having collagen or related substances as basal constituents, in the bones as well as in the soft tissues. A deficiency for this purpose results in the well-known symptoms of scurvy, such as swollen, bleeding, and ulcerated gums; loosening of teeth; weak bones; and fragility of the capillaries with resulting hemorrhages throughout the body. The widespread effects throughout the body of a deficiency of vitamin C reflects its basic role as a tissue catalyst, but the specific biochemical changes have only been partially characterized. It functions in hydroxylation reactions. Of special significance is its role in the formation of hydroxyproline, which is a constituent of collagen required for the maintenance of intercellular material. The vitamin has been ascribed a function in tyrosine metabolism and also in the conversion of folic acid to folinic acid.

There have been reports with rats and cattle that on vitamin A-deficient diets the ascorbic acid content of the tissues and blood plasma may be low and that vitamin C deficiency may thus result in species not normally having a dietary requirement. There is also much negative evidence, however, and thus claims for the practical importance of the positive findings remain to be established.

Ascorbic acid is stored to only a very limited extent in the body, and thus needs must be supplied regularly. The level in the plasma is a good measure of the current intake. The content in the white cells and platelets is an indicator of body stores. The urine excretion is used to measure the vitamin C status of the body and of its needs as indicated by the *saturation technique*. This technique is based on the fact that, when the body is depleted, an intake results first in storage and then in excretion. The intake which will keep the tissues saturated, as indicated by the urinary output, represents the maximum which can be considered of any value to the body.

11.63   Chemistry of Vitamin C

Credit for the isolation of the compound we now call ascorbic acid is due to King of the United States and Szent-Györgyi of Hungary. The chemical structure of the vitamin was worked out in 1933 by British and Swiss workers, and its synthesis was accomplished that same year. The vitamin occurs in two forms, a reduced form that is readily oxidized to a dehydro form (p. 345). Both forms are biologically active. In foods the reduced form greatly predominates, but it may reversibly oxidize to the dehydro form. The latter can be further oxidized to diketogulonic acid, an inactive compound. This reaction is irreversible. This change takes place readily, and thus the vitamin is very susceptible to destruction through oxidation, a change that is accelerated by heat. The vitamin is more stable in an acid than an alkaline medium. It is not found in dried foods and is markedly destroyed by cooking, particularly where the pH is alkaline. Cooking losses also result because of its solubility. There are also losses in storage, particularly in certain foods.

```
 OC ─────┐              OC ─────┐
  │       │              │       │
HOC       │             CO       │
  ║        O             │        O
HOC       │             CO       │
  │       │              │       │
 HC ─────┘             HC ─────┘
  │                      │
HOCH                    HOCH
  │                      │
H₂COH                  H₂COH
```

L-ascorbic acid   L-dehydroascorbic acid
(reduced form)   (oxidized form)

Vitamin C is an antioxidant. This property is made use of in the addition of the vitamin in the canning of certain fruits to prevent oxidation changes which cause darkening. Glycoascorbic acid acts as an antimetabolite for vitamin C.

In recent years, many claims have been made for large doses of vitamin C for humans. For example, vitamin C has been advocated for the prevention of colds. One recent review[93] concluded that "there is little convincing evidence to support claims of clinically important efficacy." Studies on toxicity are not without contradictions but it appears that vitamin C is relatively nontoxic except perhaps for individuals with metabolic defects such as glucose-6-phosphate dehydrogenase deficiency.

11.64 Vitamin C in Foods

One I.U. is the activity of 0.05 mg of ascorbic acid. Citrus fruits, tomatoes, green leafy vegetables, potatoes, and certain other fruits and vegetables are the principal food sources. Milk is a substantial source as drawn, but much of the vitamin value is lost in pasteurization.

11.65 Polyphenols (Flavonoids)

In 1936 Szent-Györgyi announced that there is a substance in citrus fruits, different from vitamin C, which is essential to prevent fragility of the capillaries. The substance was designated as *vitamin P*. In the past twenty-five years several papers have shown that catechol, rutin, hesperidin, chalcone, and other nonspecific polyphenols, or flavonoids, which are widely distributed in fruits and vegetables can, under certain conditions, provide some protection against capillary fragility. There is evidence that some of these compounds may have some value as a supplement to limited C intake, particularly under conditions of stress. They have not been established as essential nutrients, however, and hence on the basis of present evidence none of them should be classed as a vitamin or confused with vitamin C. They have no known role in the nutrition of farm animals. The Joint Committee on Biochemical Nomenclature has recommended that the use of the term vitamin P be discontinued.

OTHER POSSIBLE VITAMINS AND RELATED FACTORS

At every stage of the development of our knowledge of vitamins, certain factors have been under consideration which required further evidence of their reality as new vitamins. Later some of them became established as new dietary essentials, while others were proved to be old factors in a new guise or scientific mistakes. Thus today we have a long list of factors which require more definite characterization. Some have been identified chemically, but no need by higher animals established. In the case of others, there is evidence of need by these animals but a lack of chemical characterization to prove that they are actually new factors.

Lipoic acid is a compound of known structure which plays an important role in the growth of certain microorganisms. It is considered to play an essential role in the oxidative decarboxylations of α-keto acids, such as pyruvic, in carbohydrate metabolism. On this basis it has been designated as a biocatalyst. There is, however, no clear evidence for an established need in animal nutrition which enables it to be classed as a vitamin, despite several experiments with rats and chicks.

Coenzyme Q is a collective name for a number of ubiquinones, such as Q_4 and Q_{10}, which play an established role in the respiratory chain in mitochondrial systems. There is evidence that specific ones have a function in the remission of some symptoms of vitamin E deficiency, but there is no proof that justifies any being classed as a separate vitamin.

Vitamin B_{13}, obtained from distiller's solubles, is a compound of unknown structure which appears to contain orotic acid (5-carboxyl-uracil) or to yield it on decomposition. It has been found to stimulate the growth of rats, chicks, and pigs under certain conditions, but evidence remains uncertain whether it plays an essential role in an otherwise adequate diet.

Vitamin B_t is an essential growth factor for the meal worm, which has been identified as carnitine. It is widespread in vegetable and animal tissues, but there is no evidence for a dietary need by higher animals.

Vitamin B_{17} is a name sometimes given to the controversial compound, laetrile. There may be unidentified factors for chicks and poults, as indicated by impaired growth on purified diets containing all the known nutritive essentials, compared with performance where certain special products or concentrates of them are added. Thus, we have such terms as "whey factor," "fish solubles factor," and others. None of them has been isolated and identified.

NOTES

1. William Prout, Chemistry, Meteorology and the Function of Digestion, Bridgewater Treatise, William Pickering, London, 1834. Prout (1785–1850), an English physician, was a profound student of the relations of chemistry to physiology. He discovered hydrochloric acid in the gastric juice, showed that the snake excreted its nitrogen as uric acid, and demonstrated that the developing chick takes phosphorus from the yolk to form its bones.

2. F. Magendie, Sur les propriétés nutritives des substances qui ne contiennent pas d'azote, *Ann. Chim. Phys.*, (1)**3:**66–77, 1816.

3. Alex Holst (1860–1931) was trained in medicine, receiving his M.D. degree in 1892. He served as Professor of Hygiene and Bacteriology at the University of Christiana from 1892 to 1930. During this period he investigated many problems of practical hygiene as well as making his studies of "ship beriberi." Theodor Frølich (1870–1947) specialized in pediatrics and held a professorship in this field at the University of Christiana from 1922 to 1941. He made studies of Barlow's disease (infantile scurvy) and thus was led to join Holst in the studies of "ship beriberi."

4. E. V. McCollum and M. Davis, Necessity of certain lipids in the diet during growth, *J. Biol. Chem.*, **15:**167–175, 1913. Elmer V. McCollum was born in 1879 and received his B.A. and M.A. from the University of Kansas and his Ph.D. from Yale University. He began his teaching and research career as instructor in agricultural chemistry at the University of Wisconsin in 1907 and moved in 1917 to Johns Hopkins University as professor of biochemistry, where he worked until he became Professor Emeritus in 1946. His research resulted in the discovery of vitamin D as well as vitamin A and in many other fundamental contributions which have caused him to be recognized as an outstanding pioneer in the modern science of nutrition. He died in 1967.

5. T. B. Osborne and L. B. Mendel, The influence of butter-fat on growth, *J. Biol. Chem.*, **16:**423–437, 1913.

6. D. E. Ullrey, Biological availability of fat-soluble vitamins: Vitamin A and carotene, *J. Animal Sci.*, **35:**648–657, 1972.

7. S. B. Woolbach and P. R. Howe, Tissue changes following deprivation of fat-soluble A vitamin, *J. Expt. Med.*, **42:**753–777, 1925; Vitamin A deficiency in the guinea-pig, *Arch. Pathol. Lab. Med.*, **5:**239–253, 1928.

8. K. C. Hays, Comments on vitamin A, *Am. J. Clin. Nutrition*, **22:**1081–1084, 1969.

9. E. L. Jungherr and coworkers, Parotid gland lesions in experimental bovine vitamin A deficiency, *J. Dairy Sci.*, **33:**666–675, 1950.

10. J. Bitman et al., Influx of sodium thiocyanate into cerebrospinal fluid in normal and vitamin A deficient calves, *J. Dairy Sci.*, **45:**872–878, 1962.

11. E. Mellanby, Vitamin A and bone growth: The reversibility of vitamin A-deficiency changes, *J. Physiol.*, **105:**382–399, 1947.

12. L. A. Moore and coworkers, Carotene requirements for Guernsey and Jersey calves as determined by spinal fluid pressure, *J. Dairy Sci.*, **31:**533–538, 1948.

13. Fred Hale, Pigs born without eyeballs, *J. Heredity*, **24:**105–105, 1933.

14. G. E. Lamming, Hydrocephalus in young rabbits associated with maternal vitamin A deficiency, *Brit. J. Nutrition*, **8:**363–369, 1954.

15. L. D. Kamstra, A. W. Halverson, and A. L. Moxon, Effect of trace minerals and other dietary ingredients upon carotene stability in stored poultry diets, *Poultry Sci.*, **32:**352–356, 1953.

16. L. Bernstein and J. F. Thompson, Studies on the carotene-destroying processes in drying bean leaves, *Botan. Gaz.*, **109:**204–219, 1947.

17. K. A. Walsh and S. M. Hauge, Cartotene—factors affecting destruction in alfalfa, *J. Agr. Food Chem.*, **1**:1001–1004, 1953.

18. J. G. Bieri and C. J. Pollard, Studies of the cite of conversion of β–carotene injected intravenously into rats, *Brit. J. Nutrition*, **8**:32–44, 1954.

19. H. D. Eaton and associates, Effect of vitamin A depletion on liveweight, plasma and liver vitamin A and microanatomy in young calves, *J. Dairy Sci.*, **34**:386–395, 1951.

20. D. R. Hughes, P. Rietz, W. Vetter, and G. A. J. Pitt, A method for the estimation of the vitamin status of rats, *Int. J. Vet. Nutrition Research*, **46**:231–234, 1976.

21. P. Olafson, Hyperkeratosis (X-disease) of cattle, *Cornell Vet.*, **37**:279–291, 1947.

22. W. B. Bell, The relative toxicity of the chlorinated naphthalenes in experimentally produced bovine hyperkeratosis (X-disease), *Vet. Med.*, **48**:135–140, 1953.

23. W. Hansel and K. McEntee, Bovine hyperkeratosis (X-disease): A review, *J. Dairy Sci.*, **38**:875–882, 1955.

24. W. C. Russell, The effect of the curing process upon the vitamin A and D content of alfalfa, *J. Biol. Chem.*, **85**:289–297, 1929.

25. J. B. Shepherd and associates, Experimental harvesting and preserving alfalfa for dairy cattle feed, *U.S. Dept. Agr. Tech. Bull.* 1079, 1954.

26. D. Roe, Nutrient toxicity with excessive intake. 1: Vitamins, *N.Y. State J. Med.*, **66**:869–873, 1966.

27. A. A. Seawright et al., Hypervitaminosis A and deforming cervical spondylosis of the cat, *J. Comp. Pathol.*, **77**:29–39, 1976.

28. E. V. McCollum et al., Studies on experimental rickets. XXI: An experimental demonstration of the existence of a vitamin which promotes calcium deposition, *J. Biol. Chem.*, **53**:293–312, 1922.

29. A. Theiler, The osteodystrophic diseases of domesticated animals. I: The structure of the bone; atrophy; osteoporosis, osteomyelitis, *Vet. J.*, **90**:143–158, 1934.

30. J. W. Thomas, M. Okamoto, and L. A. Moore, The ulnar epiphyseal cartilage width in normal and rachitic calves and its use compared to other methods of detecting rickets, *J. Dairy Sci.*, **37**:1220–1226, 1954.

31. M. R. Haussler, Vitamin D, Present Knowledge in Nutrition, 4th ed., Nutrition Foundation, New York, 1976, p. 82.

32. A. F. Hess and M. Weinstock, Antirachitic properties imparted to inert fluids and to green vegetables by ultra-violet irradiation, *J. Biol. Chem.*, **62**:301–313, 1924.

33. H. Steenbock and A. Black, Fat-soluble vitamins. XVII: The induction of growth-promoting and calcifying properties in a ration by exposure to ultra-violet light, *J. Biol. Chem.*, **61**:405–422, 1924.

34. H. Steenbock and M. T. Nelson, Fat-soluble vitamins. XIX: The induction of calcifying properties in a rickets-producing ration by radiant energy, *J. Biol. Chem.*, **62**:209–216, 1924.

35. J. W. Thomas and L. A. Moore, Factors affecting the antirachitic activity of alfalfa and its ability to prevent rickets in young calves, *J. Dairy Sci.*, **34**:916–928, 1951.

36. G. C. Wallis, G. H. Kennedy, and Roy H. Fishman, The vitamin D content of roughage, *J. Animal Sci.*, **17**:410–415, 1958.

37. C. N. Chineme, L. Krook, and W. G. Pond, Bone pathology in hypervitaminosis D: An experimental study in young pigs, *Cornell Vet.*, **66**:387–412, 1976.

38. L. Krook, R. H. Wasserman, K. McEntee, T. D. Brokken, and M. B. Teigland, *Cestrum diurnum* poisoning in Florida cattle, *Cornell Vet.*, **65**:557–575, 1975.

39. O. B. Kasali, L. Krook, W. G. Pond, and R. H. Wasserman, *Cestrum diurnum* intoxification in normal and hyperparathyroid pigs, *Cornell Vet.*, **65**:190–221, 1977.

40. R. H. Wasserman, L. Krook, and G. Dirksen, Evidence for anti-rachitic activity in the calcinogenic plant, *Trisetum flavescens, Cornell Vet.*, **67**:333–350, 1977.

41. A. S. Brickman, J. W. Coburn, and A. W. Norman, Action of 1,25 dihydroxycholi-calcierol, a potent kidney-produced metabolite of vitamin D, in uremic man, *New Engl. J. Med.*, **287**:891–895, 1972.

42. D. Fraser and associates, Pathogenesis of hereditary vitamin D-dependent rickets, *New Engl. J. Med.*, **289**:817–822, 1973.

43. J. P. Willman et al., Prevention and cure of muscular stiffness ("stiff-lamb" disease) in lambs, *J. Animal Sci.*, **4**:128–132, 1945; F. Whiting, J. P. Willman, and J. K. Loosli, Tocopheral (vitamin E) deficiency among sheep fed natural feeds. *J. Animal Sci.*, **8**:234–242, 1949.

44. A. L. Tappel, Vitamin E, *Ann. N.Y. Acad. Sci.*, **203**:12–17, 1972.

45. J. G. Bieri, Vitamin E in Present Knowledge in Nutrition, 4th ed., Nutrition Foundation, New York, 1976, p. 98.

46. M. W. Dicks, Vitamin E content of foods and feeds for human and animal consumption, *Wyoming Agr. Expt. Sta. Bull.* 435, 1965.

47. J. C. Bauernfeind, The tocopheral content of food and influencing factors, *CRC Crit. Rev. Food Sci. Nutrition*, **8**:337–383, 1977.

48. R. H. Barnes and coworkers, Prevention of coprophagy in the rat, *J. Nutrition*, **63**:489–498, 1957.

49. R. H. Barnes and G. Fiala, Effects of the prevention of coprophagy in the rat. VI: Vitamin K, *J. Nutrition*, **68**:603–614, 1959.

50. S. I. Bechdel, C. H. Eckles, and L. S. Palmer, The vitamin B requirement of the calf, *J. Dairy Sci.*, **9**:409–438, 1926.

51. S. I. Bechdel and coworkers, Synthesis of vitamin B in the rumen of the cow, *J. Biol. Chem.*, **80**:231–238, 1928.

52. N. J. Benevenga, R. L. Baldwin, and M. Ronning, Alterations in liver enzyme activities and blood and urine metabolite levels during the onset of thiamin deficiency in the dairy calf, *J. Nutrition*, **90**:131–136, 1966.

53. F. M. Loew, A thiamin-responsive polioencephalomalacia in tropical and nontropical livestock production systems, *World Rev. Nutrition Diet.*, **23**:168–183, 1975.

54. F. D. Carroll, H. Goss, and C. E. Howell, The synthesis of B vitamins in the horse, *J. Animal Sci.*, **8**:290–299, 1949.

55. E. T. Rees Evans, C. Evans, and H. E. Roberts, Studies on bracken poisoning in the horse, *Brit. Vet. J.*, **107:**364–371, 399–411, 1951.

56. W. C. Evans, Thiaminases and their effect on animals, *Vit. Horm.*, **33:**467–504, 1975.

57. L. Sarkar and D. K. Chandhuri, Antithiamin factor in cottonseed (*Bombex malabericum*): Its isolation and characterization, *Int. J. Vit. Nutrition Research*, **46:**417–421, 1976.

58. K. Schaller and H. Holler, Thiamine absorption in the rat. IV: Effects of caffeic acid (3,4-dihydroxycinnamic acid) upon absorption and active transport of thiamine, *Int. J. Vit. Nutrition Research*, **46:**143–148, 1976.

59. H. Foy and V. Mbaya, Riboflavin, *Prog. Food Nutrition Sci.*, **2:**357–394, 1977.

60. P. J. Garry and G. N. Owen, An automated flavin adenine dinuchotide-dependent glutathione reductase assay for assessing riboflavin nutriture, *Am. J. Clinical Nutrition*, **29:**663–674, 1976.

61. E. R. Miller and coworkers, The riboflavin requirement of the baby pig, *J. Nutrition*, **52:**404–413, 1954.

62. C. A. Elvehjem and coworkers, Relation of nicotinic acid and nicotinic acid amide to canine black tongue, *J. Am. Chem. Soc.*, **59:**1767–1768, 1937. Conrad A. Elvehjem (1901–1962) was born in Wisconsin and received his bachelor's, master's, and Ph.D. degrees from the University of Wisconsin. He was appointed assistant professor there in 1930 and successively held appointments as professor, dean of the graduate school, and president (from 1958 until his untimely death). His discovery of nicotinic acid as the cause of black tongue in dogs was one of the many outstanding contributions he and his students made in the fields of the B-vitamins, trace elements, and others. He worked with a variety of species—rats, dogs, guinea pigs, chickens, farm animals—and also with microorganisms. Elvehjem was a superb teacher and trained a host of graduate students who today are occupying important positions in academic institutions, industry, and government.

63. J. H. Hopper and B. C. Johnson, The production and study of an acute nicotinic acid deficiency in the calf, *J. Nutrition*, **56:**303–310, 1955.

64. W. A. Perlzweig, F. Rosen, and P. B. Pearson, Comparative studies in niacin metabolism: the fate of niacin in man, rat, dog, pig, rabbit, guinea pig, goat, sheep and calf, *J. Nutrition*, **40:**453–469, 1950.

65. P. Ellinger and M. M. Abdel Kader, The nicotinamide-saving action of tryptophan and the biosynthesis of nicotinamide by the intestinal flora of the rat, *Biochem. J.*, **44:**285–294, 1949.

66. L. M. Henderson and L. V. Hankes, Effect of enterectomy on synthesis of niacin in the rat, *Proc. Soc. Expt. Biol. Med.*, **70:**26–28, 1949.

67. E. Kodicek, Some problems connected with availability of niacin in cereal, *Bibl. Nutrition Diet.*, **23:**86–87, 1976.

68. C. Gopalon and K. S. J. Rao, Pellegra and amino acid imbalance, *Vit. Horm.*, **33:**505–528, 1975.

69. E. R. Miller and coworkers, The pyridoxine requirement of the baby pig, *J. Nutrition*, **62:**407–419, 1957.

70. C. R. Scriver and D. T. Whelan, Glutamic acid decarboxylase in mammalian tissue outside the central nervous system and its possible relevance to hereditary vitamin B$_6$ dependency with seizures, *Ann. N.Y. Acad. Sci.*, **166**:83–96, 1969.

71. F. Lipmann (chairperson), Symposium on chemistry and function of coenzyme A, *Fed. Proc.*, **12**:673–715, 1953.

72. J. I. McKigney, H. D. Wallace, and T. J. Cunha, The influence of chlortetracycline on the requirement of the young pig for dietary pantothenic acid, *J. Animal Sci.*, **16**:35–43, 1957.

73. P. N. Achuta Murthy and S. P. Mistry, Biotin, *Prog. Food Nutrition Sci.*, **2**:405–455, 1977.

74. T. J. Cunha, D. C. Lindley, and M. E. Ensminger, Biotin deficiency syndrome in pigs fed desiccated egg white, *J. Animal Sci.*, **5**:219–225, 1946.

75. J. R. Couch and coworkers, Effect of oats, oat products and fat on the intestinal synthesis of biotin in mature fowl, *J. Nutrition*, **37**:251–261, 1949.

76. J. Scheiner and E. DeRitter, Biotin content of feedstuffs, *J. Agr. Food Chem.*, **23**:1157–1162, 1976.

77. Choline is discussed along with the vitamins of the B complex in accordance with common practice. It differs from them in being essential as a structural component of tissues rather than as a metabolic catalyst.

78. C. H. Best and W. Stanley Hartroft, Nutrition, renal lesions and hypertension, *Fed. Proc.*, **8**:610–617, 1949.

79. A. L. Neumann and coworkers, The choline requirement of the baby pig, *J. Nutrition*, **38**:195–231, 1949.

80. NCR-42 Committee on Swine Nutrition, Effect of supplemental choline on reproductive performance of sows: A cooperative regional study, *J. Animal Sci.*, **42**:1211–1216, 1976.

81. H. G. Ketola, Choline metabolism and nutritional requirement of lake trout (*Salvelinus namaycush*), *J. Animal Sci.*, **42**:474–477, 1976.

82. W. H. Griffith and H. M. Dyer, Present knowledge of methyl groups in nutrition, *Nutrition Revs.*, **26**:1–4, 1968.

83. D. S. McKittrick, The interrelations of choline and methionine in growth and the action of betaine in replacing them, *Arch. Biochem.*, **15**:133–155, 1947; The interrelations of choline, glycine and betaine in the growth of the chick, *ibid.*, **18**:437–448, 1948.

84. A. E. Schaefer and coworkers, Interrelationship of folacin, vitamin B$_{12}$ and choline: Effect on hermorrhagic kidney syndrome in the rat and on growth of the chick, *J. Nutrition*, **40**:95–111, 1950.

85. J. Farquharson and J. F. Adams, The forms of vitamin B$_{12}$ in foods, *Brit. J. Nutrition*, **36**:127–141, 1976.

86. S. E. Smith and J. K. Loosli, Cobalt and vitamin B$_{12}$ in ruminant nutrition: A review, *J. Dairy Sci.*, **40**:1215–1227, 1957.

87. A. L. Sutton and J. M. Elliot, Effect of roughage to concentrate and level of feed intake in ovine ruminal vitamin B$_{12}$ production, *J. Nutrition*, **102**:1341–1346, 1972.

88. G. W. Bigger, J. M. Elliot, and T. R. Rickard, Estimated ruminal production of pseudovitamin B_{12}, Factor A, and Factor B in sheep, *J. Animal Sci.,* **43:**1077–1081, 1976.

89. R. H. Barnes and G. Fiala, Effect of prevention of coprophagy in the rat. II: Vitamin B_{12} requirement, *J. Nutrition,* **65:**103–114, 1958.

90. V. Herbert, Vitamin B_{12} in Present Knowledge in Nutrition, 4th ed., Nutrition Foundation, New York, 1976, pp. 191–203.

91. E. C. Birney, R. Jenness, and K. M. Ayaz, Inability of bats for the synthesis of L-ascorbic acid, *Nature,* **260:**626–628, 1976.

92. J. E. Halver et al., Utilization of ascorbic acid in fish, *Ann. N.Y. Academy Sci.,* **258:**81–102, 1975.

93. R. E. Hodges, Ascorbic Acid in Present Knowledge in Nutrition, 4th ed., Nutrition Foundation, New York, 1976, pp. 119–130.

SUPPLEMENTARY LITERATURE

General

Aitken, F. C., and R. G. Hankin: Vitamins in Feeds for Livestock, Commonwealth Agricultural Bureau, Farnham Royal, Bucks, England, 1970.

Baker, D. H., Nutrient bioavailability in feedstuffs: Methodology for determining amino acid and B-vitamin availability in cereal grains and soybean meal, Proc. Georgia Nutrition Conf., 1978, pp. 1–12.

Baker, S. J.: Nutrition and diseases of the blood: The megaloblastic anaemias, *Prog. Food Nutrition Sci.,* **1:**421–460, 1975.

Brubacher, G: Biochemical studies for assessment of vitamin status in man, *Bibl. Nutrition Diet.,* **20:**31–40, 1974.

Chick, H.: The discovery of vitamins, *Prog. Food Nutrition Sci.,* **1:**1–20, 1975.

Coates, M. E.: The water-soluble vitamins and other accessory food factors, *Comp. Animal Nutrition,* **1:**136–167, 1976.

Cormier, M.: Regulatory mechanisms of energy needs: Vitamins in energy utilization, *Prog. Food Nutrition Sci.,* **2:**347–356, 1977.

Morava, E., and R. Tarjan: Vitamin absorption, *Bibl. Nutrition Diet.,* **22:**86–100, 1975.

Thompson, J. N.: Fat-soluble vitamins, *Comp. Animal Nutrition,* **1:**99–135, 1976.

Vitamin A

Cho, D. Y., et al.: Hypervitaminosis A in the dog, *Am. J. Vet. Research,* 1597–1603, 1976.

Eaton, H. D.: Chronic bovine hypo- and hypervitaminosis A and cerebrospinal fluid pressure, *Am. J. Clinical Nutrition,* **22:**1070–1078, 1969.

Gallina, A. M., C. F. Helmboldt, H. I. Frier, and S. W. Nielsen: Bone growth in the hypovitaminotic A calf, *J. Nutrition,* **100:**129–141, 1970.

Kohlmeier, R. H., and W. Burroughs: Estimation of critical plasma and liver vitamin A levels in feed lot cattle, *J. Animal Sci.*, **30**:1012–1018, 1970.

Seawright, A. A., and J. Hrdlicka: Pathogenic factor in tooth loss in young cats on a high daily oral intake of vitamin A, *Aust. Vet. J.*, **50**:133–141, 1974.

Thompson, S. Y.: Role of carotene and vitamin A in animal feeding, *World Rev. Nutrition Diet.*, **21**:224–280, 1975.

Tucker, R. E., et al.: Yellow pigments excreted by vitamin A-depleted sheep, *J. Nutrition*, **93**:518–522, 1967.

Underwood, B. A.: The determination of vitamin A and some aspects of its distribution, mobilization, and transport in health and disease, *World Rev. Nutrition Diet.*, **19**:123–172, 1973.

Wolf, G.: Retinol-linked sugars in glycoprotein syntheses, *Nutrition Rev.*, **35**:97–99, 1977.

Vitamin D

DeLuca, H. F.: Metabolism of vitamin D: Current status, *Am. J. Clinical Nutrition*, **29**:1258–1270, 1976.

Norman, A. W.: The hormone-like action of 1,25(OH)$_2$-cholecalciferol (a metabolite of fat-soluble vitamin D) in the intestine, *Vit. Horm.*, **32**:326–384, 1974.

Procsal, D. A., W. H. Okanura, and A. W. Norman: Vitamin D, its metabolites and analogs: A review of the structural requirements for biological activity, *Am. J. Clinical Nutrition*, **29**:1271–1282, 1976.

Tettelbaum, S. L.: Morphological effects of vitamin D and its analogs on bone, *Am. J. Clinical Nutrition*, **29**:1300–1306, 1976.

Wasserman, R. H., R. A. Corradino, C. S. Fuller, and A. N. Taylor: Some aspects of vitamin D action: calcium absorption and the vitamin D-dependent calcium-binding protein, *Vit. Horm.*, **32**:299–325, 1974.

Wasserman, R.: Current knowledge of vitamin D metabolism and its relationship to calcium and phosphorus metabolism, Proc. Cornell Nutrition Conf., 1976, pp. 112–117.

Vitamin E

Chow, C. K., and A. L. Tappel: An enzymatic protective mechanism against lipid peroxidation damage to lungs of rats, *Lipids*, **7**:518–522, 1972.

Horwitt, M. K.: Vitamin E: A reexamination, *Am. J. Clinical Nutrition*, **29**:569–578, 1976.

Lanmak, N., and P. Lindberg: Vitamin E and selenium deficiencies of domestic animals, *Adv. Vet. Sci. Comp. Med.*, **19**:127–164, 1975.

MacDonald, D. W., R. G. Christian, G. R. Whenhan, and J. Howell: A review of some aspects of vitamin E-selenium responsive disease with a note on their possible incidence in Alberta, *Canad. Vet. J.*, **17**:61–71, 1976.

Vitamin K

Olson, R. E.: New concepts relating to the mode of action of vitamin K, *Vit. Horm.*, **32:**483–511, 1974.

Scott, M. L.: Vitamin K in animal nutrition, *Vit. Horm.*, **24:**633–647, 1962.

Shearer, M. J., A. McBurney, and P. Barkham: Studies on the absorption and metabolism of phylloquinone (Vitamin K) in man, *Vit. Horm.*, **32:**513–542, 1974.

Stenflo, J., and J. W. Suttie: Vitamin K-dependent formation of δ-carboxyglutamic acid, *Annu. Rev. Biochem.*, **46:**157–172, 1977.

Thiamin

Miller, R. C., and associates: The influence of the thiamine intake of the pig on the thiamine content of pork with observations on the riboflavin content of pork, *J. Nutrition*, **26:**261–274, 1943.

Nishino, K., and Y. Itokawa: Thiamin metabolism in vitamin B_6 or vitamin B_{12} deficient rats, *J. Nutrition*, **107:**775–782, 1977.

Phillipson, A. T., and R. S. Reed: Thiamine in the contents of the alimentary tract of sheep, *Brit. J. Nutrition*, **11:**27–41, 1957.

Yudkin, Warren H.: Thiaminase, the Chastek-paralysis factor, *Physiol. Revs.*, **29:**389–402, 1949.

Riboflavin

McCormick, D. B.: Riboflavin, in Present Knowledge in Nutrition, 4th ed., Nutrition Foundation, New York, 1976, pp. 131–142.

Niacin

Darby, W. J., et al.: Niacin, in Present Knowledge in Nutrition, 4th ed., Nutrition Foundation, New York, 1976, pp. 162–174.

Fisher, H., H. M. Scott, and B. C. Johnson: Quantitative aspect of the nicotinic acid-tryptophan interrelationship in the chick, *Brit. J. Nutrition*, **9:**340–349, 1955.

Luce, W. G., E. R. Peo, and D. B. Hudman: Availability of niacin in wheat for swine, *J. Nutrition*, **88:**39–44, 1966.

Vitamin B_6

Gershoff, S. N.: Vitamin B_6, in Present Knowledge in Nutrition, 4th ed., Nutrition Foundation, New York, 1976, pp. 149–161.

Pantothenic Acid

Coursin, D. B.: Vitamin B_6 and pantothenic acid, *Prog. Food Nutrition Sci.*, **1:**183–190, 1975.

Folic Acid

Krumdreck, C. L.: Folic acid, in Present Knowledge in Nutrition, 4th ed., Nutrition Foundation, New York, 1976, pp. 175–190.

Malin, J. D.: Folic acid, *World Rev. Nutrition Diet.*, **21:**198–224, 1975.

Vitamin B_{12}

Grasbeck, R., and E. M. Salonen: Vitamin B_{12}, *Prog. Food Nutrition Sci.*, **2**:193–232, 1976.

VanTonder, S. V., J. Metz, and R. Green: Vitamin B_{12} metabolism in the fruit bat (*Rousettus aegyptracus*). The induction of vitamin B_{12} deficiency and its effect on folate levels, *Brit. J. Nutrition*, **34**:397–409, 1976.

Wagner, Fritz: Vitamin B_{12} and related compounds, *Annu. Rev. Biochem.*, **35**:405–434, 1965.

Wilson, T. H.: Intrinsic factor and B_{12} absorption: A problem in cell physiology, *Nutrition Revs.*, **23**:33–35, 1965.

Biotin

Bonjour, V. P.: Biotin in man's nutrition and therapy: A review, *Int. J. Vit. Nutrition Research*, **47**:107–118, 1977.

Carey, C. J., and J. G. Morris: Biotin deficiency in the cat and the effect on hepatic propinyl CoA carboxylase, *J. Nutrition*, **107**:330–334, 1977.

Ascorbic Acid

Dykes, M. H. M., and P. Merer: Ascorbic acid and the common cold. Evaluation of its efficacy and toxicity, *J. Am. Med. Assoc.*, **231**:1073–1079, 1975.

King, C. G.: Current status of vitamin C and future horizons, *Ann. New York Acad. Sci.*, **258**:540–545, 1975.

Inositol

Hawthorne, J. N., and D. A. White: Myo-inositol lipids, *Vit. Horm.*, **33**:529–573, 1975.

Chapter 12

Antibiotics, Hormones, and Other Growth Stimulating Substances

The constant effort to produce human foods from animal sources more efficiently and at lower cost to the consumer has stimulated continued search for more suitable combinations of known nutrients and for new additives which will increase the efficiency and rate of growth and the level of production of animals. These widespread efforts have led to the present use of antibiotics, hormones, and other chemicals in animal production. Thus, while these materials are not nutrients and cannot be considered as dietary essentials, it is important to understand their effects on animals, and on the meat, milk, and eggs produced.

To increase growth rate and feed utilization and to help promote good health, various drugs are added to animal feeds. These include:

1. Growth stimulants: Antibiotics, arsenicals, and hormonal compounds;
2. Disease prevention treatment: Antibiotics, antimycotics, antiprotozoals, antihelminthics, and pesticides.

The growth stimulants are distributed through feed suppliers or veterinarians. The second group of materials are used largely by veterinarians or under their direction. Pesticides have wide distribution. The use of all of the compounds is regulated by the Food and Drug Administration and the U.S. Department of Agriculture (USDA) Animal and Plant Health Inspection Service of the Federal Government.

ANTIBIOTICS

The term *antibiotic* means against life, or destructive of life. An antibiotic is a compound synthesized by a living organism which inhibits the growth of another. Realization that antibiotics would stimulate the growth rate of chicks and young pigs fed diets containing only vegetable proteins came largely from the reports in 1949 by Stokstad and Jukes of the American Cyanamid Co., Cunha and associates at the University of Florida, and McGinnis and associates at Washington State University. Moore and associates at Wisconsin had reported in 1946 that streptomycin increased the growth rate of chicks, but had not explored the observation further. In the earliest studies with vitamin B_{12}, crude sources of the vitamin, obtained as by-products of fermentations from the production of antibiotics, were used. These residues were shown to have animal protein factor (APF) activity for chicks and pigs fed all vegetable diets. As vitamin B_{12} became available in quantities suitable for farm-animal research, it was shown that fermentation residues gave greater growth responses than the pure vitamin. Stokstad and Jukes[1] demonstrated that pure chlortetracycline was responsible for the growth stimulation.

12.1 Growth Responses of Animals

Soon after the early reports of growth stimulation of chicks fed chlortetracycline (Aureomycin), it was found that young pigs, calves, turkey poults, lambs, foals, pups, mink, rats, and mice also showed growth responses when antibiotic supplements were included in the diet. It was found that oxytetracycline (Terramycin), streptomycin, and certain forms of penicillin and bacitracin stimulated growth of chicks and pigs. Growth responses to various antibiotics differed appreciably, as illustrated by the data in Table 12.1 for pigs. As might be expected from the responses, chlortetracycline, oxytetracycline, penicillin, streptomycin and bacitracin have been most extensively used in animal diets. The others shown have been used to some extent. Other antibiotics tested have included tylosin, tetracycline, oleoandomycin, erythromycin, tyrothricin, and many others, but the results have not always been favorable. There have been important differences in the way various animals respond to antibiotics. Calves show greater growth responses to chlortetracycline and oxytetracycline than to bacitracin, and growth depression occurred with penicillin. Effective levels vary from 5 to 30 g per ton of feed for the different animals and types of feed.

Increased weight gain is most evident during the period of rapid growth and then decreases. Differences between control and treated animals are greater when the diet is slightly deficient or marginal in protein, B-vitamins, or certain mineral elements. Calves show less growth response on liberal amounts of whole milk than on restricted diets of milk replacers. Pigs and poultry also

TABLE 12.1. Relative Growth Responses of Pigs Fed Various Antibiotics

Antibiotic	Index of growth	Index of ⁣feed/gain
None	100	100
Chlortetracycline	136	90
Oxytetracycline	124	94
Penicillin	111	94
Bacitracin	109	103
Streptomycin	115	94
Chloramphenicol	106	98
Polymyxin	96	100
Neomycin	93	88
Subtilin	89	130

Source: R. Braude, H. D. Wallace, and T. J. Cunha, The value of antibiotics in the nutrition of swine: A review, *Antibiotics Chemotherapy*, 3:271–291, 1953.

show greater responses on poor diets than on highest quality feeds. There is no evidence that the mature size of animals is increased by antibiotic feeding. Removal of antibiotics from the diet usually depresses the rate of gain below normal. The earlier findings with farm animals have been reviewed by Braude and associates[2] and Reid and coworkers.[3]

Rumensin, monensin sodium, an antibiotic produced by a strain of *Streptomyces cinnamonensis,* which is useful as an anticoccidial agent for broilers and lambs, is also used in beef cattle rations. This antiobiotic increases the efficiency of feed utilization by modifying rumen fermentation. The molar percentage of propionic acid is increased about 50 percent and acetic acid is lowered. Weight gain is usually not increased, but the feed required per unit of gain is lowered by about 10 percent.[4]

12.2 Other Effects of Feeding Antibiotics

Animals that respond to antibiotic feeding usually consume more feed than controls and less feed is used per unit of weight gain. Because there are fewer unthrifty animals, the growth rates are more uniform. A reduced incidence of diarrhea in young calves and pigs and fewer death losses of rabbits from enteritis have been reported. Roy and coworkers[5] found that calves deprived of colostrum were protected from scours and related infections by chlortetracycline. Diarrhea is often an important problem with calves and may lead to unthriftiness, secondary infections, and death loss. While antibiotics greatly reduce the frequency and severity of the problems, they do not prevent all scours, and they cannot be expected to replace sanitary practices and proper nutrition and management.

The use of high levels of antibiotics, 100 to 200 g per ton of feed, for short periods for poultry flocks having certain chronic infections, such as respiratory diseases, appears to stimulate recovery and bring the birds back into efficient growth or egg production sooner than otherwise possible.

The extra feed that animals receiving antibiotics consume is often sufficient to explain the faster growth rate and also the fact that these animals make more gain in weight per unit of feed consumed because a higher percentage of the total feed is available for growth. The evidence is not convincing that there is any general increase in the efficiency of feed utilization, since antibiotic-fed animals restricted to the same feed intake as the controls usually have not gained faster. Some studies have reported thinner intestinal walls in animals fed antibiotics, suggesting that more efficient absorption of nutrients may occur. However, balance studies with animals have not demonstrated that antibiotics uniformly improve digestion, absorption, or retention of nutrients, but certain vitamins and minerals may be better absorbed.

Low-level feeding of antibiotics (20 to 50 mg per 100 kg of body weight) is not an effective preventative or cure for foot rot, mastitis, or other specific infections. In beef cattle fed high grain rations antibiotics have reduced the incidence of liver abscesses. Most of the studies on the rumen bacteria of cattle receiving antibiotics support the view that low levels do not modify the normal development of rumen function in calves and that they effect little or no change in the major types of microorganisms present in the rumen of either calves or older cattle, but higher levels do cause changes.

There are differences of opinion as to whether or not the feeding of low levels of antibiotics to growing animals should be continued in the United States. During the past 25 years since antibiotics have been added to certain animal feeds to increase the rate and efficiency of weight gain some antibiotic-resistant, disease-producing organisms have emerged, but no major outbreak of disease has occurred during this period resulting from resistant organisms. Some hold the view that antibiotic feeding should be discontinued until it can be clearly proven that they are fully safe under all conditions. Some European countries do not permit addition of low levels of antibiotics to animal feeds for routine use. Many in the United States believe present usage should be continued because of its important contribution to more abundant and lower cost meat, until there is evidence that the antibiotic feeding results in a disease risk for human beings or animals.

The Food and Drug Administration (FDA) appointed a Task Force to review the usage and value of antibiotics in animal feeds and whether their use contributed to hazards to human and animal health. The Task Force was not able to reach agreement on the questions involved. Their report included the following statement:

There is a general agreement on the Task Force that the use of certain antibiotics at low levels, when administered to food producing animals, does result in selection for resistance with a resulting increase in the percentage of resistant organisms recovered from these animals. There is no solid evidence that this increase in resistant organisms in animals has caused disease problems in man which was not present prior to the development of resistance.[6]

Veterinarians in the United States and Europe have pointed out that development of resistance is mostly from therapy and not from drugs in feeds. Prolonged feeding of antimicrobial agents to animals has not been shown to be

responsible for the current high level of R (resistant) factors in the intestinal
tract of humans. It is possible that future studies will support the desirability of
removing certain antibiotics which are most effective in treating animal and
human diseases from the use in animal feeds.

The antibiotics originally used and others more recently developed are still
as effective in stimulating the growth rate as they were initially.[7] Gilliam and
Martin[8] estimated that withdrawing antibiotics from feeds of cattle, calves, and
pigs would increase the cost of production by $11.20 per 100 kg for cattle and
calves and $23.28 per 100 kg for pigs.

Economics is not the only reason to feed antibiotics and other additives.
These materials make it possible to produce more meat, eggs, and milk than
would be possible without them. Since protein deficiency is critical in many
areas increased amounts of high-quality animal protein can help to improve
human nutrition and health.

12.3 Chemical Nature of the Antibiotics

Penicillic acid was isolated in 1913 by Alsberg and Black of the USDA. It is
produced by fungi, *Penicillium puberulum* and others. It has a molecular
weight of 170.16 and its formula is $C_8H_{10}O_4$. Penicillin S potassium, the structure shown at the top of the facing page, has a molecular weight of 417, with the
formula $C_{14}H_{18}ClKN_2O_4S_2$. There are many derivatives of penicillic acid that
differ in side-chain groups and in biological activity. There are many penicil-
linases which antagonize the antibacterial action of penicillin. These are used to
treat the allergic reactions caused by penicillin.

Streptomycin, isolated by Waksman and associates in 1944, is produced
by a soil actinomycete *Streptomyces griseus.* The hydrochloride has a molecular weight of 581.58 and the formula $C_{21}H_{39}N_7O_2 \cdot HCl$. The structure reported
by Brink and Folkers in 1947 is shown in the second illustration at the top of the
facing page. Streptomycin is soluble in water and almost insoluble in alcohol,
chloroform, and ether.

Chlortetracycline (Aureomycin) was isolated from a substrate of *Streptomyces aureofaciens* by Duggar in 1948. The hydrochloride has the formula
$C_{22}H_{23}ClN_2O_8 \cdot HCl$ and a molecular weight of 515.36. Its structure is shown in
the third illustration on the facing page. It is widely used in medicine to combat
infections caused by both gram-positive and gram-negative bacteria. It is
slightly soluble in water (0.5 to 0.6 mg per milliliter). At levels of 10 to 30 mg per
kilogram of feed, chlortetracycline is a growth stimulant.

Oxytetracycline (Terramycin) was isolated from the product of *Streptomyces rimosus* in 1950. It is similar in formula to chlortetracycline, having a
molecular weight of 496.46, consisting of $C_{22}H_{24}N_2O_9 \cdot 2H_2O$. The biological
activity of Terramycin is very similar to Aureomycin. Its structural formula is
shown at the base of the facing page.

Bacitracin is an antibiotic polypeptide produced by *Bacillus subtilis* in
culture with tryptone or protein hydrolyzate broth. Bacitracin has a molecular
weight of about 1460 and the formula $C_{66}H_{103}N_{17}O_{16}S$. The identified components are ammonia, aspartic acid, cysteine, glutamic acid, histidine, leucine,

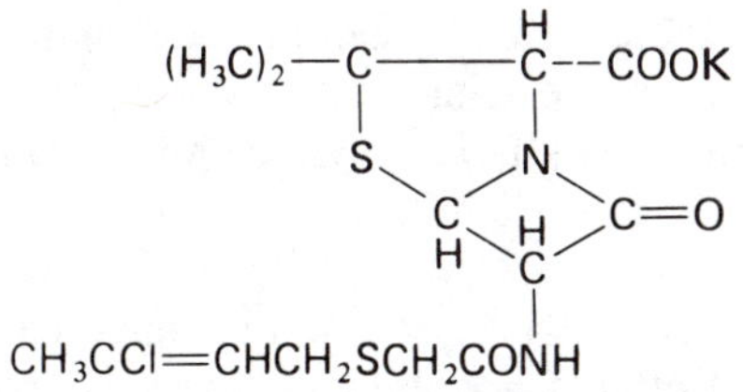

$(H_3C)_2-C$———$C--COOK$

with ring structure and S, N, $C=O$

$CH_3CCl=CHCH_2SCH_2CONH$

Penicillin S potassium

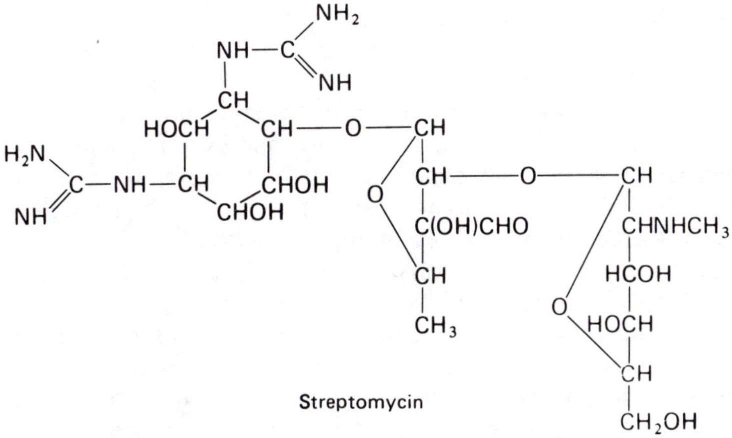

Streptomycin

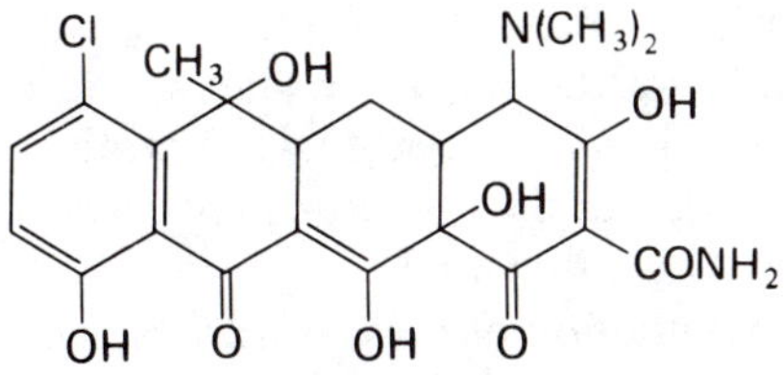

Chlortetracycline

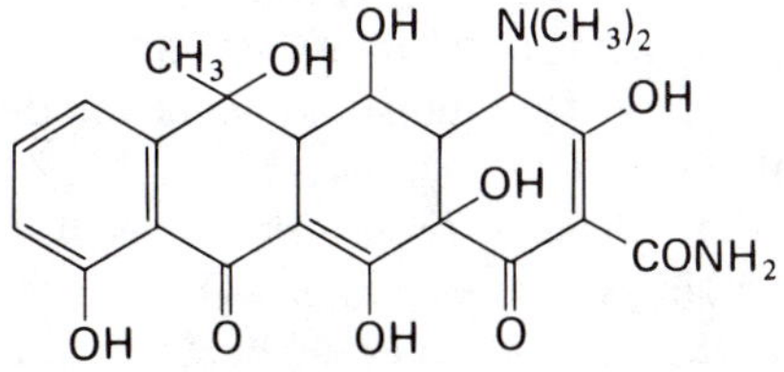

Oxytetracycline

isoleucine, lysine, and phenylalanine. It is soluble in water, alcohol, and methanol. It is less stable than the antibiotics discussed earlier, and its destruction in the digestive tract apparently explains some early studies in which no growth stimulation was observed. Zinc bacitracin is more stable than some other forms.

Many other antibiotics have been produced. Among those that have exhibited growth responses or other beneficial effects are: tylosin, flavomycin, lincomycin, monensin, neomycin, oleandomycin, spectinomycin, spiramycin, and virginiamycin. The use of antibiotics is carefully regulated by the FDA to avoid improper usage and ill effects. Since changes in regulations occur from time to time it is essential to be informed on current limitations as they appear regularly in feed trade and government publications.[9]

12.4 Mode of Action of Antibiotics

The antibiotics are drugs, not nutrients, and their effects upon the nutrition of animals are of necessity secondary. The specific method by which antibiotics exert this influence has not been fully explained, although several theories have been proposed, each of which seems to fit some of the facts but not all of them.

Antibiotics have been shown to have a number of specific beneficial effects on growing animals. On some diets they *spare nutrients* either by reducing the bacterial destruction of vitamins and amino acids by favoring bacteria that synthesize essential nutrients, or by reducing the competition of intestinal microflora of the host animal. Animals grow well on lower levels of protein or amino acids and certain B vitamins with antibiotics in the diet than without them. Antibiotics prevent the thickening of the gut wall and apparently allow better absorption of nutrients, which may explain in part, the *nutrient-sparing effect*. It has been reported that antibiotics decrease the vitamin D requirement for normal bone calcification and lowered manganese requirements for growth and prevention of perosis. Barnes and coworkers[10] found that feeding penicillin to rats increased the amount of thiamin synthesized in the large intestine. The thiamin was not absorbed, however, and only when *coprophagy* was permitted was an increased response observed. Thus, in the rat the thiamin-sparing action of penicillin was abolished when coprophagy was prevented.

Another proposed mode of action is the *inhibition of toxin-producing bacteria*.[11] Ammonia is highly toxic and since its level in the portal blood of both germfree and antibiotic-fed animals is much lower than in conventional animals it is proposed that antibiotics depress bacterial urease production. Ammonia toxicity increases the rate of destruction of mucosal cells resulting in increased thickening of intestinal walls and mucosal cell turnover. Prevention of this cellular loss could explain part of the growth response of antibiotic-fed animals. The observation that germfree rats and chicks have thin intestinal walls and do not respond to oral antibiotics is in line with this view.

Rather conclusive evidence favoring the *"disease level"* theory is presented by the studies of Forbes and Park[12] using germfree chicks. On a corn-soybean meal diet chicks hatched and reared in the absence of bacteria and fungi grew 18 to 25 percent faster than other chicks from the same hatch reared in the animal room. In the initial tests penicillin failed to elicit growth responses in either the germfree chicks or those raised in the animal room, which had not previously been used for chick rearing. When the diet was supplemented with 1 or 2 g per kilogram of lyophilized intestinal contents from chicks reared in an animal room where growth responses to penicillin were regularly obtained, a growth response to penicillin resulted. In later studies responses to penicillin were always obtained even in the absence of lyophilized intestinal material, but the growth rate was never equal to that of germfree birds. Clearly, therefore, infection of the chicks in the animal room decreased the growth rate, and the decrease was partially overcome by penicillin. Other studies have shown that conventionally reared turkey poults respond to antibiotics but that germfree poults do not respond, although they gain as fast as those fed antibiotics. Antibiotics did not increase the growth of chick embryos, which are usually entirely free of bacteria contamination.

Walton[13] has shown that antibiotics cause lesions in the cell wall making bacteria more susceptible to natural defense mechanisms of the body and to other therapeutic agents.

HORMONAL COMPOUNDS

Various drugs, many kinds of chemicals, and countless special fermentation products are being sold for use with farm animals with the claim that they will stimulate growth or in some manner improve the health or performance of the animal. Some of these claims are justified, but others are not supported by scientific evidence. In fact, many chemicals used on farms are clearly toxic if improperly used. The animal raiser must guard his animals against possible harm from exposure to chemical fertilizers, insect sprays, and similar chemicals which are commonly used. These latter materials cannot be discussed in detail here, but the student is advised to become familiar with their potential dangers so losses can be avoided.

Certain hormones have proved effective as growth stimulants. Other compounds may increase the rate of growth of animals under some conditions. The literature in this field has been comprehensively and critically reviewed by Casida and associates, and Clegg and coworkers.[14]

12.5 Thyroprotein and Goitrogens

The role of thyroxine in controlling growth and metabolism has led investigators to use thyroid-active materials to stimulate growth of body tissue and wool and secretion of milk by creating a mild hyperthyroidal state. Thyroxine or thyroprotein (iodinated casein) will increase the growth rate of young pigs and calves under some conditions. Results with broilers have been variable. In dairy cows, feeding thyroprotein usually increases milk yields (Sec. 17.3), but there are some important disadvantages to its use. Attempts to increase weaning weights of beef calves, lambs, and pigs have met with only variable success.

Thyroprotein has been reported to increase wool growth. In spite of its potential values, thyroprotein is used only to a very limited extent because of the difficulty of regulating the dosage and the uncertain responses obtained.

The goitrogens (Sec. 10.39), which interfere with thyroxine production by the thyroid gland, depress growth, and they often increase the rate of fattening. The antithyroid material, thiouracil, increases the fattening rate of pigs. If the dose is too high or if it is fed too long, growth rate and feed efficiency are markedly reduced, and because of these problems little use has been made of this goitrogen in pig feeding. In poultry, especially in combination with diethylstilbestrol, thiouracil improves finish and market quality without depressing growth rate. Goitrogens have been of little value in growing-finishing lambs. Thyroid-regulating substances seem to have little practical importance in livestock feeding.

12.6   Growth Stimulation from Hormones

Extensive use is being made of synthetic and purified estrogens, androgens, progestogens, and growth hormones to stimulate the growth and fattening of meat-producing animals. Some of these have given important increases in the rate and efficiency of gain or in the quality of the food products that result. There is concern, however, about possible harmful effects of any residues of these materials in the meat. Evidence that traces of estrogenic activity remained in the meat of cockerels implanted with diethylstilbestrol (stilbestrol or DES), a practice which had been used for a number of years to improve carcass finish and quality, led to discontinuation of such implants. There is no general agreement among scientists, however, that the amounts of estrogens found in meat following diethylstilbestrol implants might prove harmful. Many widely used natural foods, including soybeans, contain higher estrogenic activity than found in animal tissues.

Using rat-assay methods, Stob and associates[15] estimated that beef muscle and liver from implanted animals contained not more than 1 μg of hormone per 100 g of dried tissue and that lambs and chickens may contain 10 times that level. Davey and associates[16] found no measurable estrogenic activity when the stilbestrol intake was reduced from 5 to 1 mg per day the last 50 days of the feeding period. Higher doses resulted in significant traces of activity in the body fat of the lambs.

Especially with growing ruminants there is wide use of hormonal preparations to increase performance, but these materials have not increased the growth of pigs. In cattle and lambs, growth stimulation occurs when stilbestrol is implanted, and less feed is required to make a unit of gain. Implants of 15 to 30 mg for steers increase the rate of gain and decrease the feed required per unit of gain, but lower the carcass grade and reduce somewhat the amount of marbling in cattle. Undesirable side effects were noted in some of the animals given larger implants of 60 to 120 mg, such as mammary development in steers and wethers, pelvic changes in cattle, vaginal and rectal prolapse, difficult urination, and changes in the organs of the urogenital system of lambs.

Burroughs and associates[17] reported that feeding stilbestrol to fattening steers increased the rate of gain and decreased the feed needed to make a unit of gain. These findings have been confirmed by several groups of workers. Feeding 10 mg of stilbestrol daily resulted in approximately 12 percent faster gains and 10 percent less feed to make a unit of gain. It is interesting that the greatest growth stimulation in some trials occurred at the start of the study and seemed to disappear toward the end. The pelvic and mammary changes noted have been less marked than with hormone implantation. Carcass grades of the animals fed hormones appear to be slightly lower than the control steers. Wallentine and associates[18] reported that feeding 10 mg of stilbestrol per head daily to beef steers increased the rate of daily gain 7 percent and reduced the amount of feed required per 100 kg of gain by 5.4 percent. Studies of the 9 to 11 rib cuts showed that stilbestrol increased the amount of separable lean and lowered the separable fat. Chemical analyses revealed increased water and protein and less fat in the tissues. There was no residual hormone in the meat on the basis of the assays used, in contrast to implanted animals.

Studies with fattening lambs have shown that feeding 2 to 5 mg of stilbestrol daily increased the average daily gain approximately 20 percent and reduced the feed per unit of gain. Carcass grades in some trials were lower, especially on higher intakes of estrogen.

It has been well established that androgens, such as testosterone and some of its derivatives, stimulate protein anabolism in cattle and some other animals by reducing urinary nitrogen excretion. Females show a greater response to androgens than males, which might be expected since males normally make more rapid and efficient gains than females. Testosterone-implanted swine and cattle have shown no consistent increase in weight gain or feed efficiency. Fattening lambs have responded in some trials but not in others. In pregnant cattle testosterone crosses the placenta and causes anatomical alterations in female calves more drastic than those seen in freemartins.

The recommended implants for growing and finishing cattle are shown in Table 12.2.

TABLE 12.2. Hormonal Implants for Ruminants

Animals	Hormone levels
Steers	15–30 mg stilbestrol
Steers	200 mg progesterone[a]
200–450 kg	20 mg stradiol benzoate
Heifers	200 mg testosterone[a]
200–350 kg	20 mg estradiol benzoate
Suckling calves	36 mg Ralgro
	15 mg DES
Growing heifers	36 mg Ralgro
	15 mg DES

[a]These combinations are produced in pellet form and sold as *Synovex S* for steers and *Synovex H* for heifers.

Stilbestrol (DES) is manufactured in 15-mg pellets. Steer calves weighing 100–180 kg are given one pellet and those weighing 200 kg or more two pellets in the ear. Animals may be reimplanted in 100 days but there should be at least 129 days between implantation and marketing the animals to avoid tissue residues. *Ralgro* contains zeranol, a chemical derivative of the fusarium mold toxin, produced by *Gibberella zeae,* zearalenome. It acts like an estrogen, being 2.0 percent as active as DES and is not considered to be carcinogenic and no residues have been detected in animal tissues.[19] Ralgro is manufactured in 12-mg pellets. These growth stimulants are generally effective for 90 to 120 days. At one time the FDA ordered that DES be removed from the market but the Court of Appeals overturned the FDA ban and has permitted continuing use.

A full explanation of the mechanism of action of estrogens on growth of ruminants is not possible at present, but there is evidence for a small increase in secretion of growth hormone and of higher levels of plasma insulin. Whether these effects are sufficient to stimulate protein metabolism and produce the anabolic effects is uncertain.[20]

12.7 Other Growth Stimulants

The influences of many other chemicals on animal performance have been tested. Mention is made of only a small number of those that show potential value.

Copper sulfate is widely used at a level of 200 to 250 ppm of copper as an additive to pig rations in Europe, where numerous studies have shown growth responses and improved feed utilization similar to those obtained with antibiotics. Braude and associates[21] reported the results of a coordinated test at 21 different research centers in Great Britain involving over a thousand pigs. The average daily gain was 0.61 kg on the control ration compared with 0.63 kg on oxytetracycline and 0.66 kg on the copper sulfate additive. These results confirmed earlier tests in Europe. It has been suggested that copper acts on the intestinal microflora since changes have been observed, but there is no increase in the digestibility of nutrients. Responses are greatest in the absence of antibiotics, but additive effects have been noted in some tests.

Wallace[22] at the University of Florida reviewed the published research up to 1967 on feeding copper to pigs. There was an average increase in rate of gain of 8.9 percent and a 3.2 percent in feed saved in U.S. studies. The European tests showed an 8.1 percent improvement in gain and a 5.4 percent decrease in feed per unit of gain. Pigs responded to copper levels ranging from 50 to 375 ppm. Optimum levels appear to be 150 to 250 ppm in the diet. Copper was often more effective than antibiotics and the combination frequently gave an additional response. Copper carbonate, copper chloride, copper oxide, copper sulfate, and copper methionine are all effective and they are cheaper to use than the antibiotics. There does not appear to be an important effect upon reproduction.

High levels of copper are toxic. Excess copper accumulates in the liver, but not to any extent in the muscle tissues. No toxicity problems are expected at 250 ppm when the copper is well mixed in balanced diets adequate in zinc and iron. If antibiotics are removed from uses in animal feeds copper may be used without a loss in performance of growing pigs. These results have been confirmed by many other experiments.[23]

A number of other compounds have been studied as feed additives in hopes of increasing the growth rate and health of animals or enhancing feed utilization. These include arsenic compounds, enzymes, and various types of yeast and bacterial cultures, among many others.

Arsenic compounds are widely used as an aid in prevention of blackhead in turkeys and coccidiosis in chickens. They are also helpful in stimulating growth of chicks and pigs similar to the antibiotics, but dairy calves have not shown responses. The results are often more favorable under stress conditions or when chicks and pigs are somewhat unthrifty or exposed to a low "disease level." When conditions are ideal, healthy animals may show no growth response and no improvement in feed efficiency, but such conditions are seldom found in practice. Of the organic arsenicals available, arsanilic acid and 3-nitro-4-hydroxyphenylarsonic acid have been most widely studied for growing chicks and pigs. Levels of 0.002 to 0.009 percent of the complete feed are usually recommended. Arsenic compounds are toxic, and feed manufacturers take special care not to exceed the limits set by law. When birds or pigs are removed from the arsenic-containing feeds a few days before slaughter, residues do not occur in the meat.

The addition of *enzymes* to high-barley rations has increased growth rates and feed utilization by poultry in some of the Western states but not in other regions of the United States. Results with other animals have been less favorable. Physiological studies with young pigs and calves suggest that some of the digestive enzymes may be lacking or not produced in adequate amounts for efficient digestion of certain raw plant products during the first 2 or 3 weeks of life. Attempts to correct these apparent inadequacies by adding various enzymes to the feed of young pigs and calves have met with little success. Studies at Iowa with baby pigs gave small growth responses in some trials but not in others. Results in Canada were largely negative. Combs and associates[24] have reviewed these various tests and reported their own findings that diastase, pepsin, or pancreatin did not increase the performance of baby pigs.

While only negative results have been obtained with dairy calves, Burroughs and associates at Iowa reported that growing-fattening cattle gained 7 percent faster on equal feed intakes when a crude enzyme mixture was added to the feed. Digestibility of the feed was not affected. Less extensive tests at other stations have not confirmed the results, and the true role of added enzymes in livestock feeding remains undefined.

Live yeast cultures are available as an additive for cattle feeds, and they are used to a limited extent. Controlled studies at several experiment stations show they do not increase rate of gain, milk yields, or the efficiency of feed utilization.

Dried rumen cultures, which are sold with the claim that they stimulate rumen development in calves and improve feed utilization in older ruminants, have given negative results. Earlier studies at the Ohio station had demonstrated that the inoculation of young calves with fresh-cud material resulted in earlier establishment of mature-type of rumen microflora. Later tests have not shown any advantage in growth rate, feed utilization, general health, or appearance from inoculation with either fresh-cud material or dried commercial preparations.

NOTES

1. E. L. R. Stokstad and T. H. Jukes, Further observations on the "animal protein factor," *Proc. Soc. Expt. Biol. Med.*, **73:**523–528, 1950.

2. R. Braude and associates, Antibiotics in nutrition, *Nutrition Abstr. Revs.*, **23:**473–496, 1953.

3. J. T. Reid, R. G. Warner, and J. K. Loosli, Antibiotics in the nutrition of ruminants, *J. Agr. Food Chem.*, 2:186–192, 1954.

4. A. P. Raun and Associates, Effect of monensin on feed efficiency of cattle, *J. Animal Sci.*, **39:**250, 1974. D. A. Dinius, M. E. Simpson and P. B. Marsh, Effect of monensin fed with forage on digestion and the rumen ecosysten of steers, *J. Animal Sci.*, **42:**229–234, 1976.

5. J. H. B. Roy and associates, The nutritive value of colostrum for the calf. II: The effect of aureomycin on the performance of colostrum-deprived calves, *Brit. J. Nutrition,* 9:94–103, 1955.

6. L. Goldberg, Implications of environmental pharmacology and toxicology: Food sources of incidental drug exposure, *Fed. Proc.* **34:**195–196, 1975.

7. M. Finland, Relationships of antibiotics in animal feeds and salmonellosis in animals and man, *J. Animal Sci.*, **40:**1222–1236, 1975.

8. H. C. Gilliam, Jr., and J. R. Martin, Economic importance of antibiotics in feeds to producers and consumers of pork, beef, and veal, *J. Animal Sci.*, **40:**1241–1255, 1975.

9. Feed Additive Compendium, The Miller Publishing Co., Minneapolis, Minn., 1977.

10. R. H. Barnes, E. Kwong, K. Delany, and G. Fiala, The mechanism of the thiamine sparing effect of penicillin in rats, *J. Nutrition,* **71:**149–155, 1960.

11. W. J. Visek, The use of drugs in animal feeds, National Academy of Sciences, Publ. 1679, 1969, pp. 135–149.

12. M. Forbes and J. T. Park, Growth of germ-free and conventional chicks: Effect of diet, dietary penicillin and bacterial environment, *J. Nutrition,* **67:**69–84, 1959.

13. J. R. Walton, Indirect consequences of low-level use of antimicrobial agents in animal feeds, *Fed. Proc.,* **34:**205–208, 1975.

14. L. E. Casida and associates, Hormonal relationships and applications in the production of meats, milk and eggs, National Acad. Sci. National Research Council Publ. 714, 1959, pp. 1–53; M. T. Clegg and associates, *ibid.,* 1415, 1966, pp. 1–87.

15. M. Stob and associates, Estrogenic activity of the meat of cattle, sheep and poultry following treatment with synthetic estrogens and progesterone, *J. Animal Sci.,* **13:**138–151, 1954.

16. R. J. Davey, D. T. Armstrong, and W. Hansel, Studies on the use of hormones in lamb-feeding. II. Tissue assays and physiological effects, *J. Animal Sci.,* **18:**75–84, 1959.

17. W. Burroughs and associates, The effects of trace amounts of diethylstilbestrol in rations of fattening steers, *Science,* **120:**66–67, 1954.

18. M. V. Wallentine and associates, Some effects on beef carcasses from feeding stilbestrol, *J. Animal Sci.,* **20:**792–795, 1961.

19. H. C. Mussman, Drug and chemical residues in domestic animals, *Fed. Proc.,* **34:**197–201, 1975.

20. A. H. Trenkle, The mechanisms of actions of estrogens in feeds on mammalian and avian growth, in The Use of Drugs in Animal Feeds, National Academy of Sciences Publ. 1679, 1969, pp. 150–164.

21. R. Braude and associates, Effects of oxytetracycline and copper sulfate, separately and together, in the rations of growing pigs, *J. Agr. Sci.,* **58:**251–256, 1962.

22. H. D. Wallace, Effects of high level copper on performance of growing pigs, *Feedstuffs,* **40:**22–23, 1968.

23. W. F. Gipp, W. G. Pond, et al., Influence of level of copper on weight gain, hematology, and liver copper and iron storage in young pigs, *J. Nutrition,* **103:**713–719, 1973.

24. G. E. Combs, W. L. Alsmeyer, H. D. Wallace, and M. Koger, Enzyme supplementation of baby pig rations containing different sources of carbohydrate and protein, *J. Animal Sci.,* **19:**932–937, 1960.

SUPPLEMENTARY LITERATURE

Andrews, F. N., W. M. Beeson, and C. Harper: The effect of stilbestrol and testosterone on the growth and fattening of lambs, *J. Animal Sci.,* **8:**578–582, 1949.

Beames, R. M., and L. E. Lloyd: Response of pigs and rats to rations supplemented with tylosin and high levels of copper, *J. Animal Sci.,* **24:**1020–1026, 1965.

Braude, R., and T. Ryder: Copper levels in diets for growing pigs, *J. Agr. Sci., Camb.,* **80:**489–493, 1973.

Cromwell, G. L., V. W. Hays, J. R. Overfield, and B. E. Langlois: Effects of antibiotics, season and parity on reproductive performance in sows, *J. Animal Sci.,* **43:**251, 1976.

Feed and Feeding Digest, National Grain and Feed Association, P. O. Box 28328, Washington, D.C. 20005.

Hays, V. W.: The role of antibiotics in efficient livestock production, Paper at International Symp. on Nutrition and Drug Interrelations, Nutrition Science Council, Iowa State Univ., Ames, Iowa, Aug. 1976.

Jukes, T. H.: Antibiotics in Nutrition, Medical Encyclopedia, Inc., New York 1955, p. 128.

Lassiter, C. A.: Antibiotics as growth stimulants for dairy cattle: A review, *J. Dairy Sci.*, **38**:1102–1138, 1955.

National Academy of Sciences: The use of drugs in animal feeds. Hays, V. W.: Biological basis for use of antibiotics in livestock production, Publ. 1679, 1969, pp. 11–30; Bird, H. R.: Biological basis for use of antibiotics in poultry feeds, Publ. 1679, 1969, pp. 31–41. Washington, D.C.

Perry, T. W., M. T. Mohler, and W. M. Beeson: Effect of feeding different tranquilizers in combination with implanted diethylstilbestrol or oral antibiotic on fattening beef steers, *J. Animal Sci.*, **19**:533–537, 1960.

Raun, A. P., and associates: Effect of monensin on feed efficiency of feedlot cattle, *J. Animal Sci.*, **43**:670–677, 1976.

Rusoff, L. L., and associates: Effect of high-level administration of chlortetracycline at birth on the health and growth of young dairy calves, *J. Dairy Sci.*, **42**:856–862, 1959.

Swanson, E. W.: Effects of chlortetracycline in calf starter and milk, *J. Dairy Sci.*, **46**:955–958, 1963.

Symp. on drugs and feed additives, *J. Animal Sci.*, **31**:1102–1132, 1970.

Symp. on natural hormones in edible products, *J. Animal Sci.*, **46**:609–685, 1977.

Teague, H. S., and associates: Influence of diethylstibestrol implantation on growth and carcass characteristics of boars, *J. Animal Sci.*, **23**:332–338, 1964.

——, A. P. Grifo, Jr., and E. A. Rutledge: Response of growing-finishing swine to different levels and methods of feeding chlortetracycline, *J. Animal Sci.*, **25**:693–700, 1966.

Wallace, H. D., and associates: High level copper for growing-finishing swine, *J. Animal Sci.*, **19**:1153–1163, 1960.

Wieser, M. F., et al.: Intensive beef production. The effect of chlortetracycline on growth, feed utilization and incidence of liver abscesses in barley beef cattle, *Animal Prod.*, **8**:411–423, 1966.

Chapter 13

Feeding Experiments.
Feeding and Nutritional
Standards

Our previous discussion has considered the different nutrients which are required by the animal body and the metabolic changes which they undergo in
serving its various functions. A knowledge of the quantitative needs of the
body for these nutrients and of the relative value of feeds as sources of them is
the basis of scientific feeding, a knowledge which has been gained gradually by
means of research and experience over many years. An understanding of the
methods by which it has been attained and which are still being employed to
augment it is essential for the student of nutrition.

FEEDING EXPERIMENTS

Trial and experience were the means by which the art of feeding animals was
originally developed. With the establishment of specific agencies to augment
this knowledge, such as the agricultural experiment station, the feeding-trial
method naturally was adopted as the means by which current practices could
be critically tested and improved with the aid of the underlying sciences.

Feeding experiments have been carried out with farm animals during the
past two centuries to compare the value of different feeds or combinations of
feeds. More recently feeding studies have been conducted with laboratory
animals, fish, primates, and even humans to determine the value and utilization
of individual feeds and the nutritional adequacy or safety of different diets.

Feeding studies have been the primary method used to determine qualitative and quantitative requirements for most of the known nutrients and for measuring the value of foods and dietary combinations for growth, maintenance, reproduction, lactation, and even for lifetime survival.

There are many different forms of feeding trials and various measures have been used to evaluate the results, such as observations on health and general condition, live weight gain and changes in height, length and circumference of the body. The combined use of digestion studies, slaughter experiments, or balance studies along with feeding trials to measure the intake, absorption, and retention of specific nutrients add greatly to the amount of information obtained. Short-term balance experiments which have been used extensively to determine nutrient requirements, especially during growth, may lead to erroneous conclusions unless the results can be checked with feeding studies which extend over meaningful periods of time. Guidelines have been published[1,2,3] to assist researchers to plan studies which will yield the maximum amount of useful data. A feeding trial with the species in question still remains the most useful method of obtaining results which have a direct application to feeding practice. Although much has been learned there are still many unsolved problems related to the rapid and accurate evaluation of feeds[4] and the determination of nutrient requirements of animals.

13.1 Comparative Feeding Trials

In its simplest form, a feeding trial is a record of the results produced in terms of growth, milk production, or other function from a given feed or ration. Two or more rations may be compared with each other on this basis. Additional records as to the feed eaten provide a comparison of the relative amounts of the rations required to produce a unit of product, and by the use of cost figures the results may be put on a money basis. The records here obtained tell us nothing as to why one ration proved better than another, unless the poorer one was so unpalatable as to be little consumed or unless it caused absolute harm. As a further step in the interest of more specific information, individual feeds may be compared as a part of rations the other ingredients of which are held constant. Here is an old example showing that fish meal is a better protein supplement for hogs than linseed meal:

	Average daily gain, kg	Feed for 100 kg gain, kg
Ration 1		
200 kg corn		
100 kg wheat middlings	0.50	390
75 kg fish meal		
Ration 2		
200 kg corn		
100 kg wheat middlings	0.32	440
75 kg linseed meal		

This experiment gives us a specific answer as to the comparative overall effect of these two feeds, but it tells nothing as to why the fish meal was better. Was it the result of the higher percentage of protein in the fish product or a higher biological value of this protein? Was the large amount of calcium supplied by the fish meal in contrast to the very small amount present in the oil meal a factor, or did certain vitamins present in the one but not the other play a role?

It is important to know the specific nutritive quality which makes one feed better than another. For example, if the superiority of the fish-meal ration resulted entirely from the extra calcium supplied, the addition of ground limestone to the linseed-meal ration would provide a cheaper method of getting the same results. The comparison of two feeds with respect to a specific nutrient such as calcium or protein requires that all other nutritive factors be held alike and adequate in the two rations. This can never be achieved absolutely, but feeding trials can be set up in such a way as to give most of the specific information desired, as is illustrated by the modern experiments discussed later.

13.2 Feeding Trials with Laboratory Animals

Today many of the problems of nutrition are being studied with small animals, such as the rat. The processes of growth, reproduction, and lactation can be effectively investigated and the value of various feeds for these various functions determined. The much smaller cost in terms of animals, feed, and labor and the much shorter time involved for a given experiment, in view of the short life cycle of the laboratory animal, are important advantages. The influence of individual variability, a seriously disturbing factor in large-animal experimentation, can be reduced to a minimum by the use of animals of similar genetic and nutritional history, by the employment of large numbers, and by close environmental control. Slaughter for chemical and histological examination, a desirable feature of many feeding trials, presents little difficulty with small animals, compared with the economic and other considerations involved in the case of farm animals.

The laboratory animal is thus highly useful for working out many of the fundamental principles of nutrition. The results obtained in feeding trials with the small animals, however, cannot be considered to have direct application to the various species of farm animals, because of the differences in physiology and other considerations. Even here studies with small animals serve as pilot experiments, by means of which much preliminary information can be obtained more quickly and at much less cost than with the large animals and whereby it can be determined what ideas are of sufficient promise to justify the expense involved in giving them a final test with the large animals. The situation is analogous to that of an industry in which processes worked out in the laboratory are first tested on a semicommercial scale before being finally adopted. Of course, there are feeding problems which by their nature are susceptible to solution only by experiments with the farm animals themselves, but the animal industry owes much to experiments with the rat.

13.3   The Purified-diet Method

An important feature of feeding trials which has been developed along with the use of laboratory animals is the employment of purified diets. These diets consist of purified sources of the various nutrients. For example, protein is supplied as casein, purified soybean protein, or urea; carbohydrates as starch, glucose, or sucrose; fat as lard or some oil; minerals as chemically pure salts; and vitamins as pure crystalline compounds. Such a diet makes it possible to include or withdraw a given nutrient with a minimum of disturbance of any of the other nutrient relations. The influence of different levels or sources of nitrogen can be studied by including varying amounts of the pure protein or amino acids without any change in the rest of the ration, whereas the addition of a natural protein source such as meat or beans would introduce many variables, because they contain many other nutrients as well. By similar substitutions other nutrients can be subjected to specific study.

The extensive use of the purified-diet method has been a development of the last sixty years, but the idea was conceived a century and a half ago at least. As reported in 1816, Magendie fed diets of pure sugar and of pure fat to dogs to ascertain whether or not nitrogen was required in the food. Before the middle of the last century Boussingault,[5] the famous French chemist, carried on nutrition studies with various species, involving the use of diets consisting in part of purified nutrients. As later attempts were made from time to time to use this method, the discouraging result occurred that the more completely the diet consisted of purified nutrients, the less satisfactory was the effect on the animal. It was this discouraging result, however, that led to the conclusion, toward the close of the century, that there were dietary essentials unknown to the chemist and thus led to the later discovery of the vitamins and other previously unappreciated nutritive factors. There followed the intensive application of the purified-diet method by McCollum, by Osborne and Mendel, and by others, as a result of which an increasing knowledge as to essential constituents of a purified diet and new discoveries as to nutritional requirements simultaneously developed.

Thus the purified-diet method became responsible for much of our modern knowledge of nutrition, including the physiology of the vitamins, the establishment of differences in protein quality, and more exact information regarding many of the minerals. Studies of the role of an element needed by the body in small amounts can be effectively carried out only with basal diets which may be freed from it and to which it may be added in known amounts. This is only possible with purified diets, because a diet cannot be prepared from natural foods which will be free from the element in question.

The purified-diet method has limitations that should be kept in mind. The ingredients of these diets cannot be considered pure in the absolute sense. Starch, for example, cannot be entirely freed from mineral matter without breaking down its structure. The essentiality of some mineral elements, needed only in traces, may remain undiscovered because of the impossibility of eliminating them entirely from the other ingredients of the experimental diet. Some of the vitamins were identified as "impurities" in purified diets earlier assumed to consist only of known nutrients. Some of the constituents, notably protein, in purified diets may be altered from their natural state in the process of purification. The kind of pure carbohydrate used affects the significance of the results in the case of certain vitamins because of the effects of various carbohydrates on vitamin synthesis in the alimentary tract.

A completely successful purified diet for a given species cannot be prepared until all of the nutrient requirements of that species are known—which is still apparently not the case for some. The diet must be of a suitable physical nature and sufficiently palatable so that it will be consumed in large enough amounts to support fully the function under study. The method has been developed to its highest degree of usefulness in the case of the rat and chick, both because of the lesser problem involved in preparing purified nutrients on a small scale and also because of the years of experience with these species. The knowledge as to the rat's qualitative and quantitative needs is much greater accordingly, but it may not yet be complete in terms of successive generations. Successful studies are also being made with several other species of laboratory animals. As regards farm animals, the use of purified diets has contributed greatly to the modern knowledge of poultry nutrition. Other applications of the method have made important contributions to our knowledge of the nutritional needs of lambs (Fig. 13.1), older sheep (Matrone et al.),[6] cattle,[7,8] and pigs. Experiments illustrating these applications are reviewed in later chapters which deal with nutritive requirements for various body functions.

13.4 Germfree Techniques

It is evident from previous discussions regarding various vitamins that the contributions of intestinal organisms to the nutrition of the host complicate the interpretation of data on dietary requirements obtained in feeding trials. Thus, the nutrition scientist has a special interest in the techniques which have been developed for obtaining animals which are germfree at birth and for rearing them in an uncontaminated environment thereafter. Germfree means free of contamination by bacteria, yeasts, molds, fungi, protozoa, and parasites in general, that is, free of all other life. The newborn are obtained by Caesarean section and reared in specially designed apparatus by appropriate techniques, involving, of course, sterilized diets. Success has been reported with rats, rabbits, hamsters, mice, chickens, turkeys, and monkeys. Rats, mice, and chickens have been bred through successive generations. A review of the development of germfree research, the technique and the equipment used, and detailed guides for their employment are to be found in the book by Luckey.[9]

Techniques have been developed for obtaining "specific pathogen-free" (SPF) baby pigs by hysterectomy, and using them for nutrition experiments carried out under conditions which prevent infections. The techniques and special equipment used are described in a study by Schneider and Sarrett.[10]

Fig. 13.1 This sheep was reared to maturity on a purified diet which it received for 15 months. (*L. L. Madsen, C. M. McCay, and L. A. Maynard. Cornell Agr. Expt. Sta. Memoir, 178, 1935.*)

13.5 Group Feeding versus Individual Feeding

Feed records are a desirable feature of all feeding trials. Even where the feed cost of the physiological performance is not of primary concern, it is frequently essential from the standpoint of the interpretation of the results to have some record of the feed consumed. In many feeding experiments, particularly those with farm animals, the animals have been fed as a group. This is the simplest procedure from the standpoint of equipment needed and labor cost, but in many experiments it introduces complications in the interpretation of results. Such complications arise when there is a wide variability in the individual behavior within the lot, as to both production and feed consumption. The difficulty is increased when an animal, owing to accident or other unavoidable cause, has to be removed from the lot. The performance of the individual can be eliminated from consideration, but the food which it ate cannot. Individual feeding eliminates these disadvantages. It makes possible the correlation of individual performance record with the food which the animal ate. It preserves the identity of the individual. Certain species which are fed together in practice may consume somewhat less when fed individually. Thus certain workers stress this "competition in the feed lot" as being of practical importance in feeding trials with beef cattle, sheep, and hogs. Here several small groups will yield a more sensitive test than a few large ones.

While there are types of feeding experiments in which feed records of the group as a whole are sufficient or in which they may be acceptable in the interest of economy and of the use of larger numbers of animals, individual records are highly desirable in studies where only small differences are to be expected and where quantitative data are of special importance. Individual records are much more useful from the standpoint of statistical treatment (Sec. 13.10). The relative advantages of group and individual feeding are discussed by Crampton[11] in his excellent review of methods for conducting feeding trials. The review covers other topics which are presented in the following sections in this chapter.

13.6   Controlled versus Ad Libitum Feeding

When the amount of feed consumed is regulated in some way by the experimenter, the feeding is controlled, as distinguished from the ad libitum system in which each animal or group is allowed to eat all it wants. Ad libitum feeding is the most commonly used procedure in farm-animal investigations and gives unbiased results for direct practical application. By keeping records of feed intake, the results can be expressed on an efficiency basis, such as "feed required per 100 kg gain," as well as in terms of total increase in weight. This system gives unbiased results for direct practical application in terms of the feed, species, and function under study. It is subject to the limitation, however, that with certain feeds and rations differences in nutritive value may be masked by differences in palatability. Further, the method does not provide the controlled conditions required for certain purposes—for example, the determination of digestibility.

Thus, in many instances there is an advantage in using some system of controlled feeding. Early in their studies of protein quality Osborne and Mendel recognized that ad libitum feeding frequently gave rise to variable results. They raised the fundamental question: "Does one animal grow because it eats more or the other fail because it eats less?" They experimented with various procedures of controlled feeding as a means of eliminating the uncertainties here involved. In one series of studies Osborne and Mendel[12] kept the food intakes alike for each diet under study, in accordance with a prescribed schedule based upon a preliminary experiment. They were thus able to compare the growth made on different diets consumed in the same amount. Recognizing that the more rapidly growing animals might be at a disadvantage under this system in view of their increasing maintenance requirement, they carried out another series in which the food intake was adjusted in accordance with increase in weight. In another experiment Osborne and Mendel[13] allowed ad libitum feeding and selected for comparison the growth records of those animals which had consumed substantially the same amount of food under this system. The discussion presented in these papers clearly shows that a proper assessment of the effect of food consumption as such is a very important matter in any feeding experiment, and the papers are well worth reading by any student who is planning such an experiment.

13.7 Equalized Paired Feeding

This is a widely used procedure in which the feed intakes are completely controlled. In this method of comparing two rations the animals are fed alike in a preliminary period and then selected by pairs, one animal of a given pair being placed on ration A and the other being placed on ration B and both animals being given exactly the same weight of food or by some measure of food energy. The latter is accomplished by limiting the intakes of both to that of the animal consuming the lesser amount. The two animals of the pair are selected to be as nearly alike as possible in size, age, and previous history, but such equalities are not essential from pair to pair. The equalization of food intake is also limited to within the pair. This method is illustrated by the data presented in Table 13.1, obtained in an experiment in which the two rations under comparison were alike with the exception of the phosphorus carrier. Both rations contained the same amount of phosphorus and in the same ratio to calcium. It is noted that for a given pair of rats the food intakes were substantially alike over the experimental period of 35 days. When it is desired to compare three rations at the same time, the animals can be selected in trios. This may involve complications, however, in the equalization of food intake, complications which become increasingly troublesome in comparing more than three rations.

Properly conducted equalized feeding experiments have a distinct advan-

TABLE 13.1. Data from a Paired-feeding Experiment in Which Dicalcium Phosphate (*A*) and Bone Meal (*B*) Were Compared as Sources of Phosphorus for Bone Growth in Rats

| | Pair 1 | | Pair 2 | | Pair 3 | | Pair 4 | |
	A	B	A	B	A	B	A	B
Food, g	253.0	253.0	255.0	254.0	252.0	252.0	224.0	228.0
Ash in bone, %	48.44	47.43	51.63	50.64	49.77	48.91	50.81	54.23
Ash in bone, mg	191.0	157.7	190.3	154.7	179.2	166.3	162.3	171.3
Calcium in bone, %	19.81	16.04	18.59	18.20	17.78	17.63	18.20	19.26
Phosphorus in bone, %	9.9	8.5	9.3	9.2	8.9	8.8	9.0	9.6

Source: K. V. Rottensten and L. A. Maynard, The assimilation of phosphorus from dicalcium phosphate, C.P., tricalcium phosphate, C.P., bone dicalcium phosphate and cooked bonemeal, *J. Nutrition*, 8:715–730, 1934.

tage over the ad libitum method as regards the adaptability of the results to statistical treatment. Other things being equal, the larger the number of pairs or trios, the greater the reliability of the results. The data given in Table 13.1 include only four of the six pairs actually used in the experiment. The statistical analysis of the complete data revealed no certain advantage for one phosphorus carrier over the other.

The method of equalized feeding by pairs commonly referred to simply as paired feeding, has been employed to study a wide variety of problems. A paper by Mitchell and Beadles[14] gives an excellent description of the procedure and its application. The method has the advantage of eliminating the confusing effects which may arise from a variable food intake even when results are compared per unit of feed consumed. The method is subject to the criticism, however, that limiting the food intake may defeat the very object of the experiment, since a frequent effect of a nutritionally deficient ration is to decrease food consumption, with the result that the full effect of the better ration cannot express itself and that the comparison is made at a restricted food intake instead of a normal one. The force of this criticism is dependent upon the nature of the experiment. In practical tests in which palatability and level of food intake are important criteria of the relative value of the rations, equalized feeding would obviously defeat their purpose. One must be careful, however, about drawing conclusions that more fundamental differences exist where some physical factor concerned in palatability may have been solely responsible for the results obtained.

The method is not suitable for finding out *how much* superior one ration is to another for growth, because, as the animal on the superior ration increases in weight over its mate, its maintenance requirement becomes greater than that of its mate. Under these conditions, an equal food intake for both means that the larger animal must be using a larger proportion for maintenance, and less remains for growth promotion. Therefore, an absolute equality of food intake means that the quantities available for the specific function which is being used as the criterion in comparing the two rations are not equal. The faster-growing animal is penalized. Various workers have introduced modifications designed to overcome this limitation and still preserve the principle of controlled feeding. Lactation studies which adjust the feed of each animal on the basis of body weight and production represent a special type of controlled-feeding experiment.

The paired-feeding method is a useful technique but requires judgment in its application and in the interpretation of the results obtained. It would appear to find its largest usefulness in comparisons in which food consumption is not markedly restricted by the conditions imposed and in which the measure is in terms of the specific effect of the nutrient under study, as is illustrated in Table 13.1, instead of the more general measure of increase in weight. Some of the limitations of the method are brought out in the studies by Barnes and coworkers,[15] dealing with methods of measuring protein efficiency. There are many problems concerning which the use in separate experiments of both ad libitum feeding and controlled feeding will give much more information than either procedure alone. If the two give results, the validity of the conclusions is greatly enhanced thereby. Lucas[16] designed an equalized feeding system which reduced the variability of responses of lactating cows that are fed individually.

This discussion of paired feeding illustrates the fact, which holds also for other methods discussed in this chapter, that no single method is suitable for the solution of all types of nutrition problems. The effective investigator must select his method in accordance with his problem, frequently employing more than one method, and finally, he must interpret his results with a full consideration of the advantages and limitations of the method used.

13.8 Slaughter Experiments

In the previously discussed experiment presented in Table 13.1, the relative value of the two mineral supplements was measured in terms of the calcium and phosphorus content of the bones, since the growth of the animals as a whole would not have given definite information as to bone development. Such a procedure, which involves the killing of the animals and the analysis of certain specific tissues or of the body as a whole, is commonly referred to as a slaughter experiment. In many feeding trials, it is desirable to obtain more specific information regarding the effect of a given ration than is furnished by the common measures of weight and size. For example, in studies of the protein requirement for growth or of the comparative value of different protein sources, it is important to know the specific effect in terms of protein tissue formed, since the increase in the body as a whole is due to water, fat, and minerals as well as protein, the relations of which may vary.

The introduction of the slaughter method by Lawes and Gilbert has been referred to (Chap. 2). As now used it takes many forms according to the problem under investigation. To study the effect of a given diet on changes in body composition a group of like animals are selected and a part of them are slaughtered and analyzed at the start of the experiments. The others are fed a weighed and analyzed diet for a given period and then slaughtered and analyzed. The difference in their composition from that of the check animals killed at the start reveals the effect of the diet fed. The use of the slaughter method for studying protein and energy requirements is illustrated by the work of Mitchell and Hamilton.[17]

A slaughter experiment requires much more time and labor than is involved in merely weighing or measuring the animals, and in many instances difficult problems are presented in the selection of representative samples of tissues and in their preparation for analysis. For each period of observation, a sufficiently large number of animals must be examined to minimize the rather large individual variability in composition. In general, small laboratory animals are much easier to work with than the larger farm animals. Because of the economic considerations involved, work with the latter must be limited for the most part to those animals for which a return can be obtained on the carcass after the desired samples for analysis have been taken. As regards farm animals, therefore, the slaughter method has found its greatest application in studying the nutrition of beef cattle, sheep, and swine. Slaughter data may also include various measures of market value, such as dressing percentages and quality of the carcass, and such measures are frequently used in meat-production experiments to study the influence of a given ration upon the quality of the product and upon its selling price.

13.9 Experimental Designs

The statistician refers to those comparative feeding trials which are set up in such a way as to allow statistical analysis as *experimental designs*. This term is applicable to the methods previously described when they are appropriately used. In addition, there are certain specific designs with which the student should be familiar.

One is the *factorial* arrangement of treatments which may be illustrated as follows:

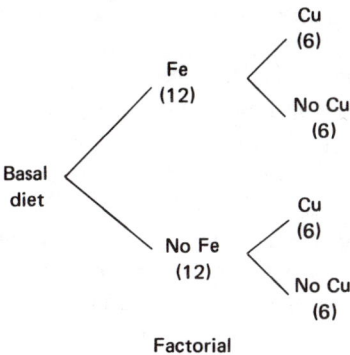

Factorial

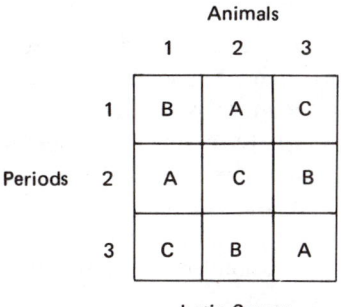

Latin Square

This design provides for testing the effects of iron and/or copper addition to a basal diet, using a total of 24 animals. Twelve are placed on the basal diet with iron, and 12 on the diet without iron. Each group of 12 is in turn subdivided into groups with and without copper. This design provides for a comparison of iron versus no iron, using 12 animals per group, and it also furnishes a compar-

ison with the same number of animals with and without copper. Further, interactions between iron and copper are brought out through statistical treatment,
answering the question, for example, as to whether iron supplementation gives
a different result in the presence of supplemental copper than in its absence.
The factorial design is considered to be of high efficiency in terms of information obtained in relation to the number of animals used. The diagram shown
represents a very simple experiment of this character. Each of the groups with
and without copper might be further broken down, e.g., into groups with and
without manganese. More animals could also be used.

There are several methods which involve the feeding of the rations to be
compared to the same animal in different periods. Thus, in the *double-reversal
system* two rations (X and Y) and two groups of animals (A and B) are used. In
period 1, group A receives ration X and group B receives ration Y. At the end
of the period the rations are reversed—group A receiving ration Y and group
B receiving ration X. One or more additional periods may be included with a
shift of groups at the beginning of each. A related method is the *latin square
design* illustrated above. Here A, B, and C represent treatments, such as feeds
for a digestibility determination, each feed being fed to each animal in a different period and thus in a different order for each animal. The essential feature
of this design is that the number of animals and the number of treatments must
be the same. It provides a high precision of measurement with a small number
of animals. This general procedure of feeding different rations to the same
group of animals in different periods has the advantage of lessening the effects
of variability among animals as compared with the method of feeding one group
one ration and the other group the other ration throughout the experiment. It
has the disadvantage, however, that there are carry-over effects when the same
animal is shifted from one ration to another. Lucas[18] has added an extra period
to this design to minimize the disadvantage.

13.10   The Use of Statistical Methods in Nutrition Experiments

In a feeding trial certain factors, such as the amounts and quantities of feed,
the time and method of feeding, and the general care and management, can be
definitely fixed. Certain other factors, inherent in the animals used, cannot be
controlled. The object of a well-planned experiment is to reduce these uncontrollable factors to minimum by giving attention, in the selection of the animals
used, to genetic and nutritional history as well as to such factors as age, size,
vigor, and the like. Even though this is effectively done, there still remain
inherent variables which cause two individuals to respond somewhat differently though treated exactly alike in an experiment.

The effect of the inherent variables cannot be measured, but the probability that the observed differences in experimental results could arise from the
uncontrollable variables alone can be estimated and taken into account. This is
done by a statistical analysis of the data obtained. Such an analysis helps the
investigator to decide whether the results from a given comparison reflect a
real difference in response to the treatments or may have occurred simply

because of inherent variations in the animals used. As has been indicated in the preceding discussions of experimental designs, statistical methods can be usefully employed in planning experiments in such a way as to make them more likely to give a definite answer to the question under study, particularly by providing results which can be statistically analyzed. Statistical methods have become an essential tool of the investigator in nutrition, and some knowledge of them is helpful to all students in this field as an aid in the evaluation of published research. It is beyond the scope of this book to attempt any presentation of them. Their application to feeding trials is discussed by Lucas.[19] A more extensive presentation is to be found in the report by Henderson.[20]

A scientist suffered confusions brought on by statistical delusions,
Though he worked 'til he lathered the data he gathered
Refused to support his conclusions.

W. W. Armistead (1916–   )

13.11   Financial Phases of Feeding Trials

It is obvious that an essential practical consideration in evaluating a ration for farm animals is its cost in terms of the return obtained for the product. Thus, in many feeding trials, records are kept of feed and perhaps of other costs and of the estimated or actual selling price of the product. Profit or loss per animal or per unit of feed thus becomes a measure of the nutritive value of the ration. While it is obvious that the financial phases of feeding operations cannot be neglected, the expression of the results in terms of dollars and cents, unless properly interpreted, may obscure rather than clarify the facts brought out by the experiment. Monetary statements are not experimental results. They are based upon factors which are not under experimental control, the same combination of which may never occur again. The relative prices of feeds and the selling price of the product vary from time to time, in fact from day to day and from place to place, according to market conditions. Clearly the product obtained per unit of feed is a much more stable and useful measure. While financial statements of feeding trials are interesting to the reader, they provide no basic or generalized measure of nutritive value, and relying upon them as a guide for practice may prove disastrous. In contrast, a statement of food consumed and product obtained provides the basic data to which the feeder can apply current prices and thus obtain a much more accurate picture of the probable financial outcome than is given by a statement of the financial results based on prices at the time and under the conditions of the experiment.

FEEDING STANDARDS AND NUTRITIONAL ALLOWANCES

Feeding standards and nutritional allowances are tables showing the amounts of food and specific nutrients which should be provided to different species for various purposes such as growth, fattening, reproduction, lactation, or strenuous exercise. They serve as guides in feeding animals and in estimating the adequacy of feed intakes and of feed supplies for groups of animals or people. To be useful these tables must be accompanied by tables listing the specific nutrient composition of the available feeds. Summaries of the N.R.C. tables of nutrient requirements and the compositions of selected animal feeds are presented in the appendix. Selected references to additional information are cited at the end of this chapter.

13.12 Early History of Feeding Standards

German workers were responsible for early development of feeding standards for farm animals. Over the past century or more many nutrition books have credited Albrecht Thaer (1752–1828) with originating the concept of *hay equivalents* as measures of relative value based on determining the materials in feeds extractable with water (and other solvents). Tyler[21] has examined the literature relating to hay equivalents and questioned Thaer's contributions. He cited a reference in 1725 in Bohemia which used *straw feed units* to compare hays. Between 1781 and 1796 workers in Germany and England conducted feeding trials to compare the value of hay with turnips, carrots, cabbage, or oil meals. Thaer only compared the value of hay and potatoes for milking cows, but did no analyses. Einhoff, Thaer's associate, isolated fiber from straw, barley, lentils, and beans by macerating them and washing with water. He assumed that the residual fiber and water were the only parts which provided no nourishment. In some cases he showed the isolated fiber was insoluble in alkali, alcohol, or acid, but these were not used in quantitative analyses. In his books (1808 and later) Thaer quoted some hay equivalent values and credited Einhoff with having defined fiber as the residue left after water, alkali, alcohol, and acid treatment, yet none of Einhoff's papers nor his immediate successors reported using any solvent except water. Boussingault (1830–1845), Henneberg and Stohman (1860), Henry (1904), and Armsby (1917) all credited Thaer with this concept and with originating hay equivalents.

Following the recognition in 1827 by Prout of protein, fat, and carbohydrate as the essential organic nutrients and the development of simple analyses by Liebig in the 1830s (Ahern[22]), Grouven made use of analyses for these nutrients to formulate in 1859 the first feeding standard for farm animals.

In 1864 Wolff devised a standard based on digestible protein, digestible ether extract, and digestible carbohydrates derived from results obtained in feeding trials. He calculated the ratio of digestible protein to digestible carbohydrate, called the *albuminoid ratio,* because he recognized that the proportion of protein in the ration affected its digestibility. This albuminoid ratio was the forerunner of the later-developed nutritive ratio (Sec. 3.17). Wolff's standards were republished annually without fundamental change until 1897, when they were modified by Lehmann to become the Wolff Lehmann standards for various classes of stock.

In 1874 Atwater brought the Wolff standards to the attention of American workers in the annual report of the Connecticut Board of Agriculture. The standards were also published in 1880 by Armsby in his book *Manual of Cattle Feeding,* which was based on Wolff's *Landwirtschafte Fütterungslehre.* As a result, the Wolff standards commenced to be used in the United States.

In 1890 Atwater proposed a feeding standard based on the "available fuel values" of feeds. These values were obtained by the use of Rubner's factors (4.1 kcal per gram for protein; 9.3 kcal per gram for fat, and 4.1 kcal per gram for carbohydrates), applied to digested nutrients. Since the Rubner factors for protein contained a deduction for urine losses of 1.25 kcal per gram of protein, Atwater's values took account of both fecal and urine losses. In 1898, Henry[23] published the first edition of his book *Feeds and Feeding,* which contained tables showing the average composition of American feeds; digestion coefficients for protein, crude fiber, nitrogen-free extract, and ether extract; and the Wolff-Lehmann standards. The standards for various classes of stock included intakes of dry matter, digestible protein, digestible carbohydrates, and digestible ether extract. These intakes were added, together with digestible ether extract multiplied by 2.4, as a sum of nutrients. The table also included nutritive ratios calculated as follows:

Digestible protein : digestible carbohydrate + digestible ether extract $\times$ 2.4

Henry considered Atwater's previously mentioned standard based on calories, but felt that their use for practical purposes was hardly warranted, "since a statement of the several nutrients themselves is more explicit and satisfactory."

In 1914 an important advance in the accuracy of the standard for dairy cows was made, as a result of many years' study by Haecker[24] of the University of Minnesota, showing that the nutritive requirements varied not only with the quantity of milk produced but also with its quality, especially its fat content. In the following words Haecker set forth a principle which should be recognized in all studies of nutritive requirements.

In order to determine the actual net nutrients required to produce a given animal product, the composition of the product should be known, as well as the composition and the available nutrients in food which is to be fed for its production, so that the nutrients in the ration might be provided in the proportions needed by the animal. Before a builder bids on a contract, he determines the quantity needed of each of the materials that are to appear in the structure. Without such specifications he would not know how much of each of the different materials would have to be provided.

Following the lead of Kellner in Germany, in 1915 Armsby set forth a feeding standard based on digestible true protein and net energy, but this system did not gain wide acceptance. The Wolff-Lehmann standard was modified in the edition of *Feeds and Feeding* published in 1915 under the authorship of Henry and Morrison.[25] The feed-intake recommendations were based on dry matter, digestible protein, TDN, and nutritive ratios. In the following years the digestible-nutrient system became used almost exclusively in the United States.

13.13   National Research Council Nutrient Requirements

As the knowledge developed regarding minerals and vitamins, recommendations regarding their quantitative requirements for various body functions were made by the investigators concerned. In 1942 the Committee on Animal Nutrition of the National Research Council (N.R.C.) of the National Academy of Sciences undertook the task of setting forth, for farm animals, a statement of quantitative needs for all the recognized nutrients. Subcommittees of experts for each class of stock were accordingly set up to review the literature. Beginning in 1944 there was thus published the Recommended Nutrient Allowances for Farm Animals, comprising separate reports for poultry, swine, dairy cattle, beef cattle, sheep, and horses. The term "allowance" was used, following the lead of the N.R.C. Food and Nutrition Board, which first adopted it for use in human dietary standards. These allowances were, in general, set higher than average determined requirements to provide a margin of safety. In 1953, the Committee on Animal Nutrition decided not to include such margins in future recommendations but to set forth intakes considered adequate for normal growth, health, and production, based on the average needs of groups of animals to achieve these results. Accordingly, it was agreed that future reports would be designated as "nutrient requirements," instead of "recommended nutrient allowances." Revised reports[26] for dairy, beef cattle, swine, horses, poultry, rabbits, foxes, mink, laboratory animals, and fish have been issued on this basis. It is believed that these N.R.C. reports, representing in each case the pooled judgment of a group of experts in the field of the species in question, should be considered the most authoritative statements of the nutritional needs of farm animals for feeding practice in the United States. Thus, where specific data on these needs are set forth in this text, they are those in the N.R.C. reports, except as important new data have become available since their publication. In applying them, the difference in concept between allowances and requirements, as discussed above, should be recognized. The N.R.C. nutrient requirement tables are given in Appendix Tables I through IX.

13.14   British Agricultural Research Council Nutrient
Requirements of Farm Livestock

These requirements are set forth in three separate reports dealing with poultry, ruminants, and pigs, each prepared by a specific working party under the general auspices of a Technical Committee of the Agricultural Research Council.[27] Each of these reports contains extensive summaries of the literature upon which the proposed requirements are based. The student will find in these reports a wealth of useful information. Frequent reference to them is made in the text under the designation A.R.C. requirements.

13.15   Usefulness and Limitations of Feeding Standards

The N.R.C. reports comprise feeding standards for total food needs and for all nutrients for which quantitative data are available. They are designed to be the best general guides for practice but should be considered subject to modification in special instances. In practical feeding operations it is frequently desirable to take economic factors into account. Thus, modifications may be called for in the interest of obtaining the rate of gain or level of milk production that seems the most economical in terms of current feed costs and the market price of the product. No standard can be a complete guide to feeding because other factors such as palatability and the physical nature of the ration must also be taken into account. The significance of these various considerations in using feeding-standard data in planning experiments and interpreting the results depends upon the nature and objective of the investigation. Further, it seems likely that environment may change nutrient requirements. As yet, there are inadequate data to adjust standards on this basis.

NOTES

1. L. L. Larson, F. G. Owen, J. L. Albright, and associates, Guidelines toward more uniformity in measuring and reporting calf experimental data, *J. Dairy Sci.,* **60**:989–996, 1977.

2. Techniques and Procedures in Animal Science Research, Am. Soc. Animal Sci., 113 N. Neil St., Champaign, Ill., 1969.

3. Report of the American Institute of Nutrition Ad Hoc Committee on Standards for Nutritional Studies, *J. Nutrition,* **107**:1340–1348, 1977.

4. Feedingstuffs Evaluation Unit, First Report, Rowett Research Institute, Aberdeen, 1975.

5. J. B. Boussingault (1802–1877), following a period of service as professor of chemistry at Lyons, France, founded the first agricultural experiment station in 1836 at Bechelbronn. Here his pioneer studies on the nutrition of various species of animals extended over many years and became models for later investigators. Boussingault ranks as one of the foremost agricultural scientists of all time. His two-volume work, Économie Rurale, published in 1843 and 1844, and dealing with soils, crops, and fertilizers as well as the nutrition of cattle, horses, hogs, and other animals, is highly worthwhile reading for the modern student.

6. G. Matrone, C. R. Bunn, and J. J. McNeill, Study of purified diets for growth and reproduction of the ruminant, *J. Nutrition,* **86**:154–158, 1965.

7. A. I. Virtanen, Milk production of cows on protein-free feed, *Science,* **153**:1603–1614, 1966.

8. R. R. Oltjen, L. L. Slyter, and R. L. Wilson, Urea levels, protein and diethylstilbestrol for growing steers fed purified diets, *J. Nutrition,* **102**:479–488, 1972. See also papers cited.

9. T. D. Luckey, Germfree Life and Gnotobiology, Academic Press, New York, 1963.

10. D. L. Schneider and H. P. Sarrett, Use of the hysterectomy-obtained SPF pig for nutritional studies of the neonate, *J. Nutrition,* **89:**43–48, 1966.

11. E. W. Crampton, Design for comparative feeding trials, in Techniques and Procedures in Animal Science Research, *Am. Soc. Animal Sci.,* 130–156, 1969.

12. T. B. Osborne and L. B. Mendel, A quantitative comparison of casein, lactalbumin, and edestin for growth or maintenance, *J. Biol. Chem.,* **26:**1–23, 1916.

13. T. B. Osborne and L. B. Mendel, The relative value of certain proteins and protein concentrates as supplements to corn gluten, *J. Biol. Chem.,* **29:**69–92, 1917.

14. H. H. Mitchell and J. R. Beadles, The paired-feeding method in nutrition experiments and its application to the problem of cystine deficiencies in food proteins, *J. Nutrition,* **2:**225–243, 1930.

15. Richard H. Barnes and coworkers, Measurement of the growth promoting quality of dietary protein, *Cereal Chem.,* **22:**273–286, 1945.

16. H. L. Lucas, A method of equalized feeding for studies with dairy cows, *J. Dairy Sci.,* **26:**1011–1022, 1943.

17. H. H. Mitchell and T. S. Hamilton, Swine type studies. III: The energy and protein requirements of growing swine and the utilization of feed energy in growth, *Illinois Agr. Expt. Sta. Bull.* 323, 1929.

18. H. L. Lucas, Extra-period, latin square change-over design, *J. Dairy Sci.,* **40:**225–239, 1957.

19. H. L. Lucas, Techniques in animal science research, Proc. Auburn Conf. on Statistics Applied to Research, Alabama Polytechnic Institute, Auburn, Ala., 1948.

20. C. R. Henderson, Design and analysis of animal husbandry experiments, in Techniques and Procedures in Animal Science Research, *Am. Soc. Animal Sci.,* **6,** 1969.

21. C. Tyler, Albrecht Thaer's hay equivalents: Fact or fiction? *Nutrition Abst.& Revs.,* **45:**1–11, 1975.

22. R. Ahern, William Prout (1785–1850): A biographical sketch, *J. Nutrition,* **107:**17–23, 1977.

23. William Arnon Henry (1850–1932) was born in Norwalk, Ohio. He received his B.S.A. degree from Cornell University in 1880 and then became professor of agriculture at the University of Wisconsin. He became dean of the College of Agriculture and director of the Agricultural Experiment Station in 1887 and held these positions for twenty years. His book, and successive editions, which bore the subtitle "A Handbook for the Student and Stockman," became the leading textbook in its field.

24. T. L. Haecker, Investigations in milk-production, *Minnesota Agr. Expt. Sta. Bull.* 140, 1914.

25. Frank Baron Morrison (1887–1958) was born in Wisconsin and received the B.S. degree from the University of Wisconsin in 1911. In 1914 he was appointed profes-

sor of animal husbandry at the university, resigning in 1927 to accept the director-
ship of the New York State Agricultural Experiment Station in Geneva. After one
year's service he moved to Ithaca to become head of the department of animal
husbandry in Cornell University, a position which he held until he retired in 1955.
He became the sole author of *Feeds and Feeding,* in the edition published in 1936
and of later editions. The last edition, published in 1956, is frequently referred to
in this text.

26. Nutrient requirements of dairy cattle, 1978; Nutrient requirements of beef cattle,
     1976; Nutrient requirements of swine, 1973; Nutrient requirements of horses, 1978;
     Nutrient requirements of sheep, 1975; Nutrient requirements of poultry, 1977;
     Nutrient requirements of rabbits, 1977; Nutrient requirements of dogs, 1974; Nu-
     trient requirements of laboratory animals, 1978; Nutrient requirements of mink
     and foxes, 1968; Nutrient requirements of trout, salmon, and cat fish, 1973. Na-
     tional Academy of Sciences, Washington, D.C.

27. Technical Committee on the Nutrient Requirements of Farm Livestock, No. 1,
     Poultry: Summary of recommendations, 1963; No. 2, Ruminants: Technical re-
     views, 1965; Ruminants: Summaries of estimated requirements, 1965; No. 3, Pigs:
     Technical reviews and summaries, 1967, The Agricultural Research Council,
     London.

SUPPLEMENTARY LITERATURE

Atlas of Nutritional Data on United States and Canadian Feeds, Publ. 1919, National
Academy of Sciences, Washington, D.C., 1971.

Black, J. L., and D. A. Griffiths: Effects of live weight and energy intake on nitrogen
balance and total N requirement of lambs, *Brit. J. Nutrition,* **33:**399–413, 1975 (see
also **30:**45, 1973).

Carpenter, K. J.: The concept of an ''appetite quotient'' for the interpretation of ad
libitum feeding experiments, *J. Nutrition,* **51:**435–440, 1953.

Duncan, D. B.: Multiple range and multiple F tests, *Biometrics,* **11:**1, 1955.

Fonnesbeck, P. V., L. E. Harris, and L. C. Kearl: Food composition, animal nutrient
requirements and computarization of diets, First International Symp., Utah State
Univ., Logan, Utah, 1978.

Gohl, B.: Tropical Feeds, F. A. O., Rome, 1975.

Heneghan, J. B.: Germfree Research, Biological Effects, Gnotobiotic Environments,
Academic Press, N.Y., 1973.

Latin American Tables of Feed Composition, Univ. of Florida, Gainesville, Fla., 1974.

Little, T. M., and F. J. Hills: Statistical methods in agricultural research, Agricultural
Extension, Univ. of California, Berkeley, Calif., 1972.

Mertens, D. R.: Principles of modeling and simulation in teaching and research,
J. Dairy Sci., **60:**1176–1186, 1977.

Proceedings of the Conference on Animal Feeds of Tropical and Subtropical Origin,
Tropical Products Institute, London, 1975.

Rechcigl, M., Jr. (ed.): CRC Handbook Series in Nutrition and Food, Sec. D: Nutritional Requirements, Vol. I, 1977; Sec. G: Vol. I–IV, 1977. Other volumes projected. Diets, culture media, food supplements.

Steel, R. G. D., and J. H. Torrie: Principles and Procedures of Statistics, McGraw-Hill Book Co., New York, 1960.

Tables of Feed Composition, *Agr. Handbook No. 8,* U.S., Dept. Agr., Washington, D.C.

Truswell, A. S.: A comparative look at recommended nutrient intakes, *Proc. Nutrition Soc.,* **35:**1–14, 1976.

United States—Canadian Tables of Feed Composition, Publ. 1684, National Academy of Sciences, Washington, D.C., 1969.

Waldo, D. R.: An evaluation of multiple comparison procedures, *J. Animal Sci.,* **42:**539–544, 1976.

Chapter 14

Maintenance

Maintenance can be defined as that state in which there is neither gain nor loss of a nutrient by the body. *Maintenance requirements* are estimates of the amounts of nutrients needed to achieve such equilibrium states (Blaxter[1]).

Whether an animal is being fed for growth, fattening, milk secretion, or other productive function, a substantial part of its food is used for supporting body processes which must go on whether or not any new tissue or product is being formed. This demand for food is referred to as the maintenance requirement, since it comprises the amount needed to keep intact the tissues of an animal which is not growing, working, or yielding any product. If this need is not met, tissue breakdown occurs, which is commonly revealed by a loss in weight and which leads to various undesirable consequences. For a considerable part of the human population, the maintenance requirement comprises the principal need for food. While this is much less true for farm animals because they are always fed for productive purposes, maintenance is an important "overhead" of the livestock business. A dairy cow weighing 500 kg and producing 20 kg of 4 percent fat milk daily uses 37 percent of its total ME requirement for maintenance, versus 23 percent at a yield of 40 kg. A beef calf weighing 200 kg and gaining 1 kg daily uses about 50 percent. The net income from the feeding operation is largely governed by the ability of the animal to consume and utilize feed in addition to its maintenance requirement. Thus the knowledge of this requirement, which the student must have in order to understand the principles underlying nutrition, has a direct practical interest. The starting point of this knowledge is the fasting catabolism.

THE FASTING CATABOLISM

A hungry people listens not to reason nor are its demands turned aside by prayers.

Seneca (4 B.C.–65 A.D.)

Possibly as many as 450 million to a billion persons in the world do not receive enough food.

(World Food and Nutrition Study, National Academy of Sciences, 1977).

The animal receiving no food, doing no external work, and yielding no product is nevertheless carrying on a variety of internal processes which are essential to life. These processes include respiration, circulation, maintenance of muscular tonus, manufacture of internal secretions, and several others. In the absence of food, the nutrients required to support these activities must come from the breakdown of body tissue itself. This destruction of body tissue is referred to as the fasting catabolism, and it can be measured in terms of the waste products eliminated through the various paths of excretion. Most of the breakdown which occurs is in response to the demand of the fasting organism for energy for its vital processes.

Total starvation leads to death. Animals and human beings can live only a few days without water (Chap. 4), but much longer without food. Lusk[2] reviewed reports on humans who fasted experimentally for 1 to 50 days and of dogs that survived starvation of up to 117 days. Body fat stores, ambient temperature, and other factors markedly influence the survival time of starving animals. Such studies have provided an understanding of nutrient needs for maintenance.

14.1 Energy Metabolism of Fasting

The energy expended in the fasting animal is represented by the fasting heat production and thus can be measured in the respiration calorimeter, or it can be obtained by one of the methods of indirect calorimetry. Its measurement provides a useful basis of reference for other phases of energy metabolism. In order that the fasting catabolism may be measured at its minimum value, it is necessary that all influences tending to increase heat production above the minimum expenditure compatible with the maintenance of life be eliminated in so far as possible. Such a minimum value is called *basal metabolism,* or *basal metabolic rate (BMR).* It has its most exact meaning in the case of humans, because it is with this species that the conditions which are essential for a true minimum value can most nearly be attained. The conditions for its measurement in man are commonly specified as follows:

1. Good nutritive condition
2. Environmental temperature of approximately 25°C
3. Relaxation on bed prior to and during measurement
4. Postabsorptive state

A good nutritive condition implies that the previous diet of the subject has been adequate, especially as regards energy and protein. A poor state of nutrition tends to decrease the heat production during fasting. The temperature of 25°C is specified as one which is above the critical, assuring that no tissue breakdown is occurring to keep up the temperature of the body, and as one below the point of hyperthermal rise where the onset of febrile conditions increases heat production. Both of these first two conditions are entirely realizable in the case of animals. The minimum muscular activity assured by the third condition, however, is obviously much less subject to control, particularly in farm animals. In addition to various miscellaneous movements, the animal may be expected to spend a variable portion of the experimental period standing and lying down. The magnitude of the influence of voluntary muscular activity is illustrated by the observations that, with the exception of the horse, the basal metabolism of different species and individuals is 10 to 15 percent greater when they are standing than while they are lying down. In making the measurements, therefore, the heat production is calculated separately for the periods of standing and lying and then computed to a standard day of 12 hours of each. The horse, an exception because of the structure of its ligaments, seems to rest as comfortably standing as lying, without any increased energy expenditure.

The fourth condition implies a state of fasting in which a long enough time has elapsed since the ingestion of food to make certain that the heat increment due to its digestion and assimilation has been dissipated. Such a condition is readily obtainable in animals with simple stomachs, but not in the case of the ruminant. In this species the anatomy and physiology of the digestive tract result in a prolonged retention of food in the rumen and a correspondingly slow passage through the tract and into the blood stream. The achievement of a truly postabsorptive state cannot be obtained except after such a prolonged period of fasting as may result in other disturbing factors which alter the normal catabolic processes. Thus the measurement of basal metabolism in the ruminant cannot have the exact significance that it does in humans.

In the ruminant a minimum value for the methane excretion is one criterion of the establishment of the postabsorptive state. Another criterion of the attainment of a basal condition is a respiratory quotient which indicates that little or no carbohydrate is being burned, a condition that is generally reached after two or three days of fast. On the latter basis the heat eliminated in the next experimental period following the attainment of a metabolism which is characterized by the nonprotein, respiratory quotient of fat (0.707) is frequently referred to as the basal metabolism. Some workers determine what is called a *standard metabolism,* which is a value obtained under specified conditions as to time after the last feeding. It is preferable from a strict point of view to refer to any value determined on a ruminant as a measure of the fasting metabolism rather than of the basal metabolism. The conditions under which such a value is obtained should be accurately defined. The term *resting metabolism* has been used to denote the heat eliminated when an animal is lying at rest, though not strictly in a thermoneutral environment or in the postabsorptive state. It is important that the significance of these various terms should not be confused.

Determinations of fasting catabolism for a given species provide a basis for studying the factors which affect this function and for comparing the metabolism in different species. They also provide a base line for measuring the effect of any superimposed factor such as muscular work, digestion, and other body activities.

14.2   Units of Reference in Fasting Metabolism

Heat production is obviously related to body size. In making use of determined values it is necessary to have some unit of reference. Rubner developed the concept, commonly referred to as the *surface-area law,* that the heat given off by all warm-blooded animals is directly proportional to their body surface and that, expressed on this basis, heat production is a constant for all species. Thus it became customary to express fasting catabolism in terms of surface rather than of weight, for example, as calories per square meter per hour. In view of the difficulties and uncertainties involved in measuring surface area, formulas were devised for computing it from weight, recognizing that surface was proportional to some fractional power of weight. Thus most of the values were really based on weight, though expressed in relation to surface area.

The data in Table 14.1 cited by Lusk suggests that basal metabolism is approximately 1000 kcal per square meter of body surface per 24 hours regardless of the size of the animal.

TABLE 14.1.   Heat Production by Fasting Animals

		kcal per day	
Animal	Body weight, kg	Per kg	Per m^2
Horse	441	11.3	948
Pig	128	19.1	1078
Human	64	32.1	1042
Dog	15.2	51.5	1039
Goose	3.5	66.7	776
Fowl	2.0	71	943
Rabbit	2.3	75.1	776
Mouse	0.018	212	1188

It is now recognized that the surface-area theory rests primarily on an empirical basis and that it does not have so general an application as previously thought. While the concept has been and still remains very useful, it is agreed that the various methods of measuring or estimating surface area give such variable results that a statement of heat elimination per unit of surface has a very limited meaning except in terms of the specific method used in obtaining the surface measure. The body surface is not a constant but varies with the position of the body. The fact that the skin is elastic causes its measurement to vary with conditions, whether measured on the live animal or after its removal.

It has now become the practice among investigators of the energy metabolism of animals to use a fractional or decimal power of weight, instead of surface area, as the unit of reference. On the basis of an analysis of a very large number of basal-metabolism data of mature animals of different species, ranging in weight from 0.02 to 4000 kg (mice to elephants), Brody and coworkers in the early 1930s found 0.734 to be the power of body weight best related to basal metabolism. The details of these studies are reviewed in Brody's textbook.[3] In 1935 the Conference on Energy Metabolism sponsored by the N.R.C. Committee on Animal Nutrition adopted the power 0.73, and it came into wide use. In 1947 Kleiber, in an extension of his earlier studies, found the power 0.756 to provide the best fit for the data he analyzed. His studies are reviewed in his book,[4] in which he proposed 0.75 as a shorter expression. Other exponents have been proposed and used, including 2/3, originally proposed for relating body weight to surface area. There is an obvious advantage in having an agreed-upon exponent for use in making statements about the metabolism of individuals and species. The N.R.C. Committee on Animal Nutrition had adopted the 3/4 power of weight as defining the metabolic size of an animal and a group in Europe accepted the same value.[5] This power has the practical advantage of rapid calculation by the use of a slide rule. An extensive table listing $W_{kg}^{0.75}$ values for weights ranging from 0.001 to 1050 kg is to be found in the report prepared by Harris.[5] A condensation of this table is presented here in Table 14.2.

Blaxter reviewed the available information and concluded that the energy expended for movement at optimal speed above the resting energy expenditure is proportional to body weight or $W_{kg}^{1.0}$. During growth metabolism increases at the expotential rate of $W_{kg}^{0.6}$ and not 3/4. Kleiber[4] considered the power function as a convenient, yet emperical way to describe the metabolism of large groups of animals for interspecies comparisons rather than as constants.

Brody originally proposed that basal metabolism could be represented by the formula:

$$BM = 70.5W^{0.734} \qquad (1)$$

and Kleiber's original formula was:

$$BM = 67.6W^{0.756} \qquad (2)$$

These two formulas differ little, particularly when one recognizes that a lower coefficient is to be expected with a higher exponent. Both authorities agreed that the basal metabolism per day for adult homeotherms may be represented by the general formula:

$$BM(kcal) = 70W_{kg}^{0.75} \qquad (3)$$

The coefficient 70 represents an average value for the kilocalories of basal heat produced per unit of metabolic size in experiments with groups of adult mammals. For a 500-kg cow the value would thus become 7402 kcal per day; and for a 50-kg sheep, 1316 kcal. These data show that heat production per kilogram is greater in the smaller animal, reflecting its relatively larger surface area and more active body mass. It should be remembered that the formulas apply only after growth is complete. Basal metabolism is highest in the newborn and gradually drops during the growth period to the figure for the adult animal, and even during adult life. Furthermore, there are species differences, as is to be expected, and there are intraspecies variables. These facts must be borne in mind in applying one of the general formulas to farm animals. The published data for farm animals have been obtained, for the most part, by the use of formula (1), with the last figure of the exponent omitted. The data for ruminants are the most variable, in part no doubt because of differences in age of the

TABLE 14.2. Metabolic Body Size $(W_{kg}^{0.75})$ for Various Body Weights

W_{kg}	$W_{kg}^{0.75}$	W_{kg}	$W_{kg}^{0.75}$	W_{kg}	$W_{kg}^{0.75}$	W_{kg}	$W_{kg}^{0.75}$
1.0	1.0	100	31.62	210	55.16	440	96.07
3	2.28	105	32.80	220	57.10	460	99.33
5	3.34	110	33.97	230	59.06	480	102.55
10	5.62	115	35.12	240	60.98	500	105.74
15	7.62	120	36.26	250	62.87	520	108.89
20	9.46	125	37.38	260	64.75	540	112.02
25	11.18	130	38.50	270	66.61	560	115.12
30	12.83	135	39.60	280	68.45	580	118.19
35	14.39	140	40.70	290	70.28	600	121.23
40	15.91	145	41.79	300	72.08	620	124.2
45	17.37	150	42.86	310	73.88	640	127.2
50	18.80	155	43.93	320	75.66	660	130.2
55	20.20	160	44.99	330	77.42	680	133.2
60	21.56	165	46.04	340	79.18	700	136.1
65	22.89	170	47.08	350	80.92	750	143.3
70	24.20	175	48.11	360	82.65	800	150.4
75	25.49	180	49.14	370	84.36	850	157.4
80	26.75	185	50.16	380	86.07	900	164.3
85	27.98	190	51.17	390	87.76	950	171.1
90	29.22	195	52.18	400	89.44	1000	177.8
95	30.43	200	53.18	420	92.78	1050	184.5

animals studied and differences in the degree to which the basal state was actually achieved. It is clear from the various studies that cattle have a higher and sheep a lower basal metabolism than is represented by the coefficient 70. The data for pigs are also on the low side. A basal-metabolism determination does not take account of the urinary loss of energy, but this is so small that this limitation does not interfere with the usefulness of the formulas discussed, for the purposes for which they are used, as described later.

14.3 Lability of Fasting Metabolism

While a properly determined value for basal metabolism is conceived as being a constant, it must be recognized that this is not true in the absolute sense. Differences in the degree of muscle tonus may exist in animals which appear entirely relaxed. The minimum influence of this tonus becomes evident during sleep. Basal metabolism is lowered by undernutrition but increased by emotional stimuli. It decreases with age. Certain internal secretions, notably that of the thyroid gland, augment heat production by increasing the heart rate, the respiration, and probably, by affecting body oxidations in other ways.

The British Agricultural Research Council (A.R.C.) metabolizable energy (ME) feeding standard uses fasting metabolism as the base to which is added the ME for productive functions. The study of Webster et al.[6] illustrates (Fig. 14.1) very well the changes in fasting metabolism as steers increase in body weight (age). Although there was no statistically significant difference between their Friesian and Angus steers they concluded that fasting metabolism is not a good basis from which to predict energy retention. These results support the view that factorially calculated requirements should be confirmed by feeding trials.

Graham et al.[7] have made similar studies with sheep. They found that 89 percent of the variance of the BMR was accounted for in the body weight term

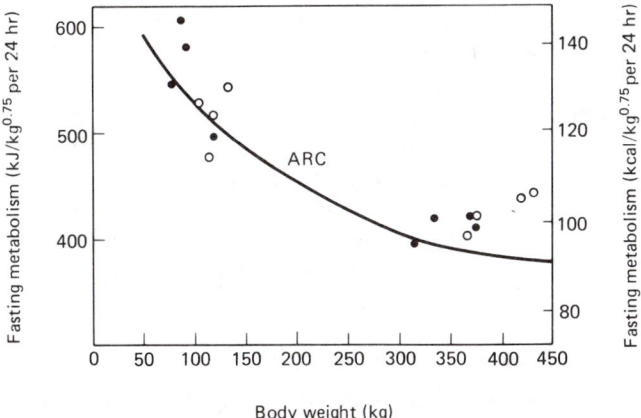

Fig. 14.1 Measured values for fasting metabolism in British Friesian (O) and Aberdeen Angus (●) steers, together with a line indicating fasting metabolism based on the Agricultural Research Council (1965). (*Courtesy of A.J.F. Webster and the Longman Group, LTD*).

($W^{0.75}$). When fat-free weight was used instead of live weight the exponent was unity ($W^{1.0}$). BMR declined about 8 percent per year of age. Feeding, growth rate, and age were about equally important in accounting for variation.

14.4 Endogenous-nitrogen Metabolism

There is a minimum essential nitrogen catabolism incident to the maintenance of the vital processes of the body, even as is the case for energy. This catabolism is measured as the minimum urinary excretion on a nitrogen-free, energy-adequate diet and called endogenous urinary nitrogen (EUN). Upon the inauguration of a nitrogen-free diet the urinary nitrogen decreases gradually. After the attainment of a postabsorptive state as regards protein, there remains "deposit protein" to be eliminated, at least in part, before the minimum endogenous value is reached. Thus the higher the level of previous nutrition, the larger the reserve of protein and the longer the time to reach the minimum level. It may be reached in a week with a rat previously on a low-protein diet, whereas on a high-protein diet four weeks or longer may be required. The minimum endogenous nitrogen is the lowest nitrogen waste of which the body is capable. It may represent a balance between destruction and synthesis—the net catabolism. A significant constituent of this net catabolism is creatinine, which arises from the breakdown of creatine in muscle action, the excretion level of which is directly related to basal energy metabolism. The principal constituent is urea, arising from the breakdown of protein and other nitrogenous compounds which yield ammonia in catabolism. A significant contribution here comes from the breakdown and reformation of those proteins which exist in a dynamic state, because reformation is not 100 percent efficient.

In order to arrive at a true value for EUN, it is essential that the animal be receiving a diet adequate in energy, because otherwise the output of urinary nitrogen may include some from body protein which has been broken down to furnish energy and thus be in excess of the value representative of the minimum essential nitrogen catabolism. Muscular activity has no appreciable influence on endogenous nitrogen so long as the energy intake is sufficient to cover it, for such activity has little, if any, influence on protein catabolism.

While the measurement of the minimum EUN metabolism is simple in theory, it is difficult in practice to obtain reliable values, particularly with certain species. Not only is a variable and, frequently, a long time required to arrive at what may be considered a constant minimum value, but it is often impossible to get animals to eat a sufficient amount of a nitrogen-free diet for any extended period. Any marked failure of adequate consumption destroys the significance of the results.

14.5 Relation of Endogenous Nitrogen to Energy Catabolism

Like basal metabolism, endogenous-nitrogen metabolism is a function of body size. Since this is true and since both represent the minimum catabolism essential to life, one would expect a relationship between them. That a relationship actually exists was first proved by Terroine and Sorg-Matter.[8] Their studies, which included mice, rats, pigeons, chicks, rabbits, and pigs, resulted in the conclusion that the *law of constant relationship of minimum nitrogen and energy output* holds for all warm-blooded animals irrespective of body weight or age. The values actually reported ranged from 2.3 to 2.9 mg of nitrogen per kilocalorie. Brody and coworkers[9] confirmed this relationship in showing that endogenous nitrogen is related to the same power of body weight in adult animals as is the basal metabolic rate, as is indicated by the following formula:

$$\text{EUN mg/da} = 146 W_{kg}{}^{0.72}$$

This formula was arrived at by the analysis of a body of data, from various sources, on minimum EUN excreted by mature animals of different species, ranging in weight from 0.02 to 500 kg.

In most of his studies, Terroine included the metabolic fecal nitrogen on his nitrogen-free diets as a part of the endogenous whole. While this fecal loss is of body origin, its level is related to food intake and not to basal metabolism, and thus the ratio may be more appropriately calculated on the basis of the urinary nitrogen only, considering that the latter is more truly representative of the same vital processes which are responsible for the energy catabolism. Using this basis, Smuts,[10] in studies with mice, rats, guinea pigs, rabbits, and pigs, obtained a value of around 2 mg of nitrogen per basal kilocalorie. Several studies with cattle and sheep were so variable that no conclusion seems justified probably because of differences in the degree to which the basal state was actually achieved. A ratio would enable one to calculate the endogenous urinary nitrogen from the basal energy value. The nitrogen value is in turn used to arrive at protein requirements.

Swanson[11] reported that the endogenous urinary nitrogen (EUN) of cattle ranging from 28- to 625-kg body weight averaged 440 mg per $W_{kg}^{0.5}$. Brody's interspecies equation cited above gives quite different requirement values for cattle than Swanson's, suggesting that data must be obtained for each animal species to have application in practice.

14.6 Mineral Catabolism in Fasting

From the discussion in Chap. 10 of the functions of mineral elements, it is evident that an active mineral metabolism continues during fasting. Differing from organic constituents of the body, however, catabolized minerals may be reutilized instead of being excreted. For example, although the red cells of the blood are constantly being destroyed, the catabolized iron is available for the resynthesis of hemoglobin.

There is, nevertheless, a constant excretion of mineral elements during fasting, as is evident from the study reported by Benedict.[12] He measured the urinary excretion of certain minerals by a man during a fast of 31 days. The outputs gradually decreased during the early days and then reached values which tended to become constant. The figures obtained for the last day's output, expressed in grams, were as follows: chlorine, 0.13; phosphorus, 0.58;

sulfur, 0.49; calcium, 0.138; magnesium, 0.052; potassium, 0.606; sodium, 0.053. The relatively large excretions of phosphorus and especially of sulfur doubtless were due in part to the breakdown of protein containing these elements as reflected by the endogenous-nitrogen catabolism. This fact suggests that the data cannot be considered to indicate the catabolism of mineral fasting alone. Had there been an adequate intake of energy and protein, certainly less sulfur- and phosphorus-containing protein would have been broken down. The large output of potassium in contrast to the low excretions of sodium, calcium, and chlorine is less readily explainable. These data from Benedict do not reflect the total excretion of all the minerals considered, since the feces are an important path of outgo for some of them.

MAINTENANCE REQUIREMENTS

Enough is as good as a feast.

John Heywood (1540).

The term *requirement* implies an exactness which it does not have, as frequently employed, and which it cannot have when used in feeding standards for practice. Strictly speaking, it is the minimum amount of a given nutrient needed to promote a given body function to the optimum in a ration adequate in all other respects; i.e., a perfectly balanced ration. Such a minimum value will not be the same for any two individuals, and thus, for this reason alone, any determined individual value or any average of such values must be increased as a practical recommendation in order that the optimum performance of all may be assured. The intakes which are just sufficient to be fully adequate in a closely controlled experiment may fail to do so under less favorable conditions in practice.

The term *minimum requirement* as employed in this text denotes an experimental average minimum value, a figure that is not suitable, however, as a practical recommendation. Despite this limitation, its determination is highly useful because it provides a base line for studying the influence of factors which increase it in practice and thus for arriving at safe and yet economical recommendations as guides for feeding operations. Such recommendations are expressed by the National Research Council (N.R.C.) requirements as was previously explained (Sec. 13.13). They are used in this text accordingly in future discussions.

14.7   The Maintenance Need for Food Energy

The energy requirement for maintenance is the minimum amount needed to keep the animal in energy equilibrium, i.e., to prevent any loss from its tissues. Thus an intake sufficient to offset the loss represented by the fasting metabolism would be the requirement under the conditions specified for measuring the latter. Expressed as net energy, it would be represented by the fasting catabolism itself, but expressed as any other measure of food energy, it would obviously be larger, since no other measure represents energy which is completely utilizable for the prevention of tissue breakdown. For example, in translating the fasting metabolism into metabolizable energy, it is necessary to choose a value for the latter which includes the fasting value plus the heat increment resulting from the food ingested, and the energy to consume the food.

14.8 Basal Metabolism as a Measure of Energy for Maintenance

Under conditions of practice, however, an intake of food energy sufficient to balance the fasting metabolism is not an adequate maintenance value because the animal is never so restricted in its activity as represented by the standard, confined conditions which are specified for the determination of the basal value. Thus one must add an *activity increment* to the basal values. There are no adequate experimental data for deciding what this increment should be. Clearly, it must vary for different animals and conditions, and any value selected for general use needs to be high enough to cover the extreme cases in order to arrive at a maintenance figure which would be adequate under all conditions of practice. The same principle applies, however, to the formulation of any generalized feeding standard. Mitchell and associates[13] have reported an experiment in which cockerels were fed to maintain constant weight, and the net-energy intake, the basal-heat production, and the excess heat produced were recorded. It was found that the excess calories produced above basal, representing the activity increment, averaged 48 percent. Mitchell proposed that the net-energy requirement for maintenance of poultry could be obtained by increasing basal metabolism by 50 percent. Limited data indicate that the increment is less in the case of cattle, sheep, and swine—perhaps of the order of 20 to 30 percent. Such increments, however, cannot be considered adequate for animals grazing widely in pasture or on the range. In the case of dairy cows, Reid and coworkers[14] found that grazing resulted in a "maintenance" requirement that was 40 percent greater than when the cows were fed in the barn. Somewhat similar findings have been obtained by New Zealand workers and others. Since animals on pasture are normally eating for purposes of production as well as for maintenance, the assessment of the cost of the grazing activity is actually not a matter of practical importance.

The additional energy required for grazing calls attention to studies such as those of Erickson and coworkers[15] which showed that the energy cost of moving the body is directly proportional to body weight. Thus there are those who feel that the power of weight used in calculating the maintenance requirement, which includes an activity increment, should be higher than the 0.75 applying to basal metabolism. In the latest reports the N.R.C. Committees have used the exponent $W^{0.75}$ as the basis for estimating the energy requirements for maintenance. The coefficients differ slightly for different species, as shown below:

Beef cattle $NE_m = 77W^{0.75}$ (kcal/day)
Dairy cattle $NE_m = 80W^{0.75}$ (adds 10% above NE_m for activity)
                $ME_m = 133W^{0.75}$
                $DE_m = 155W^{0.75}$
                $TDN_m = 35.2W^{0.75}$ (g TDN/day)
Sheep $ME_m = 98W^{0.75}$
                $DE_m = 119W^{0.75}$
                $TDN_m = 27W^{0.75}$

The energy values are expressed as kilocalories per day and W is expressed in kilograms.

These examples illustrate how basal-metabolism data can be used to arrive at the maintenance requirement. The general procedure is theoretically sound. Because of the lability of the fasting catabolism and the uncertainty as to the expenditure which should be allowed for activity, however, values so obtained need to be checked in practical feeding trials to test their reliability as bases for maintenance allowances.

The calculation of the need for maintenance by starting with the basal expenditure and adding an estimated activity increment is an example of the *factorial method* of arriving at the requirement for energy of a specific nutrient for a given function. It involves a summation of the various factors contributing to a given requirement.

When energy-maintenance values, which include the fasting metabolism and an activity increment, are expressed on any other basis than net energy, account must be taken of other losses which are deducted in arriving at this measure (see Fig. 14.2).

14.9   The Determination of Energy Equilibrium

You never know what is enough unless you know what is too much.

William Blake (1790).

The use of a respiration apparatus, or respiration calorimeter, makes possible the measurement of the effectiveness of a given ration for the maintenance of tissue integrity without slaughter of the animals. This procedure was early used by Kellner, Armsby, and others as a basis for obtaining the minimum requirement. It involves the determination of the energy balance with a ration which is just adequate to maintain weight. It cannot be expected that any such ration will result in exact energy equilibrium, but the procedures furnish specific data as to any tissue gains or losses, and the feed-energy intake, whether expressed as digestible, metabolizable, or net, can be corrected accordingly to arrive at the exact maintenance requirement.

From extensive studies on growing pigs, 16 to 196 kg in weight, involving respiration experiments carried out with the Pettenkofer apparatus, Breirem[16] developed the following formula for the net-energy requirement for maintenance:

$$NE_{kcal} = 161.05 W_{kg}^{0.56}$$

Since this formula was based on data obtained in metabolism studies, he proposed the following formula as one which included an activity increment and thus would provide a more satisfactory estimate for practice:

$$NE_{kcal} = 193.3 W_{kg}^{0.56}$$

Representative values obtained by this formula are as follows:

Weight, kg	NE, kcal
20	1034
40	1525
80	2249
150	3197

A reevaluation of the Norwegian studies[17] gave somewhat higher values. These data are of interest because the N.R.C. swine report does not list any maintenance values.

14.10   The Determination of Energy Needs from Feeding Trials

The maintenance values in the early feeding standards developed in the United States were based, for the most part, on data obtained in feeding trials. In its simplest form this method involves the determination of the amount of food required to hold adult animals at constant weight. The inclusion of a digestion trial in the course of the experimental period allows the expression of the requirement in terms of DE, TDN or ME. A study in which live weight is the sole criterion, the importance of accurate and representative data for this measure is clear. If the experiment is successful in maintaining the weight substantially constant over an extended period, a fairly accurate measure of the maintenance requirement is obtainable and a measure which is directly applicable to the conditions of practice. Allowances can be made for changes in live weight by estimating the food equivalent of the losses or gains and correcting the observed intakes accordingly. The figures proposed for this purpose by Knott and associates[18] are as follows:

Pounds gained $\times$ 3.53 = TDN required for gain
  Pounds lost $\times$ 2.73 = TDN equivalent to loss

The N.R.C. dairy cattle report lists the following values:

Energy Value of Weight Changes During Lactation

	TDN, kg	DE	ME	NE_l
		Mcal		
Weight loss	−2.17	−9.55	−8.25	−4.92
Weight gain	2.26	9.96	0.55	5.12

Such corrections can be only approximate at best because of a lack of knowledge of the kind of tissue gained or lost. As an extreme example, the change in weight might be due entirely to water, which, of course, would have no food equivalent at all. It is clear that the larger the corrections which have to be applied, the less significant become the results.

Over a period of years Lofgreen and Garrett[19] at California have conducted a series of comparative slaughter studies to measure the net energy requirements of beef cattle. By determining body composition and empty body weights initially and after a feeding period over the range from maintenance to *ad libitum* feeding it was possible to calculate the net energy required for maintenance and gain. The net-energy value of standard feeds have also been determined. Fig. 14.2 shows the relationship between daily heat produced and daily ME intake per unit of metabolic size. From this relationship it was estimated that the heat production of fasting beef cattle lies between 72 and 82 kcal per $W_{kg}^{0.75}$ with a mean of 77, the value used in the N.R.C. reports (Sec. 14.8).

14.11 Maintenance Values in Recommendations for Practice

Although there are situations under which farm animals are fed for maintenance only, for the most part they are fed primarily for productive purposes. Here maintenance values serve as basic figures to which additions are made in accordance with the level of production. Thus the experimental data previously reviewed have been used to arrive at the basic figures on which feeding standards are formulated.

The N.R.C. reports list energy requirements for maintenance of dairy cattle, beef cattle, horses and sheep, as shown in the appendix tables. Some of the values for beef and dairy cattle are summarized in Table 14.3. The dairy cattle values are for mature cows. In calculating daily rations for lactating cows energy is added to the maintenance figure to cover the milk produced, with additional amounts for pregnancy, for growth of first and second calf milking heifers and for extensive grazing activities.

The beef cattle data (Table 14.3) apply to young as well as mature females which are not gaining weight. They have special value in formulating survival rations during winter or periods of feed shortages. Comparative feed dry matter needs for maintenance, reproduction, and lactation of various animals are given in Table 16.4.

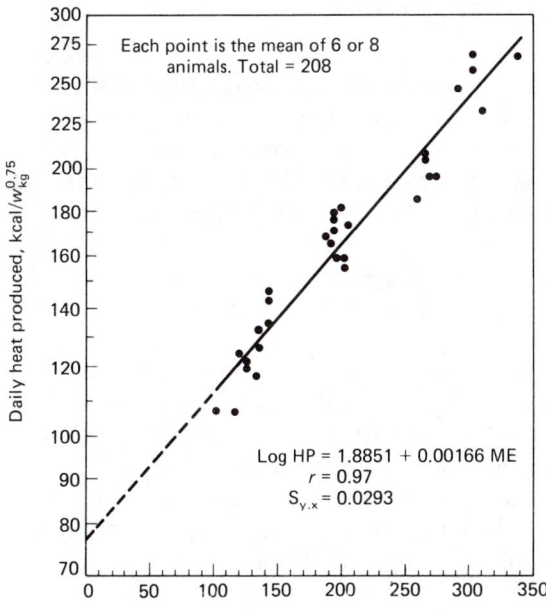

Fig. 14.2 Determination of fasting heat production. (*Courtesy of G. P. Lofgreen and W. N. Garrett, University of California.*)

TABLE 14.3. The N.R.C. Daily Energy Required for Maintenance of Beef and Dairy Heifers and Cows

	Beef Cattle				Dairy Cattle				
			ME	NE_m			DE	ME	NE_1
Body weight, kg	Daily feed, kg	TDN, kg	Mcal		Daily feed, kg	TDN, kg	Mcal		
100	2.1	1.2	4.2	2.4	—	—	—	—	—
200	3.5	1.9	7.0	4.1	—	—	—	—	—
300	4.7	2.6	9.5	5.6	—	—	—	—	—
400	5.9	3.3	11.8	6.9	5.5	3.2	13.9	11.9	7.2
500	7.2	3.9	14.1	8.1	6.5	3.7	16.4	14.1	8.5
600	8.3	4.4	16.1	9.3	7.5	4.3	18.8	16.1	9.7
700	—	—	—	—	8.5	4.8	21.1	18.1	10.9
800	—	—	—	—	9.5	5.3	23.3	20.0	12.0

Source: N.R.C. Nutrient Requirements of Domestic Animals.

14.12   The Protein Requirement for Maintenance

The discussion in Chap. 8 has shown that the need of the body for nitrogenous food, which we commonly refer to as a protein requirement, is actually a need for the building stones of protein, viz., the amino acids. It was also brought out that the figures for the protein content of foods are conventional values, calculated from nitrogen content, and thus that they include both proteins and other nitrogenous compounds calculated to a protein basis. Despite these limitations, we express the nitrogen phase of nutrition on a protein basis because it is simpler to do so and because our knowledge is insufficient for a more exact expression. No serious disadvantage is here involved providing the limitations are kept in mind.

A separate figure for a maintenance allowance is commonly utilized in feeding standards only in the case of the dairy cow. The protein need of the horse is essentially a maintenance requirement, however, because work does not involve the catabolism of protein. Maintenance values for all species are useful, nevertheless, as a base line for arriving at the overall need during production. A stated requirement assumes that the ration is adequate in energy content so that protein need be utilized only for its specific purpose.

14.13   Functions of Protein in Maintenance

The absorbed protein required for maintenance needs must make good the endogenous urinary losses and the metabolic fecal losses incident to the digestion of the ration in question and also provide for "adult growth." While the urinary losses are considered to be reasonably constant per unit of body size ($W^{0.75}$), the fecal losses are variable according to the makeup of the maintenance ration and the species. Studies have shown that in humans the fecal loss is approximately one-fourth or one-fifth as much as the urinary endogenous output. In adult ruminants on a high-roughage ration the metabolic fecal output may exceed the endogenous.

The term *adult growth* refers to the growth and renewal of hair, nails, feathers, and other epidermal tissues, a process which continues throughout life, even though the protein intake is inadequate for the maintenance of the body as a whole. As an extreme example, Mitchell and associates found that sheep fed an inadequate diet for 200 days were continuously in negative nitrogen and energy balance; yet there was appreciable growth of wool, and its content of protein was normal. This wool growth represented an increase of 0.014 kg of protein per day in the fleece per 100 kg live weight, an approximately normal rate, at the expense of the breakdown of other protein tissues of the body. In general the amount of protein required for adult growth is very small compared to the overall need. In the case of the adult rat, however, the need for the continuing hair growth is a substantial one. The same is true for feather replacement in moulting hens.

Theoretically, the minimum requirement for absorbed protein might be met by supplying the amounts needed for the above-mentioned functions in an otherwise adequate diet. Actually, a substantially larger amount is needed in practice. The minimum endogenous urinary nitrogen (EUN) represents the output of an animal in a depleted state insofar as protein nutrition is concerned. An appropriate maintenance intake should also cover the needs for maintaining a protein reserve. The absorbed intake must be large enough to cover a variable wastage in metabolism. Where the protein requirement is expressed on a total-intake basis, it must be increased to cover the fecal losses.

14.14 Biological Value and Protein Requirement

The wastage of absorbed protein in metabolism results from the fact, as discussed in Chap. 8, that body need is for specific amino acids which make up the protein or other nitrogenous tissues or compounds to be formed. The process is most efficient when all the essential amino acids are supplied at the site of synthesis in the proportions which correspond exactly to the amino acid makeup of the product to be formed. In general the amino acid mixture absorbed and carried to the tissues differs, as regards the proportions of amino acids at least, from those required for tissue synthesis. The kinds of amino acids needed for a given synthesis are taken up from the available mixture in the proportions required, and the "leftovers" are wasted in so far as protein nutrition is concerned. This loss is a very substantial one where the mixture absorbed is relatively very deficient in any one essential amino acid. The ability of a given source of protein to supply amino acids in the relative amounts needed to form the nitrogenous tissues and compounds required for body functions is referred to as its *biological value*. This statement does not apply to ruminants, insofar as the feed supply of amino acids is concerned, because of the ability of the rumen to synthesize amino acids from a variety of nitrogen compounds (Sec. 8.16). The nature of these compounds and of the varying microbiological processes which go on in the rumen do influence the kinds of amounts of amino acids which are presented to the tissues. This is reflected in differences in biological value of the total nitrogen of the feeds, and the use of the term "biological value of protein" is appropriate in the case of ruminants when the protein is conventionally calculated as $N \times 6.25$.

Studies with rats and humans, particularly, have indicated that various protein sources differ in biological value for the support of maintenance. Rat studies have shown that the amino acid needs for maintenance differ both qualitatively and quantitatively from those of rapid growth. In the latter the need is primarily for the synthesis of tissue protein, while in maintenance the need is in part to replace tissue breakdown and in part for the synthesis of simpler nitrogen compounds used up in maintenance. For these reasons biological values are different than for growth and are considered to be of a lower order.

In stating a protein requirement for any purpose, biological value must be taken into account. In the case of farm animals differences in biological value for maintenance are of lesser significance because, except in special situations, they are being fed for productive purposes. Thus the biological values for the combined function of maintenance and production are the ones of practical importance. For this reason the discussion of how biological values are obtained for various feeds and how they are used in formulating rations is taken up in the chapter on growth. In this later discussion methods for determining biological value for maintenance are also described. In the discussion of methods of arriving at protein requirements consideration is given to the bearing of biological value on the results obtained.

14.15 Estimation of Protein Requirement from Endogenous Urinary Nitrogen

It has been mentioned (Sec. 14.8) that the energy requirement for maintenance can be estimated from the fasting metabolism by the factorial method. Similarly, a protein requirement can be arrived at from EUN. Brody and associates outlined such a method based on their EUN values obtained in developing the formula,

$$\text{EUN mg/da} = 146_{kg}W^{0.72}$$

Starting with EUN, account was taken of the loss of MFN, and an assumed biological value was used to arrive at the protein requirement, and a table of requirements for different weights was constructed.

This method is best illustrated by the procedure proposed for cattle and sheep by the British A.R.C. Nutrient Requirements of Farm Livestock.[20] This procedure was developed from a detailed study of the literature to select appropriate values for the factors which should be taken into account. First, a table was developed showing, for each species, "acceptable" estimates of endogenous nitrogen excretion (UE) for animals of different weights, expressed as grams per day per kilogram$^{0.73}$. Next, additions were made for the loss of nitrogen in hair and scurf in the case of cattle (S_1), obtained from an estimate of $0.02W_{kg}^{0.73}$ g nitrogen per day; and for loss in adult fleece growth in sheep (S_2), ranging from 0.6 to 1.0 g nitrogen per day for different breeds. From a study of the rather variable published figures on losses of MFN, the figure of 5 g MFN per kilogram dry-matter intake was chosen for both species to provide the additions required to UE to cover this loss. The data in the literature on biological values for cattle and sheep fed different rations were summarized and the general estimates arrived at of 70 for cattle and 65 for sheep.

The complete formula for protein requirements includes factors to cover nitrogen needs for various productive functions; these factors will be referred to in later chapters in this text. The following formula includes the factors which relate to maintenance only, including hair or wool growth:

$$\text{TP} = (\text{UE} + S_1 \quad \text{or} \quad S_2 + \text{MFN}) \times 6.25 \times 100/\text{BV}$$

TP is "truly digestible protein," which designates digested-food protein as distinguished from digestible protein, which has taken account of MFN loss as well. Since the MFN factor, which depends on dry-matter intake, poses a problem in tabulating requirements, a formula which eliminates this factor and provides a requirement in "available protein" (AP) is used in constructing the requirement tables. AP differs from the conventional digestible crude protein by a subtraction of the protein wasted in synthesizing MFN estimated from the figure for biological value. The formula, omitting the factors for productive functions, is as follows:

$$AP = (UE + S_1 \quad \text{or} \quad S_2) \times 6.25 \times 100/BV$$

For cattle the factor S_2 drops out and BV equals 70; for sheep S_1 is omitted and BV equals 65. The tables specify "minimum requirements" for maintenance and various productive functions. The report states that the data are not to be considered as recommended levels of feeding, and that whether the latter should differ from the minimum levels presented and by how much depends on many factors, including economic consideration.

Swanson's estimates of protein requirements for maintenance were discussed in Sec. 14.5. His excellent paper discusses the advantages and limitations of various methods of estimating the protein requirements. For greater details on previous research the classic monograph of Mitchell[21] is recommended.

Differing from the situation with respect to energy needs, activity does not require consideration in calculating a protein requirement from the basic nitrogen excretion because activity does not significantly increase protein metabolism. While it is agreed that endogenous nitrogen provides the basic value for arriving at the maintenance need for protein, the uncertainties as to the constancy and general applicability of such a value, as previously discussed, suggest that any figure so arrived at should be tested in practice before being accepted as a recommended requirement.

14.16 Nitrogen-balance Data as a Measure of Protein Maintenance

The minimum protein intake in a ration otherwise complete which will keep a previously well-nourished animal in nitrogen equilibrium is a reliable measure of the requirement for the protein mixture in question. It is important that the animal be in a good state of protein nutrition at the start and that the minimum intake necessary to maintain nitrogen equilibrium in such an animal be determined. The latter is important because the animal cannot store protein appreciably and an unnecessarily high intake tends to result in equilibrium also, giving a false picture of the maintenance need. On the other hand, if the animal is in a protein-depleted state, equilibrium may be established by intake levels which fall short of maintaining the needed protein reserves. The minimum level of intake which will maintain these reserves represents the true requirement, in contrast to the smaller amount which may result in equilibrium in a depleted body. Thus, the maintenance of nitrogen balance does not always guarantee protein adequacy. The objective of a precise balance experiment should be to arrive at a protein requirement indicating the minimum amount which preserves the integrity of all protein tissues.

Around the turn of the century many German workers conducted nitrogen-balance studies with steers and dry cows. Armsby[22] summarized these various studies along with ones of his own and thus arrived at his recommendation of 0.6 lb digestible crude protein or 0.5 lb digestible true protein per 1,000 lb live weight for the maintenance of cattle. Nitrogen-balance data provided the basis for the protein intakes for maintenance which were specified by Armsby for computing rations for farm animals. It is interesting to note that, at this early date, he recognized that not all proteins were of equal value. The Recommended Dietary Allowances[23] for humans cites the results of nitrogen balance studies as the primary basis for setting the protein requirements for maintenance as well as for growth and other physiological states.

A balance experiment is a short-term measure carried out under closely controlled conditions, and thus the question always arises as to how accurately the results apply to the long term. Further, a properly determined value is a minimum figure which must be adjusted upwards to cover many conditions of feeding practice. Of course, biological value must be considered in interpreting a determined value to a different feed combination.

14.17 Determination of Protein Maintenance from Feeding Trials

The early feeding standards for dairy cattle contained recommendations for protein-maintenance requirements, based on long-term feeding trials. Intakes which, along with adequate energy, kept the nonproducing animal in good condition without loss of weight over an extended period were considered satisfactory in practice. The early extensive experiments by Hills and associates illustrate how protein requirements were thus arrived at. Their recommendations of 0.6 lb digestible protein per 1,000 lb live weight were in accord with those arrived at from nitrogen balance by Armsby.

Of course a feeding trial gives no specific information regarding the maintenance of the integrity of the nitrogenous tissues, information obtainable only by the inclusion of slaughter data, which is hardly practicable in the case of dairy cows. Nor does it give the data needed to arrive at minimum requirements. The early, extensive, long-term studies of Hills and associates[24] did provide data, however, which are still recognized as useful in arriving at data for practice.

14.18 Does the Protein Requirement for Maintenance Remain Constant During Production?

The Folin theory postulates a constant endogenous catabolism of nitrogen independent of the total protein catabolism, and the theory implies that there is a constant requirement to meet this loss, irrespective of the protein metabolism which may be taking place for the support of other body functions. It is considered that the feeding of protein to meet the needs of such a function as growth does not alter the amount required for maintenance where the latter alone is involved. Many do not agree with this concept, basing their objection on the view that the maintenance requirement in part is for certain amino acids only and that on a nitrogen-free diet the catabolism of body nitrogen compounds to furnish the acids needed results in ''leftovers'' which are wasted. It is argued that during protein ingestion, on the other hand, this wastage is decreased.

The questions here involved are of minor importance from the standpoint of feeding practice. It may be agreed that amino acids which are unsuitable or unneeded for a given productive function may serve in maintenance and thus lessen the specific intake for this purpose, but whether the gain here should be subtracted from the maintenance requirement or from the production requirement is a matter of bookkeeping.

14.19 Protein Requirements for Practice

The N.R.C. protein requirements for maintenance are shown in Table 14.4 and the appendix tables. These represent the safe minimum levels for long-time feeding of mature animals.

A requirement for practice must be high enough to meet the protein needs fully with rations of varying biological value and also to provide a protein-to-

TABLE 14.4. N.R.C. Nutrient Requirements for Maintenance (content in the dry feed)

Nutrient	Horse	Dairy cattle	Sheep	Beef cattle	Rabbit
Body weight, kg	500	500	50	500	4.5
Daily feed, kg	7.45	6.5	1.0	7.2	0.16
ME Mcal/kg feed	2.2	2.5	2.0	2.0	2.1
Protein, %	8.5	9.8	8.9	8.5	12.0
Calcium, %	0.3	0.31	0.30	0.18	—
Phosphorus, %	0.2	0.23	0.28	0.18	—
Sodium, %	0.35	0.1	0.1	—	0.2
Potassium, %	0.4	0.7	0.5	—	0.6
Magnesium, %	0.09	0.07	0.08	—	0.3–0.4
Sulfur, %	0.15	0.1	—	—	—
Iron, mg/kg	40	100	50	—	—
Zinc, mg/kg	40	40	32	—	—
Manganese, mg/kg	40	20	40	—	2.5
Copper, mg/kg	9	10	5	—	3
Iodine, mg/kg	0.1	0.1	0.1	—	0.2
Cobalt, mg/kg	0.1	0.1	0.1	—	—
Selenium, mg/kg	0.1	0.1	0.1	—	—

Source: N.R.C. Nutrient Requirements of Domestic Animals.

energy ratio which is not so wide as to depress the efficiency of the ration as a whole.

14.20   Mineral and Vitamin Needs for Maintenance

The discussion in Chap. 10 has shown that many of the mineral elements undergo a very active metabolism in connection with various processes which are essential for the normal functioning of the body in maintenance. Differing from the energy and protein metabolism, however, they are not necessarily used up and excreted in the process. It has been mentioned that the iron released from the constant breakdown of red cells is reutilized for hemoglobin synthesis. Chlorine which is secreted in the gastric juice to provide for digestion can be reabsorbed from the digestive tract and reutilized. There are other examples.

Nevertheless, for reasons only partially understood, there are regular and substantial losses of certain minerals from the body of a mature animal, as described for the fasting condition in Sec. 14.6. The maintenance of an appropriate electrolyte balance in the blood and other tissues is an important factor governing mineral conservation and excretion. The amount of a given mineral required to keep the body store intact during maintenance is readily determined by a balance experiment. This is the technique which has been used to provide the data for the mineral allowances for human adults. Few similar studies have been made with farm animals because they are fed primarily for productive purposes, and mineral-maintenance data do not have the same usefulness as a base line for arriving at production requirements as is the case for energy and protein. The report by Gallup and Briggs[25] illustrates how the balance technique can be used to arrive at the intake of a mineral required to maintain the body's supply. In this study it was found that approximately 2 g of phosphorus per 45-kg body weight was sufficient to maintain lambs in phosphorus equilibrium. The maintenance needs for the major mineral elements can also be arrived at from data on endogenous losses and on availability of the minerals in the feed. For example, the British Committee on Nutrient Requirements gives a requirement for calcium for a 454-kg bovine as 16.2 g per day on the basis of an endogenous loss of 7.3 g per day and an availability figure of 45 percent ($7.3 \div 0.45 = 16.2$).

The calcium and phosphorus requirements for maintenance, growth and lactation are compared in Table 16.5. Requirements for the other mineral elements are shown in Table 14.4. The concentration of minerals in the ration is often not different for maintenance than for growth, but the daily intakes would be lower because less food is consumed.

On the basis of our knowledge of the specific functions performed by the various vitamins, their importance during maintenance as well as for productive purposes is clear. Only limited information is available on the vitamin requirements of farm animals for maintenance since most animals are fed for growth or production. The comparative data for vitamin A in Table 15.18 show that considerably lower concentrations in the ration are adequate for maintenance than for reproduction. It has been shown that horses obtain adequate amounts of the B-vitamins from natural feeds and from synthesis by intestinal microflora for maintenance. Hard-working horses may need a dietary supplement of B-vitamins under some conditions. Not much information is available about the specific maintenance needs of nonruminants for other vitamins, but the levels are likely lower than for growth.

NOTES

1. K. L. Blaxter, Fasting metabolism and the energy required by animals for maintenance, in Festskrift til Knut Breirem, Oslo, Norway, 1972.

2. G. Lusk, The Science of Nutrition, W. B. Saunders Co., Philadelphia, Pa., 1931.

3. S. Brody, Bioenergetics and Growth, Reinhold Book Corp., New York, 1945, Chap. 13.

4. M. Kleiber, The Fire of Life, John Wiley and Sons, Inc., New York, 1961. Max Kleiber (1893–1976) trained in Switzerland in agricultural chemistry and energy metabolism. In 1929 he came to the University of California, Davis, to construct a respiration apparatus for energy metabolism studies with large animals. He developed the use of weight to the 3/4 power instead of surface area to describe energy metabolism. When isotopes became available in 1947 he began studies of metabolic processes related to mineral utilization and milk secretion, and contributed greatly to present concepts in this area.

5. L. E. Harris, Biological energy relationships and glossary, National Acad. Sci.— National Research Council Publ. 1411, 1966.

6. J. F. Webster, J. M. Brockway, and J. S. Smith, Predictions of the energy requirements for growth in beef cattle. I: The irrelevance of fasting metabolism, Animal Prod., 19:127–139, 1974.

7. N. M. Graham, T. W. Searle, and D. A. Griffiths, Basal metabolic rate in lambs and young sheep, Australian J. Agr. Res., 25:957–971, 1974.

8. E. F. Terroine and H. Sorg-Matter, Loi quantitative de la dépense azotée minima des homéothermes: Validité intraspécifique, Arch. Intern. physiol., 29:121–132, 1927.

9. S. Brody, R. C. Procter, and U. S. Ashworth, Growth and development. XXXIV: Basal metabolism, endogenous nitrogen, creatinine and neutral sulphur excretions as functions of body weight, Missouri Agr. Expt. Sta. Research Bull. 220, 1934.

10. D. B. Smuts, The relation between the basal metabolism and the endogenous nitrogen metabolism, with particular reference to the estimation of the maintenance requirement of protein, J. Nutrition, 9:403–433, 1935.

11. E. W. Swanson, Factors for computing requirements of protein for maintenance of cattle, J. Dairy Sci., 60:1583–1593, 1977.

12. F. G. Benedict, A study of prolonged fasting, Carnegie Institute Wash. Pub. 203, pp. 247–291, 1915.

13. H. H. Mitchell, L. E. Card, and T. S. Hamilton, A technical study of the growth of white leghorn chickens, Ill. Agr. Expt. Sta. Bull. 376, 1931.

14. J. T. Reid, A. M. Smith, and M. J. Anderson, Difference in the requirements for maintenance of dairy cattle between pasture and barn feeding conditions, Proc. Cornell Nutrition Conf., 1958, pp. 88–94.

15. L. E. Erickson and coworkers, The energy cost of horizontal and grade walking on the motor-driven treadmill, *Am. J. Physiol.*, **145**:391–401, 1945.

16. K. Breirem, Der Energieumsatz bei den Schweinen, *Tierernährung*, **11**:487–528, 1939.

17. C. W. Holmes and K. Breirem, A note on the heat production of fasting pigs in the range of 16 to 96 kg live weight, *Animal Prod.*, **18**:313–316, 1974.

18. J. C. Knott, R. E. Hodgson, and E. V. Ellington, Methods of measuring pasture yields with dairy cattle, *Washington Agr. Expt. Sta. Bull.* 295, 1934.

19. G. P. Lofgreen and W. N. Garrett, A system for expressing net energy requirements and feed values for growing and finishing beef cattle, *J. Animal Sci.*, **27**:793–806, 1968.

20. British A.R.C., Nutrient Requirements of Farm Livestock, No. 2, Ruminants: Technical reviews, The Agricultural Research Council, London, 1965.

21. H. H. Mitchell, Comparative Nutrition of Man and Domestic Animals, Vol. 1, Academic Press, New York, 1962.

22. H. P. Armsby, The Nutrition of Farm Animals, The Macmillan Company, New York, 1917, pp. 326–327.

23. Recommended Dietary Allowances, National Academy of Sciences, Washington, D.C., 1974.

24. J. L. Hills et al., The protein requirements of dairy cows, *Vermont Agr. Expt. Sta. Bull.* 225, 1922; J. L. Hills, The maintenance requirement of dairy cattle, *ibid.*, 226, 1922.

25. W. D. Gallup and H. M. Briggs, The minimum phosphorus requirement of lambs for phosphorus equilibrium, *J. Animal Sci.*, **9**:426–430, 1950.

SUPPLEMENTARY LITERATURE

Blaxter, K. L., and F. W. Wainman: The fasting catabolism of cattle, *Brit. J. Nutrition,* **20**:103–111, 1966.

Blaxter, K. L., and W. A. Wood: The nutrition of the young Ayrshire calf. I: The endogenous nitrogen and basal energy metabolism of the calf, *Brit. J. Nutrition,* **5**:11–25, 1951.

Close, W. H., and L. E. Mount: The rate of heat loss during fasting in the growing pig, *British J. Nutrition,* **34**:279–290, 1975.

Hutton, J. B.: The maintenance requirements of New Zealand dairy cattle, *Proc. N.Z. Soc. Animal Prod.,* **22**:12–34, 1962.

Knox, K. L., J. C. Crownover, and G. R. Wooden: Maintenance energy requirements for mature idle horses, in A. Schurch and Wenk (eds.), Energy Metabolism of Farm Animals, Jurvis Druck and Verlag, Zurich, 1970, pp. 181–184.

Meyer, J. H., and W. J. Clawson: Undernutrition and subsequent realimentation in rats and sheep, *J. Animal Sci.,* **23**:214–224, 1964.

Millward, D. J., and associates: Protein turnover, in D. J. A. Cole et al. (eds.), Protein Metabolism and Nutrition, Butterworths, London. EAAP Publ. 16, 1976.

Mitchell, H. H.: Adult growth in man and its nutrient requirements, *Arch. Biochem.,* **21**:335–342, 1949.

Papas, A.: Protein requirements of Chios sheep during maintenance, *J. Animal Sci.,* **44**:665–671, 1977.

Patle, B. R., and V. D. Mudgal: Maintenance requirements for energy in cross-bred cattle, *Brit. J. Nutrition,* **33**:127–139, 1975.

Smith, A. H., R. R. Burton, and C. F. Kelly: Influence of gravity on the maintenance feed requirement of chickens, *J. Nutrition,* **101**:13–24, 1971.

Thonney, M. L., R. W. Touchberry, R. D. Goodrich, and J. C. Meiske: Intraspecies relationship between fasting heat production and body weight: A reevaluation of $W^{0.75}$, *J. Animal Sci.,* **43**:693–704, 1976.

Trowbridge, P. F., C. R. Moulton, and L. D. Haigh: The maintenance requirement of cattle as influenced by condition, plane of nutrition, age, season, time on maintenance, type, and size of animal, *Missouri Agr. Expt. Sta. Research Bull.* 18, 1915.

Chapter 15

Growth

Growth is such a universal phenomenon that it commonly incites little curiosity, but when physiologists face the question "What is growth?" they are overwhelmed by its complexities. The fertilization of a single cell starts a multiplication and a differentiation which becomes highly varied in kind and rate in the differentiated cells yet remain coordinated and culminate in the adult. There is no complete explanation as to why the process starts, how it is coordinated during its course, or why it stops at the definite point which characterizes adult development. As expressed by Rubner[1]

Throughout the animate kingdom, from the simplest microorganisms to the most complexly organized beings, that inexhaustible power of growth which ever since the genesis of the first protoplasm in the infinite past has created the structure of the fossil remains of former ages as well as our own existence—this capacity to grow, has remained as the most remarkable phenomenon of nature, the supreme riddle of life.

Despite the complexities involved, physiological studies have produced a large body of information regarding the major processes of growth, and some knowledge of these facts is obviously essential for an understanding of the nutritive requirements involved and as to how they can be met.

THE PHYSIOLOGY OF GROWTH

15.1 The Nature of Growth

Clearly, a process as complex as growth cannot be simply defined. It is much more than an increase in size. Schloss[2] defines growth as a "correlated increase in the mass of the body in definite intervals of time, in a way characteristic of the species." This brief statement is excellent because it has very broad implications. It implies that, subject to individual variability, there is a characteristic rate of growth for each species and a characteristic adult size and development. It is considered that the maximum size and development are fixed by heredity. Nutrition is an essential factor determining whether this maximum will be reached, and an optimum nutritional regime is one which enables the organism to take full advantage of its heredity. According to the basic concept, however, the maximum development fixed by heredity cannot be exceeded through nutrition or by any other means, in the normal organism. The definition by Schloss also implies that in the growth of the organism as a whole there must be a complete and coordinated growth of all its parts. This simply stated characteristic involves a multitude of interrelated processes which are very imperfectly understood at the present time.

Optimum growth should result in an adult organism capable of optimum performance through its normal life. This is an extension to the life span as a whole of the previous statement that an optimum development is one which enables the organism to take full advantage of its heredity. Here again we are ignorant of many of the factors concerned, but it is evident that optimum growth in this sense includes much more than the rate of increase of weight and size.

True growth involves an increase in the structural tissues such as muscle and bone and also in the organs. It should be distinguished from the increase that results from fat deposition in the reserve tissues. Thus growth is characterized primarily by an increase in protein, mineral matter, and water. From the nutritional standpoint, it involves in addition a large intake of energy-producing nutrients to support the growth processes, and an adequate supply of the various vitamins concerned is also required. A minute amount of lipid material goes into the structure of each cell, but this does not represent a specific dietary requirement with the exception of the essential fatty acids (Sec. 7.18), in view of the synthesis of lipids from carbohydrate.

15.2 The Cell, the Unit of Growth

Growth takes place both by means of an increase in the number of cells, *hyperplasia,* and also through an increase in their size, *hypertrophy.* In early embryonic life both processes occur in the case of all cells. In the adult three types of cells are differentiated: the *permanent* cells, such as those in the nerves, which ceased to divide early in prenatal life and whose number has remained fixed thereafter; the *stable* cells, including those of most organs, which continued to divide for a variable but major part of the growth period but which have become fixed in the adult; the *labile* cells, composing the epithelial and epidermal tissues, which continue to divide throughout life, the process in the adult being limited to the replacement of cells worn out. All of these three types of cells undergo hypertrophy during growth, and some of them may increase in size thereafter in accordance with special physiological demands. For example, the increased muscular development which can be brought about through exercise involves a hypertrophy. The cells of the adult kidney can undergo enlargement if an increased burden is placed on this organ. It seems probable that the ability of the cells of the adult organism to hypertrophy becomes less with age.

15.3   The Course of Growth of the Body as a Whole

Conception is the starting point of growth. The discussion in the present chapter deals with postnatal growth, since it is more convenient to discuss intrauterine growth as reproduction (Chap. 16), but it should be remembered that the character of the latter has important bearings on the course of growth after birth. The evident vigor of the newborn, their content of certain nutrient reserves, and other qualities are influenced by the intrauterine nutrition. This fact is reflected in the recognized desirability of considering the diet of the mother and other factors affecting intrauterine development when selecting animals for many types of growth experiments. The percentage of the total growth period which is spent in utero differs in different species, and this also has a bearing on the nutritional and other factors concerned in postnatal development. The longer the portion of the total period spent in utero, the more advanced are the young at birth. The rat is born with its eyes closed, has no hair, does not gain the use of its legs for a considerable period, and must be nourished for a relatively long period solely by its mother's milk. In contrast, the guinea pig has a full coat of hair, its eyes are open at birth, and within a few hours it is running around nibbling leafy material. The calf, lamb, and foal

TABLE 15.1.   Equivalent Ages and Body Weight

Species	Percentage of mature weight							Mature weight, kg
	10	20	30	40	50	60	100	
	Age in months							
Horses, medium	0.5	2.5	4.3	6.7	9.5	13.0	45	636
Holstein cattle	1.8	4.8	7.1	10.0	13.4	17.4	84	636
Beef cattle	1.0	3.5	5.5	8.0	11.0	14.6	72	591
Horses, light	0.5	2.0	4.0	6.0	10.0	14.0	44	455
Swine	3.2	4.8	6.3	7.3	8.4	9.6	24	227
Sheep	0.5	1.2	2.5	3.5	4.5	6.5	24	68
Chickens	1.0	1.5	2.0	2.8	3.2	3.8	8	1.8

Source: H. R. Guilbert and J. K. Loosli, Comparative nutrition of farm animals, *J. Animal Sci.,* **10**:22–41, 1951.

resemble the guinea pig as regards their stage of development at birth, while the pig and the human baby are more like the rat.

In the various species the time that is normally spent in growth bears a rather definite relation to the length of life. The data in Table 15.1 give the age in months from birth at which definite percentages of the mature weight are achieved in the different species. While such data must be considered only approximate, they present a useful picture for comparative purposes. A comparison involving more animal species is shown in Fig. 15.1 from Brody. The different relative-growth curve for the human species during childhood is striking when compared to other mammals. Another analysis of the data on the comparative chronological age and stage of growth of various species is presented by Asdell.[3] The rate of growth is not constant, nor does its entire course follow any simple mathematical expression. There are periods of acceleration and of retardation. In the human, for example, the curve of growth is characterized by a decreasing rate during childhood, an acceleration during adolescence, and a decreasing rate thereafter.

15.4 The Growth of Parts

The growth of the body as a whole is a result of the simultaneous growth of its pairs for which the individual rates are widely variable. The classical studies of Hammond[4] with pigs showed how changes in body form and composition are brought about by differences in rate and time of growth of different body parts and tissues, as illustrated in Fig. 15.2. His conclusion that bone develops earlier than muscle, which in turn develops earlier than fat has been confirmed by more recent studies. Berg and Butterfield[5] suggested that the growth pattern in cattle was better described as bone having a low growth impetus, muscle intermediate and fat tissue a high impetus, particularly after the fattening phase begins.

Markedly different rates are exhibited by certain organs and body parts. The head of the human baby is 25 percent of its body size at birth, but only 7 to 8 percent at maturity. The brain reaches adult size early in the growth period. The eyes reach mature size at about 8 years in children. The thymus increases to puberty and then decreases. The adrenals actually lose weight for a time after birth, but this loss is balanced by an accelerated development toward the end of the growth period.

Callow[6] has published the results of a very exhaustive study of the changes in the percentages of the various tissues of the body during the growth and fattening of cattle, sheep, and pigs, including data on chemical composition. Data on the composition of 132 dairy animals, ranging from a 135-day-old fetus to a 12-year-old cow, including extensive data on calcium and phosphorus content, are to be found in a bulletin by Ellenberger, Newlander, and Jones.[7]

15.5 Measures of Growth

The growth of the body as a whole is most commonly measured as an increase in weight. Size measures, such as height, and various other body dimensions

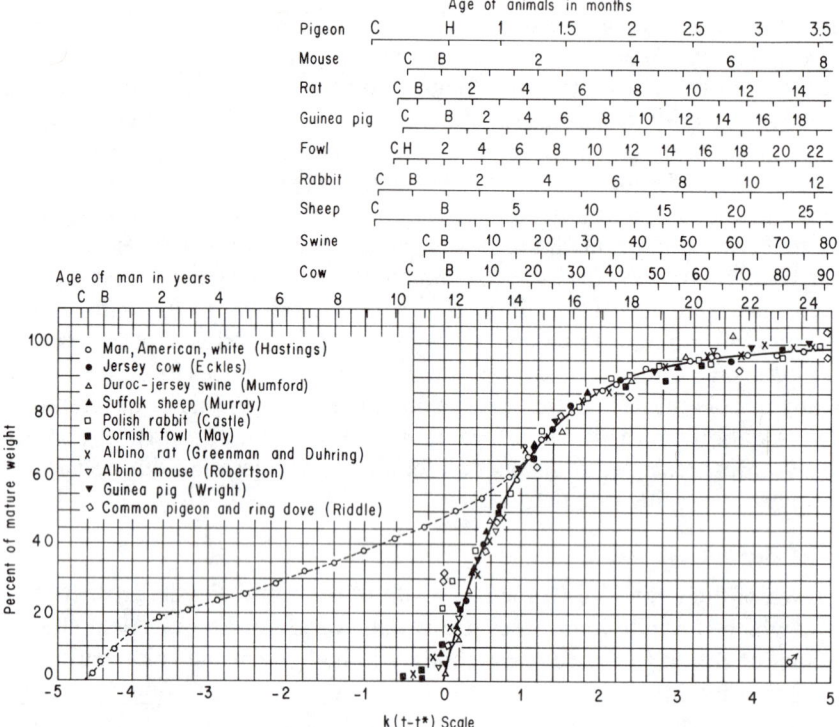

Fig. 15.1 Weight-growth equivalence of farm animals, laboratory animals, and man. (*Reproduced from Brody's Bioenergetics and Growth (1945), courtesy of the Reinhold Publishing Corp., New York.*)

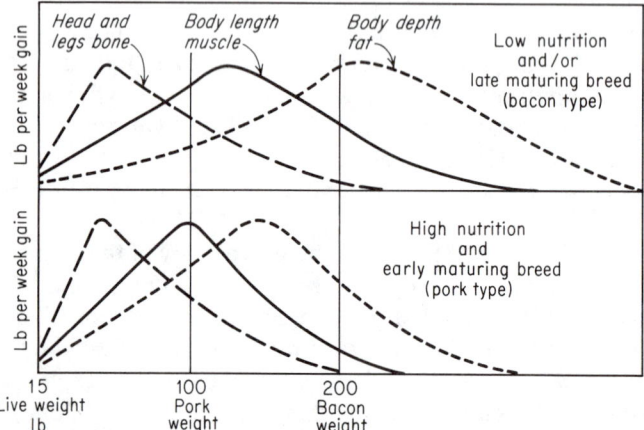

Fig. 15.2 Rates of growth of parts. The curves portray the rates of growth for the parts considered. It is noted in both charts that the rate of growth for the head and legs reaches its maximum earlier than the rate for body length and that the maximum for the latter is followed in turn by that for body depth. These three curves are considered to represent, respectively, bone growth, muscle growth and fat deposition. All these processes occur earlier where the plane of nutrition is higher. (*Courtesy of John Hammond, University of Cambridge.*)

are also frequently employed. A combination of both weight and size measures is much more useful than either alone. An animal may increase in weight through the deposition of fat without any increase in the structural tissues and organs which characterize growth. An animal which is receiving insufficient protein and energy to permit growth of its muscles and organs may still show an increase in size due to skeletal growth.

The increase in mass of the body as a whole may be expressed absolutely, as in grams per day or as a percentage of the mass at the start. While body weight is the measure most commonly used in growth experiments it is often combined with height, length, and heart girth or circumference.

Increases in weight and size are highly useful measures of growth, but they are obviously incomplete. They do not show the nature of the tissue formed, nor are they suitable measures of coordinated development. The amounts of the true growth tissue, viz., protein and the skeletal minerals, can be obtained by a balance experiment or by the slaughter procedure. The latter must be employed to record the growth of organs and other parts which provide measures of coordinated growth.

15.6 "Normal Growth"

In nutrition studies much use is made of "normal" growth data as illustrated by height-weight tables for children and by curves for increase in weight and size for farm animals. Such data are useful for comparative purposes, but their limitations must be realized. They are *averages* of the increases found for a group of individuals for which the nutritive and other factors were considered to be adequate to produce a satisfactory adult. Since the time that some of the growth data which are still used as criteria of normal growth were obtained, modern discoveries have resulted in more rapid rates of growth for various species. Some present-day investigators continue to regard their growth results as "normal" or "above normal" by comparing them with the earlier data which no longer reflect the rates that can be achieved by taking full advantage of the more recent discoveries in nutrition (Sec. 15.9). At the same time it should be recognized that increase in weight and size may be an inadequate measure of the coordinated growth which means an optimum development (Sec. 15.1). The word "normal" is much too loosely used in referring both to the state of nutrition and health and in describing growth and productive performance.

15.7 Internal Secretions and Growth

Growth is under control of certain hormones secreted by the endocrine glands. The hormone which has the most general effect is secreted by the anterior pituitary and is called the *growth hormone*. It increases anabolic processes, stimulating both muscle and bone development. Hypophysectomy, the surgical removal of the anterior lobe of the pituitary, causes growth to cease. An underactive gland results in a decreased rate of growth which is reflected in a small animal, infantile in appearance. The overproduction of the hormone by the gland causes an overgrowth of the skeleton which results in a giant in stature but a weakling physically because other growth processes do not keep pace.

Thyroxine and diiodotyrosine, secreted by the thyroid gland, have a basic effect on growth by speeding up the metabolic rate and oxygen consumption of tissues, thus increasing the energy production. Thyroidectomy, or hypofunction of the gland, stunts growth and causes a relative overgrowth of certain parts. Excess fat may be deposited. An overactive gland speeds up metabolism unduly and results in various functional abnormalities. The gland frequently becomes characteristically enlarged, a condition clinically referred to as exophthalmic goiter, which is quite different from simple goiter resulting from iodine deficiency (Sec. 10.38). Other internal secretions also play roles in growth, and there are definite interrelations in the function of certain of the endocrine glands in coordinating the processes which combine to produce normal development, physical and mental. As one example of their interrelationship, thyrotropin from the anterior pituitary stimulates the release of the hormones secreted by the thyroid gland. In turn, the level of thyroxine in the blood governs the amount of thyrotropin secreted.

The controlling roles of hormones in growth, reproduction, and lactation have stimulated research to determine whether animal production can be improved by the ingestion or injection of appropriate hormone preparations. The finding of this research has been presented in Chap. 12.

15.8   Retardation of Growth

The severe retardation or arrest of growth which results from a failure of glandular secretions is fortunately comparatively rare. Much more common are the lesser retardations which are caused by undernutrition, either in calories or in some specific essential nutrient. The nature and extent of the effect on growth are dependent upon the character and severity of the deficiency and upon the period involved. A deficiency of energy, for example, will immediately check growth in mass, while a lack of calcium may not, since its primary effect is upon bone structure rather than its size. A deficiency of certain other nutrients, such as phosphorus or thiamin exerts an indirect influence on increase in size by decreasing appetite, as well as causing direct physiological effects.

The influence of varying degrees of undernutrition on the growth process has been the object of much study. The reports of H. J. Waters at Kansas in 1910 showed that steers restricted in feed to prevent any weight gain increased in body length and height and lost their fat reserves. Refeeding demonstrated the animals were not permanently stunted. In 1915 Osborne and Mendel reported similar results with rats, but later studies suggested there might be permanent harmful effects depending on the phase of the life cycle and the nature and extent of the restrictions. Barnes et al.[8] have shown that there was permanent stunting of rats malnourished the first 3 weeks of life, but not those malnourished from week 3 to 7 of life nor those restricted during gestation.

An example of recovery from extreme underfeeding was reported by McCay and associates.[9] Rats held undersized by calorie restriction beyond the average life span for the species were able to resume growth on ad libitum feeding. This growth was reflected in organ development as well as in weight, but those severely retarded animals never reached full body size.

The major interest in the effects of growth retardation lies in the question of its influence on ultimate body development and productive life. Does any prolonged or severe retardation permanently harm any tissues or functions? The view is commonly expressed that any marked retardation is definitely undesirable. This may be true, but experiments have shown that underfeeding may prolong the growth period without affecting the ultimate size. Among the many studies with farm animals are those with identical twins at the U.S. Department of Agriculture (USDA). One animal of each pair was liberally fed for rapid growth, the other receiving a maintenance ration or a lesser allowance. Some of the restricted animals received such a ration from 6 to 12 months of age, others for shorter periods beginning at an earlier age. The rations were designed to supply adequate amounts of the essential nutrients. The only effect of the underfeeding was to prolong the overall growth period. The restricted animals attained the same final weights as the controls, without loss of meat quality.[10] Breirem and associates[11] reviewed various experiments, mostly with cattle, on the effects of undernutrition on growth rate and ultimate body size. They drew the general conclusion that, while restrictions in energy intake have a retarding effect on growth, by growing more rapidly during following periods of liberal feeding the animals are capable of recovering and may eventually reach the same size as animals well fed from birth. It was noted, however, that severe and prolonged undernutrition at an early age may result in permanent stunting.

Meyer and Clawson[12] have shown with sheep that realimentation following undernutrition resulted in an increased efficiency of feed utilization. The chemical composition of body-weight loss during starvation was similar to that of body-weight gain.

Beginning in 1960, McCance and associates at the University of Cambridge made extensive studies of severe undernutrition in growing and adult animals, specifically cockerels and pigs. The fifteenth paper of the series by Tonge and McCance[13] cites the earlier studies. The series of papers reports the results of restriction of total food intake. Young cockerels were maintained with little or no gain for 6 months and weanling pigs for 1 to 2 years, and then given ample food. In addition to resulting in growth retardation the food restriction caused marked changes in the size and structure of the bones. When an ample food supply was later provided, rapid gains resulted and the normal structure and composition of the bones were restored. Abnormalities in tooth deposition and structure resulting from undernutrition were not corrected by ample food. From the standpoint of animal husbandry practice, the question of the desirability of food restriction during growth involves considerations which differ depending on whether the animals are being fed for meat production or for later breeding and lactation performance, as shown by the discussions which follow.

15.9 Accelerated Growth Performance

It is evident that nutrition discoveries during the past fifty years have resulted in an accelerated rate of growth. A study of the inbred rat colony of the Connecticut Agricultural Experiment Station during three periods from 1910 to 1935, the era of large advances in nutrition, revealed a gradual increase in growth rate and markedly larger body size at maturity. Similar data have been reported for humans. There are convincing data from selected groups in this country and elsewhere that a physical evolution is occurring whereby children are growing to be taller and heavier than their parents. Better and more abundant food is listed as one of the causes.

Today's methods of feeding result in more rapid rates of gain of broilers, pigs, and cattle and they reach slaughter weights at younger ages on less feed and often in better condition than was possible even a few years ago. It should be recognized that improvements in breeding and management as well as in feeding practices have been concerned in these developments. Through selective breeding the potentials for faster growth and greater productivity have been increased in our domestic animals so that they can take full advantage of the improved nutrition and management. In considering these data showing increased rates of gain in weight, it is important to bear in mind that growth proper is represented by increases in protein and in skeletal development and that in this sense gains resulting from increased fat deposition do not constitute an acceleration of growth rate. The value of an increased fattening of meat animals depends upon the market for the product.

15.10 Effect of Growth Rate on Feed Efficiency and Carcass Quality

From the standpoint of animal production we are particularly interested in the influence of the growth rate on feed economy and on the final product. Extensive experiments at the Missouri Experiment Station[14] reported in a classical series of papers clearly showed that the higher planes of nutrition were more efficient in the recovery of energy and protein in the production of edible meat. Guilbert and coworkers[15] have published an important study demonstrating the superior results obtained through continuous growth in steers, brought about by supplementing the range feed during the dry season, as compared with the interrupted growth which otherwise resulted. Today's improvements over former years in the rates of growth obtained in farm animals also result in marked savings in feed consumed per unit of gain.

Several experiments have been made of the influence of plane of nutrition at different stages of growth on carcass composition. Very extensive studies at the University of Cambridge with pigs by McMeekan[16] and with lambs by Pálsson and Vergés[17] resulted in the conclusion that the proportions of bone, muscle, and fat in the carcass could be markedly influenced by altering the level of nutrition at different stages of the growth cycle. A reanalysis of the data by Elsley and coworkers[18] has indicated that the greater part of the changes noted could be attributed to the effect of level of nutrition on the deposition of fat and that there were few deviations of any importance from a uniform growth of bone and muscle under the patterns of feeding used. The authors suggest that the level of nutrition should be selected with the aim of producing whatever fat content is desired, subject to economic considerations.

Similar conclusions are drawn by Tulloh[19] in a review of data for cattle, sheep, and pigs. He found that comparisons of body composition of animals may be invalid unless they are made at the same empty-body weight. Carcass composition appears to be mainly dependent on body weight and largely independent of age.

The N.R.C. Symposium[20] and Reid's report[21] evaluated the various studies on the effect of nutrition on body composition and the methods of measuring composition. Reid pointed out that within species, breed, and sex the body composition of healthy animals maintained in positive energy balance is closely related to body weight prior to maturity and that level of nutrition has very little influence, except for a deficiency of protein level or quality and for *compensatory growth* following underfeeding. There are important specie and breed differences. The body composition of the rat, chicken, duck, and very young pig, lamb and calf can be changed more by diet than is the case of older animals. In all animals the fat and moisture content are inversely related.

Donnelly and Hutton[22] fed calves from 12 to 61 days of age diets varying from 16 to 32 percent protein at two energy levels (4.2 or 5.2 Mcal/kg) to produce gains of 610 or 830 g per day. Selected data shown in Table 15.2 illustrate the effects of different protein intakes on the tissues gained. The proportion of energy gained as fat was 0.70 on 15.7 percent protein versus 0.47 on the 29.6 percent level. Similar effects of protein intakes on the composition of gains of young pigs have been reported by Campbell.[23]

Crampton and associates have shown that restricting the feed intake of market hogs or "diluting" highly digestible rations with fibrous feeds, during the finishing period, improves the carcass for bacon by reducing fat deposition

TABLE 15.2. Effect of Protein Intake on the Composition of the Digesta-free Bodies and of the Weight Gains of Calves

	Digesta-free bodies[a]				Weight gains	
Dietary protein, %	15.7	21.8	29.6	31.5	15.7	29.6
Water, %	66.7	67.1	67.6	68.3	58.9	62.8
Fat, %	11.1	10.0	7.2	7.5	22.3	11.7
Protein, %	18.7	18.8	20.9	19.8	16.6	21.5
Ash, %	3.7	3.8	4.3	4.4	2.3	3.8

[a]The initial composition was 71.7 percent water, 3.6 percent fat, 20.0 percent protein, and 4.6 percent ash.
Source: Selected data from Donnelly and Hutton.[22]

and increasing the actual size of the muscle area. As expected, there was a decreased rate of gain and an increase in the length of the feeding period accordingly.

15.11 Rate of Growth and Productive Life

Clearly, the recent studies in nutrition have resulted in practices that have markedly increased the growth rate as measured by weight and size. It has been the general belief that this is a desirable development in terms of the life span as a whole, but there are experiments which challenge this belief. The studies by McCay and associates clearly showed that rats whose growth is severely retarded in early life, by calorie restriction only, have a longer life span than those which grow rapidly under ad libitum feeding. Their later studies showed that retarded animals were much less subject to chronic disease. They also found that, with animals which had grown normally on ad libitum diets until middle age, calorie restriction thereafter resulted in a longer life. McCay's studies have been confirmed and extended by Berg and by Berg and Simms,[24] who found that less drastic restrictions up to 800 days of age resulted in much leaner animals with somewhat less skeletal size, but improved health, female fertility, and longevity, and delayed the onset of degenerative disease.

In 1948, an experiment was started at Cornell University to measure the influence of three different levels of energy intake from birth to first calving upon growth rate and lifetime reproduction and lactation performance of Holstein cattle. Energy intakes of 62, 100, and 146 percent of Morrison's TDN standard were fed to the three groups until first calving. After first calving all animals were liberally fed a balanced ration adequate to meet their individual requirements for continuing growth, lactation, and successive reproduction. The animals were discarded only if they became sterile or if an incurable defect developed which seriously impaired function, but not because of low milk yields. Data on growth and development are shown in Table 15.3.

Heifers fed the low-energy intake were markedly retarded in early growth and development. They weighed only 72 percent as much at 1 year of age as those fed the normal ration. At first calving they weighed 100 kg less than the normal group although they were 4 months older and their calves were a little smaller at birth. Animals fed the high-energy intake grew faster and weighed 12 percent more at first calving than those fed the recommended intake. Adequate feeding after first calving permitted the stunted heifers to recover, and by the third calving they were as large as the normal group. Thus underfeeding to 32 months of age did not cause permanent stunting, but heavy feeding increased mature weight slightly. The high-level animals remained about 30 to 40 kg heavier throughout life than the others. The results with respect to reproduction and lactation are discussed in Chaps. 16 and 17.

These experiments serve to emphasize the fact that growth should be looked at primarily as a preparation for life. In humans we are interested in a healthy, productive life in which the infirmities of old age are postponed as long as possible. In breeding stock and in animals kept for milk and egg pro-

TABLE 15.3. Influence of Plane of Nutrition on Body Weights of Female Holstein Cattle

	TDN intake, % of normal		
	62	100	146
Age, month/no. of calvings	kg	kg	kg
5.5	101	144	189
12.0	198	278	337
First	384	483	548
Second	560	584	631
Third	621	627	673
Fourth	647	650	687
Fifth	671	669	695
Sixth	676	670	703

Source: J. T. Reid and associates, Causes and prevention of reproductive failures in dairy cattle. IV: Effect of plane of nutrition during early life on growth, reproduction, production, health and longevity of Holstein cows. I: Birth to fifth calving, *Cornell Agr. Expt. Sta. Bull.* 987, 1964.

duction, lifetime performance is the final measure of the success achieved in rearing these animals. The possibility must be considered to exist, particularly since there may be nutritional factors still unknown, that, from the standpoint of productive life, there are limitations in a system of rearing which relies upon a rapid increase in weight and size as the primary measure of success. However much one may doubt this possibility, it cannot be said with certainty that all the factors for growth are optimum until their influence on lifetime performance has been studied as thoroughly as their effects during the growth period itself.

ENERGY REQUIREMENTS FOR GROWTH

Previous discussions have shown that the rate and character of the body increase vary with age, as well as with species. It is evident, therefore, that a feeding standard for growth must be different for each species and must consist of a series of values corresponding to the different ages and body weights representing the growth period. Such a detailed presentation for all species is beyond the scope of a text dealing with the principles of nutrition. Rather, the physiological bases of the requirements will be considered, typical procedures for arriving at specific values will be outlined, and reference will be made to sources of detailed information for the different species which have been studied.

The total requirement for a given nutrient during growth must include the amount needed for maintenance as well as the amount required for the new tissue formed. The values given in feeding standards represent these combined requirements. Of the various nutrient needs for growth, the requirement for energy is by far the largest and primarily governs the total food allowance. It is therefore advantageous to discuss this requirement first.

15.12 Factors Governing the Energy Requirement for Growth

The maintenance component of the total energy requirement during growth increases regularly with body size, but the additional demand for the growth itself varies with the rate and with the composition of the tissue formed. Per unit of body weight, the amount of energy represented by the growth tissue formed decreases with age, reflecting the declining rate of body increase measured on a percentage basis (Sec. 15.5). But the amount of energy stored per unit of body increase becomes larger with age because of its lower water and higher fat content. While the true growth tissue contains only a trace of fat, a certain amount of fat deposition is an inevitable accompaniment of growth, and in practice a considerable amount of fattening is an integral part of growing animals for meat. Since fat contains much more energy than does protein, it is evident that the energy requirement per unit of body gain increases in accordance with its fat content. In fact, if the fattening is very rapid, the normal trend of decrease in energy stored per unit of body weight may not occur. In feeding standards for meat animals, no separate statement is made of the requirements for growth proper and for the fattening which concurrently takes place, but a distinction may be made according to the amount of fattening desired.

Except under conditions where very rapid fattening is sought, the maintenance component of the total growth requirement always markedly exceeds the portion required for the formation of new tissue. Thus the faster the growth rate, the lower the total requirement per amount of gain tends to be, but this tendency may be partially counterbalanced by the decreasing efficiency of food utilization as the intake is increased.

Since balanced rations (Sec. 12.16) involve less wastage as heat loss, they have the practical effect of decreasing the total feed required per unit of gain.

15.13 Energy Requirements by the Factorial Method

The net-energy requirement for growth may be considered to be the sum of the energy of the tissue formed plus the basal metabolism increased by an activity factor. The nutritional needs at any given period are determined by the rate of gain expected and the average body weight during the period in question. The weight gain, plus data from slaughter experiments on the composition of the gain, furnishes the figures for computing the calories required for expected gain. Such data are available for cattle, sheep, and swine. Basal metabolism data plus an activity increment provide the basis for the maintenance requirement. The sum of the calories thus obtained provides the net-energy (NE) requirement, which can in turn be translated into ME or DE by the use of appropriate factors. The method is illustrated for the pig in Table 15.4. Details of the calculations are given in the report cited in the footnote. The NE values

TABLE 15.4. Calculation of Daily Energy Requirements of Pigs

Live weight, kg	Daily live weight gain, g	Energy, kcal				Digestible energy[c] for maintenance and weight gain, Mcal
		Of maintenance[a]	Of weight gain		Of maintenance and weight gain	
			Per 100 g gain[b]	Total		
20	500	1034	220	1100	2134	3.05
40	750	1525	280	2100	3625	5.18
60	790	1914	350	2765	4679	6.68
80	790	2249	410	3239	5488	7.84
100	790	2548	480	3792	6340	9.06

[a]Calculated from Breirem's formula, $E = 196.3W^{0.56}$.
[b]Based on Hörnicke's data on the energy value of live weight gain, *Z. Tierphysiol. Tierernähr. Futtermittelk.*, **17**:28–60, 1962.
[c]Assuming that DE is 70 percent utilized for maintenance and growth.
Source: Reproduced by permission of the Agricultural Research Council of Great Britain–Nutrient Requirements of Farm Livestock, No. 3., Pigs.

for maintenance and weight gains are translated into DE values by the factor 70 percent.

Several assumptions are necessarily involved in the factorial method. The results require testing in feeding trials before being adopted in practice. A major uncertainty involved is the energy values calculated for the gains made, since these gains can vary rather widely in their proportions of fat and protein, according to genetic makeup and nutritional regime of the animal in question. Nevertheless, the method does provide data, not obtainable in any other way, which are useful for various experimental purposes, as well as for consideration along with feeding-trial data in arriving at recommendations for practice.

15.14 Energy Requirements Obtained from Feeding Trials

The data contained in the currently used feeding standards for farm animals are based primarily on the results of feeding trials. In the more critically conducted trials different groups of animals have been fed throughout the growth period at different levels of energy intake to ascertain the level that would produce economical growth and development. A feeding trial enables the statement of the requirement in terms of specific feeds or in terms of any desired measure of the energy required, by the use of appropriate methods. In the past the data for farm animals had been expressed as TDN in the United States, but the recent National Research Council (N.R.C.) reports include DE, ME, and for some species NE. Most of the energy requirement data and feed values have been calculated from the TDN values determined over the past three quarters of a century. Reevaluations are being made to check the calculated values. The British Agricultural Research Council (A.R.C.) energy requirements are expressed as ME, the system having been changed from starch equivalents (SE) in the middle of the 1960s.

TABLE 15.5. Energy Required for Growth of Dairy Cattle

Body weight, kg	TDN, kg	DE	ME	Body weight, kg	TDN, kg	DE	ME
		Mcal				Mcal	
45	1.05	4.63	3.80	204	3.08	13.58	11.14
68	1.53	6.75	5.54	227	3.21	14.16	11.61
91	1.92	8.47	6.95	250	3.34	14.73	12.08
114	2.25	9.92	8.13	273	3.48	15.35	12.59
136	2.52	11.11	9.11	295	3.61	15.92	13.05
159	2.74	12.08	9.91	318	3.75	16.54	13.56
182	2.93	12.92	10.59	341	3.89	17.15	14.06

Source: Calculated from C. H. Eckles and T. W. Gullickson, Nutrient requirements for normal growth of dairy cattle, *J. Agr. Research*, 42:603–616, 1931.

The feeding trial method of determining the energy requirements for growth is illustrated by the studies of Eckles and Gullickson with Holstein and Jersey cattle fed three levels of energy. The intakes of those which were considered to have made normal growth, as measured by weight, were then averaged to represent the requirements. These estimated daily requirements are given in Table 15.5.

The recommended energy intakes for growth that are found in the feeding standards for various species have been arrived at primarily from feeding-trial data obtained in experiments similar to the one cited above, but balance data were also considered.

From data obtained in the previously described (Sec. 14.10) growth and slaughter experiment, Garrett and associates developed the formulas shown in Table 15.6.

The N.R.C. nutrient requirements for beef and dairy cattle lists energy requirements and feed values also in net energy terms and separates net energy for maintenance (NE_m) from net energy for growth (NE_g). The study of Rattray et al.[25] illustrates the method used to determine the NE requirements of growing lambs and the energy value of the feeds using a feeding-slaughter trial.

TABLE 15.6. Equations to Estimate the Energy Requirements of Sheep and Cattle for Maintenance and Weight Gain

Sheep	Cattle
$TDN^a = 0.036W^{0.75}(1 + 2.3g)$	$TDN = 0.036W^{0.75}(1 + 0.57g)$
$DE = 76W^{0.75}(1 + 2.4g)$	$DE = 76W^{0.75}(1 + 0.58g)$
$ME = 62W^{0.75}(1 + 2.5g)$	$ME = 62W^{0.75}(1 + 0.60g)$
$NE = 35W^{0.75}(1 + 1.8g)$	$NE = 35W^{0.75}(1 + 0.45g)$

[a]TDN, W, and g are in pounds; DE, ME, and NE, in kilocalories.
Source: Garrett, Meyer, and Lofgreen, The comparative energy requirements of sheep and cattle for maintenance and gain, *J. Animal Sci.*, 18:528–547, 1959.

The following data were obtained:

Energy Value of the Diet, kcal/g DM	Requirements
DE, maintenance 3.25 ± .022	ME utilization for
DE, ad libitum 3.16 ± .014	maintenance, 65.7% gain, 40.8%
ME, maintenance 2.67 ± .018	Daily NE requirement
ME, ad libitum 2.59 ± .012	NE_m, 79.4 kcal $W_{kg}^{0.75}$
NE_m 1.75	
NW_g 1.04 ± .061	NE_g, 322.2 kcal $W_{kg}^{0.75}$

From the data obtained the energy requirements of growing lambs can be expressed as DE, ME, or NE and the amounts of feeds needed to meet the requirements can be calculated.

The daily energy requirements for normal growth of beef and dairy cattle are compared in Table 15.7. The N.R.C. values for beef cattle are slightly higher than for dairy cattle of the same body weight and making similar weight gains, probably because beef cattle usually store more fat than dairy cattle. The A.R.C. values for cattle are somewhat lower at the heavier weights than the N.R.C. values. Obviously the energy requirements vary greatly depending on the amount of weight gained, and the feeding standards take this fact into account. The N.R.C. energy requirements of various animals are summarized in Appendix Tables I through IX, and in Table 16.3.

THE PROTEIN REQUIREMENTS FOR GROWTH

Aside from water, the body increase during growth consists very largely of protein and fat. The theoretical minimum protein requirement is the amount actually stored in the body. But this is far below the actual requirement because of the wastage in digestion and metabolism. The loss in digestion can be taken

TABLE 15.7. Energy Requirements for Growth of Beef and Dairy Cattle (amount per animal per day)

Body weight, kg	N.R.C. Beef cattle		N.R.C. Dairy cattle		A.R.C. cattle	
	Gain, kg	ME, Mcal	Gain, kg	ME, Mcal	Gain, kg	ME, Mcal
100	0.70	7.5	0.75	7.2	–	–
150	0.70	10.4	0.75	9.8	–	–
200	0.70	13.8	0.75	12.3	0.75	12.3
250	0.70	15.0	0.75	14.4	–	–
300	0.70	17.1	0.75	16.2	0.75	14.7
350	0.70	19.7	0.75	17.7	–	–
400	0.70	21.7	0.75	18.8	0.75	17.4
450	0.80	24.5	0.70	19.2	–	–

Source: N.R.C. and A.R.C. reports.

account of by stating the requirement in terms of digestible protein, and data on digestibility are available for all of the common feeds. The wastage in metabolism is much less readily assessed. It is governed primarily by the efficiency with which the digested protein supplies the amino acids required for the construction of body tissue.

15.15 Amino Acid Requirements

The discussion in Chap. 8 listed the amino acids which have been found essential for the growth of rats, pigs, and chickens and described how this information was obtained. The general procedure has been to feed a diet designed to be adequate except in the amino acid to be tested and to add varying levels of it to different groups to arrive at the level which is sufficient to result in satisfactory growth. Weight increase has been the usual criterion, but in some experiments the nitrogen-balance measure has also been employed. The procedure is illustrated by the studies of Brinegar and coworkers[26] on the isoleucine requirements of pigs and by Grau and Peterson[27] in the case of chicks. Such studies measure the combined requirement for maintenance and growth.

Data on amino acid requirements for growth are presented in Table 15.8 for chicks, poults, pigs, rabbits, rats, and salmon. The values are expressed in terms of the utilizable form of each acid, usually the L form. The mixture of amino acids provided by the data in the table must be supplemented by nonspecific nitrogenous sources of the nonessential acids (Sec. 8.8). The data apply to the period of rapid growth and to the protein levels listed and assume an energy intake adequate for rapid growth. While lesser amounts of the acids are required for slower rates of growth, the proportionate needs do not change. This reflects the fact that whether protein formation is rapid or slow, it requires amino acids in the proportions represented by the protein to be formed. This means that an appropriate amino acid balance in the ration is important for efficient and optimum protein nutrition. This balance can be upset by relative excesses of one or more of the acids, as well as by deficiencies. At a very slow rate of growth, compared with a maximum rate, one would expect some differences in the proportions of acids called for, because here maintenance needs dominate, and they are somewhat different from those for growth, as previously discussed.

Another method of estimating amino acid requirements is the determination of the amino acid composition of the bodies of the species in question. Williams and coworkers[28] have done this for the essential amino acids for pigs, chicks, and rats at various stages of growth. The data show comparability within each species at different stages of growth and also close correlation among species. The requirements thus indicated are in reasonable agreement, for the most part, with those determined by growth experiments on which the data in Table 15.8 are based.

The data on amino acid requirements are being used in conjunction with those on feed composition to formulate rations for more efficient and economical protein nutrition.

TABLE 15.8. Essential Amino Acid Requirements for Growth

Nutrient	Rat	Salmon	Chick	Pig	Rabbit	Turkey
			Percent of Diet Dry Matter			
Protein	13.3	40	20	18	16	28
Arginine	0.67	2.4	1.2	0.23	0.6	1.6
Histidine	0.33	0.7	0.4	0.20	0.3	0.55
Isoleucine	0.61	0.9	0.75	0.56	0.6	1.1
Leucine	0.83	1.6	1.4	0.68	1.1	1.9
Lysine	1.0	2.0	1.1	0.79	0.65	1.5
Methionine[a]	0.67	0.5	0.75	0.56	0.6	0.87
Phenylalanine[b]	0.89	2.1	1.3	0.56	1.1	1.8
Threonine	0.56	0.9	0.7	0.51	0.6	1.0
Tryptophan	0.17	0.2	0.2	0.15	0.2	0.26
Valine	0.67	1.3	0.85	0.56	0.7	1.2
			Percent of Dietary Protein			
Arginine	1.0	6.0	6.1	1.5	3.8	5.7
Histidine	2.1	1.8	1.7	1.5	1.9	1.9
Isoleucine	3.9	2.2	4.4	4.6	3.8	3.9
Leucine	4.5	3.9	6.7	4.6	6.9	6.8
Lysine	5.4	5.0	6.1	4.7	4.1	5.4
Methionine	3.0	4.0	4.4	3.0	3.8	3.1
Phenylalanine	5.3	5.1	7.2	3.6	6.9	6.4
Threonine	3.1	2.2	3.3	3.0	3.8	3.6
Tryptophan	1.0	0.5	1.1	0.8	1.2	0.9
Valine	3.1	3.2	4.4	3.1	4.4	4.3

[a] Cystine can meet at least 50 percent of the methionine requirement.
[b] Tyrosine can meet 30 percent of the requirement.
Source: N.R.C. Reports.

They provide a basis, in both experiments and practice, for putting together combinations of feeds, the amino acid makeup of which will supplement each other (Sec. 15.23). Of course, the selection of a ration which contains amino acids in the optimum quantities and proportions as fed may be rendered less efficient by differential losses in digestion (Sec. 8.12).

The proportionate amino acid requirements can be most readily visualized by taking one of them as unity, thus providing a useful "pattern" for comparative purposes. Such a pattern is shown for pigs in Table 15.9 constructed from the data in Table 15.8. Here tryptophan is taken as unity, and thus all of the values in that table are multiplied by 6.67 to arrive at the data in Table 15.9. The calculated values for methionine and phenylalanine are, of course, subject to partial replacement as indicated in the footnotes to Table 15.8. Such a proportionality pattern is useful for comparison with a similar pattern which may be constructed for a given feed to show the extent to which the propor-

Fig. 15.3 Leucine deficiency. The top pig received a diet deficient in leucine; the bottom pig received the same diet with added leucine. (*Cornell University*).

tions of the amino acids in the feed correspond with body needs. Proportionality patterns for the requirements for different species and functions also provide a useful basis for comparative purposes.

One cannot at present visualize a protein feeding standard for practice expressed solely in terms of amino acids. But it seems clear that a knowledge of amino acid requirements and their distribution in feeds can be used to select more efficient and economical protein mixtures. In some instances, as com-

TABLE 15.9. Proportionality Pattern of Amino Acid Requirements for Swine

Arginine	1.5	Methionine	3.7
Histidine	1.3	Phenylalanine	3.7
Isoleucine	3.7	Threonine	3.4
Leucine	4.5	Tryptophan	1.0
Lysine	5.3	Valine	3.7

mercial supplies become more available and cheaper, the addition of one or more specific amino acids may become the most economical way of providing an efficient mixture with available feeds. It seems evident that with the usual feed supplies attention will need to be given to only a few amino acids which may be deficient in terms of body needs—rather than to even all the essential ones. In the case of chicks, arginine, lysine, tryptophan, and the sulfur-containing amino acids appear to be the only ones which thus need attention.

Lysine is the first limiting amino acid for pigs fed sorghum diets and threonine the second most limiting.[29] The addition of tryptophan further improved the supplemented diet for early growth but not for finishing pigs. Similar studies have been carried out with diets based on other cereal grains for pigs and poultry.

This discussion has dealt with pigs, poultry and other nonruminants. It has been recognized in ruminants that microbes synthesize as well as degrade amino acids (Sec. 8.14). The biological values of various protein sources and mixed rations for sheep have varied from about 60 to 80. Many studies have given attention to methods of improving utilization of protein and nonprotein nitrogen (NPN) by sheep and cattle. Abomasal infusion of high-quality protein or limiting amino acids has increased weight gains and growth of wool. Treating protein feeds with heat, aldehydes, or tannins to reduce the solubility of the protein reduces ammonia production in the rumen and apparent digestibility, but increases growth rate and nitrogen retention. The report of Nimrick et al.[30] illustrates the type of improvement various researchers have obtained. Many problems remain in devising practical methods of achieving maximum potential efficiency in protein and NPN utilization by ruminants. These findings and problems have been reviewed by Lewis and Mitchell.[31]

15.16 The Determination of Biological Value

The term *biological value of protein* denotes the measure of protein quality which is obtained in an animal study in which the percentage of the intake which is actually utilized is determined. The measure is sometimes expressed as the percentage of the total intake that is stored. Here losses in digestion as well as in metabolism are taken into account. The biological value, properly speaking, takes account of metabolic losses only and thus should be computed on the basis of the digested protein. This procedure measures the efficiency of the absorbed protein in supplying the amino acids needed for the synthesis of body protein, thus arriving at a figure for biological value which constitutes the more exact and preferred usage of the term. The calculation is made most simply as follows:

$$\frac{\text{N intake} - (\text{fecal N} + \text{urinary N})}{\text{N intake} - \text{fecal N}} \times 100 = \text{biological value}$$

It is evident that the data for such a calculation can be obtained from a nitrogen-balance experiment. The level of protein fed must be high enough so that marked growth will actually result as indicated by the positive balance; yet it must not be in excess of the amount needed to cause maximum growth, because an intake above this would be catabolized and excreted and thus give a biological value lower than the true one. There must be a sufficient intake of nonnitrogenous food so that the protein will not be needed as a source of energy. Other nutrients must be supplied adequately also.

15.17   The Thomas-Mitchell Method

The formula previously given measures the biological value of protein for growth purposes only. A more useful measure is one that takes account of maintenance as well. This can be accomplished by considering the metabolic and endogenous losses separately from the total fecal and urinary excretions. A method for this purpose was originated in 1909 by Karl Thomas[32] of Leipzig, who first used the term *biological value*. In so doing he had amino acid makeup in mind. It is interesting to note that Thomas developed this concept and a method for its measurement prior to the work of Osborne and Mendel which inaugurated the modern studies of protein quality. The Thomas method was modified by Mitchell[33] and this modified procedure continues to be widely used. The method is best explained by a consideration of the formula which Mitchell uses in calculating the value obtained:

$$\frac{\text{N intake} - (\text{fecal N} - \text{metabolic N}) - (\text{urinary N} - \text{endogenous N})}{\text{N intake} - (\text{fecal N} - \text{metabolic N})} \times 100$$

The feature of this formula, distinguishing it from the previous one, is its recognition of the fact that endogenous and metabolic nitrogen represent fractions which have actually been utilized by the body even though they appear as excretions. Thus, in the numerator, the fecal loss subtracted from the total intake is limited to the part actually undigested, and the urinary loss is reduced by its endogenous fraction before being subtracted. The numerator, therefore, represents the total nitrogen utilized, including the part used in maintenance as well as that built into growth tissue. Since in the denominator, also, the metabolic nitrogen is subtracted from the total fecal output, the biological value computed is the percentage of the actually digested nitrogen that is utilized. In excluding the metabolic and endogenous nitrogen from the losses, the Thomas-Mitchell method provides a measure of the efficiency of the absorbed protein for the combined functions of growth and maintenance.

The values for metabolic and endogenous nitrogen cannot be determined while the protein is under study but must be calculated from values obtained in separate periods when the animals are receiving a nitrogen-free diet. A frequent difficulty here, particularly with certain species, is that some individuals will not eat enough of the nitrogen-free diet to supply their energy needs, and thus true endogenous nitrogen values are not obtained. To obviate this difficulty a small amount of protein, such as egg or milk, which is considered to be utilized to an approximately complete degree in both digestion and metabolism, can be included in the ration. Some workers, instead of using the nitrogen-free-diet procedure, have arrived at the metabolic fecal nitrogen by the Titus extrapolation method (Sec. 8.15) and calculated the endogenous nitrogen from basal metabolism (Sec. 14.5). Others have used average values which have been reported for metabolic fecal nitrogen per unit of dry matter consumed and for endogenous nitrogen per unit of metabolic body size.

The data presented in Table 15.10 taken from a publication by Mitchell, illustrate the nature of the results obtained by the Thomas-Mitchell procedure. These figures reveal the wide differences which exist in the efficiency of various proteins, as measured by the rat, and they illustrate the extensiveness of the role that biological value can play in governing the amount of dietary protein required. The values given in Table 15.10 were determined by introducing the food in question into the basal diet in such amounts as would provide a protein level lying between 8 and 10 percent. In addition to avoiding too high levels, it is desirable to hold to approximately the same level where the data obtained with different foods are to be compared. Since the overall usefulness of a given source of protein depends on its digestibility as well as on the biological value of the absorbed fraction, the *net protein utilization* is frequently expressed as the product obtained by multiplying the biological value by the coefficient of digestibility for the source of protein in question. Hence, for milk:

85 X 0.97 (digestibility coefficient) = 82 (net protein value)

Armstrong and Mitchell[34] have compiled data from the literature giving the biological values of various feeds for growing swine which show a variability similar to that revealed in Table 15.10 for rats.

TABLE 15.10. Biological Value of the Proteins of Human Foods

Food	Biological value of protein	Food	Biological value of protein
Whole egg	94	Whole wheat	67
Milk	85	Potato	67
Egg white	83	Rolled oats	65
Beef liver	77	Whole corn	60
Beef heart	74	Wheat flour	52
Beef round	69	Navy beans (cooked)	38

Source: H. H. Mitchell, The protein values of foods in nutrition, *J. Home Econ.*, **19**:122–131, 1927.

The *nitrogen-balance index* is a measure of biological value similar to those measures obtained by the Thomas-Mitchell method in that it takes account of the same data on intake and outgo of nitrogen. It takes advantage of the fact found to hold for different species that a linear relationship exists between nitrogen intake and nitrogen balance in the region of negative and low-positive balance, and thus it is applicable to studies of biological value for maintenance and some growth. It measures the rate of change of nitrogen balance with respect to absorbed nitrogen; the higher the rate, the greater the efficiency of retention, and thus the higher the biological value. For a detailed explanation of the method and how it is carried out, the student is referred to the article by Allison[35] which also reviews other methods of measuring the nutritive value of proteins.

15.18   Protein Efficiency as Measured by Feeding Trials

As an alternative to the nitrogen-balance methods, another procedure for measuring protein quality is based on the method originally developed by Osborne, Mendel, and Ferry,[36] involving a feeding trial in which protein sources are compared in terms of gain in body weight per gram of protein or nitrogen fed. This procedure is frequently referred to as the determination of the "protein efficiency ratio" (PER). As carried out with rats to compare specific proteins or protein sources, a nitrogen-free, otherwise adequate basal diet is used in which the protein sources to be compared are included for different groups of young animals, and records are kept of growth and feed consumption. The results thus obtained are illustrated in Table 15.11 representing data taken from a study by Jones and Divine with rats in which the comparisons were made at a protein level of 9.1 percent. Examination of the data for gain per gram of protein shows that the protein of patent flour is of low quality compared with that of the other sources and that skim milk has the highest value. The superiority of whole-wheat flour over the highly milled product is indicated, though the cereal flour fell below the oil-seed flours in value.

A limitation of measuring protein efficiency in terms of body gain is that

TABLE 15.11.   Comparative Growth-Promoting Value of Proteins

Source of protein	Average food consumed, g	Average weight gained, g	Average gain per gram of protein consumed, g
Patent flour	278	19	0.75
Whole-wheat flour	342	36	1.15
Peanut flour	419	75	1.95
Cottonseed flour	455	85	2.05
Soybean flour	408	87	2.35
Skim-milk powder	560	141	2.78

Source: D. Breese Jones and J. P. Divine, The protein nutritional value of soybean, peanut and cottonseed flours and their value as supplements to wheat flour, *J. Nutrition*, 28:41–49, 1944.

the protein content of this gain may be variable. Thus some investigators have included slaughter data, where laboratory animals were involved, as a check on the results. From comparative studies which have been made, it appears that the possible errors here concerned are not likely to be of large importance in a well-planned and well-conducted experiment. However, any factor which influences the rate of growth may markedly affect the PER. For example, Ames and Brink[37] showed that growth of lambs was highest at ambient temperatures of 15 to 20°C (192 to 197 g per day) compared with 73 g at −5°C and 41 g at 40°C. PER values varied from 1.36 at maximum daily gains to 0.36 and 0.20 at the lowest and highest temperatures and lowest gains, respectively.

15.19   Net Protein Utilization (NPU) Method

This method, devised by Miller and Bender,[38] is based on a comparison of the body nitrogen content resulting from a test protein with that resulting over the same period on a nitrogen-free diet. The value is computed as follows:

$$NPU = \frac{\text{Body N content with test protein} - \text{body N content with N-free diet}}{\text{N intake}}$$

The formula measures efficiency for growth. At the close of the experiment the animals are killed, the body water determined, and the nitrogen calculated from body water. This method of arriving at nitrogen content is subject to some question, a question which can be eliminated by actual analyses of the carcasses for nitrogen. An advantage of the method is the large number of values which can be obtained over a brief test period (7 to 10 days), with a minimum of measurements. The authors list values obtained from some 25 animal and plant proteins.

15.20   Estimation of Protein Quality from Amino Acid Composition

A comparison of the quantitative distribution of the essential amino acids in a feed (Table 8.3) with the relative amounts needed by the body per unit of feed provides a method of estimating protein quality. As a composite measure of the quality of a given protein, Mitchell and Block devised a chemical rating, or score, from amino acid composition data based on the essential amino acid in greatest deficit in the protein compared with egg protein as a reference, chosen because of its top biological value. In general, there was a close correspondence between these ratios and determined biological values for growth. Later, Oser devised a similar measure based on the contribution the protein makes to all essential amino acids rather than to the one in greatest deficit. He named this measure the *essential amino acid index* (EAAI). These indices were found in general to be highly correlated with biological values. Oser[39] has described the bases of these indices and discussed their usefulness and limitations in an article which also refers to the work of Block and Mitchell. Oser points out that EAAI is useful as a tool in predicting biological value in that it permits estimates to be made for combinations of proteins or for proteins supplemented with amino acids. He notes, however, that the method is predicated on the assumption that the amino acids determined in the feed are actually available to the animal and that factors which impair the rate or degree of digestion and absorption, and thus this availability, limit the usefulness of the method for the protein sources in question. Proteins injured by heat represent a case in point.

15.21 Comparative Usefulness of Various Measures of Protein Quality

It is clear from the previous discussions that none of the methods described is free from limitations. Each has its usefulness, however, if properly carried out. The advantages and limitations of the various methods and their modifications can best be appreciated by reading the discussions of them both by Mitchell and by Barnes and Bosshardt.[40] In a review of the data on the nitrogen-balance procedure compared with those measured as the "protein efficiency ratio" in feeding trials Block and Mitchell[41] found the correlation of the results obtained by the two methods to be fairly close. Using a modification of the Oser method, Armstrong and Mitchell[34] found that values calculated for various feeds corresponded rather closely in most cases with those determined with pigs by the Thomas-Mitchell procedure.

There is no single procedure for determining protein efficiency or biological value which is best for all purposes. One method may provide more reliable and specific basic scientific information yet be less suitable than another for the solution of animal feeding problems in practice. The critical investigator can find, however, a method useful for his purpose which will yield reliable results if they are properly interpreted. Delort-Laval[42] has reviewed the criteria for biological evaluation of protein.

15.22 Comparative Protein Quality of Various Foods

On the basis of the studies with rats and the many fewer ones which have been made with pigs and chicks, certain generalizations have been made regarding the comparative biological value of individual foods and combinations. Animal products as a class are superior to foods of plant origin. Eggs stand at the top, followed closely by milk. Muscle meats, including fish, and glandular organs rank somewhat lower but above most proteins of plant origin. Most of the oil-bearing seeds and their meals have higher values than the cereal seeds. The latter do not show marked differences from one another. Milling, which removes the germ and bran, results in a lowering of the value for the resulting flours. Thus, whole wheat ranks above white flour, and the milling by-products commonly used for animal feeds above both. The lower rankings of the various feeds, compared with eggs and milk, are due to specific amino acid deficiencies. The principal deficiency in cereals and in cottonseed and linseed meals is lysine. Soybean meal and peanut meal are deficient in the sulfur-containing amino acids. A deficiency of these acids also exists in casein, with a resulting markedly lower protein quality than that shown by milk proteins as a whole. Most of these lower-ranking products mentioned have lesser deficiencies of one acid or more besides the ones mentioned.

15.23 The Supplementary Relations among Proteins

While data on the protein quality of individual proteins and foods provide important basic information, their usefulness in evaluating the combinations which occur in rations in practice is limited by the fact that, when two protein sources are combined, the resulting value is not necessarily the mean of the individual values. The explanation here is that certain proteins mutually supplement each other so that the resulting amino acid mixture has a biological value superior to that of either protein when fed alone, each liberally supplying one or more amino acids in which the other is deficient. Many studies with pigs demonstrated the favorable effects of adding animal protein sources, such as fish meal and meat or milk by-products to high cereal diets. The biological value of the combination is higher than the mean of the individual feeds. Increased growth rates and improved feed utilization confirm the higher quality of the combined protein sources.

Hegsted and coworkers[43] showed that replacing one-third of the protein in an all vegetable diet by meat protein resulted in a 10 to 15 percent increase in the biological value of the diet for rats, dogs, and humans. In the case of the chick, blood meal, a feed very deficient in isoleucine but rich in lysine, combines with gluten meal, which has a surplus of isoleucine but a deficiency of lysine, to provide a mixture having a growth-promoting value greatly exceeding that of either feed alone.

The experiments just cited also illustrate the fact that the most useful data for biological value are those obtained with specific combinations used in practice, since single feeds seldom provide the entire ration. Combinations of vegetable and animal products generally provide effective mixtures, and when seeds are supplemented with 10 to 15 percent of one of the better animal products, the combination is nearly as good as the animal source alone. In general, seeds and their products do not supplement each other, but a combination of soybeans or peanuts with cereal seeds is an exception. Wheat-germ and wheat-bran protein supplements the protein of white flour, explaining the superior growth-promoting value of whole compared to patent flour shown in Table 15.11.

The previous discussion of supplementary relationships does not apply to ruminants because the biological value of a feed or ration for these species depends in large measure on the nature and extent of the microbial action in the rumen (Sec. 8.14). It was early assumed that, since most of the feed nitrogen is converted into bacterial protein, the biological values of the protein as fed were of little significance. Many studies have shown, however, that there are real differences for certain feeds and rations, though in general less than in the case of nonruminants. Ellis and coworkers[44] have reviewed various experiments with sheep in an article which also reports their own studies. Using isolated proteins, they obtained values which ranged from 54 for gelatin to 83 for blood fibrin. As cited in Sec. 14.15, the British report on the nutritive requirements for ruminants tabulated biological values reported in the literature for a large number of feeds used in both sheep and cattle studies. From these data it was decided to use a value of 70 for cattle and 65 for sheep as general estimates in formulas for arriving at protein requirements. In addition to the kind of protein fed, there are obviously dietary factors which affect the nature of the microbial action.

15.24 Effect of Heat on Protein Quality

The heating which occurs in many commercial processes and also in home cooking can injure protein quality unless carefully controlled. For example, the following data on a menhaden fish meal subjected to high temperature in "flame drying" and on a steam-dried product were obtained at Cornell University and the processed cashew nuts were studied in Nigeria. The discarded cashew nuts were unacceptable as human food because of excess browning from overheating, but they were useful for animal feeding. In the field of human foods it was early shown that the toasting of bread and the high heating of other cereal products resulted in an impairment of protein value. Milk protein can also be injured by overheating in the manufacture of evaporated and dried milks. The high heat developed in the expeller process in removing the oil as completely as possible from oil-bearing seeds has also been found to cause injury. This has been shown to be true in the case of cottonseed meal, sunflower seed meal, soybean meal, and others. The protein of alfalfa becomes largely unavailable to animals if it is overheated and becomes browned or blackened in the stack or silo. The protein of any feed which is heat-treated in drying or processing can be injured unless the treatment is controlled. Both the temperature and the duration is involved.

	Digestibility, percent	Biological value
Flame-dried fishmeal	62	71
Steam-dried fishmeal	73	77
Cashew nuts, good[a]	84	69
Cashew nuts, discarded[a]	78	55

[a]B. Fetuga, G. Babatunde, and V. Oyenuga, Composition and nutritive value of cashew nut to the rat, *Agr. Food Chem.,* **22**:678–682, 1974.

Many studies have sought the specific cause of the injury. Certain amino acids, notably lysine, arginine, tryptophan, and histidine, are specifically involved. They are more slowly and less completely liberated in digestion. There may be some actual destruction of the amino acids, but the blocking of amino groups in such a way as to affect hydrolysis unfavorably and the combination of the acids with other chemical groups or compounds to form enzyme-resistant linkages appear to be more important causes. Much stress has been laid on the combination of the acids with reducing sugars, as occurs in the "browning reaction." Some workers have stressed the importance of the delayed digestion and absorption in lowering biological value because of the time factor in protein synthesis. Whatever may be the cause, data are available to guide feed and food manufacturers for processing their products in such a way as to avoid injury to protein quality. This is an important matter in both human and animal nutrition.

Heat also has a beneficial effect on protein quality. Nesheim and Garlich[45] reported that laying hens digested only 54 percent of the protein in raw soybeans compared with 85 percent in soybean meal. This could explain the lower energy value. Important differences have been noted among raw soybean varieties in the growth of rats, but proper heat treatment improved the value of all types.[46]

15.25 Biological Values for Different Species and Functions

Most of the data for biological value or protein efficiency have been obtained with rats. Several studies have been made with pigs and chickens, and a few with sheep and cattle. The question as to the extent to which the values obtained with one species apply to another is an important one from the standpoint of their general usefulness, particularly in view of the special suitability of the rat for determining these values. There are several experiments with fish meals indicating that the values obtained with rats tend to hold for pigs also, though this is not true for all feeds. Studies comparing the nutritive value of the protein in mixed diets for dogs, rats, pigs, and humans show striking similarities of biological values. This is explainable on the basis of the similar amino acid makeup of the tissues formed in growth. Chickens present a special case because, differing from the mammals studied, they require a dietary source of glycine. Thus, a diet could be devised, from purified nutrients at least, which would have a high biological value for the growth of mammals and yet not even maintain a chick. A relatively higher requirement for arginine also seems to be characteristic of the chick. Casein has a lower biological value for chick than for rat growth because glycine and arginine are limiting. No requirement for

TABLE 15.12. The Biological Value of Proteins for Growing Rats, Mature Rats, and Mature Humans

Protein	Growing rats	Mature rats	Mature humans
Egg albumin	97	94	91
Beef muscle	76	69	67
Wheat gluten	40	65	42
Casein	69	51	56
Peanut flour	54	46	56

histidine for maintenance in man has been clearly shown, differing from findings with rats and chicks. These various differences have been discussed in Sec. 8.8.

There are certain differences in the relative amounts of the various essential amino acids required in accordance with the function being served. This fact is illustrated for growing and mature rats in Table 15.12. These data are taken with the permission of H. H. Mitchell, from a report by Mitchell and Beadles[47] which cites the sources of data not obtained by the authors. The authors explain that wheat gluten has a much lower value for rat growth than for maintenance because of its deficiency in lysine, the requirement for which seems to be much less prominent for maintenance than for growth. Both casein and peanut flour are deficient in sulfur-containing amino acids, but their biological values indicate more intensive requirements for them for maintenance than for growth, explainable by the large need for these acids for hair growth, which is a prominent feature of protein maintenance in rats. A comparison of the data for mature rats with those for mature humans indicates that certain species differences exist. Apparently the previous explanation that wheat gluten has a relatively high efficiency for the adult rat because the lysine need is low does not hold in the case of humans.

15.26 Factors Governing Protein Needs

As commonly stated in feeding standards, the protein requirement for growth includes the amount needed for maintenance as well. The maintenance component increases with body size, but the demand per unit of new tissue formed decreases with age and body size because of the decreasing protein content of this tissue. While the total daily requirement increases with age and size, at least during early growth, it decreases per unit of weight and in relation to the energy requirement. For example, according to the N.R.C. data, a 10-kg pig requires 22 percent of protein in its feed, whereas a 60-kg pig requires only 13 percent. This relative change in the protein requirement with age means that its experimental determination should be made at different stages of the growth period. An average requirement obtained for the period as a whole would be inadequate for early growth. A figure obtained for early growth would be adequate for the entire period but wasteful during the latter part.

In view of the factor of biological value, it is evident that there can be no fixed minimum requirement except in terms of specific food sources. The fact that biological value tends to decrease with level of intake has the effect of increasing the requirement per unit of tissue gained as the growth rate is increased. The previously discussed evidence that, within rather wide limits, the wastage of energy in metabolism is decreased as the level of protein is raised suggested that the most efficient level of intake may be higher than the amount needed for its specific function as protein. All of these considerations suggest higher allowances in practice than certain minimum values which have been determined for specific combinations.

15.27   Factorial Method of Estimating Protein Requirement

The protein requirement for growth can be estimated factorially as a sum of the amount needed for maintenance plus that in the growth tissue formed, with an allowance for losses in metabolism. The net amount needed for maintenance is considered to be the endogenous nitrogen, which may be determined directly as endogenous nitrogen or calculated from the basal energy metabolism. The amount required for the tissue formed is estimated from slaughter data. The procedure is illustrated in the publication of Blaxter and Mitchell,[48] which shows the basis of the calculations and the resulting estimated requirements for fattening lambs and dairy heifers. The minimum protein requirements for the maintenance and growth of cattle and sheep, set forth for various live weights in the A.R.C. report for ruminants, were arrived at by the factorial method. For growing cattle the estimated gain in body nitrogen, as a percentage of the corresponding liveweight gain, is 2.4 percent for weights up to 600 kg; and for growing sheep, 2.5 percent up to 40 kg and then 2.4 percent to maturity. The procedure has the advantage of arriving at figures which reflect the changing needs at the various stages of the growth period in terms of animal size, rate of gain expected, and composition of the gain. In view of the assumptions involved in the various calculations, the data obtained cannot be considered, by themselves, to be reliable recommendations for practice. They are very useful, however, for consideration along with the results of more direct methods in arriving at such recommendations. They have been considered in setting up many of the N.R.C. requirements for farm animals.

15.28   Protein Requirements as Measured by Nitrogen Balance

The nitrogen-balance method provides an exact measure of the actual requirements in terms of a specific ration by determining the minimum intake which will provide maximum retention. The animals must be making the expected rate of growth during the study. The measurement should be made at a minimum of two or three times during the growth period to obtain data representing amounts which are adequate early but not wasteful later. Recommendations for practice must be set higher than the average data thus obtained to cover the needs of all individuals and to allow for the probability that the protein of some rations fed in practice will be of lower biological value than the one tested. Lassiter et al.[49] used nitrogen balances to evaluate earlier protein requirement data using diets ranging from 10 to 22 percent protein and adequate amounts of nutrients not known to be essential in earlier balance trials at Illinois. The results were in general agreement with recent feeding trial data.

15.29   Protein Requirement from Feeding Trials

In the feeding-trial method of measuring the protein requirement different levels are fed to find the minimum one which will give a maximum rate of growth.

The inclusion of slaughter tests to show the nature of the increase made provides valuable additional data which are obtainable in the case of meat animals. Many early feeding trials defined the protein requirements for various domestic animals using the different feed combinations available in specific regions. As new feeds, supplements, and management systems became available the protein requirements were reevaluated. The use of equalized feeding by pairs in arriving at the protein requirement is illustrated by the study of Becker and associates[50] with pigs.

Lofgreen and associates[51] have combined growth and nitrogen-balance studies in an investigation of the protein needs of dairy calves, in which they compared the results with those obtained by the factorial method. According to their data, the factorial method as carried out by Blaxter and Mitchell underestimates the needs of heifers during the latter part of the growth period.

While feeding trials have shown, as is to be expected from the nature of the tissue formed, that the protein needs decrease relative to energy needs as growth proceeds, few critical studies have been made of the differential quantitative requirements at different stages. Most of the experiments, especially those with cattle and sheep, have compared different percentages of the total ration or different intakes per unit of live weight for the growth period as a whole. Whatever level is selected as adequate on one of these bases, it automatically results in an increasing daily intake as weight and total feed consumption increases. More recently when supplies of vitamins, trace mineral elements, amino acids, antibiotics, and other growth-stimulating additives became economically available feeding trials were used to review the protein requirements along with determining the value of the supplements.

The N.R.C. protein requirements for growth of young animals shown in Table 15.13 in comparison with the values for maintenance, reproduction, and lactation define the requirements more accurately at different phases of growth than earlier standards. There are large differences among species of animals. The protein level required for early growth is much higher than for maintenance

TABLE 15.13. The Protein Requirement for Growth, Maintenance, Reproduction, and Lactation (percent of dry diet)

Animal	Growth		Maintenance	Reproduction	Lactation
	Early	Late			
Beef cattle	15	9	6	6	9
Dairy cattle	16	10	8	9	12
Horse	19	11	10	15	13.5
Pig	22	14	14	14	15
Sheep	10	9	8.9	9.3	11.5
Chicken	20	16	—	15	—
Rabbit	16	—	12	15	17

Source: N.R.C. Nutrient Requirements of Domestic Animals.

or reproduction and also higher than for lactation in most species. Late in the growth period the protein needed declines to the maintenance level.

15.30   Wool Production

Since the wool fiber is practically pure protein, a substantial amount of food protein is required for its production.

There are two aspects of wool growth which distinguish it from the growth of muscle. In the first place, despite a negative nitrogen and energy balance, wool growth continues at the expense of the breakdown of other protein tissues (Sec. 14.13). Secondly, wool protein has an amino acid distribution quite different from that found in muscle, calling physiologically for a different pattern for its formation. Of special note, wool protein contains, on a percentage basis, over 10 times as much cystine as does muscle protein, a difference which is only partially balanced by the higher content of methionine in muscle.

Several early studies were made to ascertain whether the feeding of cystine or sulfur as a supplement to the usual rations would improve wool production, with generally negative results. These results are explainable on the basis that microorganisms in the rumen can utilize inorganic sulfur and nonspecific sources of nitrogen to form these essential amino acids. In the studies by Thomas and coworkers[52] in which a ration lacking in the sulfur-containing amino acids and inorganic sulfur as well prevented the synthesis of these amino acids in the rumen of lambs, wool growth was interfered with also. In rats the addition of methionine to a diet deficient in the sulfur-containing amino acids produces more hair and hair richer in cystine.

Many recent studies have been concerned with the effects of nutrition on wool growth, which is clearly influenced by net-energy intake and the pattern of amino acids absorbed. Wool growth is stimulated by the intestinal absorption of casein or other high quality protein in nonbreeding adult sheep. Ingested soluble protein is largely degraded in the rumen and to be effective it must be infused into the abomasum or treated to reduce solubility to allow it to bypass the rumen. Methionine and cystine supplements also increased wool growth in some tests but not in others. Black, Faichney, and Graham[53] have reviewed the research and the problems which must be resolved before supplementation can become practical. Nimrick et al.[54] reported that methionine, lysine, and threonine were the limiting amino acids for growing lambs receiving urea nitrogen; however, the effects on wool growth were not measured.

THE MINERAL ELEMENTS

The general functions of the mineral elements required in the body have been discussed in Chapter 10. As the result of many studies fairly reliable data are available on the quantitative requirements for growth of those minerals which particularly need attention to insure that the ration will contain an adequate supply. Although some 20 mineral elements are now considered to be dietary essentials only about half of those need to be considered in evaluating feeds for domestic animals.

15.31   Factors which Influence Calcium and Phosphorus Requirements

Over 70 percent of the mineral matter of the body consists of calcium and phosphorus. Almost 99 percent of the calcium and 80 percent of the phosphorus are present in the bones and teeth. A deficiency of either element markedly affects bone development. Vitamin D is critical for calcium and phosphorus utilization in bone formation and needs to be considered along with these minerals for animals that are housed in buildings continuously. However, a dietary source of vitamin D is not necessary for any animal that is exposed to sunlight because the ultraviolet irradiation produces the vitamin in the skin (Sec. 11.15). Certain other minerals, such as copper, manganese, and fluorine, are also important in bone growth.

15.32   Measures of Skeletal Growth

The development of the skeleton cannot be measured by increase in weight, nor can its adequacy be determined by dimensional measures of the body or even of the bones themselves. Size of bone is governed largely by inheritance, and a large bone may be a very weak one if the nutrition has not been adequate. Severe nutritive deficiencies during growth manifest themselves in misshapen bones as in severe rickets, but mild deficiencies may have serious consequences without any evident early symptoms. There may be later deformities or fractures or a breakdown of the teeth, as a result of prolonged periods of mild deficiency. Even though these evident symptoms never occur, the bone development may still be inadequate, particularly as regards its content of calcium and phosphorus reserves which are normally called upon during reproduction and especially during lactation (Sec. 17.29).

The real measure of the adequacy of skeletal development is the density and strength of the bones formed as conditioned by their content of calcium and phosphorus and their histological structure. Thus the requirements for the bone-forming nutrients can be determined by slaughter experiments in which representative bones are analyzed for calcium and phosphorus or studied histologically. The measurement of density and hardness and the determination of breaking strength are useful supplementary measures. Since the ash of bone consists almost entirely of calcium and phosphorus and since this remains true no matter what the quality of the bones, the determination of the ash is more commonly used as the measure of the adequacy of bone nutrition than the more time-consuming analyses for calcium and phosphorus. Radioactive calcium and phosphorus have been found highly useful in studying bone development by an autoradiographic technique illustrated by the report of Tomlin and coworkers.[55]

The progress of bone development can be followed quantitatively in the living animal by calcium and phosphorus balances. The blood serum levels of the elements and of alkaline phosphatase and radiographs are useful supplementary measures. The application of the radiographic technique to the living animal is illustrated by the studies of Benzie and coworkers[56] with lambs.

15.33 Optimum Bone Development

In the discussion of bone growth in Chap. 10, it was pointed out that calcium and phosphorus are deposited in the bone as reserve material as well as constituents of the structural portion itself. Presumably the building of the latter has first call on the bone-forming nutrients, but where the intakes are large enough, deposition in the trabeculae doubtless occurs also. In the case of animals grown for slaughter, the state of the reserves would seem to be of minor importance, but other considerations enter for those being reared for breeding and milk production.

Early studies in Mendel's laboratory showed that an increased rate of body growth, caused by a more complete diet was accompanied by an increased rate of calcification compared with that in the slower-growing animals. When the latter reached the same mature body weight, however, there were no differences between the two groups as regards the ash content of the bones. Thus it is suggested that, if bone development keeps pace with body growth, the end result is satisfactory, even though the rate of calcification is not at its maximum. Others considered that the diet that causes the maximum rate of retention should be considered optimum until the normal store at maturity, including reserves, is attained. There is special reason for believing that this is true in the case of breeding females because of the desirability of providing them with liberal reserves before they are subjected to gestation and lactation. This must be accomplished by a diet which will cause a rapid rate of calcification throughout the growth period.

While, within rather wide limits, increasing the bone-building nutrients in the ration results in an increased storage in the bones, the percentage retention falls off markedly at the higher intake levels. This suggests that as the limit of the capacity for the deposition of reserves is approached, the process becomes increasingly less efficient. If a maximum rate of calcification is considered to be optimum, a high intake in proportion to the amount stored must be supplied. There is no present reason to believe that calcification can be overdone during growth, except by massive doses of vitamin D, nor is there evidence that excretory or other functions are unduly burdened by the ingestion of amounts of calcium and phosphorus which are large compared to the amounts stored. Duckworth and Hill[57] have presented an excellent review of the knowledge regarding the storage and mobilization of calcium, phosphorus, and other minerals in bone.

15.34 Calcium and Phosphorus Requirements Measured by Balance Studies

Balance determinations at different levels of intake ascertain the minimum level which will provide maximum retention. For a study of one of the minerals, the ration must have an adequate supply of the other and vitamin D must be provided. This method is the most reliable one which can be employed with living animals and thus has found its largest use in human nutrition studies. An example of the use of the procedure for studying the phosphorus requirements of lambs is provided by the comprehensive experiment of Preston and Pfander,[58] which included data on growth, digestibility, plasma inorganic phosphorus level, and bone development. The balance data are presented in Table 15.14. The data were analyzed to express phosphorus requirement by the equation:

$$P = 0.0194W(1 + 0.0171G)$$

where P is the daily phosphorus requirement in grams, W is body weight in kilograms, and G is daily gain in grams. Applying this equation to a 35-kg lamb gaining 180 g per day the requirement became 2.78 g per day. Plasma inorganic P levels decreased with time on the two lower levels, and bone data indicated changes typical of phosphorus deficiency.

15.35 Blood Data as Measures of Calcium and Phosphorus Needs

Inadequate bone nutrition is reflected in a lowering of the serum inorganic phosphorus and sometimes of calcium. The blood picture varies according to the specific deficiency concerned and also according to the species. A rapid rate of body growth which is accompanied by normal levels of calcium and phosphorus in the blood is highly indicative of adequate skeletal development, and these measures are frequently employed to determine the actual requirements for minerals.

Huffman and associates[59] have used this procedure for studying the phosphorus requirements of dairy cows, resulting in the following conclusions as regards growth requirements. A ration containing 0.2 percent of phosphorus caused a lowering of the blood phosphorus of calves, which persisted up to eighteen months of age. Intakes of 5.7 to 9.9 g of phosphorus per day were inadequate where the calcium-phosphorus ratio was 4:1 or wider. An intake of 10.3 g daily sufficed from 3 to 6 months of age. From 18 months to first calving 10 to 12 g daily were adequate. Similar studies by these Michigan workers have led to the conclusion that an intake of 6 to 12 g of calcium daily from birth to 2

TABLE 15.14. Phosphorus Balance in Lambs Fed Varying Phosphorus Intakes

	Phosphorus in ration, %		
	0.12	0.15	0.29
Feed intake, g/day	516	566	828
P intake, g/day	0.620	0.837	2.419
P balance, g/day	0.210	0.309	0.750

years of age is sufficient for the growth of calves. Beeson and associates[60] compared different levels of phosphorus intake for fattening lambs in terms of growth, efficiency of feed utilization, and blood-phosphorus level. They concluded that the need should be set at 5.3 g per 100 kg of live weight and that the ration, on a dry basis, should contain 0.17 percent or more of the element.

Serum alkaline phosphatase activity, which rises when bone development is inadequate, is a useful supplementary blood measure.

15.36 Calcium and Phosphorus Requirements from Growth and Bone Data

Aubel, Hughes, and Lienhardt compared rations containing various levels of phosphorus for pigs receiving ample D, using growth, blood, bone analyses, and breaking strength as measures. The data obtained for bone composition from one period of their first experiment are shown in Table 15.15.

It is noted that the calcium was held constant in all lots at a level which was certainly adequate while the phosphorus was varied. Such a scheme unavoidably involves a variation in the ratio between the two elements. The ratio of 4.4:1 existing for the ration fed Lot I may have been in part responsible for the poorer bone development, but the ample amount of vitamin D supplied in each ration would tend to overcome any such effect. The data for ash, calcium, and phosphorus clearly reveal the superiority of the intermediate level over the lower one. While the data for the highest level are still better, the differences are small. For the experiment as a whole, the bone analyses and other measures employed failed to show significant differences in favor of the highest phosphorus level. Aubel, Hughes, and Peterson[61] have reported a similarly conducted study of the calcium requirement of pigs. They found that a level of 0.25 percent in the ration was insufficient but that a level of 0.41 percent was definitely adequate.

More recently Pond, Walker, and Kirtland[62] reported that the performance of growing-finishing pigs was similar on diets containing 1.2 or 0.5 percent of calcium and 1.0 or 0.4 percent of phosphorus. The higher levels did not interfere with absorption and storage of magnesium, zinc, copper, manganese, or cobalt.

The larger the number of measures used in a given study, the better the

TABLE 15.15. Effect of Phosphorus Intake on Bone Composition of Pigs

Constituent, %	Lot I	Lot II	Lot III
Calcium in feed	0.77	0.78	0.77
Phosphorus in feed	0.18	0.33	0.59
Ash in femur and humerus	48.14	57.35	59.64
Calcium in femur and humerus	18.35	21.93	22.70
Phosphorus in femur and humerus	8.69	10.58	10.82

Source: C. E. Aubel, J. S. Hughes, and H. F. Lienhardt, Phosphorus requirements in the rations of growing pigs, *Kansas Agr. Expt. Sta. Tech. Bull*. 41, 1936.

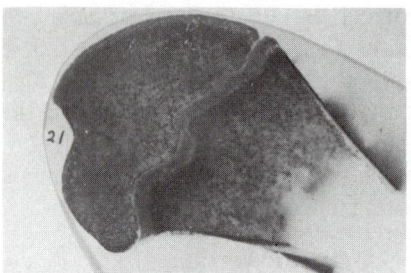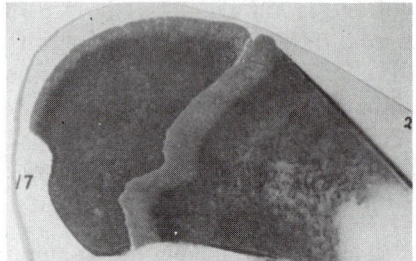

Fig. 15.4 Autoradiograms of calf femurs. The bone on the left is from a phosphorus-deficient calf. The bone on the right is from a calf fed adequate phosphorus. Note the difference in growth of new bone at the epiphyseal plate. (*Courtesy S. E. Smith, Cornell University*).

basis for arriving at requirements. Here the experiments of Wise and coworkers[63] are of special interest because they measured growth, feed efficiency, serum inorganic phosphorus and phosphatase activity, bone ash, and bone growth by autoradiography. The last measure is illustrated in Fig. 15.4. The combined measures led to the conclusion that the minimum phosphorus requirement for calves 12 to 18 weeks of age and weighing 90 to 127 kg is 0.22 percent, a figure which should be increased to 0.30 percent in practice in view of variations in animals and feedstuffs.

15.37 Estimation of Calcium and Phosphorus Requirements by the Factorial Method

Mitchell[64] has estimated the calcium and phosphorus requirements on the basis of calculated maintenance values, storage during growth as shown by carcass analyses and retention data. The factorial method provides data for the different ages and weights throughout the growth period. The calculations are necessarily based on several assumptions. By contrast, the methods previously described provide determined values for specific periods during growth with only a limited basis for translating them into values for other periods.

15.38 Calcium and Phosphorus Requirements for Practice

As is indicated by all studies of body composition, skeletal development is at a maximum early in the growth period. The requirements of both calcium and phosphorus decrease with age per unit of body weight and also per unit of dry-matter intake, but the extent of this decrease varies for the different species. The calcium requirements exceed those for phosphorus at the start, but differences become much less or nil as maturity is approached. The serious effects of phosphorus deficiency on bone growth of pigs is illustrated in Fig. 15.5.

Requirements for practice should provide for maximum skeletal development in the species in question during the period of most rapid bone growth. Lesser intakes will suffice later. The N.R.C. nutrient requirements set forth the amounts needed at the various stages of growth and production. The

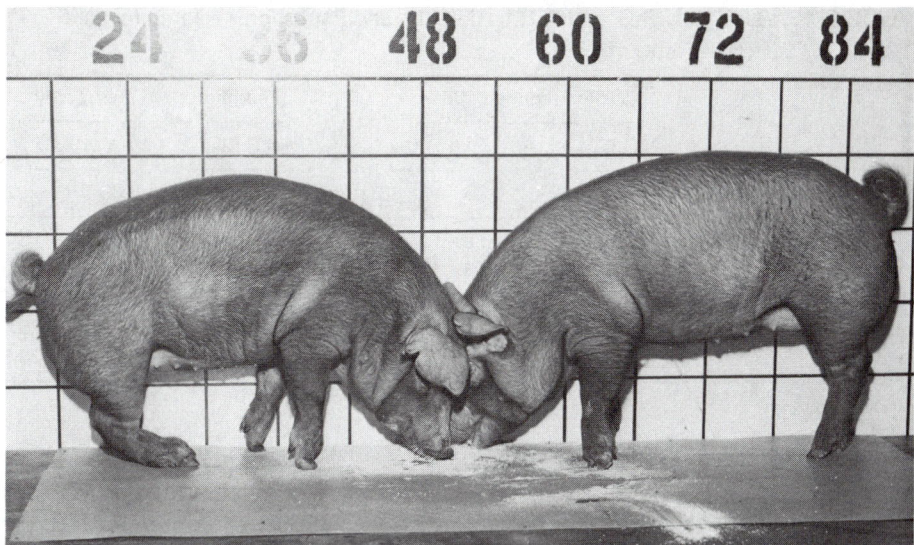

Fig. 15.5 Phosphorus deficiency. The pig on the left received a phosphorus deficient ration. Note the weak and crooked leg bones, in contrast to the condition of the pig on the right which received the same ration adequate in available phosphorus. (*Courtesy of M. P. Plumlee and W. M. Beeson, Purdue University.*)

N.R.C. requirements expressed as concentrations in the dry feed are shown in Appendix Tables I through IX. Requirements expressed as percentage of the feed intake or as grams per day can be calculated into the other measure by taking into consideration the intake of the dry feed for different weights. In practice it is not feasible to follow exactly recommendations which change several times. If one chooses a single figure, it should not be the average for the period as a whole but should be high enough to meet the needs when the maximum rate of bone growth normally occurs, even though this would be a waste later.

The daily calcium and phosphorus requirements of beef cattle at various stages of growth and making different rates of gain, as set forth by the N.R.C. and A.R.C. reports are shown in Table 15.16. It is clear that young cattle need two or three times more calcium and phosphorus for maximum growth than for a slow or moderate rate. As they mature and reach heavier body weights the difference is much less. The A.R.C. values are slightly lower than the N.R.C. values for the small cattle and appreciably higher at 400-kg body weight and above. These differences are most likely due to differences in interpretation of the published data by the committee rather than representing actual differences in requirements of the cattle. It is recognized that it is difficult to detect marginal deficiencies in these bone-forming elements over short periods of time and there is no great danger from feeding excesses of two or three times the requirements if the calcium-phosphorus ratio is suitable and solar radiation or vitamin D is provided.

TABLE 15.16. The N.R.C. and A.R.C. Calcium and Phosphorus Requirements of Cattle for Different Rates of Gain

Body weight, kg	Calcium required, g/day				Phosphorus required, g/day			
	N.R.C.		A.R.C.		N.R.C.		A.R.C.	
	0.3*	1.1*	0.33	1.0	0.3	1.1	0.33	1.0
100	14	29	10.8	27.0	11	19	5.4	12.7
200	10	25	13.8	29.7	10	19	8.1	15.2
300	13	23	17.5	33.2	13	20	12.8	19.7
400	16	19	21.9	37.3	16	19	22.0	28.6
450	16	19	21.9	37.3	16	19	22.0	28.6

*Daily gain, kg.

Nursing young never suffer from a deficiency of calcium or phosphorus, and neither does the dairy calf reared on a liberal supply of skim milk. Herbivorous animals will receive ample calcium if their roughage is one-half or more legume hay, but where grass hay is the sole roughage and particularly where it is not consumed liberally, a calcium supplement should be included unless the hay is known to contain 0.5 percent or more of this element. No concentrate mixture is rich in calcium, but a liberally fed mixture which contains 25 percent of more of some phosphorus-rich feed, such as wheat bran or one of the oil meals, will take care of the needs for the latter element unless the roughage is unusually low in it. Usually roughage alone will not suffice. Under most conditions it is more economical to make up the deficiency in the ration by adding a mineral source.

Whether or not herbivorous animals on pasture will receive enough bone-forming minerals depends upon the nature of the soil and the resulting calcium and phosphorus content of the forage. In the absence of specific information in these respects, giving the grazing animals access to a calcium and phosphorus source is a desirable procedure. These various provisions for ensuring adequate minerals in the rations of herbivora require more attention in the case of calves than in the case of lambs and colts because of the lower requirements of the latter. For a given species, the provisions demand less attention with advancing age because of the decreasing requirements.

If pigs are being fed corn or other cereal grain with the additional protein required furnished as tankage or fish meal, they will receive sufficient calcium and phosphorus. Where a vegetable-protein concentrate is used instead, additional calcium is required, and it can be supplied as a mineral supplement. Where a milk by-product is the protein concentrate, it must be liberally fed to meet the calcium requirements.

15.39   Other Minerals

The requirements of the other minerals for growth expressed as concentration in the dry diet are summarized in Table 15.17 for cattle, horses, pigs, sheep, and

TABLE 15.17. Other Mineral Requirements for Growth (amount per kilogram dry diet)

Mineral	Cattle		Horse	Pig	Sheep	Rat
	Beef	Dairy				
Magnesium, %	0.10	0.16	0.04–0.09	0.04	0.08	0.04
Sodium, %	0.06	0.1	0.35	0.1	0.1	0.06
Potassium, %	0.6–0.8	0.8	0.5	0.26	0.5	0.23
Chlorine, %	–	–	–	0.13	–	0.06
Manganese, mg/kg	10	40	40	20	40	5.6
Zinc, mg/kg	20–30	40	–	50	32	13
Iron, mg/kg	10	50	50	80	50	39.0
Copper, mg/kg	4	10	9	6	5	5.6
Cobalt, mg/kg	0.1	0.1	–	–	0.1	–
Iodine, mg/kg	0.08	0.1	0.1	0.2	0.1	0.17
Selenium, mg/kg	0.1	0.1	0.1	0.1	0.1	0.04

Source: Selected data from the N.R.C. Nutrient Requirements of Domestic Animals.

rats. More complete information is given in Appendix Tables I through IX. Allaway[65] has reviewed the interrelationships between soils, fertilizers, and the mineral composition of plants and outlined the regions in the United States where mineral deficiencies are most common. Of the nutritionally critical minerals listed in the table, potassium and iron are least apt to be deficient in the diets of farm animals.

Potassium is required for plants to grow. Thus if a plant grows at all it will nearly always contain sufficient potassium to meet the requirements of animals and people that eat the plants. Most fertilizers contain potassium to increase crop yields rather than to enhance the mineral content. Potassium deficiency has been produced on purified diets and it may occur in practice owing to metabolic upsets or illnesses that interfere with potassium utilization or cause excess losses from the body. Chickens need 0.20 percent potassium in the diet, pigs 0.26 percent, and rats[66] 0.23 percent. The higher values shown for cattle, horses, and sheep represent specie differences.

Sodium chloride. The earliest records list *common salt* as an important food item for humans and their animals. However, specific data regarding its roles in the body (Sec. 10.22) and the amounts needed for various animals has been available only recently. The most sensitive way of detecting a sodium deficiency is to measure the amount in the urine and saliva. Any excess sodium chloride consumed is rapidly excreted in the urine but on a salt-free diet the values fall to nearly zero. In salt deficiency sodium in the saliva declines and potassium increases. Murphy and Connell[67] reported a simple method of collecting saliva from cattle and sheep.

Chlorine has been known to be essential for HCl in the gastric juice for at least two centuries, but a dietary deficiency has been demonstrated only in laboratory animals. Sodium is clearly the most critical element in common salt.

The N.R.C. sodium requirements for growth (Table 15.17) vary from 0.06 percent of the dry diet for beef cattle and rats to 0.35 percent for horses. The studies of Morris and Gartner[68] and Hegstern and Perry[69] illustrate recent studies to define the requirements of cattle and sheep. The salt content of feeds and particularly of water vary greatly (Sec. 4.8) and thus the need for supplemental salt is not a universal one for animals. The requirement may be appreciably higher in hot, dry regions because of excess losses in perspiration. Providing salt as a free choice or adding 0.5 percent to the concentrate mixture are the methods most commonly used to insure adequate sodium intakes.

Magnesium is required for normal growth of all animals. Most forages and other common feeds contain sufficient magnesium to prevent a deficiency except during lactation when the requirements are highest and its utilization may be reduced (Sec. 17.32). Milk is an exception. Calves limited to milk as the only food, with no access to forages or concentrates, will suffer from magnesium deficiency after several months. The N.R.C. requirements (Table 15.17) for growth vary from 0.04 to 0.10 percent of the dry matter.

Sulfur is a dietary essential because of its role in the sulfur-containing amino acids and the vitamins thiamin and biotin. Inorganic sulfur can be utilized by rumen bacteria to synthesize methionine, cystine, thiamin, and biotin. When NPN provides an important part of the nitrogen for ruminants sulfur should be provided at a nitrogen-sulfur ratio of 1:10. A number of studies have compared the availability of different forms of sulfur for ruminants. For example, using lambs, Kahlon, Meiske and Goodrich[70] showed there was no difference in growth or nitrogen retention when calcium sulfate, sodium sulfate, DL-methionine, or the hydroxy analog supplied 0.13 to 0.15 percent of sulfur. Wool growth and sulfur retention was slightly higher on methionine hydroxy analog.

15.40 The Trace Mineral Elements

The *iron* requirement is measured as the amount needed to maintain a normal hemoglobin level and to provide an appropriate positive balance. The N.R.C. requirements for growth (Table 15.17) show 10 mg per kilogram of feed for beef calves versus 100 mg for dairy calves. It is clear that the values for dairy cattle includes a margin of safety above the minimum. Other values are 80 mg per kilogram of dry diet for pigs and chickens, 50 mg for horses and sheep, and 30 mg for rats. Milk is always low in iron (3 mg per kilogram) and animals restricted to milk only will develop iron-deficiency anemia unless extra iron is provided. The time on milk alone required for anemia to appear depends upon the level of liver reserves at birth, which varies considerably among animal species.

The blood hemoglobin levels of calves decline during the milk feeding or suckling period. While iron supplements will prevent the decline and milk replacers usually contain iron, supplements are not necessary in practice because calves begin to eat other feeds soon enough to prevent a serious enough deficiency to retard growth or injure health. Webster et al.[71] reported that weight gains and energy utilization was reduced when hemoglobin levels of calves fell below 7 g per 100 ml of blood. Iron levels of 20 mg per kilogram of dry diet resulted in only 10 g of hemoglobin per 100 ml compared with 11.9 g on a diet with 40 mg of iron. Matrone et al.[72] had earlier found that about 30 mg of iron per kilogram of feed was adequate for calves. Hibbs et al.[73] demonstrated that iron could either be fed or iron-dextran injected with good results. Ammerman et al.[74] used Fe^{59} to show that ferric oxide is much less available for calves and sheep than ferrous sulfate, ferrous carbonate, or ferric chloride. The iron in deflourinated phosphate is less available than ferrous sulfate, but at a level of 125 mg per kilogram it will prevent anemia.[75]

A special case of iron deficiency is that represented by the anemia which frequently occurs in suckling pigs. Owing to the labored breathing which is always characteristic of severe cases, this trouble was known as *thumps* long before it was discovered to be due to a lack of sufficient iron for blood formation. The trouble occurs in litters kept inside without access to soil or forage. Anemic pigs are listless and flabby, their skin becomes wrinkled, and their coats have an unhealthy appearance. As the disease progresses, the skin and mucous membranes become pale and animals become thin and weak. In advanced stages, the breathing is labored and the pigs may have a swollen appearance, especially around the head and shoulders. This anemia can be prevented or cured in its early stages by drenching the sucklings with a saturated solution of ferrous sulfate or other soluble iron salt. The weekly dosage is 1/3 teaspoonful for pigs under 1 week of age up to 1 teaspoonful at 4 weeks. A simpler and entirely effective procedure consists of the injection of iron. An injection of 100 mg of iron-dextran at 3 days of age and repeated at 21 days will maintain normal hemoglobin levels.

This nutritional anemia in pigs and the recognized deficiency of iron in the milk of all species should not lead to the suggestion that all mammals should have supplements of this mineral during the suckling period. While this anemia has been produced experimentally in calves and lambs, it does not occur in practice because these species begin to supplement their milk diet with other foods relatively earlier than does the pig. The store of iron with which they are born suffices until their needs are met adequately by grain and hay.

Copper is required, along with iron, for hemoglobin formation. It also plays many other important roles in the body (Sec. 10.32). Copper deficiency in cattle and sheep has caused serious losses in several countries. Adding copper sulfate to mineral mixtures or concentrate feeds is effective in preventing the losses. The N.R.C. requirements (Table 15.17) vary from 4 to 10 mg per kilogram of dry diet. Chickens need 4 mg and rabbits 3 mg per kilogram of diet for normal growth. While 6 ppm is the minimum amount required for growth of pigs, up to 250 ppm stimulates growth[76] similar to antibiotic supplements. Higher levels are toxic.

Iodine is a constituent of thyroxine, the thyroid hormone which controls metabolic rate (Sec. 10.39). Some regions of the United States are deficient in iodine and supplements are necessary to prevent goiter in animals and people alike. The same is true in many areas of the world. This deficiency disease primarily causes losses of newborn animals, but in humans the periods of adolescence and pregnancy are the most critical, showing up as enlargement of the thyroid, common goiter. Iodine is usually supplied to farm animals along with common salt, which is fortified to contain 0.007 percent of iodine. Iodized salt is available for human consumption. Cromwell et al.[77] reported the growing pig required 0.09 to 0.13 mg of iodine per kilogram of dry feed for normal growth. The N.R.C. iodine requirements for growth of different animals (Table 15.17) varies from 0.08 mg for beef cattle to 0.20 mg per kilogram of feed for pigs. The requirement for the rat is 0.17 mg and for chickens 0.35 mg per kilogram of feed. Soybean meal and some other feeds are goitrogenic and more iodine is needed if they are included in the diet.

Manganese deficiency has been produced on purified diets. Rojas et al.[78] reported that all calves from cows fed a low-manganese diet of natural feeds with less than 17 ppm in the dry matter had enlarged joints, twisted legs, stiffness, and a general physical weakness, while all calves were normal on a level of 25 ppm. They suggested that the requirement is about 20 ppm. Most forages and cereals contain at least that amount of the mineral element. The manganese requirement of different species of animals varies from 5.6 to 40 mg per kilogram of feed (Table 15.17). Chickens require 55 mg per kilogram and deficiencies have occurred on unsupplemented natural feeds.

Cobalt deficiency is an area problem which occurs in ruminants, but not in horses or other nonruminants on the same pasture. This is because cobalt is needed for rumen bacteria to synthesize the vitamin B_{12} necessary in metabolism. In contrast, nonruminants have a dietary requirement for vitamin B_{12} instead of cobalt. The cobalt requirement for growth is about 0.1 ppm (Table 15.17) although some reports indicate that 0.07 may be sufficent for cattle. The deficiency causes reduced appetite, weight loss, emaciation and eventual death (Sec. 10.34). Adding cobalt to the daily feed or to a mineral mixture offered free choice, or placing a cobalt bullet in the rumen are effective preventatives.

Zinc is a critical nutrient for growth of animals and to prevent dermatosis. The N.R.C. requirements vary from 20 for beef cattle to 50 mg per kilogram of feed for pigs and chicks (Table 15.17). High calcium intakes increase the zinc requirement. Grazing cattle have been shown to develop typical signs of zinc deficiency in several countries and supplements have corrected the problems as well as increasing the growth rate.

Selenium has been shown to be an essential mineral element for all animals studied to date. Its functions have been reviewed in Chap. 10. The requirement of selenium is approximately 0.1 mg per kilogram of dry feed for most animals (Table 15.17). Turkeys need twice as high a concentration and rats only half as much (0.04 mg). Ammerman and Miller[79] have reviewed the role and metabolism of selenium and its interrelationships with other nutrients. Selenium appears to be more critical for reproduction in birds and some mammals than vitamin E, and especially for very early growth of the newborn animal to prevent muscle dystrophy under grazing conditions or on all forage diets (Sec.

10.48). Heavy pellets or bullets for placement in the rumen have proved effective in preventing selenium deficiency. Adding 30 ppm of selenium to salt for feeding ad libitum has been shown to be safe for sheep and cattle.[80] Julien et al.[81] found in a field study that injecting 40 mg of selenium and 680 mg of vitamin E 20 days prepartum reduced the incidence of retained placenta in dairy cows from 51 percent in controls to 8 percent.

Schwarz and associates[82] have demonstrated that *chromium, tin, vanadium, fluorine,* and *silicon* are essential trace elements for normal growth of rats. It must be supposed that they are also dietary essentials for other animals. It appears unlikely that a deficiency of any of these elements might limit the growth or production of farm animals; however, future research may well demonstrate the need to add some of these elements to practical diets. Recent data from Finland[83] support the view that silicon in drinking water may be helpful in reducing atherosclerosis in humans.

VITAMIN REQUIREMENTS FOR GROWTH

Much has been learned in recent years about the amounts of the various vitamins needed for growth of animals, but there are still many instances in which the data are less precise than desirable. Values are not yet available for some phases of the life cycle of some animal species. In most cases adequate diets can be formulated[84] to promote maximum growth and production, but excesses of some nutrients, especially vitamins may be added to insure that a deficiency will not occur. More is known about the vitamin requirements for growth than for other phases of the life cycle, because deficiencies are more frequent in rapidly growing animals and it is easier and less expensive to measure requirements over short periods of growth than for longer periods involving maintenance, reproduction, or lactation.

15.41   Vitamin A

All higher animals which have been studied require vitamin A for growth. The N.R.C. vitamin A requirements for growth of selected animals are shown in Table 15.18. The values are expressed as international units (I.U.) per kilogram of body weight and per kilogram of dry feed. The N.R.C. reports also tabulate the amounts per animal per day. The values suggest that there are wide differences in the amounts different animal species need, which in fact is not the case. The differences primarily represent varying margins of safety reflecting the interpretations of the published data by the N.R.C. committees. The early work of Guilbert et al.[85] showed that the minimum vitamin A requirement for growth and freedom from night blindness is proportional to body weight rather than metabolic size and that it is similar for all mammalian species studied. Their value for the minimum requirement was 56 I.U. per kilogram of body weight per day. The values for growth of dairy cattle, horses, and sheep (Table 15.18) are somewhat lower and those for swine and the rat are higher than the early minimum. Eaton et al.[86] found that young dairy calves require 29 μg (97

TABLE 15.18. Vitamin A Requirements for Growth, Reproduction, and Lactation

Species	Growth	Reproduction	Lactation	Growth	Reproduction	Lactation
	Daily I.U./kg body weight			I.U./kg dry diet		
Beef cattle	60	50	75	2200	2800	3900
Dairy cattle	40	75	—	2200	3200	3200
Horses	40	50	50	2000	3400	2800
Swine	100	90	80	1500	4100	3300
Sheep	40	85	85	1000	2500	2000
Human	70	85	100	—	—	—
Rat	100	1200	1200	2000	12000	12000

Source: Calculated from the N.R.C. Nutrient Requirements of Domestic Animals.

I.U.) per kilogram of body weight per day to prevent elevated cerebrospinal fluid pressure, the most sensitive sign of vitamin A deficiency. The vitamin requirements are most conveniently expressed as the amount per unit of feed. Direct comparisons can then be made between the animal's requirement and the content in the feed so that appropriate supplements can be added. The carotene content of feeds can be related to the needs of grazing animals (Table 11.1) on the basis that 0.6 μg of β-carotene is equivalent to 1.0 I.U. of vitamin A. Such a comparison shows that fresh growing forages or bright green cured forages will satisfy the needs of ruminant animals, while bleached roughages, cereal grains, and by-product feeds are deficient. Other animals will generally need vitamin A supplements unless special care is taken to select foods that are rich in the vitamin. The requirements of other domestic animals are shown in Table 15.19 and of laboratory animals in Table 15.20, along with the other vitamins. Fig. 15.6 illustrates abnormal bone growth in vitamin A deficiency.

TABLE 15.19. Vitamin Requirements for Growth (amount per kilogram of diet)

Vitamin	Pig	Chicken	Turkey	Rabbit	Dog	Fish	Rat
Vitamin A, I.U.	2200–1300	1500	4000	580	5000	2000	4000
Vitamin D, I.U.	220–125	200	900	—	500	—	1000
Vitamin E, I.U.	11	10	10	40	48	30	30
Vitamin K, mg	2.2	0.53	0.7	—	1.4	—	0.05
Thiamin, mg	1.3–1.1	1.8	2.0	—	0.73	10	4.0
Riboflavin, mg	3.0–2.2	3.6	3.6	—	2.2	20	3.0
Pantothenic acid, mg	13–11	10	11	—	2.2	40	20
Niacin, mg	22–10	27–11	70	180	10.6	150	6.0
Pyridoxine, mg	1.5–1.1	3	4	39	1.0	10	6.0
Biotin, mg	—	0.09	0.3	—	0.1	1.0	—
Choline, g	1.10–0.90	1.30	1.90	1.20	1.20	3.0	1.0
Folicin, mg	—	0.55	0.9	—	0.18	5	—
Vitamin B_{12}, μg	22–11	9.0	3.0	—	20.0	20.0	50

Source: Selected from N.R.C. Nutrient Requirements of Domestic Animals.

TABLE 15.20. Vitamin Requirements for Growth of Laboratory Animals (amount per kilogram of dry diet)

Vitamin	Guinea Pig	Mouse	Mink	Cat	Hamster	Monkey
Vitamin A, I.U.	2160	556	3500	27700	20000	25500
Vitamin D, I.U.	0	167	—	1110	0	1590
Vitamin E, I.U.	67	22	25	150	28	—
Vitamin K, mg	2.2	—	—	—	—	0.07
Thiamin, mg	2.2	3.17	1.2	4.4	6.7	1.9
Riboflavin, mg	18	4.4	1.5	4.4	6.7	1.9
Pantothenic acid, mg	22	9.4	6.0	5.5	11	—
Niacin, mg	22	11.1	20	44	—	127
Pyridoxine, mg	3.3	1.1	1.1	2.2	6.7	32
Biotin, mg	0	—	—	—	—	0.6
Choline, g	1.7	1.3	—	3.3	—	—
Folicin, mg	6.7	—	0.5	—	—	2.6
Vitamin B_{12}, μg	—	5.6	—	—	11.0	4400

Source: N.R.C. Nutrient Requirements of Laboratory Animals.

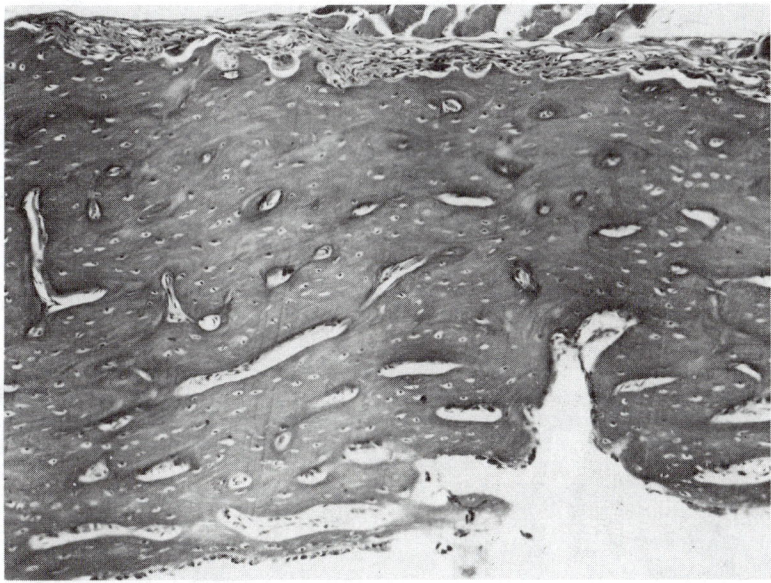

Fig. 15.6 Abnormal bone growth in vitamin A deficiency. The figure shows the parietal bone of a mink (decalcified, H & E,X 110). Near the striated muscles on the external side (top) are several multinucleated osteoclasts located in Howship's lacunae or flattened along the bone. The internal side (bottom) is lined by a single layer of mononucleated osteoblasts. This is the reverse of normal growth of the skull and explains the overcrowding of the cranium in vitamin A deficiency. (*Courtesy of L. Krook, Cornell University.*)

15.42 Vitamin D

Vitamin D plays a critical role in normal bone growth and development (Sec. 11.10) and in the utilization of calcium and phosphorus. Most animals that have access to summer sunlight daily do not need a dietary supply of vitamin D because of the activation of vitamin precursors in the skin by ultraviolet light (Sec. 11.14). In the winter months in northern latitudes there is less sunlight and it is less efficient, and thus an unreliable source. Under these conditions supplements of vitamin D are recommended as well as for animals confined indoors at any season. Fig. 15.7 shows severe rickets in a calf. The N.R.C. vitamin D requirements of beef cattle, dairy cattle, horses, and sheep are summarized in Table 15.21. Of these values the one for dairy calves of 660 I.U. per 100 kg live weight is probably the most accurate. There is no good explanation why the value for beef calves is three to four times higher. The values for vitamin D per unit of feed were calculated from the N.R.C. feed intakes listed for calves of different body weights during early growth. Nonfortified milk is a poor source of vitamin D and will not meet the requirements of the young animal. Milk replacers and dry feeds for young animals are fortified in the vitamin. Exposure to sunlight or the consumption of high quality sun cured hay should supply enough vitamin D. The requirements of other animals are shown in Tables 15.19 and 15.20. Note that no requirement can be demon-

Fig. 15.7 Severe rickets produced on a ration lacking in vitamin D. Note humped back, enlarged joints, and buckling of front legs. (*Courtesy of S. I. Bechdel, Pennsylvania State University.*)

TABLE 15.21. The N.R.C. Vitamin D and E Requirements for Growth of Cattle, Horses and Sheep

| | Vitamin D, I.U. | | Vitamin E, I.U. | |
Animal	Per kg body weight	Per kg feed	Per kg body weight	Per kg feed
Beef cattle	17–32	275	45–180	15–60
Dairy cattle	6.6	230–500	12	600
Horses	6.6	167–250	0.33	15
Sheep	5.5–6.7	112–140	1.6	180–280

Source: N.R.C. Nutrient Requirements of Domestic Animals.

strated for growing guinea pigs and hamsters unless there is also a deficiency of calcium and/or phosphorus or the ratio between them is very abnormal.

15.43 Vitamin E

Vitamin E deficiency has been discussed in Sec. 11.17. All animals have a dietary requirement for this vitamin. The amounts needed for nonruminant domestic animals are shown in Table 15.19, for the laboratory animals in Table 15.20 and for cattle, horses, and sheep in Table 15.21. The values for most animals vary from 10 to 67 I.U. per kilogram of dry diet. There is no recognized reason for the higher values for the cat, sheep, and dairy cattle. While vitamin E is a dietary essential for young nursing lambs and calves, it seems to have no further practical importance as a dietary supplement in the nutrition of older animals. Even in young animals selenium has been shown to be more effective in preventing some deficiency signs, such as muscle dystrophy than vitamin E (Sec. 10.47). A reevaluation of the vitamin E requirement in the presence of an adequate amount of selenium appears to be in order.

15.44 Vitamin B Group

The information presented in Chap. 11 has indicated that experiments have failed to show any dietary need by cattle and sheep for the B group of vitamins after the rumen begins to function. The recognition of the need for various of these factors by swine and poultry was followed by many studies to establish the quantitative requirements for normal growth. The common procedure in these studies has been to feed a basal diet lacking in the vitamin under study but otherwise adequate and to add increasing levels of the vitamin to different groups. In this way the amount resulting in optimum growth without any appearance of characteristic deficiency symptoms was established. The commercial availability of synthetic sources of the vitamins has been largely responsible for the development in this field. On the basis of the studies made, quantitative requirements have been established for all the B-vitamins which appear to be of practical importance in swine and poultry feeding. The N.R.C. requirement for nonruminant domestic animals and laboratory animals are presented in Tables 15.19 and 15.20. More complete information is given in Appendix Tables III, IV, VI, VII, and VIII. The effects of riboflavin deficiency on young pigs are illustrated in Fig. 15.8.

Young horses are assumed to require a dietary source of the B-vitamins before their synthesis by the microflora begins. Deficiencies of several of the B-vitamins have been produced experimentally. The thiamin requirement has not been established, but 3 mg per kilogram of feed will maintain feed intakes, weight gains, and normal levels in the skeletal muscles. Thiamin is synthesized by the intestinal microflora and at least a quarter of the free thiamin in the cecum is absorbed. The riboflavin requirement of the horse for growth is not known, but 2.2 mg per kilogram of diet is adequate for maintenance. Niacin is not a dietary essential because of intestinal synthesis. Pyridoxine, pantothenic acid, biotin, vitamin B_{12}, and folic acid are synthesized in the lower digestive tract of the horse. Information on requirements is lacking.

The establishment of these quantitative requirements does not mean that appropriate amounts of the vitamins as such should be added to rations. The feed sources of them have been discussed in Chap. 11. For the most part body needs are easily met by properly selected rations of common feeds. Thiamin deficiency is not encountered in either pigs or chickens in practice because their customary rations contain large proportions of whole grains and milling by-products rich in the vitamin. Chicks require special supplements of riboflavin during the early weeks, but this does not appear to be the case for pig rations. The need for niacin as such is modified by the tryptophan content of the diet (Sec. 11.35). There appears to be no need for special sources of niacin in either poultry or hog rations which contain protein in the amount and quality recommended. There is no evidence at present that special supplements to otherwise good diets are needed in the case of any of the other B-factors with the exception of B_{12}. Rations lacking in animal protein need a special source of this vitamin. Such a source is provided by the B_{12} feed supplements now in common use. The feed sources of the various factors which are essential for growth have been discussed in detail in Chap. 11.

Previous discussions (Chap. 11) have shown that newborn ruminants have a dietary need for certain B-vitamins which are later provided by rumen synthesis. For this reason and because of possible health benefits from extraliberal intakes early in life, it has been proposed by some that dairy calves should receive supplementary B-vitamins, as well as A and D, from birth. Most experiments have failed to find any value in such a practice. Various studies in this field are reviewed by Erb and associates[87] in a report of their own experiments which failed to reveal any benefits from massive doses of nicotinic acid and vitamins A and D. Commercial feeds for young calves contain supplements of vitamins A and D as insurance against a deficiency.

15.45 Other Vitamins

Vitamin K is synthesized by microorganisms in the digestive tract in amounts adequate to cover the requirements of ruminants and some other animals (Sec.

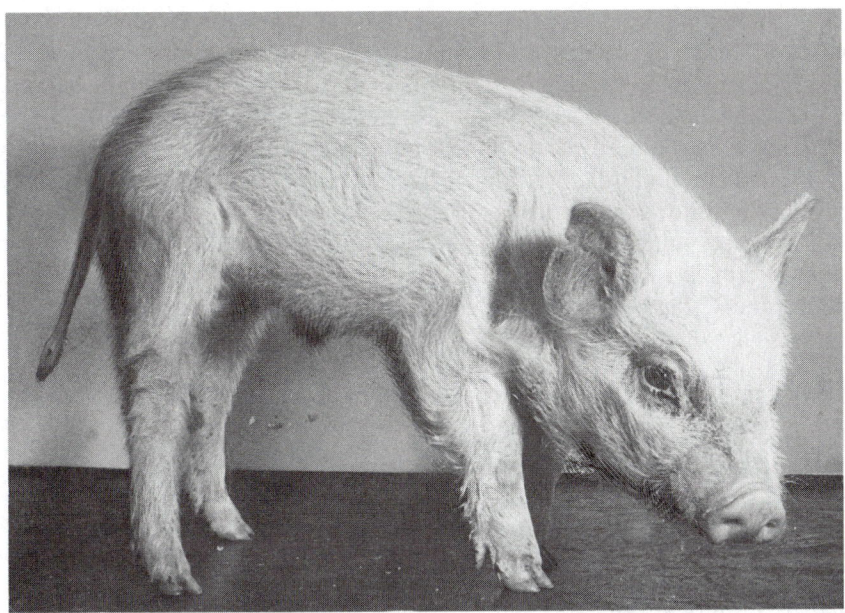

Fig. 15.8 Riboflavin deficiency. Top pig received no riboflavin; bottom pig received an adequate supply. (*Pictures furnished by R. W. Luecke, Michigan Agricultural Experiment Station.*)

11.22), such as the horse, hamster, rabbit, and rat. The requirements of other animals are shown in Tables 15.19 and 15.20. Some intestinal synthesis of vitamin K occurs in guinea pigs and monkeys; however, this may not be adequate to satisfy the entire need.

Vitamin C is required in the diet of humans, subhuman primates, guinea pigs, bats, and certain birds, fish, and perhaps reptiles (Sec. 11.59). The N.R.C. reports list the following requirements for growth: monkey, 1270 mg per kilogram of dry diet; guinea pig, 222 mg per kilogram of diet; and fish, 100 mg per kilogram of diet. The recommended dietary allowances for humans are 35 mg for infants and 45 mg per day for older children and adults.

NOTES

1. Max Rubner, Das Problem der Lebensdauer und seine Beziehung zum Wachstum, R. Oldenbourg, Munich, 1908, p. 81. Translation given in Lafayette B. Mendel, Abnormalities of growth, *Am. J. Med. Sci.,* **153:**1–20, 1917.

2. Ernst Schloss, Pathologie des Wachstums, S. Karger, Berlin, 1911, p. 4.

3. A. Asdell, Comparative chronological age in man and other mammals, *J. Gerontology,* **1:**224–226, 1946.

4. John Hammond, Pigs for pork and pigs for bacon, *J. Roy. Agr. Soc.,* **93:**1–15, 1932.

5. R. T. Berg and R. M. Butterfield, Growth patterns of bovine muscle, fat and bone, *J. Animal Sci.,* **27:**611–619, 1968.

6. E. H. Callow, Comparative studies of meat. II: The changes in the carcass during growth and fattening and their relation to the chemical composition of the fatty and muscular tissue, *J. Agr. Sci.,* **38:**174–199, 1948.

7. H. B. Ellenberger, J. A. Newlander, and C. H. Jones, Composition of the bodies of dairy cattle, *Vermont Agr. Expt. Sta. Bull.* 558, 1950.

8. R. H. Barnes and associates, Maternal protein deprivation during pregnancy or lactation in rats and the efficiency of food and nitrogen utilization of the progeny, *J. Nutrition,* **103:**273–284, 1973.

9. C. M. McCay, M. F. Crowell, and L. A. Maynard, The effect of retarded growth upon the length of life span and upon the ultimate body size, *J. Nutrition,* **10:**63–79, 1935.

10. C. F. Winchester and P. E. Howe, Relative effects of continuous and interrupted growth on beef steers, *U.S. Dept. Agr. Tech. Bull.* 1108, 1955; C. F. Winchester and N. R. Ellis, Delayed growth in beef cattle, *ibid.,* 1159, 1957.

11. K. Breirem, A. Ekern, and T. Homb, Relation of nutrition of the young animal to subsequent fertility and lactation. III, *Fed. Proc.,* **20:**275–283, 1961.

12. J. H. Meyer and W. J. Clawson, Undernutrition and subsequent realimentation in rats and sheep, *J. Animal Sci.,* **23:**214–224, 1964.

13. C. H. Tonge and R. A. McCance, Severe undernutrition in growing and adult animals: 15, The mouth, jaws and teeth of pigs, *Brit. J. Nutrition,* **19:**361–372, 1965.

14. C. R. Moulton, P. F. Trowbridge, and L. D. Haigh, Studies in animal nutrition. III: Changes in chemical composition on different planes of nutrition, *Missouri Agr. Expt. Sta. Research Bull.* 55, 1922; and earlier papers here cited.

15. H. R. Guilbert and coworkers, The importance of continuous growth in beef cattle, *California Agr. Expt. Sta. Bull.* 688, 1944.

16. C. P. McMeekan, Growth and development of the pig with particular reference to carcass quality, *J. Agr. Sci.,* **30:**276–343, 387–436, 511–569, 1940; **31:**1–49, 1941.

17. H. Pálsson and Juan B. Vergés, Effect of the plane of nutrition on growth and the development of carcass quality in lambs: Part I. The effects of high and low planes of nutrition at different ages, *J. Agr. Sci.,* **42:**1–92, 1952; Part II. Effects on lambs of 30-lb carcass wt., *ibid.,* **42:**93–149, 1952.

18. F. W. H. Elsley, I. McDonald, and V. R. Fowler, The effect of plane of nutrition on the carcasses of pigs and lambs when variations in fat content are excluded, *Animal Prod.,* **6:**141–154, 1964.

19. N. M. Tulloh, The carcass compositions of sheep, cattle, and pigs as functions of body weight. Symposium on carcass composition and appraisal of meat animals, *Comm. Sci. Inds. Research Organization,* **5:**1–18, 1964, Australia.

20. N.R.C. Symposium, Body Composition in Animals and Man, Publ. 1598, National Academy of Sciences, Washington, D.C., 1968.

21. J. T. Reid, Body composition of animals: Interspecific, sex and age peculiarities and the influence of nutrition, in Festskrift til Knut Breirem, Oslo, Norway, 1972, pp. 213–238.

22. P. E. Donnelly and J. B. Hutton, Effects of dietary protein and energy on the growth of Friesian bull calves. I: Food intake, growth and protein requirements. II: Effect of level of feed intake and dietary protein content on body composition, *New Zealand J. Agr. Research,* **19:**289–295, 409–414, 1976.

23. R. G. Campbell, The response of early weaned pigs to various protein levels in a high energy diet, *Animal Prod.,* **24:**69–75, 1977.

24. B. N. Berg, Nutrition and longevity in the rat. I: Food intake in relation to size, health and fertility, *J. Nutrition,* **71:**242–254, 1960; B. N. Berg and H. S. Simms, Nutrition and longevity in the rat. II: Longevity and onset of disease with different levels of food intake, *ibid.,* **71:**255–262, 1960.

25. P. V. Rattray, W. N. Garrett, and H. H. Meyer et al., Net energy requirements for growth of lambs age three to five months, *J. Animal Sci.,* **37:**1386–1389, 1973.

26. M. J. Brinegar and coworkers, The isoleucine requirement for the growth of swine, *J. Nutrition,* **42:**619–624, 1950.

27. C. R. Grau and D. W. Peterson, The isoleucine, leucine and valine requirements of chicks, *J. Nutrition,* **32:**181–186, 1946.

28. H. H. Williams and coworkers, Estimation of growth requirements for amino acids, by assay of the carcass, *J. Biol. Chem.,* **208:**277–286, 1954.

29. R. S. Cohen and T. D. Tanksley, Jr., Limiting amino acids in sorghum for growing and finishing swine, *J. Animal Sci.,* **43:**1028–1034, 1976.

30. K. Nimrick, A. P. Peter, and E. E. Hatfield, Aldehyde-treated fish and soybean meals as dietary supplements for growing lambs, *J. Animal Sci.,* **34:**488–490, 1972.

31. D. Lewis and R. M. Mitchell, Amino acid requirements of ruminants, in D. J. A. Cole and associates (eds.), Protein Metabolism and Nutrition, Butterworths, London, 1976, pp. 417–424.

32. Karl Thomas, Über die biologische Wertigkeit der stickstoff Substanzen in vershiedenen Nährungsmittel, Arch. Anat. u. Physiol., Physiol. Abt., 1909, pp. 219–302.

33. H. H. Mitchell, A method of determining the biological value of protein, J. Biol. Chem., 58:873–903, 1924; H. H. Mitchell, W. Burroughs, and J. R. Beadles, The significance and accuracy of biological values of proteins computed from nitrogen metabolism data, J. Nutrition, 11:257–274, 1936.

34. D. G. Armstrong and H. H. Mitchell, Protein nutrition and the utilization of dietary protein at different levels of intake by growing swine, J. Animal Sci., 14:49–68, 1955.

35. J. B. Allison, Biological evaluation of proteins, Physiol. Revs., 35:664–700, 1955.

36. T. B. Osborne, L. B. Mendel, and E. L. Ferry, A method of expressing numerically the growth-promoting value of proteins, J. Biol. Chem., 37:223–229, 1919.

37. D. R. Ames and D. R. Brink, Effect of temperature on lamb performance and protein efficiency ration, J. Animal Sci., 44:136–140, 1977.

38. D. S. Miller and A. E. Bender, The determination of the net utilization of proteins by a shortened method, Brit. J. Nutrition, 9:382–388, 1955.

39. B. L. Oser, An integrated essential amino acid index for predicting the biological value of proteins, in A. A. Albanese (ed.), Protein and Amino Acid Nutrition, Academic Press, Inc., New York, 1959, Chap. 10.

40. R. H. Barnes and D. K. Bosshardt, The evaluation of protein quality in the normal animal, Ann. N.Y. Acad. Sci., 47:273–296, 1946.

41. R. J. Block and H. H. Mitchell, The correlation of the amino acid composition of the proteins with their nutritive value, Nutrition Abstr. & Revs., 16:249–278, 1946.

42. J. Delort-Laval, Biological criteria of protein evaluation, in D. J. A. Cole et al. (eds.), Protein Metabolism and Nutrition, Butterworths, London, 1976, pp. 233–247.

43. D. M. Hegsted and coworkers, A comparison of the nutritive value of the proteins in mixed diets for dogs, rats and human beings, J. Lab. Clin. Med., 32:403–409, 1947.

44. W. C. Ellis and coworkers, Nitrogen utilization by lambs fed purified rations containing urea, gelatin, casein, blood fibrin and soybean protein, J. Nutrition, 60:413–425, 1956.

45. M. C. Nesheim and J. Garlich, Digestibility of unheated soybean meal for laying hens, J. Nutrition, 88:187–192, 1966.

46. J. T. Yen, T. Hymowitz, and A. H. Jensen, Utilization by rats of a protein from a trypsin-inhibitor varient soybean, J. Animal Sci., 33:1012–1017, 1971.

47. H. H. Mitchell and J. R. Beadles, Biological values of six partially purified proteins for the adult albino rat, J. Nutrition, 40:25–40, 1950.

48. K. L. Blaxter and H. H. Mitchell, The factorization of the protein requirements

of ruminants and of the protein value of feeds, with particular reference to the significance of the metabolic fecal nitrogen, *J. Animal Sci.*, **7**:351–372, 1948.

49. J. W. Lassiter and coworkers, Protein levels for pigs as studied by nitrogen balance, *J. Animal Sci.*, **15**:392–399, 1956.

50. D. E. Becker and coworkers, Levels of protein in practical rations for the pig, *J. Animal Sci.*, **13**:611–621, 1954.

51. G. P. Lofgreen, J. K. Loosli, and L. A. Maynard, Comparative study of the conventional protein allowances and theoretical requirements, *J. Animal Sci.*, **10**:171–183, 1951.

52. W. E. Thomas and coworkers, The utilization of inorganic sulfates and urea nitrogen by lambs, *J. Nutrition*, **43**:515–523, 1951.

53. J. L. Black, G. J. Faichney, and N. M. Graham, Wool growth responses to increased absorption of amino acids, in D. J. A. Cole et al. (eds.), Protein Metabolism and Nutrition, Butterworths, London, 1976, pp. 478–479.

54. K. Nimrick et al., Quantitative assessment of supplemental amino acids for growing lambs fed urea as the sole nitrogen source, *J. Nutrition*, **100**:1301–1306, 1970.

55. D. H. Tomlin and coworkers, Autoradiographic study of growth and calcium metabolism in the long bones of the rat, *Brit. J. Nutrition*, **7**:235–252, 1953.

56. D. Benzie and coworkers, Studies of the skeleton of the sheep. IV: The effects of interactions of dietary supplements of calcium, phosphorus, cod-liver oil, and energy, as starch, on the skeleton of growing blackface wethers, *J. Agr. Sci.*, **54**:202–221, 1960.

57. J. Duckworth and R. Hill, The storage of elements in the skeleton, *Nutrition Abstr. & Revs.*, **23**:1–17, 1953.

58. R. L. Preston and W. H. Pfander, Phosphorus metabolism in lambs fed varying phosphorus intakes, *J. Nutrition*, **83**:369–378, 1964.

59. C. F. Huffman and associates, Phosphorus requirements of dairy cattle when alfalfa furnishes the principal sources of protein, *Michigan Agr. Expt. Sta. Tech. Bull.* 134, 1933.

60. W. M. Beeson and associates, The phosphorus requirement for fattening lambs, *J. Animal Sci.*, **3**:63–70, 1944.

61. C. E. Aubel, J. S. Hughes, and W. J. Peterson, Calcium requirements of growing pigs, *J. Agr. Research*, **62**:531–542, 1941.

62. W. G. Pond, E. F. Walker, and D. Kirtland, Weight gain, feed utilization and bone and liver composition of pigs fed high or normal Ca-P diets from weaning to slaughter weight, *J. Animal Sci.*, **41**:1053–1056, 1975.

63. M. B. Wise, S. E. Smith, and L. L. Barnes, The phosphorus requirements of calves, *J. Animal Sci.*, **17**:89–99, 1958.

64. H. H. Mitchell, The mineral requirements of farm animals, *J. Animal Sci.*, **6**:356–377, 1947.

65. W. H. Allaway, The effect of soils and fertilizers on human and animal nutrition, *U.S. Dept. Agr. Information Bull.* 378, 1975.

66. J. G. Bieri, Potassium requirements of the growing rat, *J. Nutrition,* **107**:1394–1398, 1977.

67. G. M. Murphy and J. A. Connell, A simple method of collecting saliva to determine the sodium status of cattle and sheep, *Aust. Vet. J.,* **46**:595–602, 1970.

68. J. G. Morris and R. J. W. Gartner, The sodium requirements of growing steers given an all sorghum grain ration, *Brit. J. Nutrition,* **25**:191–205, 1971.

69. I. Hagstern, T. W. Perry, and J. B. Outhouse, Salt requirements of lambs, *J. Animal Sci.,* **40**:329–334, 1975.

70. T. S. Kahlon, J. C. Meiske, and R. D. Goodrich, Sulfur metabolism in ruminants. II: In vivo availability of various chemical forms of sulfur, *J. Animal Sci.,* **41**:1154–1160, 1975.

71. A. J. F. Webster et al., Energy exchange of veal calves fed a high-fat milk replacer diet containing different amounts of iron, *Animal Prod.,* **20**:69–75, 1975.

72. G. Matrone et al., A study of iron and copper requirements of dairy calves, *J. Dairy Sci.,* **40**:1437–1447, 1957.

73. J. W. Hibbs et al., Occurrence of iron deficiency anemia in calves at birth and its alleviation by iron dextran injection, *J. Dairy Sci.,* **46**:406–410, 1967.

74. C. B. Ammerman et al., Utilization of inorganic iron by ruminants as influenced by form of iron and status of the animal, *J. Animal Sci.,* **26**:404–410, 1967.

75. E. T. Kornegay, Availability of iron contained in defluorinated phosphate, *J. Animal Sci.,* **34**:569–572, 1972.

76. R. Braude and T. Ryder, Copper levels in diets for growing pigs, *J. Agr. Sci.,* **80**:489–493, 1973.

77. G. L. Cromwell, D. T. H. Sihombing, and V. W. Hays, Effects of iodine level on performance and thyroid traits of growing pigs, *J. Animal Sci.,* **41**:813–818, 1975.

78. M. A. Rojas, I. A. Dyer, and W. A. Cassatt, Manganese deficiency in the bovine, *J. Animal Sci.,* **24**:664–667, 1967.

79. C. B. Ammerman and S. M. Miller, Selenium in ruminant nutrition: A review, *J. Dairy Sci.,* **58**:1561–1577, 1975.

80. D. E. Ulberg et al., Selenium supplementation of diets for sheep and beef cattle, *J. Animal Sci.,* **46**:559–565, 1977.

81. W. E. Julien, H. R. Conrad, and A. L. Moxon, Selenium and vitamin E and incidence of retained placenta in parturient dairy cows. II: Prevention in commercial herds with prepartium treatment, *J. Dairy Sci.,* **59**:1960–1962, 1976.

82. K. Schwarz, Recent dietary trace element research exemplified by tin, fluorine and silicon, *Fed. Proc.* **33**:1748–1757, 1974.

83. K. Schwarz et al., Inverse relation of silicon in drinking water and atherosclerosis in Finland, *Lance,* **5**:538–539, 1977.

84. M. Rechcigl, Jr. (ed.), Handbook Series in Nutrition and Food, Section G: Vol. 1, Diets, Culture Media and Food Supplements, C.R.C. Press, Cleveland, Ohio, 1977.

85. H. R. Guilbert, C. E. Howell, and G. H. Hart, Minimum vitamin-A and carotene requirements of mammalian species, *J. Nutrition,* **19:**91–103, 1940.

86. H. D. Eaton et al., Reevaluation of the minimum vitamin A requirement of Holstein male calves based upon elevated cerebrospinal fluid pressure, *J. Dairy Sci.,* **55:**232–237, 1972.

87. R. E. Erb and associates, Observations on efficiency of vitamin supplements for new-born calves, *J. Animal Sci.,* **8:**425–431, 1949.

SUPPLEMENTARY LITERATURE

Alternate sources of protein for animal production, National Academy of Sciences, Washington, D.C., 1973.

Babatunde, G. M., et al.: Effect of plane of nutrition, sex and body weight on the chemical composition of Yorkshire pigs, *J. Animal Sci.,* **26:**718–726, 1967.

Conrad, J. H., and L. R. McDowell (eds.): Latin American Symp. on Mineral Nutrition Research with Grazing Ruminants, Animal Science Dept., Univ. of Florida, Gainesville, 1978.

Elsley, F. W. H.: Limitations to the manipulation of growth, *Proc. Nutrition Soc.,* **35:**323–337, 1976.

Folman, Y., and associates: Compensatory growth of intensively raised bull calves. III: Restricted feeding and breed differences, *J. Animal Sci.,* **39:**788–795, 1974.

Koong, L. J.: A new method for estimating energetic efficiencies, *J. Nutrition,* **107:**1724–1728, 1977.

Machlin, L. J., and associates: Effects of a prolonged vitamin E deficiency in the rat, *J. Nutrition,* **107:**1200–1208, 1977.

Meyer, J. H., and W. N. Garrett: Efficiency of feed utilization, in Techniques and Procedures in Animal Science Research, American Society of Animal Science, 1969.

Millward, D. J. and P. J. Garlick: The energy cost of growth, *Proc. Nutrition Soc.,* **35:**339–349, 1976.

National Research Council: The effect of genetic variance on nutritional requirements of animals, National Academy of Sciences, Washington, D.C., 1975.

Spears, J. W., C. J. Smith, and E. E. Hatfield: Rumen bacterial urease requirement for nickel, *J. Dairy Sci.,* **60:**1073–1076, 1977.

Testing and Treating Mineral Disorders in Dairy Cattle, Centre for Agricultural Publishing and Documentation, Wagningen, 1973.

Underwood, E. J.: Trace Elements in Human and Animal Nutrition, 4th ed., Academic Press, New York, 1977.

Young, L. G., and associates: Selenium and vitamin E supplementation of high moisture corn diets for swine reproduction, *J. Animal Sci.,* **45:**1051–1060, 1977.

Chapter 16

Reproduction

While it has long been appreciated that regular and normal reproduction is the essential basis of a successful animal industry, including commercial milk production, it is only within comparatively recent years that the various aspects of this primary physiological function have received detailed study. The reproductive function is conditioned by a long series of distinct but interrelated physiological events in which the body as a whole, as well as the sex organs, is concerned. A failure of any one event to take place normally can result in temporary or permanent failure of the function as a whole. The economic importance of temporary failure has been stressed by Marshall and Hammond[1] as follows:

Low fertility and sterility of a temporary nature, because of their prevalence, are the cause of much greater loss to the breeding industry than infertility of a more permanent kind which occurs less frequently, although the latter, because of its striking effects, generally attracts most attention.

It is self-evident that nutrition must play at least a general role in the development and functioning of the organs of reproduction, but its significance is much larger than this. It is apparent also that, although there are no substances needed by the reproductive organs which are not needed by the tissues, the metabolic pathways followed by some of the nutrients provided by the bloodstream differ from those which have been identified in other organs. To understand these functions some knowledge of the physiology of reproduction is required. This discussion which immediately follows deals with mammalian reproduction and its nutritive requirements. At the close of this chapter the avian process is very briefly considered from the standpoint of the nutritive needs for egg production.

PHYSIOLOGY OF REPRODUCTION

The sexual organs reach their full development and become functional at an age which varies with the species, the breed, and the nutrition of the individual. The development of these organs is a rather gradual process controlled by secretions from the pituitary, the gland which also secretes a substance controlling body growth (Sec. 15.7). These organs become functional before body growth is completed.

16.1 Spermatozoa Production

Sexual maturity in the male is characterized by the full development of the testicles and certain accessory glands, and by the production of viable sperm which become motile when mixed with the secretions of the accessory organs. As well as producing the sperm, the testes also secrete androgenic hormones, including testosterone, which cause sex drive and the development of male characteristics, and promote the development and functioning of the accessory glands. These glands consist of the seminal vesicles, Cowper's glands, and the prostate, providing secretions which are mixed with the sperm on ejaculation. Malnutrition can delay the development of sexual maturity and, after maturity, can lessen the amount of sperm produced and their viability and motility. It can even cause the production of viable sperm to cease entirely. Specific nutrients are here involved, as is discussed later.

16.2 Ovulation and Fertilization

In the female the functional development of the ovary is followed by a recurring cycle of events, an early stage of which is characterized by the onset of heat, or estrus. The ovary contains a large number of minute eggs, each enclosed in a *Graafian* follicle. As the animal comes in heat, one or more of these follicles enlarge, the egg is matured and liberated and then passes down the *Fallopian* tubes, or oviducts, to the uterus. Fertilization normally occurs shortly after the eggs enter the oviduct provided the animal has been served during the heat period. With the shedding of the egg, the cavity of the Graafian follicle becomes filled with the corpus luteum or yellow body. While this yellow body is present, no more follicles are ripened and thus no more eggs are matured. If fertilization has occurred, the corpus luteum persists during the ensuing pregnancy in most species (the mare is an exception); thus no new fertilization can take place until its termination. If pregnancy does not occur, the corpus luteum usually degenerates after a brief period, allowing a new follicle to mature as the start of a new cycle. The ovarian changes of the estrous period are accompanied by cyclic changes in the epithelial lining of the vagina. A study of these changes in the rat, dog, and guinea pig by the vaginal-smear technique[2] has proved useful as an indicator of the regularity of estrus and of the ovarian processes involved, as influenced by various nutritional and other factors. Less well-defined cyclic, vaginal changes have been noted also in the cow, sow, and ewe.

16.3   Hormonal Control

Each of the events which comprise the estrous cycle is under hormonal control. Three hormones arising from the anterior pituitary gland, namely, the *follicle stimulating hormone* (FSH), *luteinizing hormone* (LH), and *prolactin,* control the activity of the gonads. Secretion rates of FSH and LH are in turn controlled by a gonadotrophin releasing hormone (GnRH) which originates in the hypothalamus, a part of the brain-stem. Similarly, the secretion rate of prolactin is governed by a prolactin inhibitory hormone (PIH) and by a stimulatory principle, both of which are of hypothalamic origin.

LH is responsible for testosterone secretion by the interstitial cells interspersed among the tubular elements of the testes. LH and testosterone are mainly responsible for spermatogenesis within the tubules, testosterone being specifically required for the reduction division resulting in the production of secondary spermatocytes. FSH is necessary for production of an androgen binding protein by the Sertoli cells within the tubules which causes an accumulation of testosterone and other androgens in close proximity to androgen-dependent cells in the germinal epithelium. Apparently, both LH and FSH are required for the production of normal numbers of spermatozoa.

Follicular development within the ovary is controlled by FSH and LH. LH stimulates steroidogenesis and results in the production of progesterone and testerone by the cells of the *theca interna* in the follicle walls. Under the influence of FSH, testosterone is converted to estradiol by the granulosa cells which surround the egg within the follicle. The resulting high circulating levels of estrogen trigger the release of large amounts of LH from the pituitary during estrus; this LH results in ovulation and corpus luteum development. LH is the major hormone responsible for progesterone secretion by the corpus luteum in the cow. In the rat, and possibly the ewe, prolactin also plays a role in corpus luteum maintenance and progesterone secretion.

The hypothalamic hormones are small peptides, the pituitary hormones are larger peptides, and the ovarian and testicular hormones are lipid-soluble steriods. The peptide hormones interact with receptor molecules on the cell membrane and may not need to enter the cell to cause their effects. The steroids, however, are thought to pass readily through the cell membrane and first interact with receptor molecules within the cytoplasm.

In addition to evoking the behavioral manifestations of estrus, estradiol promotes the growth of the vaginal epithelium and helps to develop the uterine mucosa. Progesterone, along with estrogen, is responsible for development of the glandular system in the uterus that is required for implantation of the embryo and the maintenance of pregnancy. The corpus luteum also produces relaxin, a peptide hormone that acts to relax the pelvic ligaments and enlarge the channel through which the fetus is expelled at the end of gestation.

The foregoing, very incomplete discussion of the roles of hormones in reproduction serves to indicate the intricate nature of the various reciprocal hormonal actions involved at different stages of the cycle. Some of the hormones discussed here also play roles in mammary development and function (Sec. 17.2). Many of the effects of nutrition on the reproductive process are mediated through the endocrine system, as reviewed by Leatham.[3] His concluding paragraph states:

The development, composition and normal functioning of the reproductive system is dependent on adequate nutrition. Only gradually are data being acquired which are pertinent to the elucidation of the nutritional-gonadal relationship.

16.4 Estrogenic Activity of Feeds

One of the hormones concerned in the estrous cycle and in the maintenance of pregnancy is estradiol produced by the developing follicles. Following observations that ewes grazing on subterranean clover showed impaired fertility, evidence was obtained that an estrogen present in the forage was responsible. There were morphological changes in the animals indicative of excessive estrogen levels. These findings, reported by Curnow and associates in Australia,[4] have been confirmed by workers in England and the United States who have shown that subterranean clover and red clover contain genistein and biochanin A, isoflavones having estrogenic activity. Other phytoestrogens include coumestrol and formononetin. Formononetin, the most hazardous isoflavone affecting sheep fertility, is converted to other estrogenic compounds in the rumen. These include equol and methyl equol. Although these compounds clearly cause infertility in sheep in certain areas, their role as infertility agents in cattle has not been clearly established.

16.5 Fetal Growth

The fertilized egg is nourished for a short time by secretions from glands of the uterus, and during this time it develops the placenta by which it becomes attached to the uterine walls. Following this implantation, it receives its nourishment from the maternal blood through its placenta and umbilical cord. The blood vessels are among the first permanent structures in the embryo, providing for the circulation of nutrients and the removal of waste products through interchange with the maternal blood. This interchange, including the oxygenation of the fetal blood, occurs in capillaries in the placenta. While the fetus receives most of its nutrients preformed, it certainly carries on some synthetic functions in connection with its growth.

Expressed arithmetically, the growth of the fetus takes place at an increasing rate throughout the gestation period. More than half of the period elapses before the weight of the fetus equals that of its membranes, whereas, at term, the placenta make up only about 20 percent of the total weight of the products of conception. Most of the growth takes place in the last third of the gestation period, as is illustrated by the curves in Fig. 16.1 taken from the studies of Mitchell and associates[5] with swine. These workers slaughtered pregnant gilts in groups of one to three at weekly intervals from the fifth to the sixteenth week of gestation and determined the nutrients stored in the fetus and the placenta. The data were corrected to a standard litter of eight and treated mathematically to provide curves showing the increase in nutrient storage over the gestation period.

It is evident from Fig. 16.1 that approximately half of the protein and more than half of the energy storage occurs in the last quarter of pregnancy. Additional data showed that even larger proportions of the calcium and phosphorus are stored toward the end of the period. In accordance with the general picture in growing organisms, the percentage of water in the fetus decreases with age. On a dry-matter basis, protein makes up about two-thirds of the products of conception, a figure which shows little change over the gestation period. Neither do the percentages of fat or iron change markedly, but the calcium and phosphorus contents make up an increasing percentage with age. It is therefore clear that the quantitative demands for nutritive material are small in early pregnancy and that they progressively increase to become several times as large toward the close of the period.

In multiparous animals, the larger the number of fetuses the smaller the individuals tend to be, owing to crowding, and there is frequently a marked difference in size among the individuals of a litter. In animals which may give birth to one or more young, multiple births do not produce so large individuals as do single births.

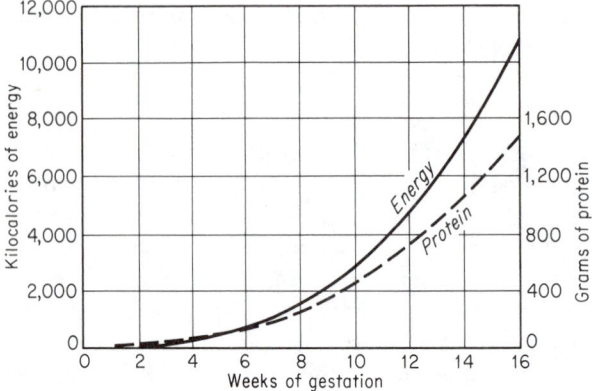

Fig. 16.1 Gross energy and protein in products of conception at different stages of gestation—sow. (*Courtesy of Mitchell, Carroll, Hamilton and Hunt.*)

For each species there is a certain duration of pregnancy which is recognized as normal. Its termination is associated with the degeneration of the corpus luteum and is probably under the influence of both nervous and endocrine factors. The delivery of young is followed by the expulsion of the placenta, or afterbirth. Level of nutrition has some influence on fetal size, but genetics is more important. Specific nutritional deficiencies may cause abnormal fetal development and reduce livability.

NUTRITIVE REQUIREMENTS

Nutritional factors play vital roles in the various physiological events which occur in the attainment of sexual maturity and in the course of the reproductive process. As is detailed later in the discussions of the requirements for the various nutrients, specific deficiencies can result in injury and even complete failure at specific stages in the reproductive process. More commonly, however, the troubles that occur in practice apparently result from multiple deficiencies which reflect general undernutrition caused by rations inadequate in amount as well as in quality.

16.6 Importance of Plane of Nutrition

Undernutrition delays puberty in both the male and the female in most species and, if severe, it may cause retrogressive changes in the sex organs after they are fully developed. It is a well-recognized fact that half-starved animals are relatively infertile. If the severely undernourished female becomes pregnant, the drain on her body by the developing young may result in permanent damage. The death of the fetus in utero or the birth of a weak animal, perhaps prematurely, may also occur. In the male, undernutrition decreases the number and vigor of the sperms and may cause cessation of spermatogenesis.

Casida[6] has reviewed various experiments with farm animals on the effects of undernutrition. In his summary he concludes that the restriction of food intake during the growing period delays the occurrence of puberty in both sexes of cattle and swine and reduces the production of eggs immediately after puberty in swine and sheep and of sperm in cattle, sheep, and swine. He notes that potential fertility in the three species was not greatly affected, however. Nutritional inadequacies are reflected in altered secretion rates of both the peptide and steroid hormones. For example, decreased ability of the corpus luteum to respond to LH following a restricted energy intake in cattle has recently been demonstrated by Apgar et al.[7]

Chow and Lee[8] reported that restricting the feed allowance of female rats during gestation and lactation by 25 to 50 percent caused permanent stunting of the young from which they did not recover. This suggests that adequate nutrition during gestation may be even more critical than previously suspected.

The practical importance of adequate nutrition during gestation has been demonstrated with various species. For example, the bad effects of general undernutrition are illustrated by the excellent experiment with ewes by Thomson and Thomson.[9] They reported studies of a high versus a low plane of nutrition during the latter half of pregnancy and during lactation on reproduction and lactation performance. The high plane provided a ration excellent in amount and quality. The low plane supplied little more than half the calories and nutrients supplied by the high plane. The low plane produced small, weak lambs, most of which died at birth or soon after—in contrast to vigorous lambs and few deaths on the high plane. The milk yield on the high plane was approximately 90 kg in a 13-week lactation period compared with 50 kg on the low plane.

A number of experiments with cattle fed different planes of nutrition have shown that low energy levels delay development and the onset of sexual maturity, but tend to allow higher reproductive efficiency than overfeeding. Early studies in Europe, reviewed by Breirem, Ekern, and Homb[10] reported that Red Danish cattle averaged 4.3 calvings on recommended feeding levels compared with 4.6 for underfed and 3.4 for those overfed by 25 to 30 percent. The culling rate for sterility was 50 percent for heavily fed animals versus 38 percent for those restricted.

In the Cornell experiment (Sec. 15.11), where wide differences in plane of nutrition were compared, marked influences were noted in the age at which heifers reached sexual maturity. The data from this study in Table 16.1 show that underfed heifers were 20.3 months of age when the first estrus occurred compared with 11.2 months for those on the recommended energy level and 9.2 months on heavy feeding, but there was no difference in the weight of the different groups. The heavy-fed heifers were less fertile than the others since fewer of them conceived at the first mating. One heifer on the high level failed to conceive after 13 breedings. She was not included in calculating average matings per conception. In subsequent reproductive cycles there were no ap-

TABLE 16.1. Influence of Plane of Nutrition on Reproductive Performance of Holsteins

Measure	TDN intake (percent of normal)		
	62	100	146
Number of heifers	33	34	34
Age at first estrous, months	20.2	11.2	9.2
Weight at first estrous, kg	303	265	277
Conceived to first mating, %	79	68	58
Matings per first conception, no.	1.55	1.41	1.48
Age at first calving, months	32.0	28.5	27.9
Weight at first calving, kg	384	483	548
Weight of first calf, kg	36	39	41
Requiring help at calving, %	45	26	24
Delivering living calves, %	87	88	94
Matings per second conception, no.	1.71	1.76	2.09
Matings per third conception, no.	1.90	1.64	1.90
Culled for sterility, %	6	12	20

parent differences in ease of conception, except that 20 percent of the cows were culled because of sterility for the high group versus 6 percent among the low.

The stunted heifers weighed 20 percent less than normal at first calving, but their calves weighed only slightly less. About twice as many had difficult first calvings, but as the result of veterinary assistance the number of live calves produced was about the same for all groups. Following refeeding the low-plane heifers performed in a perfectly normal manner.

Overfeeding during the growth period may also be disadvantageous. Studies with dairy cattle have indicated that feeding 40 to 50 percent more TDN than called for by feeding standards, while hastening the onset of puberty, tends to result in breeding troubles later and a shorter productive life accordingly. With gilts, Casida has reported that what is gained by increased ovulation rate on full feeding may be lost by increased embryo mortality.

A condition of extreme fatness appears to be deleterious to reproduction. The ovaries may become so infiltrated with fat as to hinder the development of the follicles, with a consequent irregularity or cessation of estrus which results in delay or failure in breeding. There may be also such an excessive amount of fat in the reproductive tract that, even if the egg is matured and fertilized, it may fail to reach the uterus and become implanted properly. Extreme fatness has also been noted to interfere with the production of fertile sperms in the male and lessens his desire to mate. In attributing reproductive failure to extreme fatness, it must be recognized that in some instances the condition may be merely an accompaniment of some specific deficiency or other cause.

While the nutritive intake is qualitatively of equal importance in both sexes, obviously the quantitative requirements are vastly greater for the female. Thus most of the following discussion deals with the needs of the mother. Her nutrition must have the double object of producing normal offspring and of protecting her own tissues, for on an inadequate ration the mother sacrifices her bones and other tissues to nourish her fetus.

As is indicated by the curves in Fig. 16.1 the last quarter of the gestation period is the time of critical importance. It has been clearly proved that a system of feeding which takes full account of the increased needs at this time is much more effective than one which supplies the same amount of feed over the period as a whole but at the same intake level throughout. The beneficial effects of meeting fully the current needs toward the end of the period are reflected not only in more vigorous young but also in a higher potential level of milk secretion by the mother. It is during the last part of the period of gestation that the formation of secretory cells in the udder is most active (Sec. 17.1). Inadequate nutrition at this time limits this process and thus lessens the milk-secreting capacity that is developed.

16.7 Energy Requirements

While the effects of general undernutrition previously mentioned are undoubtedly the result of more than one deficiency, it is evident that the energy supply is perhaps the most critical factor. Experiments with rats have shown that a deficiency of energy alone results in a delay in the opening of the vagina, a prolongation of the period between this event and the first estrus, and an irregularity or cessation of the estrus cycle. The studies by Asdell and Crowell[11] showing these effects included observations on sexually mature rats which were held at constant weight far below normal size by energy restriction. Under these conditions, the cycles were highly irregular. When the animals were given an increment of energy sufficient to cause some growth, regular cycles occurred until the animals reached a weight where growth ceased because the total energy intake was needed for maintenance. Here sexual activity ceased also. Thus it was shown that neither growth nor sexual activity takes precedence over the other.

The energy requirement for reproduction consists of the energy stored in the new tissue formed plus the energy expended in the process. The tissues formed include the fetus and its membranes, the enlargement of the uterus, and the mammary development. The energy content of these tissues at different stages provides the basic figures for estimating the nutritive requirements over the gestation period. Mitchell and associates in their studies, previously cited, with swine computed the daily increase in energy and other nutrients in the products of conception. These data are reproduced in Table 16.2. It is noted that the energy storage is very small during the early weeks. While 272 kcal are

TABLE 16.2. Computed Daily Rate of Increase in Weight and Energy Content and Computed Daily Deposition of Nutrients in the Uteri of Pregnant Gilts

Week of gestation	Total weight, g	Gross energy, kcal	Crude protein, g	Ash, g	Calcium, g	Phos- phorus, g	Iron, mg
1	27	1.6	0.54	0.028	0.0001	0.0011	0.28
2	49	5.9	1.5	0.126	0.0018	0.0074	0.71
3	71	12.5	2.7	0.30	0.0081	0.022	1.24
4	91	21.0	4.2	0.57	0.024	0.048	1.84
5	111	32.0	5.9	0.93	0.055	0.087	2.50
6	131	45.0	7.7	1.38	0.109	0.142	3.2
7	150	59.0	9.6	1.93	0.194	0.215	4.0
8	169	76.0	11.8	2.6	0.32	0.31	4.8
9	187	94.0	14.0	3.3	0.50	0.42	5.6
10	205	115.0	16.0	4.3	0.74	0.56	6.5
11	224	137.0	19.0	5.2	1.05	0.72	7.4
12	242	160.0	21.0	6.2	1.46	0.91	8.3
13	259	186.0	24.0	7.4	1.97	1.13	9.2
14	277	213.0	27.0	8.7	2.60	1.38	10.2
15	294	242.0	30.0	10.1	3.37	1.67	11.2
16	312	272.0	33.0	11.7	4.29	1.98	12.3

Source: H. H. Mitchell, W. E. Carroll, T. S. Hamilton, and G. E. Hunt, Food requirements of pregnancy in swine, *Illinois Agr. Expt. Sta. Bull.* 375, 1931.

stored daily during the last week, computation shows that the average daily deposition for the period as a whole is only 104 kcal. No data are available for estimating the energy stored in the mammary growth, but except in a first pregnancy, it should not exceed 10 percent of that in the uterine products. On this basis a daily intake of 115 kcal could be considered as the net-energy requirement for reproduction in a sow producing a litter of eight pigs. That these calculations are only approximations is clear, but the final value obtained is useful for comparison with the estimated net-energy requirement for maintenance.

The gilts studied by Mitchell weighed around 90 kg. Armsby gives the net-energy requirement for maintenance of a pig of this weight as 1.99 Mcal. Even on the basis of this figure, which is probably too low, the average daily need for reproduction itself adds only 6 percent to the maintenance requirement and less than 15 percent during the last week of gestation when the demand is greatest. Approximately the same relations should hold for the requirements in terms of digestible nutrients. The preceding calculations are useful to show that even for the sow, in which multiple birth involves a relatively higher reproductive performance than for other farm mammals, the additional energy needs above maintenance are very small until the last part of the gestation period. The data are not particularly useful, however, for arriving at allowances for practice because of other factors involved. Animals are usually bred before they have reached their full growth, and thus allowances during gestation must take account of growth needs as well. Mature animals frequently begin their gestation in rather poor flesh because of the previous lactation and need additional feed accordingly. Lamond[12] suggested that cattle have a target weight needed for high fertility. Animals below that weight and still losing show reduced fertility.

Grazing cattle often will not eat enough mature poor-quality forage of low protein content and digestibility to maintain their body weight even when the supply is abundant and sometimes the amount available is limited.[13] Unless supplements of energy, protein, and often minerals are provided reproductive levels will be greatly reduced. Moe and Tyrrell[14] reported that the metabolizable energy requirements for dairy cows was increased only slightly the first half of gestation, but at term it was 75 percent above the maintenance requirement. Their data are slightly higher than the N.R.C. values in Appendix Table I.

Thus, in practice, most pregnant animals must be given a sufficient energy allowance to enable them to gain some weight during the period as a whole, with special attention given to the last quarter when the specific needs are substantial. The aim should be to have the animals in good flesh at parturition without being too fat.

16.8 Protein Requirement

Controlled studies with rats have shown that a low-protein diet causes a cessation of estrus and that, if fertilization occurs, fetal resorptions or the birth of premature, dead, or weak offspring results. These findings are illustrated by the study of Nelson and Evans.[15] Other work by these authors has indicated that in the absence of dietary protein the reproductive failure is due to a lack of ovarian hormones. In the male rat a lack of protein results in limited testes growth and sperm formation and even in the absence of testosterone secretion, as a result of a reduction in the level of circulating pituitary gonadotropic hormones, according to the review by Leathem.[16] Studies with other species also indicate reproductive failures when the protein nutrition is inadequate, although the evidence is less specific than for the rat. The quality as well as the quantity of the protein has been shown to be important. Stuart[17] has reviewed the evidence indicating that protein deficiency in human reproduction can cause inability to conceive, toxemia of pregnancy, and immature newborn.

Rations which are fully adequate in protein for maintenance, and for growth if this function is not completed, should be adequate also for conception and the initiation of fetal growth. The additional needs here are negligible. Since, however, the dry matter of the products of a conception consist largely of protein, it is evident that there is an increased need as fetal growth proceeds. The data in Table 16.2 show that the quantitative need does not become of large importance until the last half of pregnancy, during which the daily storage increases rapidly. The average daily figure for the gestation period as computed from these data is approximately 14 g, but it is more than double this figure during the last 2 weeks. Calculations similar to those made for net energy (Sec. 16.7) indicate that the additional requirements for digestible protein for the products of reproduction, as a percentage of maintenance needs, are considerably higher than the energy increments called for. In other words, pregnancy in the sow increases the need for protein much more than for energy. Similar calculations, based on meager data, indicate that the same is true for the cow.

As in the case of energy, the protein intake during pregnancy should be

TABLE 16.3. Protein and Energy Requirements

Item	Cattle		Horse	Pig	Sheep	Chicken	Rabbit	Rat
	Beef	Dairy						
Protein, % of dry diet								
Maintenance	8.5	8.5	8.5	14	8.9	–	12	4.4
Growth	15–10	16	16	22	10.0	20	16	13.3
Reproduction	9	9	11	14	9.3	15	15	13.3
Lactation	9–11	12	14	15	11.5	–	17	13.3
Energy, Mcal/kg dry diet								
Maintenance	1.90	1.4	2.3	–	2.0	2.85	2.1	4.0
Growth	2.50	2.6	3.1	3.36	2.2	2.90	2.5	4.0
Reproduction	1.90	1.3	2.5	3.17	2.1	2.85	2.5	4.0
Lactation	2.00	2.6	2.8	3.17	2.4	–	2.5	4.0

Source: N.R.C. Nutrient requirements of domestic animals. Energy values are expressed as ME for beef cattle, pig, sheep, chicken and rat; DE for the horse and rabbit; NE for dairy cattle.

TABLE 16.4. Dry Matter Intakes for Maintenance, Reproduction, and Lactation

Species (weight, kg)	Daily intake per animal, kg		
	Maintenance	Reproduction	Lactation
Beef cattle, 500	7.2	8.6	11.8
Dairy cattle, 500	6.5	8.5	19.7[a]
Horse, 500	7.5	8.5	9.8
Pig, 110–250	1.8	2.0	4.0–5.5
Sheep, 70	1.2	2.1	2.5
Chicken, 1.8	0.07	0.11[b]	—
Rabbit, 4.5	0.16	?	0.55
Rat, 0.3	0.013	0.019	0.033

[a] For 30 kg of 4% milk.
[b] For 60% production.
Source: N.R.C. Nutrient Requirements of Domestic Animals.

set higher than the sum of the needs for maintenance and for the formation of the products of reproduction. For animals bred before they have reached their full growth, the further growth needs must be covered. The intake should also be sufficient to take advantage of the special ability of the pregnant animal to store protein in its own body as well as in the products of conception. This is important because early in lactation a negative-nitrogen balance frequently occurs despite very liberal protein nutrition at the time. Thus the dam's body protein is drawn upon to supply a part of the protein secreted in the milk. If advantage has been taken of her capacity to build reserves during gestation, this can occur without harm. These facts were first established in studies with women by Hunscher and associates.[18]

The N.R.C. requirements for reproduction in dairy cattle, beef cattle, swine, horses, sheep, chickens, rabbits, rats, and humans are given in Appendix Tables I through IX. The N.R.C. requirements summarized in Table 16.3 do not show an increase in the energy concentration in the diets of animals for gestation above those for maintenance and only for the horse, sheep, rabbit, and rat are higher protein levels shown. However, additional feed intakes are recommended for all animals during the last quarter of gestation, the extra feed being adjusted to the condition of the animal. The following increases in energy and total feed are listed: women, 15 percent more kcal; rabbits, 19 percent more DE; dairy cows up to 40 percent more feed; beef cows and sheep, 20 percent more, and horses, 7.5 more feed. For women the protein allowance is increased from 46 to 75 g for pregnancy. Table 16.4 gives more complete data on feed intakes.

16.9 Effects of Calcium and Phosphorus Deficiencies

The observations in the phosphorus-deficient areas (Sec. 10.2) throughout the world are in agreement that reproductive troubles are very common and that they have caused very large losses in the animal industry. Conclusive proof for cows is furnished by the extensive studies of Theiler and associates[19] carried out in the phosphorus-deficient area in South Africa. The studies included observations on 200 animals over a period of two years. In groups in which the phosphorus-deficient pasture was supplemented by bone meal or other phosphorus sources, the calf crop was approximately 80 percent in contrast to a figure of approximately 51 percent in the control group. Similar evidence has been presented from various areas in the United States and other countries of the world. The most frequently observed specific trouble is irregularity or cessation of estrus, corresponding to the commonly reported finding in rats. It has become apparent, however, that other deficiencies, such as lack of vitamin A, protein, or trace minerals are also concerned in some of the reproduction troubles in phosphorus-deficient areas.

Calcium deficiency can upset the normal reproductive performance, particularly by decreasing the number of viable young in the case of multiple births. Severe deficiency results in intrauterine death of some of the fetuses in rats and pigs, possibly because of a lack of tone in the uterine muscle. Calcium deficiency is much less of an area problem than is the case for phosphorus. In fact, grazing cattle seldom show calcium deficiency.

Though of primary importance, regularity of breeding and normal pregnancy and parturition do not constitute complete proof that the calcium and phosphorus nutrition is adequate for reproduction. Despite a normal birth there may be pathological changes in the osseous system of the newborn as a result of mineral deficiency in the diet of the mother. Of equal importance, the mother's bones may be depleted to supply the minerals in the skeleton of the fetus. This has been shown to occur in various species of animals fed rations low in calcium and phosphorus. To the extent that it involves only the reserves of the minerals in the bones (Sec. 10.10), no structural injury is caused, but since, in the lactation to follow, the demand for these minerals is so large that losses from the bones cannot be prevented despite the most liberal nutrition (Sec. 17.29), it is clearly desirable to husband the reserve in the bones during gestation. It has been shown with sheep by Fraser and associates[20] and with swine by Evans[21] that losses from the bones which occur on a deficient ration can be prevented by increasing the intakes of the bone-forming minerals. Similar findings have been reported from studies with women and with rats. Severe or continued depletion of the bones results in osteomalacia (Sec. 10.12).

The fact that the mother's bones can be sacrificed in the interests of the fetus provides a means of protecting the offspring of a first pregnancy from serious skeletal defects. X-ray studies have shown that calcification is best in infants from mothers showing the highest retentions of calcium and phosphorus during pregnancy. In a study with sows fed a ration deficient in calcium, Davidson[22] found that a calcium-deficient ration did not produce an immediate effect because of the store in the maternal body. In successive farrowings there was an increase in the number of pigs born weak or dead and a decrease in the number reared to weaning. There was a serious reduction and eventual failure of the milk supply. Several investigators have reported fetal rickets, gross hypoplasia of the enamel, and defective dentine in infants from mothers under-nourished in bone-forming minerals during pregnancy.

16.10 Calcium and Phosphorus Requirements

In addition to providing for the growing fetus, the calcium and phosphorus intakes during gestation must be sufficient to meet the maintenance require-ment of the mother and also to build up reserves in her bones in so far as this is possible. Data as to the amounts of these minerals present in the products of conception at birth are of limited value in arriving at the actual intake needs, for several reasons. For most species there is a lack of information as to the maintenance requirement, and there is no reliable figure as to the percentage retention which can be assumed where it is desired to ensure storage in the mother's bones to the fullest extent possible, as well as in the fetus. The efficiency of utilization is recognized to be low under these conditions. Further, the daily requirement for fetal growth cannot be based upon the average stor-age in products of conception, for almost all of this storage takes place in the last half of gestation and especially in the last fifth. This fact is clearly brought out in Table 16.2. The intake chosen for the period as a whole must be that which will be optimum for the last days of gestation, or provision must be made for increasing a lower initial intake in accordance with increasing storage. There are data for cattle and sheep showing that the storage of calcium and phosphorus during pregnancy follow the same course as indicated for swine in Table 16.2. Calcium deficiency in gestation may result in spontaneous fractures of long bones, parturient paresis (Sec. 17.37), or lameness of sows.

While extensive data are available for rats, few studies have been made with farm animals providing information as to the minimum intakes of calcium and phosphorus which can be relied upon to be optimum over the gestation period alone. In a very extensive balance experiment with swine, Evans com-pared intakes of approximately 0.6 and 0.04 percent of calcium in a basal ration adequate in other nutrients including vitamin D. On the low level of intake only 5 g of calcium was retained during gestation, although five times this amount was found in the products of conception at the close. A large depletion of the sow's skeleton therefore occurred. There was a lack of milk secretion following farrowing. The high level resulted in a storage in gestation greatly in excess of the demand for fetal growth, demonstrating the ability of the maternal organism to build up her reserves when the dietary supply is sufficiently large. The striking effect of an inadequate dietary supply of calcium on the bones of the gilt, as observed by Hogan, is shown in Fig. 16.2.

Table 16.5 compares the N.R.C. calcium and phosphorus requirements of several species of animals for maintenance, growth, reproduction, and lacta-tion. The percentages of these elements required in the feed is the same or only slightly higher for reproduction than for maintenance, and considerably lower than for maximum early growth or lactation. Following are the recommended levels of calcium and phosphorus, respectively for reproduction, as percent of the dry diet: dairy cows, 0.34 Ca and 0.21 P; beef cows, 0.18 and 0.18; swine,

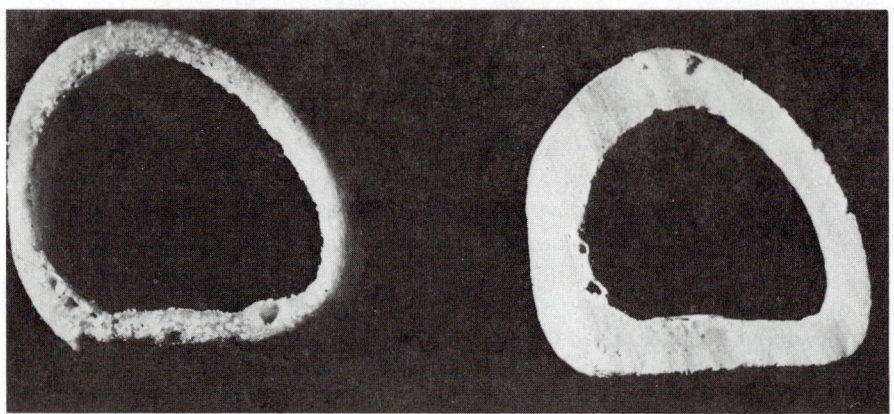

Fig. 16.2 Cross section of metacarpal bones. The bone on the left was taken from a gilt on a low-calcium ration. Note the thin-walled, spongy, porous condition compared to the bone on the right, produced on an adequate ration. The poor bone contained only half as much calcium. (*Courtesy of A. G. Hogan, University of Missouri.*)

TABLE 16.5. The Calcium and Phosphorus Requirements of Animals for Maintenance, Growth, Reproduction and Lactation

Animal	Maintenance	Growth	Reproduction	Lactation
	Calcium requirements, % of diet			
Beef cattle	0.18	0.50	0.18	0.39
Dairy cattle	0.23	0.43	0.34	0.53
Swine	—	0.80	0.75	0.75
Horse	0.3	0.70	0.50	0.50
Sheep	0.27	0.45	0.24	0.48
Rabbit	—	0.4	0.45	0.75
Rat	—	0.56	0.67	0.67
Chicken	—	1.0	2.75	—
Turkey	—	1.2	2.25	—
	Phosphorus requirements, % of diet			
Beef cattle	0.18	0.38	0.18	0.36
Dairy cattle	0.18	0.33	0.21	0.39
Swine	—	0.60	0.50	0.50
Horse	0.20	0.50	0.35	0.35
Sheep	0.25	0.25	0.23	0.34
Rabbit	—	0.22	0.37	0.5
Rat	—	0.44	0.56	0.56
Chicken	—	0.7	0.6	—
Turkey	—	0.8	0.75	—

Source: N.R.C. Nutrient Requirements of Domestic Animals.

0.75 and 0.50; horses, 0.33 and 0.25; sheep, 0.24 and 0.23; rats, 0.67 and 0.56; rabbits, 0.45 and 0.37; turkeys, 2.25 and 0.75; and chickens 2.75 and 0.75. These levels have been used extensively in practical production of animals with favorable results.

16.11 Iron

The studies by Mitchell showed that the sow producing a litter of eight stored 580 mg of iron in the products of conception. During the last week of pregnancy, the daily storage was 12.3 mg (Table 16.2) and the average figure for the period as a whole was 5.5 mg. Even the latter figure is greatly in excess of the maintenance requirement. Studies with women have also indicated that the demands for fetal growth are much greater than those for maintenance. It is probable that, for all species, the maintenance requirement must be increased two or three times to cover the needs of gestation. This is readily understandable on the basis that the principal iron metabolism in maintenance results from the breakdown and resynthesis of hemoglobin, which involves no loss of iron from the body and thus no requirement for replacement (Sec. 10.28), whereas gestation calls for the building of new tissue as in growth.

A large store in the newborn is needed to be drawn upon for blood formation during the suckling period, when milk, which is low in iron is the principal food (Sec. 10.27). The problem is critical in young pigs, where anemia always develops during the suckling period unless iron is provided. Pond and associates[23] reported that injection of iron-dextran into sows at 100 days of gestation did not influence hemoglobin levels or help prevent anemia in suckling pigs. Intramuscular injection of 150 mg of iron as iron-dextran into suckling pigs at 10 days of age resulted in higher hemoglobin levels than oral doses of 400 mg of ferric or ferrous oxide or sulfate. Despite the very large increase in the iron requirement which accompanies pregnancy, there is no evidence that a practical problem exists for other farm animals.

16.12 Iodine

The occurrence of goiter in farm animals as a result of a deficiency in the diet of the mother during gestation has been discussed (Sec. 10.38). The increase in iodine metabolism during pregnancy is indicated by the fact that the blood level is doubled at this time. In areas where goiter troubles occur, the most practicable way of supplying the need for iodine is the use of iodized salt. It is a common recommendation that cattle, sheep, and horses receive 1 percent of stabilized iodized salt (containing 0.0076 percent iodine) in the grain portion of their rations and that swine be fed this salt as 0.5 percent of the grain ration. The N.R.C. report for swine mentions evidence that the requirement is approximately 4.4 μg per kilogram body weight. Where goiter troubles in mammals have been experienced, feeding iodized salt during the last three-quarters of the gestation period only has given protection. The need for supplementary iodine at other periods in the life of the animal and in areas where goiter troubles are not evident has been discussed (Sec. 10.39).

16.13 Selenium

Selenium has been shown to be an important element for reproduction and early growth. The role of selenium in preventing white muscle disease in lambs and calves has been reviewed (Sec. 10.48). The effective way to prevent early death of newborn animals from this disease is to provide adequate amounts of selenium and vitamin E during gestation. These nutrients are transferred through the placenta to protect the developing fetus and provide limited reserves. Buchanan-Smith et al.[24] found that satisfactory reproduction in ewes fed a purified diet was obtained only when both selenium and vitamin E were provided. The fertility of rams was not affected by the basal diet. Necrosis was evident in skeletal and cardic muscles of ewes that died, but no pathology was noted in reproductive tracts or fetuses of the ewes. More recent studies by Whanger et al.[25] showed that all lambs from ewes fed a purified diet without supplements of selenium and vitamin E during gestation and early lactation developed white muscle disease, while only a few lambs were affected when either nutrient was provided. Vitamin E appeared to be more effective in preventing white muscle disease than selenium alone. Their results suggest that 0.1 ppm of selenium in the dry diet is not always an adequate level. Sang-Hwan et al.[26] reported that the selenium requirement of reproducing ewes and their lambs fed a practical type diet was about 0.12 percent of the dry diet.

Julien et al.[27] confirmed earlier reports that selenium deficiency may be one cause of retained placentas in cows. When they changed the diet to increase the daily selenium intake from 0.23 mg to 0.92 mg the incidence of retained placentas declined from 38 percent to zero.

16.14 Other Mineral Elements

Magnesium deficiency during gestation and early lactation may result in *grass tetany* in ruminants (Sec. 10.19). The nutritional relationships for dairy cows are discussed in Sec. 17.32. Harmon et al.[28] have shown that a level of 0.04 percent of magnesium in the dry diet is adequate for gestation in sows. The N.R.C. reports (Table 15.17) show that beef cattle require 0.1 percent, dairy require 0.07 percent, and sheep require 0.08 percent magnesium in the diet.

Zinc is a critical mineral element for reproduction. Soon after it was shown that zinc deficiency was the cause of parakeratosis in pigs (Sec. 10.45) it was discovered that a deficiency in pregnant rats caused congenital malformations in the skeleton and organs. Difficult deliveries and death of many females occurred. Apgar[29] has shown that injection of 900 μg on day 18 of gestation gives normal parturitions. Fosmire et al.[30] reported that zinc levels of 11 mg per liter of drinking water prevented the difficulties.

All of the other mineral elements required for growth are also essential for successful reproduction. However, the amounts needed have not been determined for many of them. It is safe to assume that a diet which is adequate for growth will also be satisfactory for reproduction. Research is being published regularly which will fill the gaps in our knowledge about the elements that are critical for reproductive functions. Values for growth are shown in Table 15.17.

16.15 Vitamin A

The hind (cow) also calved in the field, and forsook it, because there was no grass. And the wild asses did stand in the high places, they snuffed up the wind like dragons; their eyes did fail, because there was no grass.

Jeremiah 14:5–6.

In all species, in so far as studied, the reproductive process is dependent upon an adequate supply of vitamin A. This is true for both sexes. The generalized effect of this vitamin on the epithelial tissues explains its role in reproduction. In the male a deficiency causes a degeneration of the germinal epithelium of the testes, which results in a decrease in spermatogenesis and its eventual cessation if the deficiency is prolonged and severe. In the female the vaginal epithelium becomes cornified, which may result in irregularity of estrus and delayed breeding. The major reproductive disturbance in the female, however, occurs during the latter part of gestation, resulting in an abortion or the birth of weak, blind, or dead offspring. These results stem from a keratinization of the vaginal epithelium and degeneration of the placenta. Injury to the epithelium has been reported to cause congenital malformations in the soft tissues and bones. Blindness in calves is due to constriction of the optic nerves caused by stenosis of the optic foramen.

The symptoms which occur in farm animals differ somewhat according to the severity of the deficiency, and some may be more prominent in one species than another. In cattle, Hart and Guilbert[31] have reported the birth of dead or weak calves with frequent retention of the placenta, a condition which simulates infectious abortion. Severe diarrhea resembling white scours was present in the weak, newborn calves. These results occurred in animals, negative to the blood test for abortion, which had been maintained for an extended period on dried-up range. Following a failure during gestation, the animals commonly did not come in heat again until they had access to green feed. The disastrous results which occur in dairy cows from the continuous feeding of low-grade grass hay in contrast to the performance on high-quality legumes have been clearly shown by research and practical experience to be lack of vitamin A.

In sows a disturbance of the estrus cycle and the farrowing of premature, weak, or dead pigs have been reported by Hughes and associates[32] (Fig. 16.3). In sheep the lambs die in utero or soon after birth. In a study by Miller and associates[33] with ewes depleted to night blindness, 65 percent of them conceived, but every lamb was born dead or died within 48 hours.

16.16 Vitamin A Requirements

Early studies by Guilbert and associates demonstrated that the carotene requirements for growth were related directly to body weight (Sec. 15.40). Presumably this relationship holds throughout life, but more vitamin A activity is required for rapid growth than for prevention of symptoms. The requirement for reproduction is also higher than for the prevention of clear deficiency.

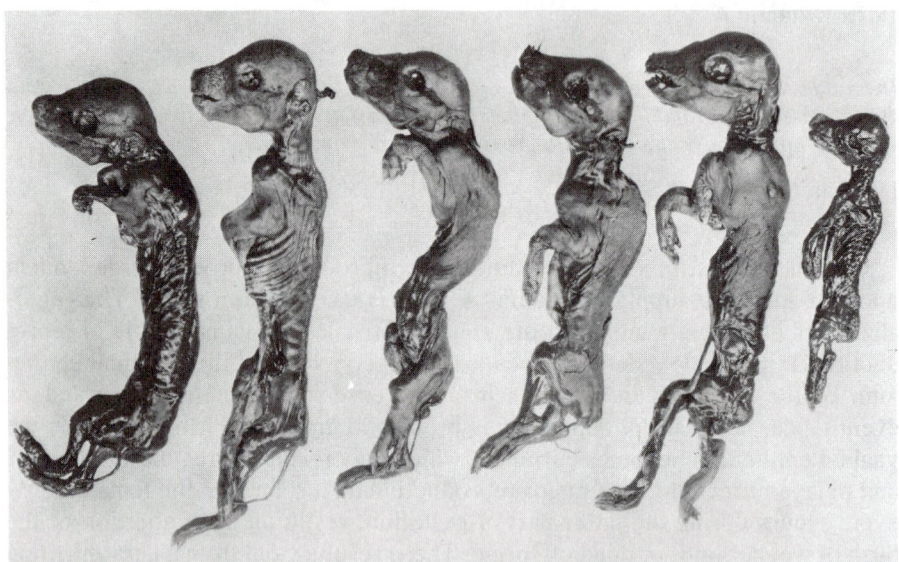

Fig. 16.3 Fetuses removed from sow in advanced vitamin A deficiency. These fetuses were obtained on a postmortem examination 81 days after the sow was due to farrow. They reveal advanced stages of resorption. (*Kansas Agr. Expt. Sta. Tech. Bull. 23, courtesy of authors.*)

The approximate requirements for vitamin A have been summarized in Table 15.18 to compare the levels for maintenance, reproduction, and lactation from the N.R.C. reports (see appendix tables). The values are also shown per unit of body weight. It is seen that the requirements are higher for reproduction than for maintenance whether the values are expressed per unit of feed or body weight. If the requirements are proportional to body weight and similar for all mammals one would expect the differences between the different animals to be much less than shown in the table. It is possible that some of the N.R.C. reports allow a larger margin of safety than others. There is no problem in meeting the needs of grazing animals if reasonably good pasture is available. The needs of sows may be met by liberal inclusion of bright green alfalfa in the grain mixture when they are not on good pasture, and yellow corn can make a substantial contribution unless its vitamin value has been depleted in storage. For animals in confinement vitamin A should be added.

Fortunately, the ability of animals to store vitamin A protects them from reproductive failure during short periods on feed deficient in this factor. The vitamin intake of the mother has little influence on the store in the newborn. Experiments have shown that massive intakes are required to increase the limited placental transfer that normally occurs. A liberal intake does result in a substantial liver store in the mother, which in turn provides the newborn with a colostrum and milk that are richer in the vitamin than otherwise. Thus the situation with respect to the placental and mammary transfer of A is the opposite of that for iron (Sec. 17.34).

16.17   Vitamin D

This vitamin is needed during fetal growth, as well as during body growth, to ensure adequate calcium and phosphorus assimilation. In the case of rats, most studies have reported an increased calcium and phosphorus retention during pregnancy and an increased content of the minerals and of the vitamin in the newborn where liberal intakes of the factor have been provided. Several studies have been carried out with women. In some cases the calcium and phosphorus balances have been improved by additions of the vitamin, in others not. It is generally agreed that the newborn infant from a mother who has received a liberal intake of the antirachitic factor is less susceptible to rickets, because of the storage of the vitamin in the fetus. In the case of the cow, however, Eaton and associates[34] did not find any significant increased placental transfer from extra vitamin D supplements.

It is clear that the vitamin is needed for normal reproduction in farm animals, but rigid experimental conditions are required to demonstrate this need, as is illustrated by the studies of Wallis.[35] By keeping cows out of the light for a long period and by using a ration in which molasses beet pulp replaced hay, he was able to produce deficiency symptoms in the course of lactation, followed by the birth of rachitic calves. No recommendations are made in the N.R.C. reports as to vitamin D requirements for reproduction in cattle, sheep, or horses, since sun-cured roughages or usual exposure to sunlight provides all that is needed. The vitamin requirements for other animals for reproduction are shown in Table 16.6. In the case of swine an intake of 275 I.U. per kilogram of dry feed, or 550 I.U. per day, is proposed for bred gilts and sows. For dairy cows not exposed to sunlight a daily intake of 5000 to 6000 I.U. will prevent deficiency signs in newborn calves. Data are not available for

TABLE 16.6. Vitamin Requirements for Reproduction (amount per kilogram of diet)

	Pig	Chicken	Turkey	Rabbit	Rat
Vitamin A, I.U.	4,100	4,000	4,000	>1160	12,000
Vitamin D, I.U.	275	500	900	–	–
Vitamin E, I.U.	11	–	25	40	33
Vitamin K, mg	2.2	–	–	0.2	–
Thiamin, mg	1.5	0.8	–	–	2.8
Riboflavin, mg	4.0	3.8	–	–	4.4
Pantothenic acid, mg	16.5	10	3.8	–	8.9
Niacin, mg	22.0	10	16.0	–	–
Pyridoxine, mg	–	4.5	–	–	0.67
Biotin, mg	–	0.15	–	–	–
Choline, g	–	–	–	–	1.0
Folicin, mg	–	0.35	–	–	–
Vitamin B_{12}, µg	14.0	3.0	800.0	–	5.6

Source: N.R.C. Nutrient Requirements of Domestic Animals.

other species. Under conditions of practice, sunlight or the kind of ration that is satisfactory for growth will certainly take care of the vitamin D requirements of farm animals during reproduction.

16.18   Vitamin E

The effects of a deficiency of vitamin E have been described (Sec. 11.17). The deficiency signs related to reproductive failure include fetal death and resorption or birth of weak young in the rat, mouse, hamster, and guinea pig. Poor hatchability of eggs occurs in poultry. There is testicular degeneration in the rat, dog, hamster, rabbit, pig, fish, and male chicken, all irreversible except in the chicken. In contrast there appear to be no specific effects on reproduction in cattle, sheep, and goats and the evidence is contradictory in horses. Gullickson et al. reported a study with cattle in which both males and females were fed a ration of rice straw, and an E-low grain mixture over three generations, no deleterious effects were noted on the estrus cycle or on spermatogenesis. Thirty services produced 25 conceptions, 19 normal parturitions, and no abortions. In other studies sheep and goats had apparently normal reproduction on rations which resulted in failure in rats. An extensive study with bulls used in artificial insemination showed no advantage from feeding wheat germ oil, a rich source of vitamin E, on semen volume, quality or fertility. Injection of large doses of vitamin A, D, and E at drying off and freshening, a practice some people recommend had no effect on reproduction, productivity, or health in a large field test in New York.[36]

A primary sign of vitamin E deficiency in the newborn or suckling young is muscle dystrophy which is preventable by supplying lambs with 11 mg of vitamin E per kilogram of body weight plus 0.1 mg of selenium per kilogram of diet. Ames[37] has shown that the vitamin E requirement for reproduction in the rat increases substantially with age.

It has been established that placental transfer takes place and can be increased somewhat when a ration low in E is supplemented by a concentrated source of the vitamin. A higher level of E in the blood and liver of the newborn results. Thus, authorities recommend that the ration during pregnancy contain a liberal supply of the vitamin. This can readily be accomplished by the appropriate selection of natural feeds. This does not obviate, however, the need for an adequate supply of E in the lactation ration (Sec. 17.38).

The vitamin E–selenium relationships have been discussed in Sec. 11.18.

16.19   B-vitamins

It is obvious that all of the B-vitamins required for growth are also required for reproduction because the production of a fetus is involved. The N.R.C. vitamin requirements for reproduction are summarized in Table 16.6. The requirements are adequate to permit normal development of the fetus and to give body reserves in the young as well as in the mother for the stress of parturition and early lactation, or for high hatchability of the eggs of chickens. Quantitative data for all B-vitamins for all species are not available. Additional data are shown in appendix tables. There appears to be no uniform pattern in comparing the requirements for growth and reproduction. It is clear that more research is needed to define more accurately the requirements of different animals for the various stages of the reproductive cycle.

There have been reports of unidentified factors needed specifically for reproduction, but the possibility remains that the deficiencies noted may have been due to inadequate intakes of one of the known factors for which quantitative requirements for reproduction have not been clearly established.

EGG PRODUCTION

Differing from mammals, which nourish the embryo inside their bodies, give birth to living young, and nurse them, birds produce eggs which contain sufficient nutrients for the embryo to develop outside the body and no special food is required after hatching. In the hen the egg-formation phase of reproduction has been extended into a continuous process, aside from the molting period, whereby egg production has become a tremendous industry as a source of human food, as well as serving in the propagation of the species.

The egg of the hen is made up approximately as follows: yolk, 31 percent; albumen or white, 59 percent; and shell, 10 percent. The albumen contains, approximately, water, 88 percent; protein, 11 percent; and carbohydrate, 1 percent. The yolk consists of water, 48 percent; protein, 17 percent; fat, 33 percent; carbohydrate, 1 percent; and ash, 1 percent. The development of the egg starts in the ovary where the yolk portion is formed. Here there are many ova, each enclosed in a follicle. The yolk is deposited in concentric layers, and when the process is completed, the follicle bursts and the yolk, surrounded by a membrane, passes into the oviduct. Here the albumen is put on and finally the shell, each process requiring several hours. During its passage through the oviduct, the developing egg is fertilized if sperm are present. After shell formation is completed, the egg passes out through the vent. Under proper temperature conditions, the fertilized egg develops into the chick in 21 days. The reproductive process in the cock is similar to that in mammals.

16.20 Nutritive Requirements for Egg Production

The hen ranks with the dairy cow in her productive performance. In a year she may produce up to four times as much dry matter as is contained in her body. Clearly an intensive metabolism and very large nutritive requirements are involved. The egg has the following proximate composition: water, 66 percent; protein, 13 percent; fat, 10.5 percent; ash, 10.5 percent. These figures reveal the fact that, in addition to the energy requirement, there are large demands for protein and especially for mineral matter. There are also important requirements for various vitamins. Besides the large nutritive demands for the formation of the egg as such, there are additional requirements for the production of an egg that will hatch and yield a strong chick. The magnitude and complexity of the demands for the intensive egg production which characterizes the modern commercial practice have made its nutrition a specialized field, which has been fully reviewed by Scott et al.[38]

The nutritional needs above maintenance depend upon the number and size of the eggs produced. These requirements are commonly expressed per unit of the total feed intake. On this basis the protein requirement is similar to that during the latter part of growth (15 percent). This protein must be of high quality, as indicated by the fact that egg protein stands at the top of all protein sources studied, in terms of biological value (Table 15.8). The average egg contains approximately 2 g of calcium and 0.12 g of phosphorus. Nearly all of the calcium is in the shell, which contains 94 percent of calcium carbonate, while the phosphorus is concentrated in the yolk, principally combined with protein. Owing to this need for shell formation, the calcium requirement of the laying hen is over twice that of the growing chick, per unit of feed, and an even greater excess of the need of any mammal for any purpose.

The intensity of the metabolism involved is reflected in a doubling of the calcium level in the blood serum during the laying period. A 2-kg hen which lays an egg daily requires for its egg alone twice as much calcium as the child, many times the hen's size, needs for growth. A deficiency of the element results in thinner shells, a marked depletion of the bones of the hen, and a lowering of egg production. The same effect on the hen and upon her production occurs from a lack of phosphorus. Though this element is needed in much smaller amounts than is calcium, no mammal has as high a requirement, per unit weight, for any purpose. Some mobilization of calcium and phosphorus from the bones during heavy egg production is a normal physiological process even as is the case for lactation (Sec. 17.29). This fact emphasizes the importance of building up the reserves in the growing pullet and of continuing a high level of feeding at all times in the case of the hen.

The iron requirement of the laying hen is very large in proportion to her maintenance need, as is evident from the fact that the average egg contains 1.1 mg of this element. Manganese is another mineral element of special importance from the standpoint of eggshell strength and hatchability.

All of the vitamins needed by chicks for growth are also required for laying and breeding hens. In the case of some of them the amounts needed for maximum egg production are not sufficient to produce eggs of high hatchability. This is true in the case of riboflavin, pantothenic acid, biotin, and folacin.

The N.R.C. nutritive requirements for breeding hens are set forth in Appendix Table VI. It is noted that data are lacking for some of the vitamins and trace minerals. To ensure an adequate supply of all of these nutrients in a laying mash capable of high production, the supplementation of mixtures of common feeds with calcium, phosphorus, iodized salt, manganese, zinc, riboflavin, vitamin B_{12}, vitamin A, and vitamin D is generally recommended.

16.21   Effect of Ration on Nutritive Value of Eggs

While the mineral content of eggs, except for iodine and manganese, is not influenced by the nature of the diet, there are marked effects in the case of several of the vitamins. This is particularly true of vitamin A, vitamin D, and riboflavin. The kind of ration which will result in the best production and hatchability is also the kind that provides eggs of the highest nutritive value for human consumption.

NOTES

1. F. H. A. Marshal and J. Hammond, Fertility and animal breeding, 6th ed., *Ministry Agr. Engl. Fish. Bull.* 39, 1945.

2. C. R. Stockard and G. N. Papanicolaou, The existence of a typical oestrous cycle in the guinea-pig—with a study of its histological and physiological changes, *Am. J. Anat.,* **22:**225–265, 1917.

3. J. H. Leathem, Nutritional effects on hormone production, Proc. Seventh Bien. Symp. on Animal Production, *J. Animal Sci.,* **25:**68–82, 1966 (supplement).

4. D. H. Curnow, T. J. Robinson, and E. J. Underwood, Estrogenic action of extracts of subterranean clover, *Aust. J. Expt. Biol. Med. Sci.,* **26:**171–180, 1948.

5. H. H. Mitchell, W. E. Carroll, T. S. Hamilton, and G. E. Hunt, Food requirements of pregnancy in swine, *Illinois Agr. Expt. Sta. Bull.* 375, 1931.

6. L. E. Casida, Effect of feed level on some reproductive phenomena of cattle, sheep, and swine, in Reproductive Physiology and Protein Nutrition, Rutgers Univ. Press, New Brunswick, N.J., 1959, pp. 35–44.

7. J. Apgar, D. Aspros, J. E. Hixon, R. R. Saatman, and W. Hansel, Effect of restricted feed intake on the sensitivity of corpus luteum to LH in vitro, *J. Animal Sci.,* **41:**1120–1123, 1975.

8. B. F. Chow and C. J. Lee, Effect of dietary restriction of pregnant rats on body weight gain of the offspring, *J. Nutrition,* **82:**10–18, 1964.

9. A. M. Thomson and W. Thomson, Lambing in relation to the diet of the pregnant ewe, *Brit. J. Nutrition,* **2:**290–305, 1948; Effect of diet on milk yield of the ewe and growth of her lamb, *ibid.,* **7:**263–274, 1953.

10. K. Breirem, A. Ekern, and T. Homb, Relation of nutrition of the young animal to subsequent fertility and lactation, *Fed. Proc.,* **20:**275–283, 1960.

11. S. A. Asdell and Mary F. Crowell, The effect of retarded growth upon the sexual development of rats, *J. Nutrition,* **10:**13–24, 1935.

12. D. R. Lamond, The influence of undernutrition on reproduction in the cow, *Animal Breeding Abstr.,* **38:**359–372, 1970.

13. J. H. Tops, Effect of energy and protein deprivation on the performance of beef cattle, in A. J. Smith (ed.), Beef Cattle Production in Developing Countries, Univ. of Edinburgh, 1976, pp. 204–215.

14. P. W. Moe and H. F. Tyrrell, Metabolizable energy requirements of pregnant dairy cows, *J. Dairy Sci.,* **55:**480–483, 1972.

15. M. M. Nelson and H. M. Evans, Relation of dietary protein levels to reproduction in the rat, *J. Nutrition,* **51:**71–84, 1953.

16. J. H. Leathem, Male reproductive system and protein nutrition in Reproductive

Physiology and Protein Nutrition, Rutgers Univ. Press, New Brunswick, N.J., 1959, pp. 12–22.

17. H. C. Stuart, Effect of protein deficiency on the pregnant woman and fetus and on the infant and child, *New Engl. J. Med.,* **236:**507–513, 537–541, 1947.

18. H. A. Hunscher and associates, Metabolism of women during the reproductive cycle. V: Nitrogen utilization, *J. Biol. Chem.,* **99:**507–520, 1933.

19. A. Theiler, H. H. Green, and P. J. du Toit, Studies in mineral metabolism. III: Breeding of cattle on phosphorus deficient pasture, *J. Agr. Sci.,* **18:**369–371, 1928; Phosphorus in the live stock industry, *Union S. Africa J. Dept. Agr.,* **8:**460–504, 1924.

20. A. H. H. Fraser, W. Godden, and W. Thomson, The effect of a calcium-deficient diet on pregnant ewes, *Vet. J.,* **89:**408–411, 1933.

21. R. E. Evans, Protein and mineral metabolism in pregnant sows on a normal or high calcium diet, compared with a calcium-deficient diet, *J. Agr. Sci.,* **19:**752–798, 1929.

22. H. R. Davidson, Reproductive disturbances caused by feeding protein-deficient and calcium-deficient rations to breeding pigs, *J. Agr. Sci.,* **20:**233–264, 1930.

23. W. G. Pond, J. H. Maner, and J. K. Loosli, Parenteral iron administration to sows during gestation or lactation, *J. Animal Sci.,* **20:**747–750, 1961.

24. J. G. Buchanan-Smith and associates, Effect of vitamin E and selenium deficiencies in sheep fed a purified diet during growth and reproduction, *J. Animal Sci.,* **29:**808–815, 1969.

25. P. D. Whanger, P. H. Weswig, J. A. Schmitz, and J. E. Oldfield, Effects of selenium and vitamin E deficiencies on reproduction, growth, blood components and tissue lesions in sheep fed purified diets, *J. Nutrition,* **107:**1288–1297, 1977.

26. Oh Sang-Hwan, A. L. Pope, and W. G. Hoekstra, Dietary selenium requirement of sheep fed a practical-type diet assessed by tissue glutathione peroxidase and other criteria, *J. Animal Sci.,* **42:**984–992, 1976.

27. W. E. Julien and associates, Selenium and vitamin E and incidence of retained placenta in parturient dairy cows, *J. Dairy Sci.,* **59:**1954–1959, 1976.

28. B. G. Harmon, T. C. Liu, A. H. Jensen, and D. H. Baker, Dietary magnesium for sows during gestation and lactation, *J. Animal Sci.,* **42:**860–865, 1976.

29. J. Apgar, Effects of zinc deficiency and zinc repletion during pregnancy on parturition in two strains of rats, *J. Nutrition,* **107:**1399–1403, 1977.

30. G. J. Fosmire, S. Greeley, and H. H. Sandstead, Maternal and fetal response to various suboptimal levels of zinc intake during gestation in the rat, *J. Nutrition,* **107:**1543–1550, 1977.

31. G. H. Hart and H. R. Guilbert, Vitamin-A deficiency as related to reproduction in range cattle, *California Agr. Expt. Sta. Bull.* 560, 1933.

32. J. S. Hughes, C. E. Aubel, and H. F. Leinhardt, The importance of vitamin A and vitamin C in the ration of swine, *Kansas Agr. Expt. Sta. Tech. Bull.* 23, 1928.

33. R. F. Miller, G. H. Hart, and H. H. Cole, Fertility in sheep as affected by nutrition

during the breeding season and pregnancy, *California Agr. Expt. Sta. Bull.* 672, 1942.

34.   H. D. Eaton and associates, The placental transfer and colostral storage of vitamin D in the bovine, *J. Dairy Sci.,* **30:**787–794, 1947.

35.   G. C. Wallis, Some effects of a vitamin D deficiency on mature dairy cows, *J. Dairy Sci.,* **21:**315–333, 1938.

36.   D. A. Hartmen, R. P. Natzke, and R. W. Everett, Injectable vitamins A, D, and E: A field study, *J. Dairy Sci.,* **59:**91–96, 1976.

37.   S. R. Ames, Age, parity and vitamin A supplementation and the vitamin E requirement of female rats, *Am. J. Clinical Nutrition,* **27:**1017–1025, 1974.

38.   M. L. Scott, M. C. Nesheim, and R. J. Young, Nutrition of the Chicken, 2nd ed., M. L. Scott and Assoc., Ithaca, N.Y., 1976.

SUPPLEMENTARY LITERATURE

Cunha, T. J., A. J. Warnick, and M. Koger (eds.): Factors Affecting Calf Crop, Univ. of Florida Press, Gainesville, 1967.

Fosmire, G. J., S. Greeley and H. H. Sandstead: Maternal and fetal response to various suboptimal levels of zinc intake during gestation in the rat, *J. Nutrition,* **107:**1543–1550, 1977.

Greenhalgh, J. F. D., and associates: Coordinated trials on protein requirement of sows. I: A comparison of four levels of dietary protein in gestation and two in lactation, *Animal Prod.,* **24:**307–321, 1977.

Greep, R. O., M. A. Koblinsky, and F. S. Jaffe: Reproduction and Human Welfare: A Challenge to Research, M.I.T. Press, Cambridge, Mass. 1976.

Hafez, E. S. E. (ed.): Adaption in Domestic Animals, Lea and Febiger, Philadelphia, Pa., 1968.

Hafez, E. S. E. (ed.): Reproduction in Farm Animals, Lea and Febiger, Philadelphia, Pa., 1974.

Hidiroglou, H., M. Ivan, and K. J. Jenkins: Influences of barley and oat silages for beef cows on occurrence of myopathy in their calves, *J. Dairy Sci.,* **60:**1905–1909, 1977.

Johnson, A. D., W. R. Gomes, and N. L. Vandemark: The Testis. I: Development, Anatomy and Physiology. II: Biochemistry. III: Influencing Factors, Academic Press, New York, 1970.

Libal, G. W., and R. C. Wahlstrom: Effect of gestation metabolizable energy levels on sow productivity, *J. Animal Sci.,* **45:**286–292, 1977.

Louca, A., A. Mavrogenis, and M. J. Lawlor: Effect of plane of nutrition in late pregnancy on lamb birth weight and milk yield in early lactation of chios and awassi sheep, *Animal Prod.,* **19:**341–349, 1974.

Nalbandov, A. V.: Reproductive Physiology of Animals and Birds, W. H. Freeman Co., San Francisco, Calif., 1976.

Robinson, J. J.: The influence of maternal nutrition on ovine foetal growth, *Proc. Nutrition Soc.*, **36:**9–16, 1977.

Terrill, C. E.: Proc. Symp. on Management of Reproduction in Sheep and Goats, Am. Soc. Animal Sci., July 1977.

Vaccaro, L. P. de: Some aspects of the performance of purebred and crossbred dairy cattle in the tropics. I: Reproductive efficiency in females, *Animal Breeding Abstr.*, **41:**571–591, 1973; II: Mortality and culling rates, *Ibid.*, **42:**93–103, 1974.

Williams, R. B.: Trace elements and congenital abnormalities, *Proc. Nutrition Soc.*, **36:**25–32, 1977.

Wilson, J. G.: Teratogenic effects of environmental chemicals, *Fed. Proc.* **36:**1698–1703, 1977.

Chapter 17

Lactation

The cow is the foster mother of the human race. From the day of the ancient Hindoo to this time have the thoughts of man turned to this kindly and beneficent creature as one of the chief sustaining forces of human life.

W. D. Hoard (1836–1918)

A 600-kg cow producing 10,000 kg of milk in a year secretes in this milk approximately eight times as much dry matter as is present in her entire body. The highest recorded milk production of a cow in the United States of over 23,000 kg in a year was more than 15 times the dry matter in her body. While less subject to direct measurement, it is apparent that milk secretion in the sow nursing a large litter also represents a noteworthy physiological performance, and the same is true for the mare and dairy goat. The studies of Macy and coworkers[1] show that the human organism is capable of producing an astonishing output of milk. It is clear that the metabolism of lactation is tremendous. While this metabolism includes many processes such as the digestion, absorption, circulation, and mobilization of nutrients, it particularly involves the functioning of the mammary glands. As an introduction to a consideration of nutritional requirements for lactation an understanding of the physiological processes concerned is essential.

17.1 The Mammary Glands

The glands usually occur in pairs, the number of pairs varying with the species. They are modified cutaneous glands which make their appearance early in embryonic life but reach their full development only after a normal parturition. As far as is known they have no function other than milk secretion, for they can be removed at any stage of the life cycle without any observable harmful effect from their absence. A diagram showing the structure of the functioning udder is presented in Fig. 17.1.

The glands are present in a rudimentary form at birth and undergo little development until puberty, at which time a marked growth occurs. Thereafter there are periodic changes which are correlated with the ovarian cycle. Histological studies show that at each estrus there is some duct growth in the gland, and an occasional secreting cell may be formed, which explains the fact that a watery secretion has been obtained before pregnancy. With the onset of gestation, there is a large increase in growth which involves the production of ducts, alveoli, and secreting cells. A secretory activity thus develops which results in an accumulation in the gland of products making up the colostrum. With the withdrawal of the secretion following parturition, its quantity gradually rises for a period which varies in different species and then gradually falls until lactation ceases. As cessation occurs, the gland shrinks enormously, owing to the involution of the alveoli and a decrease in size of the ducts, and remains quiescent until another pregnancy starts a renewed growth and secretion.

17.2 Hormonal Control

The physiological mechanisms controlling these various events are still incompletely understood, but it is recognized that the activities of the ovaries, placenta, pituitary, and mammary glands are interrelated. While certain neurological mechanisms are of basic importance, the essential stimulus for the growth and functioning of the gland is hormonal rather than nervous, and the present evidence indicates that at least a dozen hormones are concerned, directly or indirectly. Mention can be made here of only the principal ones as presently understood. The interaction of several results in mammary growth. In the main, estrogens are responsible for duct growth, and progesterone for alveolar growth, but neither of these hormones can produce its maximum effect unless somatotropin and prolactin from the pituitary gland are present. At full development of the gland, prolactin brings about the initiation of secretion. Maximum secretory activity is dependent on the presence of adrenocorticotropin (ACTH). The gradual decline in production, which follows the peak attained after parturition (Sec. 17.5), results from a gradual cessation of secretion by the cells built up during the previous pregnancy and the lack of new cellular growth.

For a detailed review of present knowledge and its experimental basis, the student is referred to the book by Schmidt.[2]

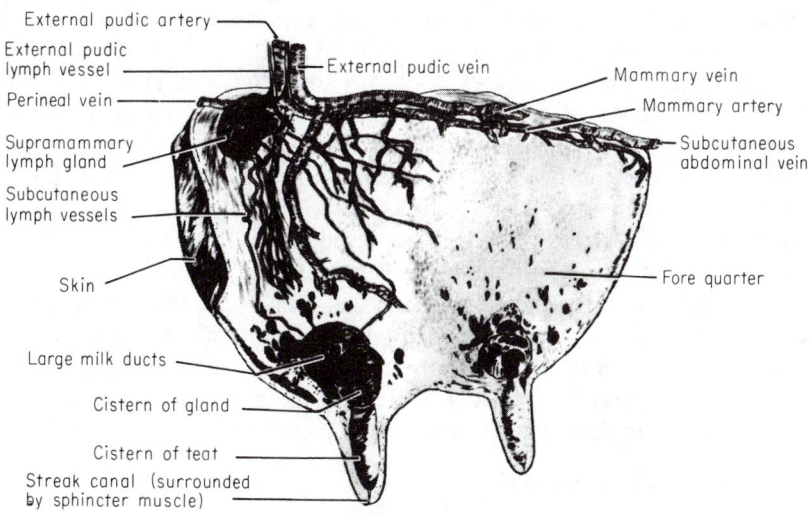

External pudic artery
External pudic lymph vessel
Perineal vein
External pudic vein
Mammary vein
Mammary artery
Supramammary lymph gland
Subcutaneous abdominal vein
Subcutaneous lymph vessels
Skin
Fore quarter
Large milk ducts
Cistern of gland
Cistern of teat
Streak canal (surrounded by sphincter muscle)

Fig. 17.1 Cross section of functioning udder of the cow. (*Courtesy of C. W. Turner, University of Missouri.*)

17.3   Influence of Iodinated Proteins on Milk Secretion

Reference was made in Sec. 15.7 to the role of thyroidal products in stimulating metabolism. Commercial sources of highly active iodinated casein have been developed for applications in animal production, dating particularly from the initial work of Reinecke and Turner[3] who developed the method of producing an iodinated casein of high biological activity. In later studies the Missouri workers and several others have shown that the feeding of this or similar products results in significant increases in milk yield and even larger effects on fat yield, in cows after the peak of lactation.

Schmidt and associates[4] published the results of extensive experiments and reviewed the earlier research. Their studies confirmed and extended the earlier data in showing an increase in milk yield and fat test when thyroprotein was fed, but the increases did not persist long even with continued feeding of the thyroprotein. In later lactation production was even lower than the controls and for the full lactation there was no significant difference in either milk yield or fat test. There were large losses in body weight and increases in heart rate, respiration rate, and body temperature. Treated cows had longer calving intervals and required more services per conception than the controls, but there were no other differences in reproduction or general health of the cows.

17.4   Induction of Lactation by Synthetic Estrogens

Following an initial finding by DeFremery in Holland that udder growth could be induced in virgin goats by anointing the udder with a salve containing estradiol benzoate, Folley and associates in England found that use of diethylstilbestrol caused udder development and a secretion of milk as well.

Reports of other workers followed, describing the induction of lactation in cows by injection or inunction, or both, of the synthetic estrogens. Next, it was shown by Folley and associates and by Hammond and Day[5] that the subcutaneous implantation of tablets of stilbestrol or hexestrol was a simple and more effective procedure. Hammond and Day reported on the treatment of cows and heifers that had failed to get in calf. While the treatment was not successful in every case and while sometimes the volume of secretion was low, most animals produced 1180 to 3000 kg over periods of 40 to 64 weeks. More recent research by Smith and Schanbacher[6] has shown that injecting estradiol-17β and progesterone for 7 days induced lactation in about 60 percent of the treated cows and milk production was about 70 percent of the best previous records of the cows. The availability of methods for measuring blood levels of the various hormones involved should lead to a fuller understanding of the factors controlling the initiation and regulation of lactation.[7]

17.5   The Course of Milk Secretion

The normal lactation curve of the cow in relation to energy intake and body weight is shown in Fig. 17.2. The time involved in reaching the peak depends upon inherited factors and upon the condition of the cow prior to calving and how she is fed and managed thereafter. The rise in secretion following parturition increases faster than food intakes. The impulse to produce milk is so strong at the start of lactation that the animal readily draws on her own reserves for a short time to produce milk. However, unless current nutrition balances milk output, the maximum yields cannot be sustained more than a few weeks on body reserves. Previous nutrition, which determines the cow's reserves at calving, is thus concerned. Weight loss in early lactation presents no problem in otherwise healthy cows unless the loss becomes excessive and if the losses are replaced in later lactation and the dry period.

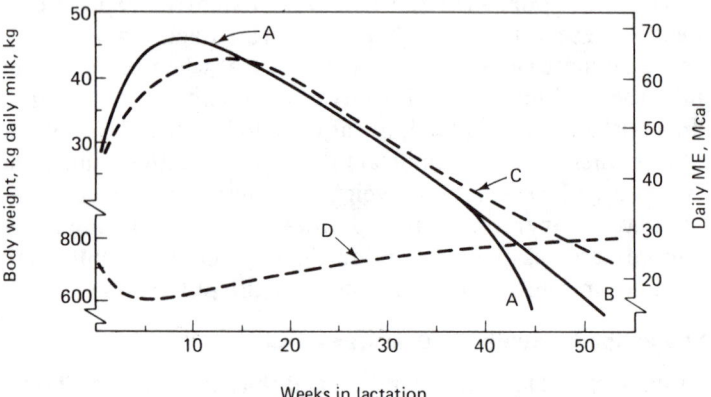

Fig. 17.2   Smoothed lactation curve of a cow showing the relative relationships of milk yield (A) for 12 months calving, (B) 15 months calving, (C) metabolizable energy intake, and (D) body weight.

Following the peak, there is a regular decline in yield such that the curve is of the descending exponential type, each month's yield being a constant percentage of that of the preceding month. *Persistency* is the term used to denote the measure of this rate of decline, which varies with the individual and in different lactations and which is accelerated at the twenty-second week after a new conception. The onset of a new pregnancy is thus a determining factor in the length of the lactation of the cow, both because of this accelerated decline and because of the necessity of giving her a rest period before another lactation. The fact that many cows will continue to secrete some milk right up to parturition if milked regularly shows that there is no physiological mechanism for absolutely stopping the process before the event. Cows which remain unbred may continue to secrete milk at a decreasing rate for two or three years or even longer.

Failure to remove the milk regularly and completely from the gland lessens its activity and brings about cessation, a fact which finds practical application in the drying off of animals. The onset of a new pregnancy during lactation results, after a period, in a more rapid decline in the secretion than otherwise occurs. In animals nursing young, lactation is usually artifically terminated at weaning, since failure to remove the secretion stops the process. Underfeeding during the declining period of secretion has an immediate effect in lowering the output, in contrast to its lesser influence at the start of lactation, but no system of feeding will counteract in any way the normal decline.

The lactation curve of the goat is similar to that of the cow, and milk-secretion studies with sheep indicate a similar one for this species also. In the case of women, however, the peak is reached much later, or there is a plateau instead of a peak. The increase after parturition may continue to the twenty-eighth week, and the secretion may persist at or near its maximum level to the fortieth week.

With lactations of substantially equal length, the yield of the cow generally increases during the first four of five. The growth of the animal as a whole is a factor during the first three lactations, and there is markedly greater hypertrophy of the gland during the second pregnancy than during the first, with smaller increases in succeeding pregnancies. Persistency decreases in succeeding lactations. Since this is true, it is apparent that the higher yields which are obtained in succeeding lactations must be due to a greater secretion during the first part of the lactation. Expressed another way, it may be said that the level at which secretion begins increases to maturity but that the total yield for the lactation is not proportional to this rise because of a declining persistency factor.

The longer the dry period, the greater the persistency in the next lactation. Shortening the dry period before a second lactation lowers yield to a much greater extent than shortening it before a later lactation. This is readily explainable on the basis of the greater growth of body and gland that takes place before the second lactation than later. The importance of an adequate dry period to build up nutrient reserves is well understood.

17.6 Milk of Different Species

The chemical composition of the milk of various species is presented in Table 17.1. Data on other minerals and on the vitamins found in cow's milk are discussed later in the chapter. It should be emphasized that the figures for individual animals may vary widely from these average values. This is certainly true for cows of different breeds and for individuals within the breed as is discussed later (Sec. 17.19). Doubtless the variations have become greatly accentuated in this species as a result of selection, but they must be expected

TABLE 17.1. Percentage Composition of Milk of Different Species

Species	Water	Protein	Fat	Lactose	Ash	Calcium	Phos- phorus	kcal
Baboon	85.8	1.6	5.0	7.3	0.3	—	—	84
Bison	82.2	5.1	7.5	4.4	0.80	—	—	115
Camel	87.2	3.7	4.2	4.1	0.75	—	—	76
Cat	82.2	9.1	3.3	4.9	0.51	—	—	101
Cow	87.2	3.5	3.7	4.9	0.71	0.121	0.095	73
Cow	85.0	3.9	5.5	4.9	0.7	—	—	92
Deer	75.0	8.7	10.4	4.4	1.49	—	—	163
Dog	75.4	11.2	9.6	3.1	0.73	0.325	0.222	164
Donkey	90.2	1.8	1.3	6.3	0.4	—	—	47
Elephant	68.0	3.1	19.6	8.8	0.50	—	—	233
Giraffe	72.4	5.8	13.5	3.4	0.9	—	—	171
Goat	86.5	3.6	4.0	5.1	0.80	0.131	0.104	79
Guinea pig	84.2	8.1	3.9	3.0	0.82	0.173	0.126	94
Human	87.5	1.0	4.4	7.0	0.21	0.035	0.013	
Kangaroo	75.4	11.5	9.1	2.4	1.60	—	—	158
Mare	89.0	2.7	1.6	6.1	0.51	0.100	0.060	54
Monkey	87.8	2.1	3.9	5.9	0.26	—	—	71
Porpoise	38.0	11.0	49.0	1.2	0.60	—	—	520
Rabbit	64.8	15.8	14.4	2.7	2.07	0.636	0.436	233
Rat	76.9	8.7	9.3	3.7	1.40	0.349	0.272	150
Reindeer	41.6	10.0	20.9	2.5	1.4	—	—	209
Ewe	80.1	5.8	8.2	4.8	0.92	0.250	0.166	127
Sow	80.4	5.4	8.3	5.0	0.85	0.252	0.151	126
Water buffalo	83.0	3.8	7.4	4.9	0.78	0.180	0.120	109
Water buffalo	78.6	5.9	10.4	4.3	0.8	—	—	163
Whale	42.8	12.2	42.3	1.3	1.42	0.300	0.193	465
Zebra	86.2	3.0	4.8	5.3	0.7	—	—	82

Source: Various sources in the literature. The data for human milk are expressed as grams per 100 ml. With the exception of this milk, the data for calorie content were calculated by the authors. The data are expressed as gross kilocalories per 100 g, calculated by Atwater's specific values for milk, viz., protein, 5.65; fat, 9.25; and lactose, 3.9. L. A. Maynard, *J. Nutrition*, **28**:443–452, 1944.

to occur to a certain extent in all species. The data presented in the table are useful, nevertheless, to indicate the nutrients involved in milk secretion and to bring out certain differences which exist among the species. It is noted that the milk of the sow and ewe contains considerably more dry matter than the milk of any of the other species of farm animals and that this is reflected in a higher energy value and ash content. Particularly noteworthy is the much higher content of calcium and phosphorus. Clearly per unit of product, the nutritive requirements for milk secretion in the sow and ewe are markedly greater than in the cow, goat, or mare. It is also interesting to note that the least variable constituent for all species is lactose. The same is true among individuals and also for different samples from the same individual, as is brought out in later discussions for the cow.

Rabbit's milk contains about 15 percent of protein and 2.0 percent of ash, considerably higher than any other species listed. The especially high fat content of the milk of the porpoise and the whale and their low water content present striking contrasts to the other mammals.

A very comprehensive compilation of the comparative composition and properties of human, cow, and goat milk, prepared for the Food and Nutrition Board by Macy, Kelley, and Sloan,[8] has been published by the National Research Council (N.R.C.). The tables include data for the various vitamins, trace minerals, amino acids, and other constituents.

Studies have been made of milk production in the sow and ewe over the lactation period by weighing the young before and after suckling at short intervals. Such an experiment has been reported for sows by Barber and coworkers[9] and for ewes by Coombe and associates.[10]

THE SECRETION OF THE MILK CONSTITUENTS

17.7 The Composition of Blood and Milk

Most of the organic constituents of milk arise from specific synthetic processes of the mammary gland, representing products which are not found elsewhere in nature. As an aid to a consideration of the physiology involved and the chemical changes which take place, representative data are presented in Table 17.2 on the composition of the milk and of the blood plasma of the cow. For convenience in later discussions, the different milk constituents are listed opposite the corresponding blood constituents. Several other constituents not shown in the table are present in both fluids, such as amino acids and other nonprotein nitrogen compounds, free and combined cholesterol, additional mineral elements, and the various vitamins. A study of the figures in the table makes it evident that milk has a quantitative composition very different from that of the blood plasma from which it is made and that there are qualitative differences as well. The two fluids are isotonic.

Certain milk constituents, including lactose, casein, and some of the fatty acids, are clearly synthetic products of the gland, while others, such as the

TABLE 17.2. Comparative Composition of Blood and Milk of the Cow

Blood plasma		Milk	
Constituent	Percent	Constituent	Percent
Water	91.0	Water	87.0
Glucose	0.05	Lactose	4.90
		Casein	2.90
Serum albumin	3.20	Lactalbumin	0.52
Serum globulin	4.40	Lactoglobulin	0.20
Neutral fat	0.06	Neutral fat	3.70
Phospholipids	0.24	Phospholipids	0.10
Calcium	0.009	Calcium	0.12
Phosphorus	0.011	Phosphorus	0.10
Sodium	0.34	Sodium	0.05
Potassium	0.03	Potassium	0.15
Chlorine	0.35	Chlorine	0.11
Citric acid	Trace	Citric acid	0.20

minerals and vitamins, pass in directly from the bloodstream. There are various organic constituents for which there is a lack of information on whether they enter directly or arise as products of the gland's metabolism.

The milk of the individual cow tends to be of constant composition for most constituents, but there are periodic fluctuations, notably of fat, and there are characteristic changes over the course of the lactation. With the normal decline in yield, the percentage of fat rises, and so does the protein to a lesser degree. In contrast, the lactose declines slightly, and for the maintenance of osmotic relations, its decline is balanced by a rise in chlorides. These same changes tend to occur when the yield is subjected to an abnormal drop as the result of sickness, off-feed, or other disturbing factors.

In general, the nature of the ration has little influence on the percentage composition of milk. Special cases in which certain specific feeds have been found to affect the fat percentage are mentioned in Sec. 17.13. It has been shown that prolonged undernutrition, which results in a drop in milk yield and in body condition, can also result in a lowering of the percentage of *nonfat solids* (NFS). The lowering occurs primarily in the protein fraction. The evidence indicates that the cause is a deficient intake of energy rather than of protein or any other specific nutrient. The concentration of certain vitamins in the milk is markedly influenced by the nature of the feed. By contrast, the percentages of the major mineral elements are not affected by commonly fed diets or other factors (Sec. 17.15).

17.8 Mechanism of Milk Secretion

A complex series of reactions is involved in the secretion process, viz., the passage of blood constituents into the cells, the synthesis of secretory granules, the movement of the granules into the lumen of the alveoli, the control of water content of the secretion in the alveoli, and the passage of the product into the ducts.

Milk secretion is a continuous process. As a result of the accumulation of milk in the udder, intermammary pressure gradually increases. This increasing pressure is considered to result eventually in a declining rate of secretion. Here lies the basis for the advocacy of frequent milking as a means of obtaining larger yields. Several studies, such as those by Schmidt,[11] have shown that the rate of milk secretion by high-producing cows does not decline significantly for 12 hours but does decline thereafter. Intermammary pressure is the responsible agent for stopping the secretion when milk is left in the udder as a means of drying off the cow.

When the milking act is initiated, a small quantity of milk can be immediately removed, then there is a lag period followed by a large inflow of milk into the teat and gland cistern of the udder. One says that the cow "has let down" her milk. The stimulation of the sensory nerves in the skin and teats results in the release of oxytocin from the posterior lobe of the pituitary, which in turn causes a contraction of the myoepithelial cells, specialized muscle cells surrounding the alveoli, whereby the milk is forced out of the alveoli and ducts.

Maintenance of lactation depends upon effective milking management to stimulate oxytocin release and prevention of epinephrine release, which causes vasoconstriction and prevents oxytocin from entering the mammary gland. When a cow or goat is frightened milk ejection is greatly reduced.

17.9 Methods of Studying the Secretion of Milk Constituents

A useful method for the study of milk secretion is the analysis of the blood before and after passing the gland. Data obtained in this way have provided important information on the blood precursors of the milk constituents, but they cannot establish quantitative relations between blood changes and the constituents secreted unless the blood volume passing through the gland is measured also so as to arrive at the blood-milk ratio. The blood-milk ratio can be obtained indirectly by measuring the uptake of an element, such as calcium, which is known to be absorbed, and analyzing the milk for the element as a means of arriving at the milk volume corresponding to the blood change. It is recognized that the disturbance of the cow incident to the drawing of blood samples and the possible slowing down of secretion accordingly limit the value of quantitative data.

Perfusion of the isolated udder was early used to study the secretion of milk constituents but became subject to the criticism that the rapid deterioration of the gland tissues resulted in unreliable data. Later procedures have overcome this limitation. Harwick and Linzell[12] have described a method which kept the gland alive and secreting milk for 12 and, occasionally, 24 hours. Mammary tissue slices and cell-free extracts of mammary tissues are also employed for in vitro studies.

During the past years radioactive traces have been widely used to study the blood source of various milk constituents. The radioactive atom is introduced into the source to be studied, which is, in turn, injected into the blood or used in a perfusion experiment, with a resulting large increase in specific information. The perfusion technique has the advantage that the tagged compound in question enters the gland as such, whereas such a compound injected into the general circulation may be subject to metabolism in other tissues, such as the liver, with the result that the radioactive element may not enter the gland in the same combination as injected. In either case the identification of the element in a milk constituent shows that the compound into which it was introduced is a precursor, but it remains uncertain whether glandular action alone was involved.

17.10 The Secretion of Lactose

It is generally accepted that glucose is the principal blood precursor of lactose. In 1906 Kaufmann and Magne took samples simultaneously from the jugular and mammary veins of a milking cow and found that the mammary blood contained 18 percent less glucose, whereas similar samples from a dry cow showed no difference. Considering the jugular blood as representative of the supply of the gland, they suggested that the lactose of milk was made from the glucose of the blood. Somewhat later Foa, an Italian physiologist, obtained, by perfusion experiments, specific evidence that the gland can use glucose to make lactose. Latest estimates suggest that at least 85 percent of the lactose and 50 percent of the glycerol are derived from glucose.

There are many experiments using isotope-labeled glucose which have shown that it is the principal precursor of lactose. Such evidence has been presented by Dimant and associates[13] in a perfusion experiment. Similar evidence has also been obtained, without exception, in several studies in which the labeled glucose has been injected into the bloodstream. Injections of labeled propionate, butyrate, and bicarbonate have also shown a definite, though small, incorporation of the labeled carbon into lactose. It is generally believed that these compounds first serve as precursors of glucose outside the gland. The various experiments dealing with the blood source of lactose and with the pathway by which glucose is changed into lactose in the gland are reviewed by Ebner and Schanbacher.[14]

The steps by which glucose from the blood is changed to lactose in the gland, under the action of specific enzymes, is as follows:

Glucose + ATP $\longrightarrow$ glucose-6-phosphate + ADP
Glucose-6-phosphate $\longrightarrow$ glucose-1-phosphate
Glucose-1-phosphate + UTP $\longrightarrow$ UDP-glucose + phosphate
　　　　　　(uridine triphosphate)
UDP-glucose $\longrightarrow$ UDP-galactose
UDP-galactose + glucose-1-phosphate $\longrightarrow$ lactose-1-phosphate + UDP
Lactose-1-phosphate $\longrightarrow$ lactose + phosphate

17.11   The Secretion of Protein

In 1920 Cary of the U.S. Department of Agriculture showed that the free amino acid nitrogen of the blood plasma suffered a large drop in passing the gland, from which it was concluded that milk protein is formed from plasma amino acids. Confirmatory evidence was later obtained by others using the same technique and also by perfusion experiments. Later studies have dealt with specific amino acids labeled with C^{14}. As regards the essential acids, which obviously must arise from the blood rather than from gland synthesis, it has been shown that the plasma amino acids are the principal precursors at least. Based on studies by isotopic methods, it is apparent that at least half of the nonessential amino acids in milk protein also arise from the same acids of the plasma, but the probability that some of them may be synthesized within the gland from other blood precursors is a reasonable one. Kleiber and his students at California have shown, by isotope studies, that glucose, acetic acid, and propionic acid are precursors of certain nonessential acids found in casein. The experiments with propionic acid are reported by Black and Kleiber[15] in a paper in which possible metabolic pathways are discussed.

Mepham and Linzell[16] demonstrated by simultaneously measuring blood flow, arteriovenous differences for all amino acids, milk yield, and milk nitrogen that the uptake of amino acid nitrogen was sufficient to provide all of the nitrogen of milk protein synthesized in the mammary gland. The uptake and output of the essential amino acids and glutamic acid were about equal, serine uptake was less than output, and the other amino acids were variable. Larson and Jorgensen[17] reviewed the later research on milk protein synthesis.

17.12   The Secretion of Fat

The fatty acids in milk occur almost entirely as triglycerides, in contrast to the situation in the plasma where they are present largely as phospholipids (Table 17.2). After studies had failed to confirm earlier evidence that milk fat arose from plasma phospholipids, the work of Maynard and associates[18] with cows showed that the triglycerides of the blood were precursors. This finding was confirmed by others. The data did not provide any quantitative information, however, on the extent to which milk fat was thus formed or on the origin of specific fatty acids, particularly the short-chain ones which do not occur in the blood plasma.

With the demonstration of Bloch and Rittenberg in 1945, using the isotope tracer technique, that fatty-acid chains could be built up from acetate, considerable amounts of which are continually being produced in body metabolism, attention was centered on this compound as a possible source of milk fat. In this connection the findings that the arterial blood of the cow and goat contain substantial amounts of acetic acid as an absorbed product of rumen activity (Sec. 6.9) seemed particularly pertinent.

Direct evidence that the gland could make fatty acids from acetate commenced to appear around 1950 as a result of the studies of Folley, Popjak, and their associates at the National Institute for Research in Dairying in England. In vitro evidence was first obtained by the use of mammary gland slices. The most conclusive evidence was produced by the use of carbon-labeled acetate, in studies with lactating ruminants, by the above workers and by Kleiber and his associates at the University of California.

The use of labeled fatty acids and the development of gas chromatographic methods for the quantitative determination of specific ones have added greatly to our knowledge of their secretion into milk. There is conclusive evidence that the gland takes up palmitic and stearic acids from the blood plasma, and also lauric, myristic, and oleic if they are present. Milk fat is mostly triglycerides. Phospholipids comprise less than 1.0 percent of the total milk fat and is mostly associated with the milk-fat globule membrane. Circulating chylomicra and low-density lipoproteins are the primary blood fractions providing fatty acids to milk fat. Since food fat contains almost no C_4 to C_{14} fatty acids these are all synthesized as well as part of the C_{16} acids.

Milk fatty acid synthesis is a very complex process only partially understood. According to the present theory fatty acid synthesis involves a source of substrates and associated enzymes for conversion to acetyl-CoA and NADPH in the cell and its conversion to malonyl-CoA and its orderly addition to a "primer" (acetyl-CoA and/or butyryl-CoA) until the fatty acid is released from the fatty acid synthetase complex. Glucose is the carbon source for fatty acid synthesis in nonruminants, while in ruminants acetate and β-hydroxybutyrate from rumen fermentation are the main carbon sources. The details of present knowledge are reviewed by Bauman and Davis.[19]

17.13   Variation in the Secretion of Fat

Fat is the most variable constituent of milk. In addition to its variation among breeds and individuals, its percentage in the milk of a given animal varies from milking to milking, from quarter to quarter, and increases progressively during the milking process. In a study with a Guernsey cow, Van Slyke found that the first fraction drawn, consisting of about 18 percent of the whole, contained only 1 percent of fat, whereas the final fraction of 30 percent contained 10 percent of fat. It is very difficult to find an explanation for these large variations.

Periodic changes in fat percentage are inversely correlated with milk yield. A lowering of yield which results from an abrupt change in conditions, such as weather, surroundings, or even a change in the milker, is frequently accompanied by a rise in fat content. The sharply lowered yield which results during a period of off-feed is generally accompanied by a rise in fat percentage. This inverse relationship is also shown in the course of the lactation, for as the decline in yield progresses following the peak, the percentage of fat tends to rise.

There is a marked seasonal variation in the percentage of fat, the maximum occurring in winter and the minimum in summer. That environmental temperature is the primary cause of this variation is indicated by controlled studies in which it has been shown that within certain limits fat percentage increases regularly with drop in temperature. From one lactation to another there is no consistent change in fat percentage. While an increase in milk yield may be expected during the first four or five lactations, a cow which has a low-fat test as a heifer will not better it materially in succeeding years.

17.14 Fat Percentage and Composition as Influenced by Feed

The previously described large variations in fat percentage early led investigators to study the possibility of increasing this percentage through feeding. The early literature is filled with contradictory data and conclusions. Most of these experiments were of short duration and failed to take account of the nondietary factors which cause large variations. Studies have produced some authentic cases of increased fat percentage through feeding. Such an effect has been caused by large intakes of soybeans (18 percent fat). Purdue workers briefly reported persistent increases when the grain mixture contained 25 to 50 percent of soybeans. Byers and associates reported such an increase from feeding 4.5 kg of the beans daily as a supplement to alfalfa hay. Cornell workers found that the fat percentage increased from 3.5 to 4.5 and that the increase persisted for the 30-day test period when soybeans made up the entire concentrate portion of the ration. On the contrary, the feeding of soybeans at the lower levels which would probably represent the maximum in practical rations has not resulted in significant increases in fat percentage. Early European studies showed that such increases could be obtained by feeding coconut meal or palm-kernel meal, high in fat, an effect not produced by other feeds of similar fat content. Later work has confirmed these findings. In studies by Storry[20] and associates a basal diet providing an intake of 105 g fat, a low intake compared with that of usual dairy rations, was supplemented with coconut, red-palm, or peanut oil to provide intakes of fat of 379, 455, and 379 g, respectively. The fat content of the milk was increased from 3.63 percent to 3.97 percent in the case of the coconut supplementation and from 3.63 to 3.99 percent in the case of the palm-oil ration. There was no increase in percentage where peanut oil was added. The milk yield tended to be higher with all supplements and significantly so in the case of the peanut oil. The three supplements had markedly different effects on the distribution of the various fatty acids in the milk fat produced. Liberal intakes of whole cottonseed (22 percent fat) will also increase the fat percent or prevent its decline in warm weather. Adding vegetable oils to a concentrate mixture to supply an equivalent level of fat will not increase the fat test. With the above exceptions there is no practical ration which will substantially increase the fat content of cow's milk. Feeding thyroprotein (Sec. 17.3) causes large increases in the fat percent, but it appears to be of short duration.

Patton and associates[21] at Pennsylvania State University first reported that feeding methionine hydroxy analog (MHA) modified fat metabolism of lactating cows. Of the many reports on MHA feeding some have observed increased fat percentages or yields, but others have not. Chandler et al.[22] found a response on a concentrate with 12.5 percent protein, but not with 15.5 percent. They have summarized much related research. Clearly the practical usefulness of methionine hydroxy analog has not been resolved.

There are, however, certain feeds and combinations of them which have practical effects in lowering the fat content of milk. Following the initial observations by Powell[23] some thirty years ago that a limited intake of roughage fed in a finely ground state to dairy cows resulted in a lowering of the fat content of the milk, several other workers have noted a similar effect, particularly when the roughage was cut to less than half the normal amount and the concentrate mixture increased accordingly. Problems with abnormally low milk-fat percentages of 1.5 to 2.0 may occur in dairy cows as the result of excessively high intakes of concentrates and restricted amounts of roughage. Heat-treated concentrate and expanded corn or sorghum grain seems to be a contributing factor. Pelleting of concentrate feeds results in a small decrease of the order of 0.1 percentage unit in fat test, according to Bishop and associates[24] and more recent studies by others. Grazing of rapidly growing, low-fiber pasturage may also lower the fat percentage of milk.

Any ration which causes a marked lowering of the milk-fat percentage produces changes in the movements of the reticulo-rumen and in the physical and chemical composition of the digesta. The production of acetic acid is lowered and that of propionic increased, and their concentrations in the blood are changed accordingly; these changes are associated with the lowering of the fat percentage. As has been discussed (Sec. 17.12), the three blood precursors of milk fat are acetic acid, triglycerides, and β-hydroxybutyric acid. A decrease in acetic lessens the synthesis of the short-chain fatty acids, while a decrease in triglycerides lowers the production of the C_{16} and higher acids. Since propionic acid has an antiketogenic action its increase in the blood lessens the concentration of β-hydroxybutyrate, which is also a precursor of lower fatty acids. Propionic acid, which is a glucose former, tends to increase the glucose content of the blood, which may be of significance because of reports that the intravenous infusion of glucose lowers the content of C_{18} acids in the milk fat. The experimental evidence for these blood changes and their significance, outlined above, have been reviewed by Storry and Rook[25] in a paper which reports studies of the interrelationships between plasma levels and amounts and proportions of fatty acids secreted, with a diet low in roughage and high in flaked maize. The diet lowered the milk-fat percentage from a normal 4 percent on the control diet to minimal values of about 1.5 percent. The lowering was accompanied by a fall in acetate, β-hydroxybutyrate, and triglycerides in the blood plasma, which recovered on restoration of the control diet. The changes in the distribution of the various fatty acids in the milk, resulting from the experimental diet, compared with those of the control diet, are also reported.

A specific lowering of the fat percentage results from the feeding of cod-liver oil. This peculiar response was first observed by Golding and associates[26] in connection with studies of the influence of the oil on the vitamin content of the milk. It has since been confirmed in several laboratories. Feeding as small amounts as 50 ml per day has caused a lowering, but larger and more consistent effects are observed with intakes of 110 to 170 ml. A 30 percent decrease in the fat level has been noted. The specific factor in the oil which is responsible is found in the saponifiable fraction, but it has not been identified. The deleterious effect of the oil is eliminated by hydrogenation. This property of lowering the fat percentage is not shared by fish oils generally. Clearly, vegetable oils do not exert the effect.

Storry et al.[27] in England fed lactating cows 300 g per day of cod-liver oil or 450 g of protected cod-liver oil–casein powder mixed in the concentrate. The free cod-liver oil depressed the fat percentage from 3.8 to 3.0 and the fat yield, but the protected oil had no adverse effect. Milk yield was not affected. Total rumen VFAs were not changed; however, the free cod-liver oil increased the proportion of propionate and decreased the proportion of butyrate. The protected oil caused no changes. The marked effect of free cod-liver oil on rumen VFAs reduced the supply of acetate and β-hydroxybutyrate to the mammary gland and indirectly the synthesis of C_{4-18} fatty acids. Both the free and the protected cod-liver oil contributed C_{20-22} fatty acids. The yields of the major fatty acids are shown below:

Effects of Supplements on Fatty Acids in Milk (g/day)

Fatty acid	Control	Basal + free cod-liver oil	Basal + protected cod-liver oil
C_{4-14}	242	174	254
C_{16}	280	196	263
C_{18}	226	178	229
C_{20-22}	0	34	35

Frobish and Davis[28] have proposed the theory that the low-milk-fat syndrome results from alterations in propionate metabolism brought on by increased rumen propionate production coupled with a decrease in the amount of vitamin B_{12} available for its metabolism. They presented some evidence to support their theory, however, since only a part of the cows injected with vitamin B_{12} responded with increased fat test or yield this may not be the full explanation. The pathway they proposed is shown in Fig. 17.3. The effective way to cure or prevent the syndrome remains to feed adequate amounts of coarse forage.

With the development by Australian researchers of methods of protecting unsaturated oils from hydrogenation in the rumen in order to provide maximum levels of polyunsaturated fatty acids in body and milk fat of cattle, several groups have reported marked changes in the amounts of $C_{18:2}$ fatty acids.[29] There was generally no effect on the fat percentage or yield of milk. Years ago Maynard and McCay[30] showed that feeding a solvent extracted concentrate mixture reduced the milk yield of cows about 20 percent in a 5-week period in comparison to a standard feed containing about 6 percent of fat, with no effect

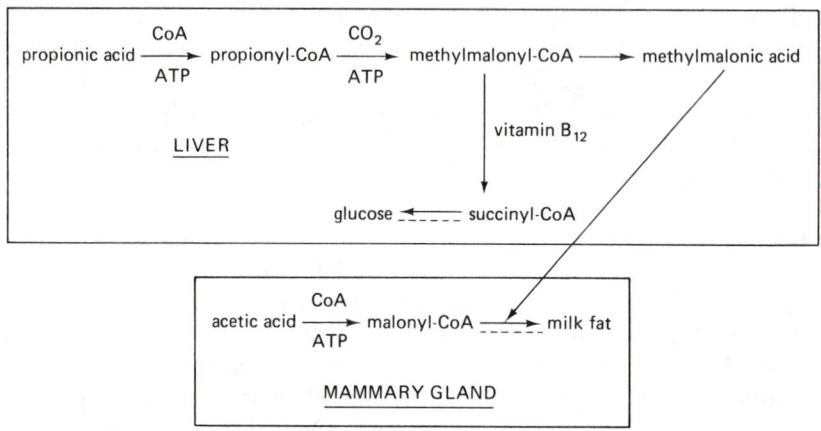

Fig. 17.3   Pathway of propionic acid metabolism in animal tissues.

on the fat content of the milk. Later studies comparing solvent extracted versus expeller type oil meals to supplement the protein in cereal grain based concentrate mixtures led to the conclusion that about 4.0 percent of fat was needed for maximum production. It was observed, however, that feeding more of the low-fat concentrate mixture usually resulted in equal amounts of milk. Thus it appears that except for highest levels of milk yields, where the cow's feed capacity may be limiting, the cost of fats versus other energy sources should govern the use of supplemental fats for lactating dairy cows.

17.15   The Secretion of the Mineral Elements

Over thirty mineral elements have been found in milk, many of them present in traces. Thus, the list is more than double the number which have been found to be essential nutrients. They all enter from the blood plasma, but, as is illustrated in Table 17.2 for five of the major ones, the concentrations of many of them in the milk are very different from their concentrations in the blood. Milk contains approximately thirteen times as much calcium, ten times as much phosphorus, and five times as much potassium, but only one-seventh as much sodium and one-third as much chloride as does blood plasma. How the gland selectively secretes its minerals presents a physiochemical problem which is largely unexplained.

Many of the minerals in milk, such as calcium, potassium, sodium, chlorine, iron, nickel, magnesium and phosphorus are affected very little by the dietary intake. Other elements including zinc, cobalt, copper and manganese may be lowered by a deficient diet. Milk iodine is highly variable depending on the dietary intake. Milk is well protected against toxic levels of cadmium and mercury. The same appears to be true for selenium.[31] While there is a positive correlation between dietary intake and milk concentration, the transfer to milk is very inefficient. Daily intakes of 0.99, 2.02, 3.02 or 6.01 mg per cow resulted in milk concentrations of 0.0127, 0.0156, 0.0150 and 0.0190 ppm of selenium. Sow's milk doubled when 0.1 ppm was added to the diet, but the reserves of the young pig weaned at 4 weeks are rapidly depleted.[32] This explains why selenium deficiency occurs in weanling pigs even when their dams are fed selenium and vitamin E supplements at standard levels.

17.16 Pigments

Milk contains both fat-soluble and water-soluble pigments. Of the fat-soluble, carotenoid group of pigments, which are synthesized by plants but not by animals, *carotene* is the principal one found in the milk of the cow. Chlorophyll is destroyed in the digestive tract, and this is presumably true to a certain extent also for xanthophyll, only a small amount of which gets into the milk. The occurrence of carotene in milk is limited primarily to the bovine species. The milk of the sheep, goat, sow, water buffalo, and camel has little or none, and women's milk is nearly colorless. The reason for these breed and species differences is not known, but when there is no pigment in the blood plasma, the milk is also free. The principal water-soluble pigment of milk is *riboflavin*. Since carotene is the precursor of vitamin A, and since riboflavin is also a vitamin, the factors which govern their secretion in milk are best discussed later where the vitamin requirements for lactation are considered.

17.17 Colostrum

The first product for the mammary gland following parturition, the colostrum, is richer in total solids and total ash, much richer in protein, and lower in lactose than normal. Per unit of total solids, colostrum has approximately twice as much protein, the same amounts of fat and ash, but only one-third as much lactose. The proteins, which make up approximately 15 percent of the product in the case of the cow, consist principally of globulin and albumin, in marked contrast to their very low content in normal milk. This lactoglobulin provides immune bodies transferred from the blood, which are in turn ingested by the newborn, and thus play an important role in disease resistance early in life. As the product drawn from the gland assumes the composition of normal milk in the course of a few days following its first withdrawal, the immunizing properties disappear. Table 17.3 shows the comparative values.

Colostrum has nutritive values that are of special importance for the newborn because of its richness in certain vitamins and iron compared with the later normal milk. For example, Sutton and associates[33] found that the first-milking colostrum of a group of pasture-fed cows of different breeds to be approximately ten times as potent in carotene, six times as potent in vitamin A, and three times as potent in riboflavin as the milk obtained at the twentieth milking. The vitamin A value of the colostrum of cows fed poor roughage during the gestation is much less than of those on pasture or dry roughage rich in carotene. Colostrum is several times richer in iron than is normal milk and also somewhat richer in thiamin and vitamin D. Both its special nutritive

TABLE 17.3. Composition of Colostrum and Normal Milk

Species	Total solids, %	Fat, %	Protein, %	Lactose, %	Ash, %	Specific gravity
Cow[a]						
Colostrum	23.9	6.7	14.0	2.7	1.11	1.056
Milk	12.9	4.0	3.1	5.0	0.74	1.032
Mare[b]						
Colostrum	25.2	0.7	19.1	4.6	0.72	0.076
Milk	11.3	2.0	2.7	6.1	0.50	1.035

[a]D. B. Parrish et al., Properties of the colostrum of the dairy cow. V. Yield, specific gravity and concentrations of total solids and its various components of colostrum and early milk, *J. Dairy Sci.*, **33**:457–465, 1950.
[b]D. E. Ullrey et al., Composition of mare's milk, *J. Animal Sci.*, **25**:217–222, 1966.

values and its immunizing properties indicate why it is so important for the newborn animal to receive the colostrum.

After the cow has been dried off in preparation for the next calving, the gland continues to secrete a fluid which is similar to colostrum, particularly rich in globulins. During the last two weeks there is a tremendous increase in the immune globulins. These globulins are immunologically similar to those in the maternal blood, and it is considered that they arise directly from the blood rather than as a synthetic product of the gland.

17.18 Abnormal Milk Constituents

It has been mentioned that some of the normal constituents of milk are apparently merely filtration products. Among these there are substances, such as urea, which represent useless products and which apparently pass into the milk in small amounts instead of being excreted entirely through the usual channels, because the membrane is not a perfect barrier. This fact raises the question as to the extent to which such substances as drugs and other nonnutritive or toxic substances may pass into the milk in health and in disease. Early studies showed that the healthy gland is highly protective against the passage of foreign substances in harmful concentrations. Neither heavy metals, such as mercury and arsenic, nor volatile organic substances, such as alcohol, ether, and chloroform, pass into the milk in toxic amounts. The same was found true for salicyclic acid, aspirin and related compounds, and various alkaloids such as morphine, atropine, and quinine.

It is noteworthy that dairy cows can tolerate levels of fluorine in the diet up to 55 ppm over long periods without showing toxic effects. Higher levels up to 109 ppm, which caused bone and teeth changes and reduced milk yields, increased the fluorine content of the milk only from 0.036 to 0.198 ppm, still well below the level that would cause harmful effects in milk-fed animals.[34] Feeding 150 ppm for a period as short as a month did not depress milk yield.[35] There is concern about residues of aflatoxin,[36] and pesticides in milk and other foods. Aflatoxin was not destroyed by pasturation or storing milk at 4°C for 17 days, and 45 percent of it remained after 120 days storage of frozen products. Thus it is important to keep milk safe from these residues.

Knowles[37] has reported studies of some 35 drugs, in terms of dosage, and resulting blood plasma and milk concentrations, with particular reference to human lactation. He concluded that nearly all products ingested by the mother are excreted in her milk—in considerable amounts in the case of some drugs, but in only minute amounts for others. He recommends curtailment of unnecessary medication during lactation. Increasing numbers of dairy cows are being treated during their dry period with intramammary infusion of antibiotics to aid mastitis prevention. Johnson and associates[38] studied milk samples from treated cows and found no residues of penicillin or cloxacillin for cows treated 9 days or more before calving.

In the case of the cow it is recognized that certain essential oils may pass into milk to the extent of causing an odor and taste and that the poisonous principles of certain plants, such as white snake root and rayless goldenrod, may be secreted in sufficient amounts to render the milk harmful. Certain antibiotics in the feed, if present in excessive amounts, may pass into the milk. Market milk sometimes contains sufficient antibiotic activity to interfere with its use in producing cheese. This most often arises from cows which have been treated for mastitis or generalized infections when the milk is not discarded for the specified three to four days following the treatment. Antibiotics are not recommended as additives to feeds for dairy cows because of the possible residue problem in milk.

In diseases of the udder, particularly those characterized by inflammation, the membrane becomes much more permeable. The milk itself changes in the direction of the composition of blood with a resulting increase in protein and salts and a decrease in lactose. The protein fraction contains more albumin and less casein. With a more permeable gland, the danger from the entrance of harmful foreign substances is greatly increased. In diseases not affecting the udder, the primary effect is on yield of milk rather than on its composition.

THE ENERGY REQUIREMENT

Unlike other glands of the body the mammary gland contributes no advantage to the animal, but places tremendous demands when secreting maximally. At the start of lactation which occurs along with the stress of parturition, there is a drastic increase in the demand for nutrients and in metabolic rate and a redistribution of the blood supply. Frequently animals are unable to adjust their metabolism quickly enough to prevent acute metabolic disturbances such as ketosis, parturient paresis, or grass tetany, which may cause death of the animal.

In addition to her maintenance requirement, the lactating animal must receive sufficient nutrients to supply those secreted in her milk and to cover the wastage involved in the process. It is obvious that a separate requirement can be stated for lactation only for those animals for which the current milk yield and its composition are known, i.e., for those used in commercial milk production. Thus we have specific knowledge of the nutritive needs for milk secretion only in the case of the dairy cow and sow. Nevertheless, the information gained in studies with these species has established principles which are useful in estimating the needs of others as well. The following discussions deal principally with the cow.

17.19   Variations in Milk Composition

While the milk of the individual cow tends to remain constant in composition, aside from fluctuations and cyclic changes in fat content, there are wide differences among individuals within a given breed as well as among the breeds themselves. Data are given in Table 17.4 on the average composition of milk of different fat contents. These data show that there is a regular increase in protein and ash with fat but that the lactose tends to decrease rather than otherwise. The large rise in energy value reflects the increasing fat level, primarily.

The data clearly show that the energy and protein requirements for milk production must be based on the composition of the milk. With new methods now available it is easy and inexpensive to determine the fat and protein content of milk for estimating the energy and protein requirements.

From data on the composition of milk of different fat contents Gaines[39] devised an equation for calculating energy value from fat content. He proposed the following formula for expressing milks of different fat contents in terms of 4 percent fat milk:

4 percent milk = $0.4M + 15F$

where M equals weight of milk and F equals weight of fat contained in it. By this calculation, 30 kg of 5 percent milk is equivalent on an energy basis to 34.5 kg of 4 percent milk, while 40 kg of 3 percent is equivalent to only 32 kg.

TABLE 17.4.   Composition of Cow's Milk as Related to Fat Content

Fat, %	Protein, %	Lactose, %	Ash, %	Energy, kcal per 100 g	Total solids, %
3.0	2.7	4.90	0.67	62	11.27
3.5	2.9	4.89	0.68	68	11.97
4.0	3.1	4.87	0.70	73	12.67
4.5	3.3	4.85	0.72	79	13.37
5.0	3.5	4.83	0.74	85	14.07
5.5	3.7	4.82	0.75	91	14.77
6.0	3.9	4.81	0.76	96	15.47

Source: The data for fat, protein, lactose, ash, and total solids in this table were compiled by A. C. Dahlberg, Cornell University, and are reproduced with his permission. The data for energy content were calculated by the authors using the appropriate Atwater coefficients. (See footnote, Table 17.1.)

This calculation is widely used in experimental work to put yields of milk of different fat contents on a common basis—"fat-corrected milk."

Tyrrell and Reid[40] have shown that Gaines' formula is inaccurate for equating the energy value of cow's milk when the fat content is depressed below 2.5 percent. They have devised a formula based on both fat and solids-not-fat contents which is more accurate under such conditions, as follows:

SCM(kg) = 12.3(F) + 6.56(SNF) – 0.0752(M)

where SCM equals solids-corrected milk having 750 kcal per kilogram; F, SNF, and M equal fat, solids-not-fat, and milk, respectively, expressed in kilograms.

17.20   Estimation of Energy Requirements from Metabolism Studies

Both direct and indirect calorimetry have been used to arrive at the energy requirements for milk production and to express the results either as NE or ME per unit of milk of a given gross calorie content. The essential feature here is the determination of the percentage of the energy intake above the maintenance requirement which appears in the milk. In setting standards the requirements are commonly expressed per unit of milk of a given fat content since its kilocalorie content closely follows the fat percentage (Table 17.4). Lactating cows are fed a ration selected to be adequate for the expected milk production, and data are obtained on the gross energy intake and the various energy losses, including the energy secreted in the milk. Later, in the dry, nonpregnant animal the energy intake and losses are measured to arrive at the maintenance requirement. The subtraction of this requirement from the gross energy intake, with corrections for the energy represented by the gain or loss of weight during the lactation trial, gives the energy actually used for milk production. The percentage of this energy which appears in the milk represents the efficiency of lactation. Such was the general method used in the studies from which the average figure of 70 percent efficiency on the basis of metabolizable energy was arrived at (Chap. 9). The details of such an experiment are presented in the paper by Flatt and associates,[41] who cite the many earlier studies. The assumptions made regarding the application of the maintenance requirement determined in the dry cow and the factors used for correcting for tissue gain or loss can markedly affect the reliability of the efficiency coefficients thus arrived at (Moe and Tyrrell[42]).

17.21   Energy Requirements as Determined in Feeding Trials

The energy requirements for milk production which are specified in the commonly used standards have been derived principally from feeding trials. The general procedure here has been to record the feed intakes and production of groups of cows differently fed and to select as the requirements those intakes which were found to give the best production over extended periods and, at the same time, maintain the weight of the animal. By the subtraction of a maintenance requirement determined after the cows were dry, or upon other animals, the net intakes required for the milk produced were obtained. In the United States most commonly the resulting data had been expressed as TDN, either based upon digestion trials run as a part of the experiment or else calculated by the use of the average coefficients for digestibility.

Haecker, who was the first to recognize that the food requirements must vary in accordance with the composition of the product, began in 1897 a series of investigations of nutrient requirements for milk production which extended over many years. His publication in 1914[43] set forth TDN requirements in accordance with the milk produced. The requirements, expressed as TDN, DE, ME and NE per unit of milk of various fat contents, are set forth in the N.R.C. report. The Morrison standards list ranges for TDN and ENE on the basis that, in practice, the relation between the cost of feed and the selling price of milk should be taken into account in determining the energy intake to be provided. The methods used in Europe for stating the energy requirements for milk production have been discussed and compared with United States standards by Breirem.[44]

The British Agricultural Research Council (A.R.C.) standards use ME (Sec. 13.14).

The aim throughout lactation should be to prevent any marked loss of body weight, but fattening should be avoided except for the purpose of restoring any large loss which has occurred early in lactation, because milk secretion tends to decline if marked fattening occurs.

A limitation in the usefulness of the feeding trial as a method of determining the energy required for lactating cows is their ability to replace the body fat used for milk secretion with water and thus to resist body weight changes. Later in lactation the reverse process occurs, as illustrated by the studies of Wagner and Loosli.[45] Energy-balance studies seem to be the only accurate method of estimating the requirements for lactation.

In practice it is found impossible to get the milking cow with highest genetic potential for milk yield to consume enough total food during the peak of her lactation to prevent loss of body fat. Such cows are able to secrete considerable milk while in negative energy balance at the expense of body energy stores, but after a few weeks milk yield decreases rapidly until equilibrium is reached. The practice in the United States of allowing cows all the concentrates they will consume during the first 2 or 3 months of lactation has resulted in higher peak yields, but total milk for the full lactation is not always increased and care must be exercised to prevent digestive disturbances and related health problems.[46]

It now seems clear that high-yielding cows do not require more energy per unit of milk secreted. Although digestion studies (Sec. 3.8) show large decreases in apparent digestibility of energy and decreasing TDN values for some types of high-grain rations fed at four to five times maintenance for high-yielding cows compared with maintenance levels, metabolism studies indicate lower heat increment losses largely compensate so that the NE required for each unit of milk is similar.

The level and efficiency of milk production is influenced not only by the genetic ability of the animal and current nutritional status but also by the plane of nutrition the animal received during growth. In the Cornell experiments to determine the influence of plane of nutrition during growth on the lifetime performance of Holstein cattle (Secs. 15.11 and 16.5), important differences in level and efficiency of milk production were observed, as summarized in Table 17.5.

After the first lactation, the cows that had been grown on the low plane of nutrition and were greatly stunted until they were refed at first calving produced more milk during each subsequent lactation than those grown on the high level. The underfed heifers generally performed as well or slightly better than those grown on the recommended energy intake. The average differences are sufficient to be of financial importance to the dairy farmer and the advantage becomes great when consideration is given to the much larger expenditure for feeds to grow the heifers on the high plane. The results agree with European studies which showed that cattle grown on high energy intakes produced less milk and had shorter productive life spans than those fed lower intakes. Experiments at Tennessee by Swanson have shown that overfattened heifers have low lactational performance in later life.

17.22 The Importance of Food Fat in Lactation

Over thirty years ago Cornell workers extracted most of the fat from a then commonly used grain mixture for dairy cattle, replacing it by an isodynamic amount of starch, and found that a substantial drop in milk production resulted.[47] The extracted mixture contained only 0.66 percent of fat compared with a level of 5.8 percent in the unextracted. These initial studies were followed by several others in which the higher fat mixture always showed an advantage. The results have been confirmed by others, e.g., Leroy and Bon-

TABLE 17.5. Plane of Nutrition During Growth and Lactation Performance

	Plane of nutrition					
	62[a]		100[a]		146[a]	
Lactation	FCM, kg	Efficiency[b]	FCM, kg	Efficiency	FCM, kg	Efficiency
First	4010	53	4120	50	4185	48
Second	4672	53	4767	53	4424	47
Third	4981	53	5088	55	4882	50
Fourth	5288	54	5027	52	4852	49
Fifth	5631	55	5700	58	4875	47
Sixth	5626	55	5180	52	5114	49

[a]Percent of Morrison's TDN standard.
[b]Relative milk yield per unit of metabolic body size.

net.[48] It requires a carefully controlled experiment to produce significant differences in production where the differences in fat intake are small, which explains why a few studies have failed to confirm the findings reported above.

Supplementary studies have indicated that the beneficial effects of the rations of higher fat content were not the result of an increased intake of fat-soluble vitamins or of a possible improvement of the digestibility of the ration as a whole. No evidence was obtained that a specific kind of fat was responsible. The probable explanation is that the higher fat intake resulted in a more effective utilization of the digested energy for milk production. Such an explanation is supported by the later findings that increasing the fat level in the diet increases its efficiency by diminishing the heat increment. This explanation is also suggested by the Cornell findings that the difference in milk production could generally be compensated for by feeding a larger amount of the low-fat grain mixture. Thus, in practice, the question may be purely an economic one as to whether a grain mixture of higher fat content would cost more than the extra feed otherwise required, provided the cow's digestive apparatus can handle the larger amount of the lower energy feed required for top production.

Adding whole cottonseed (22 percent fat) or ground soybeans (18 percent fat) to the ration of lactating cows will increase the fat content of the milk and help overcome the fat depression which occurs during warm weather or when course roughage intakes are low. Mixing oils from the same sources with the concentrate feed does not give the same response unless the oils are protected. The higher energy concentration contributed by the extra fat helps to maintain energy intakes during warm seasons when feed intakes may be depressed. In poultry the feed intake is regulated by heat increment as well as by energy density. Extra fat not only adds extra energy per volume of diet, it also decreases heat increment and increases net energy per unit of metabolizable energy.

17.23   Energy Requirements of Other Species

In the absence of any measures of the amount of milk being secreted, no attempt is made to specify energy requirements for animals nursing young, separate from their total needs covering maintenance as well. On the basis of feeding trials, standards had been proposed earlier to cover the overall requirements of sows, ewes, mares, and beef cows during lactation. The N.R.C. requirements of these species are summarized in Appendix Tables II, III, IV, and V. In practice their total feed allowances are regulated in accordance with their apparent needs by giving sufficient amounts to keep them in good flesh during the nursing period. The energy requirements of various domestic animals for lactation are shown in Table 16.3 in comparison with maintenance, growth, and reproduction.

A study of the energy requirements of lactating sows which included data on feed consumption, body weight changes, milk yield, and milk composition has been reported by Elsley.[49] Similar studies with ewes have been published by Gardner and Hogue.[50]

THE PROTEIN REQUIREMENT

The minimum protein requirement for milk production is the amount secreted in the milk plus any catabolized as a specific accompaniment or result of the secretory process. In translating this minimum value into an actual requirement, account must be taken of the wastage in digestion and metabolism. The wastage in metabolism, however, is much less readily accounted for, since this is dependent upon the quality of the protein in terms of its amino acid makeup.

17.24   The Biological Value of Protein for Lactation

The differences in the biological values of various proteins and combinations of them for growth, which modify the intakes required for this function according to the sources used, have been mentioned (Sec. 15.23). That similar differences exist as regards milk secretion is to be expected, since there is no evidence that the mammary gland has special powers for the synthesis of amino acids not possessed by other tissues of the body. Insofar as farm animals are concerned, sows are the only ones for which differences in protein quality for lactation have been demonstrated. Here it appears that the kind or combinations of proteins that are efficient for growth are also efficient for milk production.

The evidence that protein quality is of much less importance for ruminants and the reasons therefore have been previously discussed. Carefully conducted feeding trials with milking cows have failed to reveal significant differences in efficiency among the protein combinations that show such differences in the case of nonruminants. They have shown, however, that urea and other simple nitrogen compounds can replace a part of the protein otherwise required for milk production. The nitrogen-balance procedure, which provides the most exact measure of the biological value of protein for growth, cannot be relied upon to produce significant data from lactation studies because there are variables that cannot be controlled. It seems probable that the nature of the protein fed is of some importance in the case of lactating ruminants, but it appears to be a minor consideration in selecting their rations. By the use of protected proteins it should be possible to define the limiting amino acids for lactating ruminants and to increase the efficiency of production by improving protein quality.

17.25   Protein Requirements as Measured by Feeding Trials

The protein requirements for milk cows recommended by current standards are based primarily on the results of feeding trials. Using combinations of feeds which are recognized to be satisfactory for milk production in other respects, the object has been to determine the protein intake which would certainly prove adequate for maximum production. Long-time feeding trials are essential for this purpose because an animal can keep up its production for an extended period, particularly in the first half of lactation, at the expense of its own tissues.

Haecker, whose long-time, pioneer studies have been referred to (Sec. 13.1), concluded that, allowing 318 g of digestible protein per 454 kg live weight for maintenance, an additional intake representing 138 percent of that secreted in the milk was adequate for satisfactory production and condition. He rejected a lower figure, because he felt that it did not keep the animals in the best condition. In setting up a standard for practice, Haecker increased the figure of 138 percent to 175 percent to allow a "factor of safety."

Many other experiments have been conducted to determine the protein required for milk production. For example, Harrison and Savage[51] compared concentrate mixtures containing 12, 16, 20 and 24 percent protein fed with corn silage and timothy-clover mixed hay during complete lactations. The 16 percent feed gave as good results as higher levels, but the 12 percent was inadequate. Based on Haecker's maintenance figure the 16 percent supplied 128 percent of the protein in the milk and the 20 percent concentrate 150 percent.

The 1971 N.R.C. dairy cattle report used 300 g of digestible protein as the base for maintenance and values for smaller and larger body weights were calculated at the same rate per unit of weight to the 3/4 power. The standard added 150 percent of the protein in the milk. Feeding trials estimate the total protein needed for maintenance and milk combined and the proportions assigned to maintenance and milk will vary inversely.

Van Horn and Zometa[52] reviewed recent research and summarized their own studies. Fig. 17.4 shows the milk yields and dry matter intakes in response

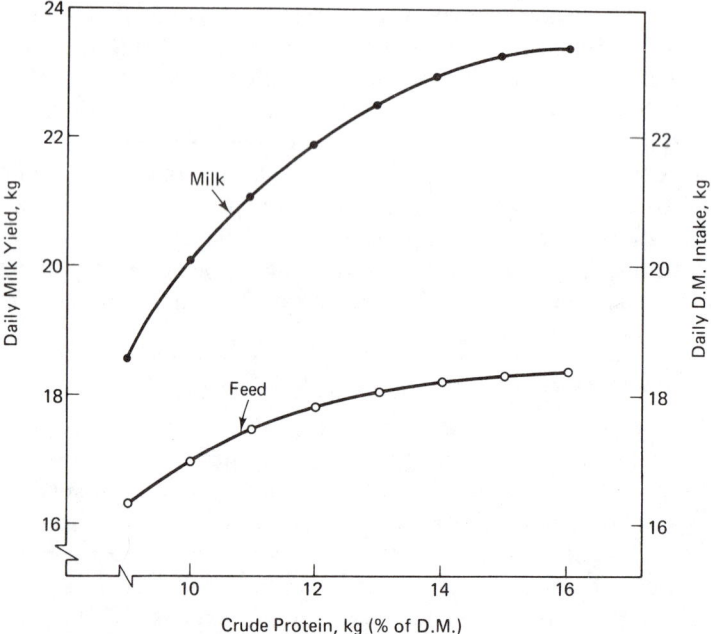

Fig. 17.4 Daily milk yield and dry matter intake as influenced by protein level. (*Courtesy of H. H. Van Horn, University of Florida.*)

to varying levels of protein in the complete mixed ration when soybean meal was the main supplemental protein source. Other protein feeds were also studied. Levels of protein above 13 to 14 percent may not be profitable in practice even though cows of high genetic merit may respond with extra milk.

17.26 Nitrogen-balance Data as Measures of Adequate Protein Nutrition

The fact that milk production can take place at the expense of body tissue has caused various workers to study the nitrogen balance as a further measure of the adequacy of intakes which were giving satisfactory production. It has generally been found that satisfactory production and nitrogen equilibrium can be maintained, at least over short periods, on intakes considerably below those recommended on the basis of feeding trials.

When the protein intake is reduced below the requirement, there may be a drop in production or its level may be maintained at the expense of the body and be reflected in a negative balance. Undoubtedly both take place over any long period on an inadequate protein intake. Such was found to be the case in a supplementary study of the ration containing the 12 percent protein grain mixture used by Harrison and Savage. Clearly, nitrogen-balance data alone are not an adequate measure of a satisfactory protein intake for milk production, but they provide useful supplementary information in connection with feeding trials.

17.27 Protein Requirements for Practice

Based on a maintenance requirement of 0.3 kg of digestible protein daily for a 500-kg dairy cow with requirements for other weights calculated on the basis of the 3/4 power, the N.R.C. standard furnishes 135 to 145 percent of the protein in the milk as an addition to the maintenance requirement. The additional amounts required increase as the fat content of the milk increases, as is shown in a table in the committee's report. Despite the evidence that lower total intakes have resulted in excellent production in several experiments, liberal allowances seem wise as general recommendations because of the indications that they may produce somewhat more milk under many conditions. From the standpoint of practice, however, the possible advantage of the higher level of intake may be overbalanced in many situations by the extra cost of high-protein feeds. The N.R.C. requirements of various species for lactation in comparison with growth, reproduction, and maintenance are summarized in Table 16.3. Greater details are given in the appendix tables and the original N.R.C. reports.

THE MINERAL REQUIREMENTS

Aside from the elements supplied by common salt, the minerals which most often require consideration in feeding rations to promote milk secretion are calcium and phosphorus. In some areas lack of copper and cobalt may limit performance. The other minerals which occur in milk are generally supplied adequately by the commonly used feeds.

17.28 Calcium and Phosphorus

Compounds of calcium and phosphorus make up approximately 50 percent of the ash of milk, and thus its secretion requires a liberal supply of them in the ration. A cow producing 8,000 kg of milk during her lactation secretes in it approximately 9.7 kg of calcium and 7 kg of phosphorus, and at the peak of her production the daily calcium output may exceed 50 g, with a somewhat smaller figure for phosphorus. These figures, however, do not represent the total requirements of the lactating animal because of the needs for maintenance, as well as for the pregnancy which normally occurs in the course of the lactation, and because of the fact that there is a large wastage of calcium and phosphorus in metabolism.

17.29 The Cycle of Calcium and Phosphorus Metabolism

Owing primarily to the pioneer and extensive work of Forbes and associates,[53] begun in 1912 at the Ohio Experiment Station, it has come to be recognized that the natural and significant unit of time in the calcium and phosphorus metabolism of the dairy cow is the annual cycle of lactation and gestation. By means of balance experiments, Forbes found that the most liberal feeding of calcium and phosphorus would commonly not meet the current needs of liberally producing cows during the first part of the lactation but that, toward the end of the lactation and particularly during the dry period, the earlier losses from the body ceased and were replaced by a storage of the elements. A similar cycle has been found to occur in lactating women, and there is evidence that the same is true for the rat and dog. Probably this depletion and restoration of the bone reserves are common occurrences in all mammals during the lactation cycle.

By continuous balance studies over the entire cycle of lactation and gestation both Ellenberger and coworkers[54] and Forbes and coworkers have shown that liberally producing cows may be in negative calcium and phosphorus balance for extended periods early in lactation and still end the cycle with adequate body reserves of the minerals, as a result of storage later in the lactation and during the dry period. Data from one of the Vermont studies are presented in Fig. 17.5. This chart gives the record of a mature Ayrshire cow from the beginning of lactation over a period of 58 weeks until calving. She was in milk for 47 weeks and produced 5115 kg during this period. Her ration consisted of timothy hay, corn silage, and grain during the winter, but during the summer the silage was discontinued and fresh-cut grass was largely substituted for the hay. The average daily intakes of calcium and phosphorus were approximately 45 and 60 g, respectively.

The weekly balances presented in the upper part of the chart, while showing considerable fluctuations, reveal losses of the elements, particularly of calcium, during the early weeks, in contrast to storages which regularly oc-

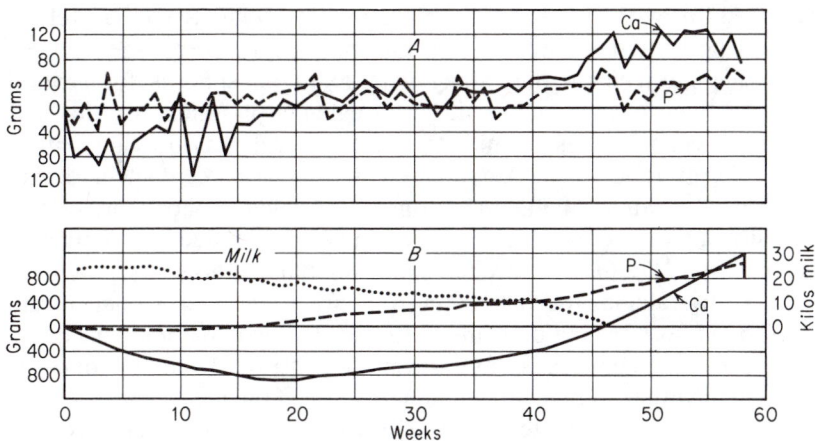

Fig. 17.5 Calcium and phosphorus balances throughout the lactation cycle. (A) weekly balances of calcium and phosphorus, (B) Milk yield and cumulative balances of calcium and phosphorus. (*Courtesy Ellenberger, Newlander, and Jones, Vermont Arg. Exp. Sta.*)

curred during the last half of the period. The results over the period as a whole are better shown in the lower part of the chart, where the cumulative balances are presented. It is noted that the cow had lost calcium for 20 weeks and did not regain her losses completely until the forty-sixth week, a point which happened to coincide with the close of the lactation. During the following weeks, the calcium store accumulated rapidly. The curve for phosphorus shows a net loss through 12 weeks, followed by a storage which was accelerated as a decline in milk flow became more rapid. The dips in both curves at the end represent a subtraction of the minerals in the calf and placenta. Clearly, negative balances early in lactation do not necessarily mean that the ration is inadequate in calcium and phosphorus for the cycle as a whole, and it would appear that the utilization of reserves early in lactation is a normal process, not harmful to the animal provided the losses are not too great and provided they are fully made good later.

This cycle of calcium and phosphorus depletion and repletion is of special interest in high-producing lactating dairy cows as it relates to milk fever (*parturient paresis*). The role of vitamin D in regulating calcium absorption and mobilization and in maintaining homeostatic blood plasma levels has been reviewed (Sec. 11.17). Milk fever results from a sudden decline in blood calcium which causes paralysis and death of the cow unless she is treated by intravenous injection of calcium or suppressing milk secretion by inflating the udder with air to build up intramammary pressure sufficient to stop milk secretion.

Based on earlier research with laboratory animals, Boda and Cole[55] postulated that if cows were subjected to a calcium-deficient diet shortly before parturition it would cause a "compensatory hypertrophy of the parathyroid glands such that at the initiation of lactation the increased calcium drain would be compensated for by the increased mobilization of calcium from the skeletal reserves" and prevent the critically low plasma calcium levels (Sec. 10.8). They fed a calcium-deficient diet along with extra phosphorus for several weeks before calving to give widely different calcium-phosphorus ratios and intakes. There were no cases of milk fever on the calcium-low diet with a calcium-phosphorus ratio of 1:3.3 compared with a 10 percent incidence in cows receiving a 1:1 ratio and 30 percent on the 6:1 calcium-phosphorus ratio. It was also shown that massive doses of vitamin D would reduce the incidence of milk fever (Sec. 17.36). Researchers at Iowa State University[56] demonstrated that a low-calcium diet consisting of corn silage and a concentrate mixture of ground corn, rolled oats and cottonseed which contained 0.12 percent calcium prevented all cases of milk fever. In contrast, 37 percent of cows fed the usual feed with 0.55 percent calcium developed the syndrome.

Yarrington et al.[57] reported that a low-calcium diet increased the parathyroid hormone activity in the parathyroid gland and increased the mobilization of the skeletal calcium.

A similar cycle occurs in sheep, as is indicated by the extensive studies of Benzie and coworkers[58] of the skeletal changes in calcium and phosphorus during lactation. In the case of calcium, for example, with a daily intake of about 5 g there was a loss of 6.5 percent from the skeleton at mid-lactation, a loss that was fully replaced later. With an intake of 2 g, on the other hand, there was a loss of 18.2 percent which was not replaced two months after the end of lactation. Replacement did occur, however, if the intake was raised to 5 g at midlactation.

17.30   Effects of Calcium and Phosphorus Deficiencies

If, owing to an inadequate ration, the demands for calcium and phosphorus during lactation are in excess of the reserve supply, or if the losses are not made good, both the animal and her production eventually suffer. With rations which are extremely low in either of the minerals, the bones may become so impoverished in them as to break, destroying the usefulness of the animal. In less severe situations, the bones may become progressively weakened in succeeding lactations due to incomplete restoration of the losses, and thus it becomes increasingly difficult for the animal to keep up her milk flow. The production may fall off more rapidly than normal in a given lactation, or it may fail to reach previous levels in succeeding lactations. The effect of small deficiencies of calcium and phosphorus may not become evident until after two or three or more years, the essential effect being to shorten the productive life of the animal.

Striking evidence of the effect on production of inadequate calcium and phosphorus nutrition has come from studies in the phosphorus-deficient areas. In South Africa[59] the feeding of bone meal to cows on deficient pasture increased the milk production by 40 percent, while in Minnesota[60] the addition of phosphorus increased the yield by 50 to 146 percent. Severe calcium deficiency causes serious effects, as reported from Florida by Becker and cowork-

ers.[61] Owing to the very low content of this element in the roughages, broken
hips and ribs were not an uncommon occurrence in the lactating animals. When
the calcium intake was raised by the addition of bone meal, the yield, per
lactation, increased by 50 percent and the cows became more persistent producers. When they were slaughtered at the close of the experiment, tests of the
bones revealed an excellent state of mineral storage. Depleted bones from an
animal on the calcium-deficient ration are shown in Fig. 17.6.

To what extent a less severe and thus unnoticed bone depletion may limit
milk production and productive life, by reason of rations inadequate in calcium
and phosphorus, is unknown. In fact, it is possible that, even with the best
mineral nutrition we know how to provide, productive life may be shortened
by failure to meet the physiological demands of extremely high production. It
seems important to reevaluate the calcium and phosphorus utilization of high-

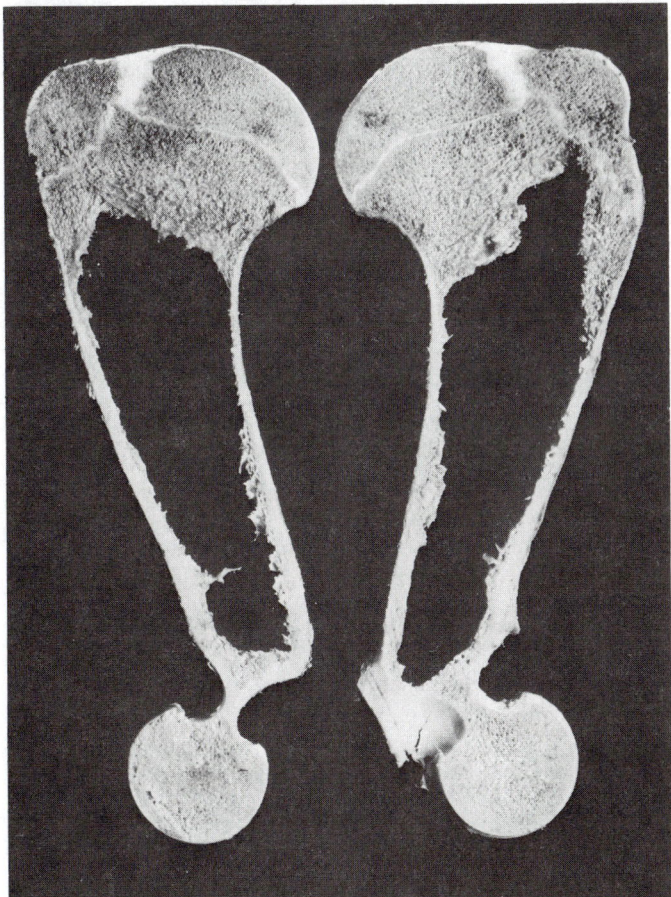

Fig. 17.6   Humeri depleted by a ration low in calcium. There bones were taken from a first calf
heifer. (*Courtesy of R. B. Becker, Florida Agr. Exp. Sta.*)

yielding dairy cows receiving liberal intakes of these mineral elements and of vitamin D to determine whether early lactational losses might be lowered or eliminated and to evaluate long-time effects on production and health.

The extent of the losses from the bones which can occur early in lactation without immediate detriment to production or the bones themselves, provided the losses are made good later, depends upon the state of the reserves at the start. It is clear that, in considering the entire cycle as a unit, we cannot ignore its various parts. The early losses of calcium and phosphorus can be kept at a minimum by liberal intakes at this time; in fact, some investigators have reported that cows producing from 27 to 36 kg a day have been held in equilibrium either by the use of natural feeds rich in the minerals or by the addition of mineral supplements. While the majority of the experimental results show that such success is unusual, it is desirable to feed the minerals liberally so that excessive losses will certainly be avoided.

17.31   Calcium and Phosphorus Requirement

Annual balance experiments such as those previously described, slaughter experiments as to the state of the bones after several lactations, and long-time feeding experiments provided the basis for the early United States standards. A troublesome factor here was the wide differences in the results of the various investigations, which are now considered to be due, in the case of calcium at least, to the ability of an animal to adapt to a low intake by its more effective utilization. In the "adapted" animal there is probably both a lowered output of endogenous calcium and a higher net absorption. Thus a lactating cow which had been accustomed to a low calcium ration would show a more favorable balance than one which had not. Recommendation for practice, however, must be made adequate for "unadapted" animals.

A method of arriving at calcium and phosphorus requirements for a given body weight and milk production takes account of the endogenous outputs, the amounts secreted in the milk, and the estimated percentage utilization of the supply in the feed. For example, in the report of the British A.R.C. nutrient requirements, the calcium requirement of a 454-kg animal producing 30 kg of milk containing 1.28 g calcium per kilogram is arrived at as follows, assuming an availability of 45 percent for the calcium in the feed:

7.3 g endogenous Ca ÷ 0.45 = 16.2 g for maintenance
38.9 g milk Ca   ÷ 0.45 = 85.0 g
   Total daily requirement   101.0 g

A possible error arises from the choice of the figure for utilization since reported figures vary rather widely with different rations and levels of feeding. It is believed, however, that the one selected is a reasonable one for general use. The United States N.R.C. Committee has taken into account calculations such as presented above, and also the results of long-term feeding and balance experiments, in arriving at the requirements as set forth in Appendix Table I. The report specifies, for example, for maintenance of a 450-kg cow, 18 g of calcium and 14 g of phosphorus, with additions during the last 2 or 3 months of gestation. For such a cow producing 30 kg of 4 percent fat milk, 2.7 g of calcium and 2.0 g of phosphorus are added to the maintenance figure for each kilogram of milk produced. Using these figures, it can be calculated that the cow would need 99 g of calcium per day compared with 101 g shown above. It is not possible to be certain which of these values is the more accurate.

The N.R.C. calcium and phosphorus requirements for lactation in various animals, in comparison with maintenance, growth, and reproduction are summarized in Table 16.5. Although there is a large increase in the daily requirement for these mineral elements for milk synthesis there is no uniform increase in their percentages in the feed above that needed for growth or reproduction. Lactating animals consume considerably more feed during lactation than when not producing milk so that the daily intake of minerals would be higher accordingly, with the same percentage in the feed. The considerable variation among animal species (Table 16.5) may represent, in part, a lack of precise knowledge about the requirement and the inclusion of variable margins of safety to insure adequacy. The accuracy of requirements of calcium, phosphorus and magnesium are largely determined by the low accuracy of estimates of their availability.

17.32 Magnesium

Grass tetany (staggers) resulting from low blood magnesium levels is an important cause of losses among grazing cattle and sheep in some areas. Early treatment of affected animals by injection of magnesium salts can prevent death loss. Hypomagnesemic tetany has been related to environmental factors, chemical composition of the herbage, and physiological status of the animal. In cattle and sheep the tetany and death usually occur in early lactation in the cooler season when the grass is high in quality. Tetany can be prevented by providing an adequate amount of magnesium to maintain blood plasma levels above 2.0 mg per 100 ml. This is sometimes difficult because of the highly variable utilization of the mineral element (Sec. 10.19).

House and Mayland[62] reviewed the research on methods of supplying additional magnesium to ruminants. Dolomitic limestone, magnesium oxide, carbonate, and sulfate have all been used. Magnesium utilization has been reported to vary from 7 to 33 percent. The oxide was better utilized than magnesium from dolomite. Magnesium alloy bullets were effective, but the rate of decomposition was variable. Frye and associates[63] found that beef cows would consume adequate magnesium when a mineral mixture fed ad libitum contained one-third of a palatible ingredient such as cottonseed meal, dry sugarcane molasses, ground corn, or corn distillers dried solubles along with equal parts of trace mineralized salt and magnesium oxide. The N.R.C. magnesium requirements are 0.2 percent of the dry diet for dairy cows, 0.18 precent for beef cows and 0.06 percent for sheep.

17.33   Other Minerals

As is indicated in Table 17.2, milk contains considerable amounts of both sodium and chlorine. Studies have been cited in Sec. 10.22 showing that milk secretion is decreased by depriving lactating cows of salt and that the requirement for cows producing around 5000 kg per year is about 30 g per day in addition to that supplied in the feeds used. The addition of 1 percent of salt to the grain mixture serves the double purpose of providing this needed mineral and increasing the palatability of the mixture. Giving cows free access to salt enables those who crave more to get it. This is obviously important when little or no grain is being fed.

While potassium is the mineral occurring in milk in the largest amount (Table 17.2), it is always abundantly supplied in the feeds of plant origin commonly consumed. As the lactating animal enters a new gestation, an iodine supplement is needed in iodine-deficient areas. While the evidence of specific need is less clear, lactating animals, in areas where copper and cobalt have been found deficient in terms of growth and health, should receive supplements of those minerals.

Offering a mineral mixture containing common salt, calcium, phosphorus and the needed trace minerals free choice will usually meet the requirements of dairy cattle. Some cows will consume considerably more than they need and others too little. The surest way to meet the requirements is to mix the mineral into a complete feed since the appetite of cattle is not well regulated by requirements. For grazing animals not receiving a supplement there is no other option except free-choice feeding of minerals.

THE ROLES OF VITAMINS IN LACTATION

Vitamins are important in lactation both as essential nutrients for the physiological process involved and as components of the secretion itself.

17.34   Vitamin Content of Milk of Different Species

All the recognized vitamins are found in the milk of all species, insofar as studies have been made. A condensation of a table published by Hartman and Dryden[64] in their review of the vitamin content of milk and various milk products is presented in Table 17.6. This publication, consisting of 119 pages with 1,212 citations from the literature and dealing almost entirely with cow milk and its products, contains a wealth of information of special interest in terms of vitamin nutrition for lactation, the vitamin values for milk products, and their nutritive values in human nutrition. The values for cow milk not shown in this table are presented in later sections of this chapter.

17.35   Vitamin A Value of Milk

This value is of special importance in the nutrition of lactating animals because in all species the value is dependent on the feed supply of vitamin A per se or

TABLE 17.6. Vitamin Content of Milk

Species	Vitamin A, I.U./liter	Thia-mine	Ribo-flavin	Nicotinic acid	Vitamin B_6	Vitamin B_{12}	Ascorbic acid
				mg/liter			
Cow	1560	0.4	1.75	0.9	0.6	0.004	21
Buffalo	2024	0.51	1.05	1.42	. . .	0.003	21
Goat	2074	0.40	1.84	1.87	0.07	0.001	15
Sheep	1460	0.69	3.82	4.27	. . .	0.006	43
Dog	8667	0.05	6.11	7.8	0.08	0.007	. . .
Guinea pig	1834	0.59	2.60	11.10	. . .	. . .	333
Horse	800	0.30	0.33	0.58	0.21	0.001	100
Human	1898	0.16	0.36	1.47	0.10	0.0003	43
Pig	1036	0.70	2.21	8.35	0.40	0.002	140
Rat	4333	1.49	1.12	18.1	0.79	0.028	8
Whale	7194	1.16	0.96	20.4	1.10	0.008	70

Source: A. M. Hartman and L. P. Dryden, Vitamins in milk and milk products, A.D.S.A. Publ., 1965.

of carotene, apart from a contribution which may be temporarily made by body reserves. The cow on its natural herbivorous ration receives vitamin A only in the form of carotene. A portion of the carotene ingested is secreted in milk as such, and a portion is transformed into vitamin A and so secreted. The more yellow the milk and butter, the larger the amount of carotene present, but this is not a true measure of vitamin A value for it gives no information as to the amount of the vitamin present as such. Jersey and Guernsey milk has much more color than Holstein milk because these cows convert a smaller proportion of their carotene intake into the vitamin. Most of the vitamin value of their milk is due to carotene, whereas Holstein milk contains less of the pigment and more of the colorless vitamin. As a result, when the cows are fed the same ration, Holstein butterfat has fully as high a vitamin A value as that from Jerseys or Guernseys despite the marked difference in color. The difference in degree of carotene conversion is also reflected in larger amounts of pigment in the adipose tissue and skin secretions. The extent of the conversion varies among individuals as well as among breeds. Those species, such as the ewe, goat, and water buffalo, which secrete a colorless milk make a complete conversion, and thus the vitamin value of their fat may be very high though no color is present.

The vitamin A value of milk can vary widely according to the feed supply. The amount found in the milk may be several times as great on feeds high in carotene as on feeds which contain very little. Of the natural feeds, pasture results in the richest milk, but nearly as large a potency can be obtained by feeding properly cured alfalfa, dried grass, or silage. Much higher potencies are also produced by feeding concentrated sources of the vitamin, notably shark-liver oil. Despite the large differences in the vitamin A potency of milk according to the nature of the diet, the percentage of the intake that appears in the milk is very small. In the various experiments carried out, the percentage recovery has seldom exceeded 3 percent.

A comprehensive study by the U.S. Department of Agriculture showed that milk produced in the winter had an average of 1200 I.U. of vitamin A per liter and summer milk had 1900 I.U. Much of the summer milk produced on pasture had 2500 to 3000 I.U. per liter. The vitamin A values averaged 24,000 I.U. per kg and 39,900 I.U. for winter and summer butter, respectively.[65] Neither pasteurization nor irradiation decreases the vitamin A value of milk. Butter does not lose vitamin A or carotene in good commercial storage. The vitamin A value of colostrum is highest immediately after calving, 4 to 25 times that of normal milk. This level declines rapidly so that by the third to the tenth day after parturition it reaches the level of normal milk. In the United States low-fat milk is fortified to contain 2,000 I.U. of vitamin A and 400 I.U. of vitamin D per quart (2,100 and 420 I.U. per liter).

17.36   Vitamin A Requirements

It is clear that the animal nursing young should receive a liberal intake of vitamin A in order that its milk may contain an adequate amount for the nutrition of its offspring, at least until they are able to obtain this vitamin from supplementary foods. This is particularly true because the newborn generally has little reserve of the vitamin in its body even though the mother may have been fed liberally during pregnancy. The latter feeding does provide stores in the body of the mother which will be drawn upon for her milk and which lessen accordingly the necessity of large intakes during lactation. Unless the cow is fed for an extended period on very poor roughage, there should be no practical problem in meeting the needs of the suckling offspring, particularly when the calf is given access early to green, leafy roughage. The same should be true for lambs and colts. In the case of pigs, which are dependent on their mother's milk for a relatively longer period, the proper feeding of the mother may be of greater importance. Her needs should be met by selecting her ration in accordance with the same principles mentioned in the discussion of reproduction (Sec. 16.16). The needs of the dairy calf, which is early changed from whole milk to skim milk or a "milk replacer" have been discussed (Sec. 15.40).

The mature animal requires vitamin A for various body functions, but whether this need increases for the specific function of lactation apart from the demand for secretion in the milk has not been definitely proved. Clearly, the first effect of a deficient diet is a lowering of the vitamin content of the milk. One would expect any eventual effect on production itself to be accompanied by signs of deficiency in the lactating animal. Several experiments have clearly shown that feeding massive doses of vitamin A as a supplement to rations containing adequate amounts for reproduction does not increase milk yield, though greatly increasing the level of the vitamin in milk.

Swanson and Associates[66] demonstrated that lactating cows produced as much milk on vitamin A-deficient rations as on adequate intakes of the vitamin or with daily supplements of 50,000 I.U. The vitamin A requirement for lactation is lower than for normal reproduction, and there is no advantage from larger intakes except to increase the concentration of the vitamin in the milk.

The N.R.C. requirements for lactation which, in the case of beef cows, ewes, and sows, are considered to provide an adequate level in the milk for the nutrition of the young are set forth in the appendix tables. For dairy cows the N.R.C. report states that the amount specified for reproduction (Sec. 16.16) should suffice for maximum milk production, though higher intakes would increase the level in milk.

17.37   Vitamin D

The early studies of Forbes (Sec. 17.29), showing that the negative calcium and phosphorus balances at the height of lactation could not be eliminated by increasing the content of these minerals in the ration, were made before vitamin D was known. With the discovery of its role in the improvement of calcium and phosphorus retention, investigations were undertaken with the expectation that the addition of this vitamin to the ration would do away with the negative balances previously noted. Some initial experiments with goats supported this viewpoint, but more extensive studies with cows gave only negative results. Sun-cured hay, cod-liver oil, irradiated yeast, or any other source of vitamin D, including direct irradiation, was found ineffective in preventing the losses of calcium and phosphorus which occurred in the liberally producing cow early in lactation, although the vitamin was absorbed and its level in the milk increased. Extensive studies of this question were reported by Hart and associates.[67]

Wallis[68] has clearly demonstrated that dairy cows require vitamin D for lactation. Cows kept out of the sunlight and fed a ration in which molasses-beet pulp replaced the hay and which was otherwise practically devoid of the vitamin eventually developed striking deficiency symptoms. These experiments cannot be considered to cast doubt, however, on the earlier findings of Hart and associates that, under normal conditions of feeding and management, supplementary vitamin D will not materially benefit the calcium and phosphorus metabolism of the milking cow.

Hibbs and Conrad[69,70] reported that continuous feeding of a concentrate containing about 70,500 I.U. per kilogram reduced the incidence of milk fever (parturient paresis) in cows with previous histories from 60 to 26 percent but it had no beneficial effect in cows with no previous milk fever. In earlier studies they had shown that massive doses of 20 to 30 million I.U. for 3 to 8 days prepartum reduced the incidence of milk fever; however, there was risk of toxicity with longer feeding. Earlier, Boda and Cole had prevented milk fever by feeding a calcium deficient diet prepartum. Others have confirmed these results (Sec. 17.29). Littledike[71] used alternate periods of milking and nonmilk-

ing to show that initiation of milking resulted in hypocalcemia, hypophosphatemia, and hypermagnesemia which increased parathyroid hormone
concentrations.

More recently the Ohio researchers[72] observed in 17 private dairy herds
intramuscular (I.M.) injection of 10 million I.U. of vitamin D_3 about one week
prepartum reduced the incidence of milk fever of cows with previous histories,
but not in cows with no previous milk fever. Lower calcium and phosphorus
intakes prepartum tended to reduce the incidence further to about 10 percent.
Estimated calcium intakes varied from 29 to 113 g per cow per day and the
calcium-phosphorus ratios were from 1:1 to 3:1.

Gast et al.[73] obtained total protection by intravenous (I.V.) administration
of 0.1 mg of 1-hydroxy vitamin D_3. Intramuscular injection of even larger doses
gave incomplete protection. There was a rapid increase in serum calcium from
the I.V. treatment. No evidence of hypervitaminosis D was found with doses
up to 3.0 mg.

In a condition sometimes confused with milk fever, referred to as "downer
syndrome" cows show similar symptoms, but fail to respond to standard treatments and most of them die. Julien et al.[74] reported that cows fed a 15 percent
protein ration during the dry period had an incidence of 25 to 35 percent of
downer cows versus none in the cows receiving only 8 percent of protein.
Phosphorus levels between 0.29 and 0.70 percent of the diet and calcium-phosphorus ratios of 2:0.1 versus 1:1 had no influence on the incidence. No suggestions about the specific cause or treatments were presented and blood studies
revealed no important differences, thus the problem remains unsolved.

The N.R.C. recommendation for sows is 200 I.U. of vitamin D per kilogram of feed. There appears to be no basis for specific recommendations in the
case of the other farm animals. There is no reason for believing that the needs
exceed those during growth and reproduction or that otherwise good rations
require any supplementation, particularly in view of the role of sunlight.

No milk produced from natural feeds is a rich source of vitamin D in terms
of the needs of the growing young. Vitamin D values ranging from 5 to 44
U.S.P. units per liter were found in Guernsey milk, while for Holsteins the
range was from 3 to 28 units. The highest value occurred in the summer and
the lowest in winter, and there was a close correlation between the hours of
sunshine and the vitamin level, indicating that sunlight was the principal factor
involved in the variations. There was little difference in the potencies of the
butterfats of the two breeds; thus the Guernsey milk was richer because of its
higher fat content.

It is recognized that sunshine, by producing the vitamin in the animal's
body through irradiation, is the sole factor responsible for the increase in
potency of summer milk. Growing pasture grass, which plays such a large role
in increasing the vitamin A value of milk is actually devoid of vitamin D. In
fact, while the vitamin D potency of milk is somewhat influenced by the nature
of the feed, no system of feeding other than massive intakes of the vitamin has

been found effective in practice. Very large intakes of vitamin D concentrate or irradiated ergosterol produce relatively small change in the milk. Feeding vitamin D supplements to the human mother will not cure rickets in the suckling infant.

17.38 Vitamin D Milk

The best milk that can be produced in summer or on any ration of natural feed falls far short of meeting the requirements of children for protection against rickets.

The recognition of this fact and the discovery of methods of enriching milk to the effective level by direct irradiation or by feeding irradiated yeast to the cow resulted in the production and use of vitamin D milk. This development received the approval of medical authorities and nutritionists because, despite the knowledge of other effective methods of preventing rickets, the disease persisted unduly. It was felt that, if the vitamin were adequately supplied in a product which formed a considerable part of the daily diet of children, a more certain way of ensuring the needed intake would be provided.

The standard for vitamin D milk calls for a minimum level of 400 I.U. per quart (420 I.U. per liter). While this level can be obtained either by irradiating the milk or by feeding irradiated yeast to the cow, vitamin D milk is produced by the direct addition of a concentrate of vitamin D_2 or D_3.

17.39 Vitamin E

Cow milk normally contains 20 to 35 mg of vitamin E per gram of fat. The level can be increased by feeding tocopherols. It drops on rations very poor in the vitamin. Experimentally, muscular dystrophy has been produced in suckling rats by feeding a diet devoid of the vitamin. In the Cornell studies with suckling lambs previously described (Sec. 11.18) the blood serum, colostrum, and milk of mothers of the lambs which developed dystrophy were low in vitamin E. It has also been reported that pigs from sows on rations experimentally designed to be low in the vitamin exhibited muscular incoordination. Further studies in this general area, with particular reference to practical applications are needed. Reference has been made in Sec. 11.20 to the wide distribution of total tocopherols in livestock feeds.

Krukovsky and associates[75] have shown that milk high in tocopherol content is more resistant to the development of oxidized flavors, and the levels of the vitamin in the milk were found to vary with the character of the ration. Thus tocopherols are nutrients of importance from the standpoint of the stability and quality of market milk.

Goering et al.[76] have shown that added tocopherol helps to prevent oxidative changes in milk that has a high unsaturated fat content from feeding protected safflower oil. Flavor changes in milk remains a very active field of research.

17.40 Thiamin

The physiological needs for thiamin increase in accordance with the amount of milk secreted because of its role in energy metabolism and because of its content in the milk. These needs present no dietary problem in cattle and sheep because of rumen synthesis. The N.R.C. requirement for sows is set at 1.0 mg per kilogram of feed. In nonruminants the thiamin content of the milk secreted drops when the ration is deficient in the vitamin. There is a low ceiling, however, above which the level cannot be raised no matter how much thiamin is added to the diet. The nature of the feed has little influence on the thiamin content of the milk of ruminants. Commercial pasteurization commonly destroys 10 to 20 percent of the level in cow milk, though this loss can be reduced below 5 percent by careful control. Milk is not a rich source of thiamin in terms of human needs, a liter furnishing less than one-quarter of the recommended allowance for an active man.

17.41 Riboflavin

The story with respect to riboflavin is similar to that for thiamin. Ruminants need no dietary source, but sows do require a supply in their feed. The N.R.C. requirement for the sow is 3.0 mg per kilogram of feed. For those species requiring it in the diet, the level in the milk varies markedly according to the feed supply. The nature of the feed does have some influence on the level in cow milk also. In contrast to thiamin, cow milk is a rich source of riboflavin in terms of human needs, a liter as drawn containing enough to meet the full daily allowance of a physically active man. The milk which reaches the consumer, however, may have a markedly lower value because of exposure to sunlight. Bottled milk in clear glass has been found to lose as much as three-fourths of its riboflavin in 2 hours in bright sunlight.

17.42 Other B-Vitamins

The content of milk of several other B-vitamins is shown in Table 17.6. In addition to those listed, cow milk contains in milligrams per liter: biotin, 0.03; folic acid, 0.003; and pantothenic acid, 3.5. The levels of these other B-vitamins are little influenced by the feed, but some of them are variable according to breed, stage of lactation, and season. Dietary intakes are of no importance in the case of ruminants. The N.R.C. has set forth requirements for the sow in the cases of niacin, B_{12}, B_6, choline, biotin, folic acid, and pantothenic acid. The quantitative needs for the others have not been established.

17.43 Ascorbic Acid

In those species requiring vitamin C in the diet, its level in the milk depends on the dietary supply. Its level in cow milk is affected somewhat by breed and seasonal influences but not by the ration. Only a small part of the total amount secreted is found in the market milk that reaches the consumer. As drawn from the udder the product contains 2.0 to 2.5 mg per 100 ml, practically all in the reduced form. In contrast, Stewart and Sharp[77] have reported a survey showing that market milk in consumers' homes or retail stores contained 0.58 mg per 100 ml (0.34 mg in the reduced form and 0.24 mg as dehydroascorbic acid). Thus approximately three-fourths is lost in pasteurization and other marketing operations. Exposure to light is an even more destructive factor in the case of ascorbic acid than of riboflavin. Stewart and Sharp found that reconstituted powdered whole milk had twice as much vitamin C as the pasteurized fresh product. The losses from pasteurization and light exposure, which customarily occur in the processing and marketing of fresh milk, can be reduced to a minimum by appropriate procedures.

17.44   Unidentified Lactation Factors

There have been various reports in the literature during the past thirty years of factors required for lactation in addition to those definitely established as essential for growth. None of these "alleged factors" have been definitely identified, and thus their existence remains in doubt.

NOTES

1. I. G. Macy and associates, Human milk flow, *Am. J. Diseases Children,* **39:**1186–1204, 1930.

2. G. H. Schmidt, Biology of Lactation, W. H. Freeman and Co., San Francisco, Calif., 1971.

3. E. P. Reinecke and C. W. Turner, Formation in vitro of highly active thyroproteins, their biologic assay and practical use, *Missouri Agr. Expt. Sta. Research Bull.* 355, 1942.

4. G. H. Schmidt and associates, Effect of thyroprotein feeding on dairy cows, *J. Dairy Sci.,* **54:**481–492, 1971.

5. J. Hammond, Jr., and F. T. Day, Oestrogen treatment of cattle: Induced lactation and other effects, *J. Endocrinol.,* **4:**53–82, 1944.

6. K. L. Smith and F. L. Schanbacher, Hormone induced lactation in the bovine. I: Lactational performance following injections of 17β-estradiol and progesterone, *J. Dairy Sci.,* **56:**738–743, 1973.

7. R. E. Erb et al., Hormone induced lactation in the cow. IV: Relationships between lactational performance and hormone concentrations in blood plasma, *J. Dairy Sci.,* **59:**1420–1428, 1976.

8. I. G. Macy, H. J. Kelley, and R. E. Sloan, The composition of milks, *National Acad. Sci. National Research Council Publ.* 254, 1953.

9. R. S. Barber, R. Braude, and K. G. Mitchell, Studies on milk production of large white pigs, *J. Agr. Sci.,* **46:**97–118, 1958.

10. J. B. Coombe, I. D. Wardrop, and D. E. Tribe, A study of milk production of the grazing ewe, with emphasis on the experimental technique employed, *J. Agr. Sci.,* **54:**353–359, 1960.

11. G. H. Schmidt, Effect of milking intervals on the rate of milk and fat secretion, *J. Dairy Sci.,* **43:**213–219, 1960.

12. D. C. Hardwick and J. L. Linzell, Some factors affecting milk secretion by the isolated mammary gland, *J. Physiol. London,* **154**:547–571, 1960.

13. C. Dimant, V. R. Smith, and H. A. Lardy, Lactose synthesis in the mammary gland perfused with 1-C^{14} glucose, *J. Biol. Chem.,* **201**:85–91, 1953.

14. K. E. Ebner and F. L. Schanbacher, Biochemistry of lactose and related carbohydrates, in B. L. Larsen and V. H. Smith (eds.), Lactation: A Comprehensive Treatise, Vol. II, Academic Press, New York, 1974, pp. 77–113.

15. A. L. Black and M. Kleiber, The transfer of carbon from propionate to amino acids in lactating cows, *J. Biol. Chem.,* **232**:203–209, 1958.

16. T. B. Mepham and J. L. Linzell, A quantitative assessment of the contribution of individual plasma amino acids to the synthesis of milk proteins by the goat mammary gland, *Biochem. J.,* **101**:76–83, 1966.

17. B. L. Larson and G. N. Jorgensen, Biosynthesis of milk proteins, in B. L. Larsen and V. R. Smith (eds.), Lactation, *op. cit.,* pp. 115–146.

18. L. A. Maynard and associates, Studies of the blood precursor of milk fat, *Cornell Agr. Expt. Sta. Mem.* 211, 1938; L. Voris, G. H. Ellis, and L. A. Maynard, The determination of neutral fat glycerol in blood with periodate: Application to the determination of arteriovenous differences in blood fat, *J. Biol. Chem.,* **133**:491–498, 1940.

19. D. E. Bauman and C. L. Davis, Biosynthesis of milk fat, in B. L. Larson and V. R. Smith (eds.), Lactation, *op. cit.,* pp. 31–75.

20. J. E. Storry, J. A. F. Rook, and A. J. Hall, The effect of the amount and type of dietary fat on milk fat secretion in the cow, *Brit. J. Nutrition,* **21**:425–438, 1967.

21. R. A. Patton, R. D. McCarthy, and L. C. Griel, Jr., Observations on rumen fluid, blood serum and milk lipids of cows fed methionine hydroxy analog, *J. Dairy Sci.,* **53**:776–780, 1970.

22. P. T. Chandler and associates, Protein and methionine hydroxy analog for lactating cows, *J. Dairy Sci.,* **59**:1897–1909, 1976.

23. E. B. Powell, One cause of fat variation in milk, *Proc. Am. Soc. Animal Prod.,* 1938, pp. 40–47; Progress report on the relation of the ration to the composition of milk, *J. Dairy Sci.,* **24**:504–505, 1941.

24. S. E. Bishop and associates, Effects of pelleting and varying grain intakes on milk yield and composition, *J. Dairy Sci.,* **46**:22–26, 1963.

25. J. E. Storry and J. A. F. Rook, The effects of a diet low in hay and high in flaked maize on milk-fat secretion and on the concentrations of certain constituents in the blood plasma of the cow, *Brit. J. Nutrition,* **19**:101–109, 1965.

26. J. Golding, K. M. Soames, and S. S. Zilva, The influence of the cow's diet on the fat-soluble vitamins of winter milk, *Biochem. J.,* **20**:1306–1319, 1926.

27. J. E. Storry, P. E. Brumby, A. J. Hall, and B. Tuckley, Effects of free and protected forms of codliver oil on milk fat secretion in the dairy cow, *J. Dairy Sci.,* **57**:1046–1049, 1974.

28. R. A. Frobish and C. L. Davis, Theory involving propionate and vitamin B$_{12}$ in the low-milk fat syndrome, *J. Dairy Sci.,* **60**:268–273, 1977.

29. T. R. Wrenn et al., Increasing polyunsaturation of milk fats by feeding formalde-hyde protected sunflower-soybean supplement, *J. Dairy Sci.*, **59**:627–635, 1976.

30. L. A. Maynard and C. M. McCay, The influence of a low-fat diet upon fat metab-olism during lactation, *J. Nutrition*, **2**:67–81, 1929.

31. T. W. Perry, R. C. Peterson, and W. M. Beeson, Selenium in milk from feeding small supplements, *J. Dairy Sci.*, **60**:1698–1700, 1977.

32. D. C. Mahan, A. L. Moxon, and M. Hubbard, Efficacy of inorganic selenium supplementation to sow diets on resulting carry-over to their progeny, *J. Animal Sci.*, **45**:738–746, 1977.

33. T. S. Sutton, R. G. Warner, and H. E. Kaeser, The concentration and output of carotenoid pigments, vitamin A, and riboflavin in the colostrum and milk of dairy cows, *J. Dairy Sci.*, **30**:927–932, 1947.

34. D. A. Greenwood and associates, Fluorosis in cattle, Special Rep. 17, *Utah Agr. Expt. Sta.*, 1964.

35. J. W. Suttie and D. L. Kolstad, Effects of dietary fluoride ingestion on ration intake and milk production, *J. Dairy Sci.*, **60**:1568–1573, 1977.

36. L. Stoloff and associates, Stability of aflatoxin in milk, *J. Dairy Sci.*, **58**:1789–1793, 1975.

37. J. A. Knowles, Excretion of drugs in milk: A review, *J. Pediatrics*, **66**:1068–1082, 1965.

38. M. E. Johnson et al., Persistence of antibiotics in milk from cows treated late in the dry period, *J. Dairy Sci.*, **60**:1655–1661, 1977.

39. W. L. Gaines, The energy basis of measuring milk yield in dairy cows, *Univ. Illinois Agr. Expt. Sta. Bull.* 308, 1928.

40. H. F. Tyrrell and J. T. Reid, Prediction of the energy value of cow's milk, *J. Dairy Sci.*, **48**:1215–1223, 1965.

41. W. P. Flatt, C. E. Coppock, and L. A. Moore, Balance studies with lactating, non-pregnant dairy cows consuming rations with varying hay to grain ratios, in K. L. Blaxter (ed.), Energy Metabolism, Proc. Third Symposium, *European Assoc. Animal Production Publ.* 11, Academic Press, Inc., New York, 1965, pp. 121–130.

42. P. W. Moe and H. F. Tyrrell, Efficiency of conversion of digested energy to milk, *J. Dairy Sci.*, **58**:602–610, 1975.

43. T. L. Haecker, Investigations in milk-production, *Minnesota Agr. Expt. Sta. Bull.* 140, 1914.

44. K. Breirem, Nutrition and lactation in domestic animals and particularly in the cow, *Ann. Nutrition Aliment.*, **11**:3–4, A3–A32, 1957.

45. D. G. Wagner and J. K. Loosli, Studies on the energy requirements of high-producing cows, *Cornell Agr. Expt. Sta. Memoir* **400**:1–40, 1967.

46. G. W. Trimberger and associates, Effects of liberal concentrate feeding on health, reproductive efficiency, economy of milk production and other related responses of the dairy cow, *N.Y. Food and Life Sciences Bull.* 8, 1972.

47. L. A. Maynard and C. M. McCay, The influence of a low-fat diet upon fat metab-olism during lactation, *J. Nutrition*, **2**:67–81, 1929.

48. A. M. Leroy and J. Bonnet, Influence de la teneur en matière grasse de la ration sur la production de nature grasse des vaches laitières, *Ann. Agron.,* **17**:455–476, 1947.

49. F. W. H. Elsley et al., The effect of level of feed intake in pregnancy and in lactation upon the productivity of sows, *Animal Prod.,* **11**:225–241, 1969.

50. R. W. Gardner and D. E. Hogue, Studies on the TDN requirements of pregnant and lactating ewes, *J. Animal Sci.,* **22**:410–417, 1963.

51. E. S. Harrison and E. S. Savage, The effect of different planes of protein intake upon milk production, *Cornell Univ. Agr. Expt. Sta. Bull.* 504, 1932; E. S. Harrison, E. S. Savage, and S. H. Work, The effect of different planes of protein intake upon milk production II: Further comparisons of 16-, 20-, and 24-percent mixtures, *ibid.,* 578, 1933.

52. H. H. Van Horn and C. A. Zometa, Utilization of protein sources for lactation, Proc. Florida Nutrition Conf., Univ. of Florida, Gainesville, 1978.

53. E. B. Forbes and F. M. Beegle, The mineral metabolism of the milch cow. I. *Ohio Agr. Expt. Sta. Bull.* 295, 1916; E. B. Forbes and associates, The mineral metabolism of the milch cow. II, *ibid.,* 308, 1917; E. B. Forbes, J. O. Halverson, and L. E. Morgan, The mineral metabolism of the milch cow. III, *ibid.,* 330, 1918.

54. H. B. Ellenberger, J. A. Newlander, and C. H. Jones, Calcium and phosphorus requirements of dairy cows. I: Weekly balances through lactation and gestation period, *Vermont Agr. Expt. Sta. Bull.* 331, 1931; Calcium and phosphorus requirements of dairy cows. II: Weekly balances through lactation and gestation periods, *ibid.,* 342, 1932.

55. J. M. Boda and H. H. Cole, The influence of dietary calcium and phosphorus on the influence of milk fever in dairy cattle, *J. Dairy Sci.,* **37**:360–372, 1954.

56. K. D. Wiggers, D. K. Nelson, and N. L. Jacobson, Prevention of parturient paresis by a low-calcium diet prepartum: A field study, *J. Dairy Sci.,* **58**:430–431, 1975.

57. J. T. Yarrington et al., Effects of a low calcium prepartal diet on calcium homeostatic mechanisms in the cow: morphologic and biochemical studies, *J. Nutrition,* **107**:2244–2256, 1977.

58. D. Benzie and coworkers, Studies of the skeleton of the sheep. II: The relationship between calcium intake and resorption and repair of the skeleton in pregnancy and lactation, *J. Agr. Sci.,* **48**:175–186, 1956. III: The relationship between phosphorus intake and resorption and repair of the skeleton in pregnancy and lactation, *ibid.,* **52**:1–12, 1959.

59. A. Theiler, H. H. Green, and P. J. du Toit, Phosphorus in the live stock industry, *Union S. Africa J. Dept. Agr.,* **8**:460–504, 1924.

60. C. H. Eckles, T. W. Gullickson, and L. S. Palmer, Phosphorus deficiency in the rations of cattle, *Minnesota Agr. Expt. Sta. Tech. Bull.* 91, 1932.

61. R. B. Becker, W. M. Neal, and A. L. Shealy, Effect of calcium-deficient roughages on the milk yield and bone strength of cattle, *J. Dairy Sci.,* **17**:1–10, 1934.

62. W. A. House and H. F. Mayland, Magnesium and calcium utilization in sheep treated with magnesium alloy rumen bullets or fed magnesium sulfate, *J. Animal Sci.,* **42**:506–514, 1976.

63. T. M. Frye, J. P. Fontenot, and K. E. Webb, Jr., Relative acceptability of supplemental magnesium oxide mixtures by beef cows, *J. Animal Sci.,* **44:**919–926, 1977.

64. A. M. Hartman and L. P. Dryden, Vitamins in milk and milk products, Am. Dairy Sci. Assoc., 1965.

65. Anon., Vitamin A in butter, *U.S. Dept. Agr. Misc. Publ.* 571, 1945.

66. E. W. Swanson and associates, Milk production of cows fed diets deficient in vitamin A, *J. Animal Sci.,* **27:**541–548, 1968.

67. E. B. Hart and associates, Dietary factors influencing calcium assimilation. XI: The influence of cod-liver oil on calcium metabolism of milking cows, *J. Biol. Chem.,* **84:**359–365, 1929; XII: A study of the influence of hays cured with varying exposure to sunlight on the calcium metabolism of milking cows, *ibid.,* **84:**367–376, 1929; XIII: The influence of irradiated yeast on the calcium and phosphorus metabolism of milking cows, *ibid.,* **86:**145–155, 1930.

68. G. C. Wallis, Some effects of vitamin D deficiency on mature dairy cows, *J. Dairy Sci.,* **21:**315–333, 1938; Vitamin-D deficiency in dairy cows, *South Dakota Ag. Expt. Sta. Bull.* 372, 1944.

69. J. H. Hibbs and H. R. Conrad, Milk fever in dairy cows. VII: Effect of continuous vitamin D feeding on incidence of milk fever, *J. Dairy Sci.,* **59:**1944–1946, 1976.

70. J. H. Hibbs and H. R. Conrad, Milk fever in dairy cows. VI. Effect of three prepartal dosage levels of vitamin D on milk fever incidence, *J. Dairy Sci.,* **43:**1124–1129, 1960.

71. E. T. Littledike, Relationship of milk secretion to hypercalcemia in the dairy cow, *J. Dairy Sci.,* **59:**1947–1953, 1976.

72. W. E. Julien et al., Milk fever in dairy cows. VIII: Effect of injected vitamin D_3 and calcium and phosphorus intake on incidence, *J. Dairy Sci.,* **60:**431–436, 1977.

73. D. R. Gast, J. P. Marquardt, N. A. Jorgensen, and H. F. Deluca, Efficiency and safety of 1-hydroxy vitamin D_3 for preventation of parturient paresis, *J. Dairy Sci.,* **60:**1910–1920, 1977.

74. W. E. Julien, H. R. Conrad, and D. R. Redman, Influence of dietary protein on susceptibility to alert downer syndrome, *J. Dairy Sci.,* **60:**210–215, 1977.

75. V. N. Krukovsky, J. K. Loosli, and F. Whiting, The effect of tocopherols and cod liver oil on the stability of milk, *J. Dairy Sci.,* **32:**196–201, 1949.

76. H. K. Goering, and associates, Effect of feeding protected safflower oil on yield, composition, flavor and oxidative stability of milk, *J. Dairy Sci.,* **59:**416–425, 1976.

77. A. P. Stewart, Jr., and P. F. Sharp, Vitamin C content of market milk, evaporated milk, and powdered whole milk, *J. Nutrition,* **31:**161–174, 1946.

SUPPLEMENTARY LITERATURE

Adams, R. S.: Variability in mineral and trace element content of dairy cattle feeds, *J. Dairy Sci.,* **58:**1538–1548, 1975.

Baldwin, R. L., and S. Louis: Hormonal actions on mammary metabolism, *J. Dairy Sci.,* **58:**1033–1041, 1975.

Bingham, E. W., and H. N. Farrel, Jr.: Phosphorylation of casein by the lactating mammary gland: A review, *J. Dairy Sci.*, **60**:1199–1207, 1977.

Broderick, G. A., L. D. Satler, and A. E. Harper: Use of plasma amino acid concentration to identify limiting amino acids for milk production, *J. Dairy Sci.*, **57**:1015–1023, 1974.

Broster, W. H.: Protein energy interrelationships in growth and lactation of cattle and sheep, *Proc. Nutrition Soc.*, **32**:115–122, 1973.

Burroughs, W., D. K. Nelson, and D. R. Mertens: Protein physiology and its application in the lactating cow: The metabolizable protein feeding standard, *J. Animal Sci.*, **41**:933–944, 1975.

Clark, J. H., H. R. Spires, and C. L. Davis: Uptake and metabolism of nitrogenous components by the lactating mammary gland, *Fed. Proc.*, **37**:1233–1239, 1978.

Clark, J. H., H. R. Spires, R. G. Derrig, and M. R. Bennink: Milk production, nitrogen utilization and glucose synthesis in lactating cows infused postruminally with sodium caseinate and glucose, *J. Nutrition*, **107**:631–644, 1977.

Dunkley, W. L., N. E. Smith, and A. A. Franke: Effects of feeding protected tallow on composition of milk and milk fat, *J. Dairy Sci.*, **60**:1863–1869, 1977.

Gordon, F. J.: The effect of protein content on the response of lactating cows to level of concentrate feeding, *Animal Prod.*, **25**:181–191, 1977.

Gregory, M. E.: Nutritive value of milk and milk products: Water-soluble vitamins in milk and milk products, *J. Dairy Research*, **34**:169–181, 1967.

Larson, B. L., and V. R. Smith (eds.): Lactation: A Comprehensive Treatise, Academic Press, New York, 1974, Vols. 1, 2, and 3.

Linzel, J. L., and M. Peaker: Mechanism of milk secretion, *Physiol. Rev.*, **51**:564–577, 1971.

Madsen, A. O.: A comparison of some suggested measures of persistency of milk yield in dairy cows, *Animal Prod.*, **20**:191–197, 1975.

Mahan, D. C.: Effects of feeding various gestation and lactation dietary proteins sequences on long-term reproductive performance in swine, *J. Animal Sci.*, **45**:1061–1072, 1977.

McCarty, K. S., and K. S. McCarty, Jr.: Early mammary gland responses to hormones, *J. Dairy Sci.*, **58**:1022–1032, 1975.

Mendelson, C. R., and associates: Lipoprotein lipase and lipid metabolism in the mammary gland, *J. Dairy Sci.*, **60**:666–676, 1977.

Miller, W. J.: New concepts and developments in metabolism and homeostasis of inorganic elements in dairy cattle: A review, *J. Dairy Sci.*, **58**:1549–1560, 1975.

National Research Council: The effect of genetic variance on nutritional requirements of animals, National Academy of Sciences, Washington, D.C., 1975.

Papas, A.: Protein requirements of lactating chios ewes, *J. Animal Sci.*, **44**:672–679, 1977.

Peaker, M.: Recent advances in the study of monovalent ion movements across the mammary epithelium: Relation to onset of lactation, *J. Dairy Sci.,* **58:**1042–1047, 1975.

Phillips, M. C., and G. M. Briggs: Symposium on milk and dairy products for the American diet, *J. Dairy Sci.,* **58:**1751–1763, 1975.

Pickard, D. W.: Calcium requirements in relation to milk fever, in W. Haresign, H. Swan, and D. Lewis (eds.), Nutrition and the Climatic Environment, Butterworths, London, 1977, pp. 113–122.

Wood, P. D. P.: Algebraic models of the lactation curves for milk, fat and protein production, with estimates on seasonal variation, *Animal Prod.,* **22:**35–40, 1976.

Chapter 18

Work Production

Horsepower was safer when only the horse had it.

Will Rogers (1879–1935)

Many different kinds of animals, such as the elephant, camel, yak, ox, mule, donkey, water buffalo, reindeer, llama, alpaca, and even the moose, are used for work in various countries throughout the world. However, in the United States, the primary working mammals are the human, the dog, and the horse (and his relatives). This chapter deals mostly with these three classes but studies on work should not be limited to them. Knowledge of energy expenditure and nutrient requirements for various activities of other animals such as the cow, sheep, and pig can also be useful in the formulation of rations and in the evaluation of management practices. For example, how much energy does a range animal expend when grazing or traveling to water? What's the effect of exercise on carcass quality?

The number of horses in the United States increased dramatically in the 1960s and 1970s and it appears that the industry will continue to thrive as long as leisure time and money for recreation are available. Of course, the large increase in horse numbers is primarily due to the increase in pleasure horses but the number of race horses has also increased and many of the pleasure horses perform a substantial amount of work in various activities at shows, hunts, and trail rides. Some horses still work on ranches, as police horses, and in rodeos. Even the draft horse, which decreased dramatically in numbers with the increased use of farm tractors and trucks, is making a modest comeback. Much of the comeback is due to increased use of the draft animal for recreation or hobbies but some of the increase is also due to the desire by some persons to use draft horses on small farms.

The number of dogs has also increased dramatically and again most of the increase is due to pleasure animals. The number of racing dogs (greyhound and sled), guard dogs, and police dogs has increased substantially but they still remain a relatively small but important fraction of the total population.

18.1 Nutrients Involved in Muscle Activity

The muscles are the agencies by which mechanical work is performed. In their contraction, nutrients are catabolized. The early physiologists considered that the muscle was broken down in the process, and since the muscle was known to consist mostly of protein, the assumption arose that protein yielded the energy for the work done. Such was Liebig's view. For many years he taught that work production involved an increased excretion of nitrogenous end products and, therefore, required an increased intake of protein in accordance with the amount of work performed. Because of his eminence, Liebig's views continued to be accepted, though apparently not based upon experimental results, even after Voit showed in 1860 that work could be performed by a dog without increasing the protein catabolism.

Fig. 18.1 Many young people devote a great deal of time and effort to their horses. Horses have provided an excellent form of physical activity. Shown above is Ms. Stacy Lowe with her horse A Bit OK. (*Courtesy of J. E. Lowe, Ithaca, New York.*)

In 1866, Fick and Wislicenus ascended a Swiss mountain 1956 m high after having abstained from nitrogenous food for 17 hours and measured their urea output before and during the ascent. They found no considerable increase in the output while the work was being performed, and their calculations revealed that the total nitrogen excretion could account for only a fraction of the catabolism which must have occurred to furnish the energy needed for the work done. In 1879, Kellner showed that, as long as the total amount of feed of the work horses was ample, the protein catabolism was not increased by work. If, however, the feed was restricted and the work was increased to the point where the animal lost flesh, then a larger nitrogen excretion occurred. Thus it gradually came to be accepted that the muscle is not broken down in work and that its energy is normally supplied by nonnitrogenous food but that, if the food supply is insufficient, body protein as well as fat may be drawn on.

Although energy is by far the principal food need for muscular performance, protein and various other nutrients are also concerned. Phosphorus compounds and calcium and magnesium ions are minerals directly involved. Various B-vitamins serve as cofactors in the metabolic reactions which take place.

18.2   The Chemistry of Muscle Action

The contractile process is not completely understood but the actin and myosin filaments are of special importance. One hypothesis is that a nerve impulse causes a release of Ca^{++} ions which in turn releases the inhibition of actin-

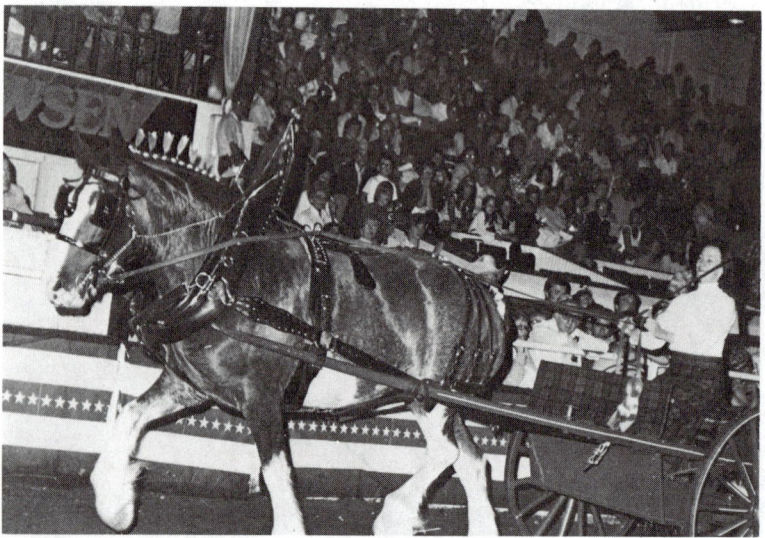

Fig. 18.2   The draft horse is also making a comeback. Many draft horses are used for recreational purposes. Shown above is Collessies Queen Foot Print of Starlane Farms being driven by Ms. Sue Quick at the New York State Fair. (*Courtesy of Mr. and Mrs. D. Flinn, Starlane Farms, Trumansburg, N.Y.*)

myosin interaction effected by troponin. Ca^{++} ions also stimulate the ATPase activity which is in the projection or head part of myosin. This enzyme catalyzes the conversion of ATP to ADP and H_3PO_4. Because ATP has a high-energy bond, mechanical energy and heat energy are released in the conversion of ATP to ADP. The actin-myosin fibers then separate and slide across each other to a new position, that is, they contract. When nervous stimulation is stopped, the concentration of Ca^{++} ions decreases. The troponin inhibition reappears and the muscle fibrils return to the resting or steady state.

Muscle contains only a small amount of ATP as such but the supply is renewed from phosphocreatine, another compound containing a high-energy phosphate bond, which is present in muscle in relatively large amounts and serves as a labile reservoir of ATP. The breakdown of phosphocreatine may be represented as follows:

$$
\begin{array}{l}
\quad\quad\quad O \\
\quad\quad\quad \| \\
HN \sim P-OH \\
\quad\quad | \\
\quad\quad OH \\
C = NH \\
\quad | \\
N-CH_2COOH \\
\quad | \\
CH_3
\end{array}
$$

Creatine phosphokinase
(CPK)

Phosphocreatine

Phosphocreatine + ADP $\rightleftharpoons$ ATP + creatine

The process is reversible, and the reformation occurs during the resting period of the muscle following contraction. The muscle is thus "recharged" to maintain the level of phosphocreatine. This recharging, however, calls for a supply of ATP which can come from the breakdown of glucose. Supported by oxygen from the blood, the glucose is completely catabolized to carbon dioxide and water yielding 36 moles of ATP per mole of glucose.

The muscle, like other cells, also has the capacity, to a limited extent, to break down glucose anaerobically and this is important in peak muscular activity. The glucose is broken down in the glycolytic pathway. The end product is lactate, whereas under aerobic conditions glycolysis stops at the pyruvate state where pyruvate is converted by oxidation to acetyl coenzyme A, which enters the tricarboxylic cycle for complete catabolism. The anaerobic process stops at the lactate stage because oxygen is required for further breakdown. The net yield of ATP is represented by the following equation:

Glucose + 2 phosphate + 2 ADP $\longrightarrow$ 2 lactate + 2 ATP

Thus it is seen that the anaerobic breakdown of a molecule of glucose yields only 2 ATP, whereas the complete aerobic breakdown yields nearly 20 times as much, namely, 36. Further, the anaerobic process is of limited duration because the accumulation of lactic acid causes the muscle to go into an inactive state called "rigor." Nevertheless, the process is an important one because it makes possible short periods of tremendous exertion not possible by the aerobic process alone, due to the inability of respiration to supply all the oxygen needed. When this occurs, the deficit of oxygen is spoken of as the *oxygen debt*. This provision of nature enables a human being or an animal to engage in exertion eight or ten times as strenuously as would be possible if all of the oxygen had to be supplied currently. A well-trained athlete cannot take in more than about 4 liters of oxygen per minute, yet he can temporarily perform work which would require oxygen at the rate of 30 liters per minute since it is possible for him to go into oxygen debt. The horse can, for a few seconds, perform work at least five times the equivalent of its maximum oxygen supply.

If the state of oxygen debt is long continued, the accumulation of lactic acid and perhaps other products results in muscular fatigue. This is transitory if the level of performance and the oxygen supply are such as to enable recovery to proceed. This effect of oxygen debt cannot explain the fatigue which gradually develops where the steady state is maintained by an oxygen supply adequate to serve the level of performance. Here the structure of the muscle itself and the nerve action may be affected. There are several chemical and physical explanations which, however, remain only partially understood.

Lactate production varies with level of work. Studies with human beings indicate that small or no changes in rate of lactate production are seen when work requires oxygen consumption below 50 percent of capacity. At high, submaximal loads, lactate production increases for the first few minutes of exercise. At maximal load, a constant production of lactate occurs. The lactate may diffuse out of the muscle and be transported via the blood to the liver where gluconeogenesis occurs when two lactate molecules are converted to glucose (Cori cycle). Some lactate may also be taken up by various tissues where it enters the tricarboxylic acid cycle via pyruvate. Training or conditioning decreases the level of lactate production.

18.3 Cellular Source of Energy

Thus, although all energy is ultimately traced to the oxidation of compounds such as glucose and fatty acids, the immediate source of energy for muscle contraction is ATP.

The cellular source for ATP production depends on many factors such as duration and intensity of exercise, cell type, condition, and metabolic state of the individual. Glucose is the primary source for short-term exhausting or maximal efforts, but fatty acids become the predominant source during long-term, low-intensity work. Free fatty acids have been shown to contribute 70 to 80 percent of the energy expenditure in human athletes under such conditions. Migratory species such as the salmon, eel, and birds utilize fats almost exclusively during sustained work. For example, the spawning salmon uses lipid reserves whereas the glycogen content of liver actually increases. Human athletes doing submaximal work who have adapted to endurance exercise obtain

Fig. 18.3 Although many Americans suffer from a lack of exercise interest in physical fitness has greatly increased in recent years. Shown above is the start of a race sponsored by the Finger Lakes Runners Club. (*Courtesy of Ithaca Journal*).

more of their energy from fat and less from carbohydrates than untrained individuals. Goodman et al.[1] reported that both fat and carbohydrate were used as sources in unconditioned and conditioned horses trotting for 1/2 hour but that the conditioned horses used fatty acids more efficiently than the unconditioned animals.

Studies by Lindholm and Saltin[2] and Carlsson et al.[3] have also demonstrated that a significant amount of free fatty acids is taken up by the muscle of exercising horses. Weil[4] suggested that the working muscle in horse preferentially extracts oleic and perhaps linolenic acid. Several studies have shown that dogs utilize free fatty acids.[5]

Other compounds such as alanine and the ketones, acetoacetate and 3-hydroxybutyrate, are also used as fuel but their contributions appear to be relatively minor.[6]

The horse[7] has three major muscle fiber types: (1) slow twitch, oxidative, (2) fast twitch, glycolytic, and (3) fast twitch, oxidative-glycolytic fibers. The fast-twitch, glycolytic fibers are adapted for rapid contractions and obtain energy by anaerobic metabolism. Slow-twitch fibers are the opposite extreme. They have slow contractions and obtain energy by aerobic metabolism. The slow-twitch fibers have a higher myoglobin content, more lipids, more blood flow, and a lower ATPase activity of the myosin than the fast twitch fibers. The fast twitch, oxidative-glycolytic fiber have characteristics intermediate between slow and fast twitch and can vary from primarily aerobic to primarily anaerobic.

Lindholm et al.[8] reported that horse muscle contains a higher content of glycogen than does human muscle and that the glycogen content and depletion rate of horse muscle varies with fiber type and type of exercise. The slow-twitch and fast-twitch oxidative-glycolytic fibers were depleted after exercise at slow trotting speeds with the former showing the faster rate of depletion. When glycogen was depleted, exercise capacity was impaired. The glycogen content of fast twitch glycolytic fiber was not depleted at low-speed exercise until the other fibers were depleted, but glycogen content of fast twitch glycolytic fibers was depleted at maximal or submaximal high-intensity exercise, and the length of time required for depletion was more than 30 minutes.

18.4   Dietary Source of Energy

What is the best dietary source of energy for work production? High-carbohydrate diets are frequently recommended for human athletes.[9] It is suggested that high-carbohydrate diets increase work performance and are more efficient for the production of muscular work than high-fat or high-protein diets. The explanation for the beneficial effect is that carbohydrates yield more energy per liter of oxygen and also promote greater buildup of muscle glycogen during the training period. Increased muscle glycogen stores appear to increase endurance. Saltin and Karlsson[10] concluded that in humans, when work requires between 65 and 89 percent of maximal oxygen uptake for times of 45 to 200 minutes, the depletion of glycogen in the working muscles seems to be the limiting factor. At work loads demanding more than 90 percent of maximal oxygen uptake, work time is reduced and muscle glycogen depletion does not limit performance. At low work rates, exercise can continue for many hours and glycogen stores may not ever be depleted.

Several recent experiments indicate that high-fat diets can be equal to or superior to high-carbohydrate diets in many situations. The performance of racing pigeons over distances of 200 miles or more was improved with the addition of fat in the diet.[11] Endurance was improved in swimming rats,[12] but not in running rats,[13] by increasing the intake of fat. Sled dogs fed high-fat seal meat performed better than when fed the seal meat plus cereal. Sled dogs fed diets containing 61 percent fat and no carbohydrate had higher plasma-free fatty acids after strenuous exercise than dogs fed diets containing 34 percent fat and 38 percent carbohydrates.[14] It was concluded that feeding a diet high in fat and low in carbohydrate enhanced the ability of the animal to mobilize body fat for periods of exercise. Increased mobilization would be advantageous during endurance events. High-fat diets have not been studied extensively in working horses but several recent studies have demonstrated that horses can efficiently digest fat. Also, it has been suggested that the addition of fat may improve the performance of horses on endurance rides. Furthermore, the increased caloric density of a high-fat diet would decrease the total amount of dry matter required and reduce gastrointestinal fill. High-fat diets might also provide a method of feeding a horse a high amount of energy with a reduced risk of founder. Founder (laminitis) is the result of the inflammation of the sensitive laminae of the feet. The sensitive laminae are leaflike structures that are lined up vertically around the region between the hoof wall and coffin bone. In founder, the sensitive laminae separate from the insensitive laminae of the hoof wall and the horse becomes lame. There are several causes of founder but overfeeding of soluble carbohydrate is probably the most common.

Carbohydrate loading has been used by many athletes, particularly those competing in events requiring sustained muscular activity, such as marathon races, long distance ski races, and bicycle races. For example, one study credited glycogen loading with improving the time of marathon racers by 8 or more minutes.[15] Briefly, the procedure is to deplete the glycogen stores and then provide excess carbohydrate. The body overcompensates in response to the glycogen depletion and stores more than the normal amount of glycogen. For example, an athlete may work strenuously and consume diets high in fat and protein during days 6, 5, and 4 prior to the competition. This combination greatly depletes glycogen stores. The athlete consumes high levels of carbohydrates during the three days prior to the event and glycogen content is greatly enhanced. However, glycogen loading may have disadvantages, such as increased water retention, intestinal cramps, and electrocardiographic abnormalities.

Glycogen loading does not appear to have many advantages for horses and dogs. It would be impractical to feed high-protein, high-fat, no-carbohydrate diets to horses and, even when such a diet was fed, the contributions of intestinal microfloral such as propionate would probably help maintain glycogen stores. Furthermore, high glycogen stores may result in muscle disorders such as tieing-up or *azoturia* (rhabdomyolysis). For example, azoturia was frequently reported in draft horses fed high levels of grain on Sunday though they were not worked. On Monday morning, when the horses were taken out for work, the muscles would stiffen and the urine would be coffee-colored because of the presence of myoglobin released during the breakdown of muscle cells. The condition was commonly called Monday morning disease. Much more work is needed on the causes of the condition but increased glycogen storage and its subsequent breakdown to lactic acid has been implicated. No studies evaluating the effect of muscle glycogen content on performance in endurance horses are available but the glycogen content does not appear to be of significance for race horses. The studies of Lindholm et al.[8] discussed earlier demonstrated that glycogen would not be depleted in short-term races. Muscle disorders have also been reported in dogs in which it was attempted to increase glycogen stores.[16] Kronfeld and Van Soest[17] concluded that that which is the best source of energy, predominantly carbohydrate or fat, "appears to depend on the severity and duration of the work bouts, the species, and probably on other conditions." It should also be stressed that dietary manipulation should not be used as a substitute for proper training. Training or conditioning can increase size and number or mitchondria per cell, increase aerobic capacity, increase the glycogen content of muscle, increase turnover rate of adipose tissue fatty acids, increase efficiency and rate of utilization of free fatty acids, decrease lactate production, and, thus, increase the efficiency of energy utilization by muscles.

When discussing high-fat diets, it must be remembered that the term *high fat* has different meanings for different species. Most horse rations normally contain less than 4 percent fat; thus a diet containing 15–20 percent fat would be considered high fat by horse nutritionists. Most human diets contain at least 10 percent fat and levels of up to 20 percent are not unusual. Some authorities suggest that for good cooking, at least 20 percent of the energy in the food should come from fat.[18] A common recommendation is that the diet of human athletes should have 30 to 35 percent of the total calories in the form of fat.[19]

18.5 Amount of Energy Required

The amount of energy needed depends on the amount of work performed and the efficiency with which the work is done. Nervous animals may require more energy to complete a task than calm animals. The ability of the rider or driver, condition and training of the animals, fatigue, environmental temperature, and constituents in the diet all influence efficiency of utilization.

Increasing the speed beyond a certain point decreases the net efficiency with which work is done. It is evident that a human uses up much more energy in running 100 m at top speed than in running at a trot. The energy cost to a human being of walking at different speeds has been found to be as follows in terms of kilocalories per kg per hour: 5 km per hour, 2; 6.5 km per hour, 3.4; and 8.5 km per hour, 9.3. Zuntz and associates found that approximately 15 percent more energy was required by the horse for locomotion at 5.8 km per hour than at a rate of 4.6 km. At a trot nearly twice as much energy was expanded as at a walk. They concluded that the horse is most efficient when working at a speed between 4 and 5 km per hour. Though gross efficiency increases with load and speed, a heavy load at low speed is more efficient than a light load at high speed.

Of all the forms of work investigated by Zuntz and associates, the ascent of a moderate grade appeared to be the most efficient on the net basis, but the efficiency decreased as the grade became steeper. Draft up a grade was performed less efficiently than draft alone, and as the grade increased from 1.5 percent to 8.5 percent, the net efficiency decreased from 31.3 to 22.7 percent.

Training increases working efficiency. When the horse or human attempts an unaccustomed task, many unnecessary muscles are brought into play which are not used when skill in performing the work has been acquired.

Estimates of energy required to perform various activities of horses are limited. Some values are listed by the N.R.C. subcommittee on horse nutrition. Standing does not increase the energy requirement above that of resting in the horse because of the stay apparatus. That is, the horse can "lock" his leg and, thus, decrease cost of standing. Galloping may increase the energy requirement 15 times and severe strenuous effort may increase it 20 to 30 times that of resting. Thus, a horse galloping 2 or 3 hours a day may have a daily energy need two or more times that of a resting horse.

Feeding standards are useful to indicate how the feed requirements vary in accordance with amount of work performed, but in practice, they can serve only as general guides. Having selected a suitable ration, it should be fed in accordance with the amounts needed to keep the horse in good working condition rather than as arbitrarily specified allowances.

More studies have been conducted on energy requirements for humans than for horses and estimates of the energy requirements for various human

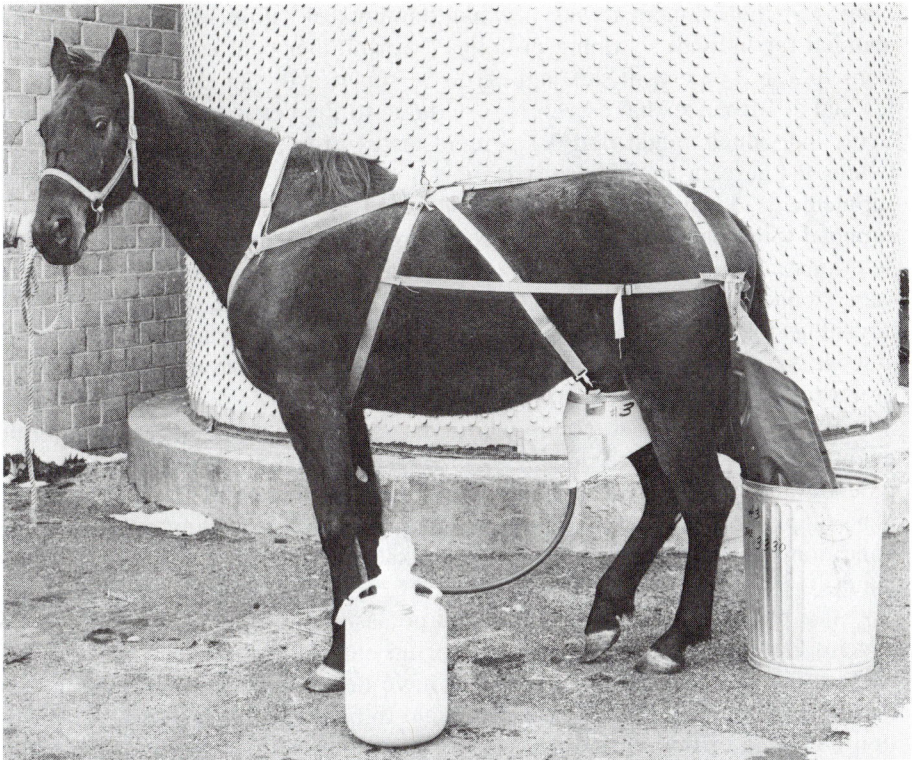

Fig. 18.4 There has been an increase in research on horse nutrition in recent years. Shown above is a method for conducting digestion trials with horses. (*Vandernoot, Fonnesback and Lydman, J. Animal Sci. 24:691–696, 1965. Courtesy of New Jersey Agr. Expt. Sta., Rutgers— the State University.*)

activities are available in many books on human nutrition. For example, tennis playing can increase the energy expenditure three to four times that of resting, long distance running seven to eight times, and sprinting ten to twelve times that of rest. But, of course, the time of the sprint is so short that it would have an insignificant impact on the total daily energy requirement if it was the only work performed.

18.6 Minerals and Exercise

Relatively few studies have been conducted on the effect of work on the requirements of minerals other than salt and electrolytes. The kidney can regulate urinary sodium losses but significant amounts of sodium are lost in sweat. Horses may lose more than 60 grams of sodium chloride per day when actively exercised. Losses could be even greater in hot, humid climates. Losses of 15 or more grams have been reported in humans doing heavy work. It is assumed that the salt requirements of horses can easily be met by allowing free choice access to salt. Even in humans, most losses can easily be recovered in a single meal. But the frequent intake of small amounts of saline solutions is recommended for individuals doing prolonged work under heat stress. Saline solutions (0.3 to 0.4 percent salt) appear to be superior to salt tablets as gastrointestinal disturbances have been reported due to the consumption of the latter. Some individuals might find 0.3 percent tastes "too salty." In solutions containing 0.1 percent salt, the salty taste is not detected. The value of potassium supplements during competition is still being debated. Potassium is usually supplied to humans in the form of commercially prepared isotonic electrolyte solutions containing sodium, potassium, and carbohydrates. One recent review[20] concluded that "there is yet no conclusive evidence of a physiological advantage [of isotonic solutions of electrolytes] over water, saline solutions, or glucose syrup drinks in improving performance." However, many athletes prefer to use the electrolyte solutions.

Electrolytes are also commonly given to horses, particularly during long events such as endurance rides. Carlson and Mansmann[21] reported that horses on endurance rides may develop synchronous diaphragmatic flutter or thumps. (The diaphragm flutters in synchrony with heart beats.) They suggested the condition was probably associated with alkalosis, hypocalcemia, and hypokalemia.

The plasma changes of sodium and potassium during exercise depend on several factors. Increases, decreases, or no changes have been reported. Increases in both sodium and potassium have been reported in marathon runners.[22] The increase in sodium was thought to be the result of hemoconcentration but some of the potassium increase was thought to be due to efflux from intracellular stores. No change in concentrations of sodium, potassium, or chloride were observed in horses ridden 35 miles in 6 hours but, as discussed earlier, Carlson and Mansmann[21] observed hypokalemia in horses ridden 100 miles in 12 to 14 hours.

Potassium deficiency results in decreased endurance times in swimming

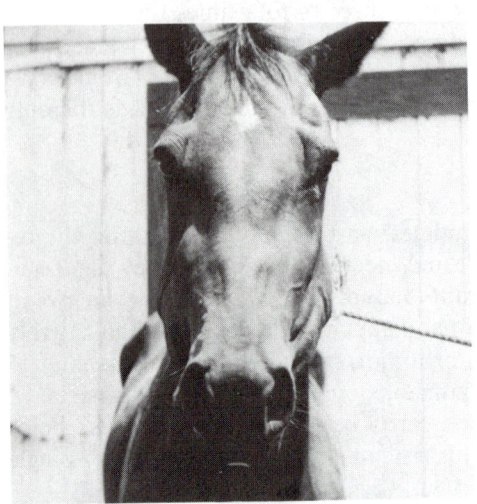

Fig. 18.5 Nutritional secondary hyperparathyroidism in a horse with calcium deficiency. The concentration of calcium necessary in the diet is not increased because the greater feed intake of working animals supplies the needed minerals. However working horses fed marginal deficient diets are more likely to develop calcium deficiency than non-working horses because of the greater rate of bone turnover. (Sec. 10.12). (*Cornell University.*)

rats and disruption of carbohydrate metabolism in the muscle. Severe muscular dystrophy has been reported in potassium-deficient rabbits.[23]

Calcium urinary losses are increased and bone changes take place (disuse osteoporosis) during periods of inactivity such as a prolonged bed rest but the losses may be due to absence of longitudinal pressure on the bones rather than physical inactivity.[24] Does exercise result in an increased calcium need above that to correct for sweat losses? Dalin and Olsson[25] reported that cross-country runners had a greater bone mineral content than the control group in certain bones but not in others. They concluded that physical training seemed to have little effect on the amount of bone mineral in the axial loaded skeleton but that it might be important for the appendicular skeleton. They also tested the effect of training on male office employees for a 3-month period. The training did not result in an increased bone mineral content in spite of an 11 percent increase in oxygen capacity. Thus it was concluded that physical training would not cause a rapid increase in bone mineral in healthy normal subjects. Studies on calcium retention in exercised dogs have been contradictory. Exercise did not increase net calcium deposition in horses or ponies exercised for various periods, but it did increase the rate of calcium turnover in the bone. It is concluded that, if the diet contains an adequate percent of calcium for maintenance, the increased need for calcium due to sweat losses will be corrected by the increased intake of the diet due to the greater energy needs. However, working animals fed a diet containing a marginal or low level of calcium could be more susceptible to calcium deficiency and bone problems than nonworking animals fed the same diet.

The active phosphorus metabolism which occurs during muscular activity has directed attention to the question of an increased requirement for this mineral. During World War I, Emdem conducted experiments on German soldiers with acid phosphate drinks, with apparently beneficial results. Their use became popular, accordingly, particularly with athletes. Later studies, however, have increased the skepticism regarding an effect of high-phosphate intake on muscular performance. Harvey and associates[26] have reported that hard work has no effect on the calcium and phosphorus balance in horses. Phosphorus is lost in sweat but at least in the horse the quantities are probably not as great as those of calcium.

Exercise frequently causes an increase in blood phosphorus in horses and humans. Hammel et al.[27] reported a decrease in blood phosphorus in sled dogs which they suggested may reflect in part decreased kidney tubular resorption associated with aciduria. They also pointed out, however, that, unlike hypocalcemia, hypophosphatemia has little or no clinically observable effects.

There may be significant amounts of iron lost in human sweat, but most diets contain enough iron to compensate for the loss.

PROTEIN REQUIREMENTS

While it is now accepted, as discussed at the beginning of the chapter, the protein is not the normal fuel of muscular work, some still adhere to the view that protein catabolism is increased during the work even though there is an ample supply of nonnitrogenous nutrients. To many, it is inconceivable that the muscle cells are entirely resistant to wear, and they believe that destruction and renewal must occur. It is stated that such a destruction may occur and yet not be reflected in an increased excretion because of a reutilization of the catabolic products. Such a viewpoint is very difficult to prove or disprove. There are experiments in which an increased output of urinary nitrogen has been recorded during work and others in which no such increase has been found. At least some of the positive experiments are inconclusive because of the uncertainty as to whether the intake of nonnitrogenous nutrients was adequate to meet the needs for energy. A well-conducted experiment with draft animals (bullocks) has been reported by Kehar and coworkers,[28] in which severe work did not increase the urinary nitrogen output.

Any increase in protein catabolism during work is certainly small. Harvey and associates[29] have reported that 818-kg Percheron horses fed protein at a maintenance level remained in positive nitrogen balance when working at rates as high as 1.27 hp per hour for over 4 hours daily. Forbes[30] states that experiments with humans confirmed earlier ones in showing that the protein needs are not measurably increased above maintenance by muscular activity, despite the popular ideas of athletes and hard workers. In studies with three miners over a 32-week period, Kraut and Lehmann[31] found that the minimum nitrogen intake to keep the men in positive nitrogen balance was the same with and without work, viz., 7 to 8 g daily. They noted, however, that at hard work there was a decrease in working capacity as well as psychic depression when the intake fell below 9 to 10 g daily. This intake is below the N.R.C. allowance for a sedentary man. Other studies have shown that, as the customary intakes of hard workers are markedly reduced in quantity and quality, though not down to the maintenance level, working capacity suffered. Experience in this country has shown that intakes of protein, particularly of animal origin, have important psychic effects on work capacity at levels far higher than any demonstrated physiological need.

Of course, if the working athlete is increasing muscle mass, the protein requirement is increased. Protein intakes appear to be adequate for mature working horses and humans when 10 percent of the gross energy in the diet is in the form of protein. Quality of protein is also important but animal protein is not essential as a reasonable mixture of amino acids can be supplied by vegetable protein. In fact, several top athletes are vegetarians. In most cases, benefits of the steak-training table appear to be primarily psychological. But such benefits should not be ignored. Davidson et al.[18] concluded, "Many distinguished athletes are more than normally sensitive people. The wise trainer will humour their dietary fancies and see that they have the food they like."

18.7 Vitamins

There is no doubt that exercise increases the total requirements of many vitamins. The need for the B-vitamins involved in energy metabolism is increased because of the greater energy intake. But there is little evidence to support the conclusions that the amounts of vitamins needed per unit of energy intake is increased or that oversupplementation with vitamins may improve performance. Numerous studies have failed to show a benefit from vitamin supplementation to diets containing adequate or normal levels of vitamins. Thus, most nutritionists consider that the increased needs for vitamins would be met when the total intake of a nutritionally balanced diet is increased. There is no evidence that the requirements for fat-soluble vitamins are increased during exercise. As discussed earlier, excessive amount of vitamins A and D could be harmful.

High doses of vitamin C are frequently recommended for human and equine athletes but there is little evidence to support the recommendations. Vitamin C has been recommended for race horses that are "nose-bleeders," that is, during the race, the small blood vessels would rupture. The recommendation was not based on experimental evidence but rather on the fact that vitamin C deficiency in other species caused increased capillary fragility. However, studies failed to show any difference in vitamin C blood levels between bleeders and nonbleeders. Furthermore, other drugs are now available that effectively treat the condition and it appears unlikely that vitamin C is involved in "nose-bleeders."

Vitamin C did not improve the performance of sprinters or long distance runners eating adequate diets.

18.8 Water

Dehydration is probably one of the most common problems of athletes working in hot, humid environments. Horses on endurance rides are particularly prone to dehydration. Hinton[32] suggested the simplest and most obvious way to prevent dehydration is to allow horses to drink whenever they wish. But he also pointed out that this is not a universally recommended practice. In fact, it is often recommended that the horse should not be allowed to drink appreciable amounts of water during events, advice Hinton condemned as being misleading and potentially harmful. He stated that there is no danger in allowing horses water when they are hot providing that they are offered water frequently and not allowed to become too thirsty. He recommended that horses be given the opportunity to drink every hour or so. Cold water can be given without harmful effects. If a horse is extremely thirsty, the water should be rationed initially, perhaps a gallon or so every 15 to 20 minutes.

The greatest joy of those who are steeped in work and who have succeeded in finding new truths and in understanding the relations of things to each other, lies in work itself.

Carl von Voit (1831–1908)

NOTES

1. H. M. Goodman et al., Determination of energy source utilized by the light horse, *J. Animal Sci.*, **37**:56–62, 1973.

2. A. Lindholm and B. Saltin, The physiological and biochemical response of standardbred horses to exercise of varying speed and duration, *Acta Vet. Scand.*, **15**:310–324, 1974.

3. L. Carlsson et al., Concentration and turnover of the free fatty acids of plasma and concentration of blood glucose during exercise in horses, *Acta Physiol. Scand.*, **63**:434–441, 1965.

4. H. Weik, Effect of work on individual free fatty acids in plasma of the horse, *Zentrabt. Vet. Med.* 17A:712–718, 1970 (as abstracted in *Nutr. Abst. Rev.* **41**:912, 1971).

5. W. A. S. Shaw, et al., Interrelationship of FFA and glycerol turnovers in resting and exercising dogs, *J. Appl. Physiol.* **39**:30–36, 1975.

6. R. M. McGilvery, The use of fuels for muscular work, *Proc. 2nd Int. Sym. Biochem. Exer.* 12–30, 1973.

7. However, muscle types can vary greatly among vertebrates. For a review of muscles of domestic birds, see K. Kiessling, Muscle structure and function in the goose, quail, pheasant, guinea hen, and chicken, *Comp. Biochem, Physiol.* **57B**:287–292, 1977.

8. A. Lindholm, H. Bjerneld, and B. Saltin, Glycogen depletion pattern in muscle fibers of trotting horses, *Acta Physiol. Scand.* **90**:475–484, 1974.

9. C. F. Consolazio and H. L. Johnson, Dietary carbohydrate and work capacity, *Amer. J. Clin. Nutrition* **25**:85–90, 1972.

10. B. Saltin and J. Karlsson, Muscle glycogen utilization during work of different intensities, *Adv. Exp. Med. Biol.* **11**:289–300, 1971.

11. H. M. Goodman and P. Greminger, Effect of dietary energy source on racing performance in the pigeon, *Poultry Sci.* **98**:2058–2068, 1969.

12. D. Tollenar, Effects of mineral and vitamin supplementation on swimming times and other parameters related to performance of rats on a low calorie regimen, *J. Nutrition* **90**:441–448, 1966.

13. D. Tollenar, Dietary fat level as affecting running performance and other performance-related parameters of rats restricted or non-restricted in food intake, *J. Nutrition* **106**:1539–1547, 1976.

14. D. S. Kronfeld et al., Hematological and metabolic responses to training in racing sled dogs fed diets containing medium, low, or zero carbohydrate, *Amer. J. Clin. Nutrition* **30**:419–430, 1977.

15. P. Slovic, What helps the long distance runner run? *Nutrition Today,* **10**(3):18–21, 1975.

16. D. S. Kronfeld, Diet and performance of racing sled dogs, *J. Am. Vet. Med. Assoc.,* **162**:470, 1973.

17. D. S. Kronfeld and P. J. Van Soest, Carbohydrate nutrition, *Comp. Animal Nutrition,* **1**:23–73, 1976.

18.   S. Davidson, R. Passmore, J. F. Brock, and A. S. Truswell, *Human Nutrition and Dietetics,* 6th ed., Churchill Livingstone, England, 1975.

19.   E. R. Buskirk, Diet and athletic performance, *Postgraduate Med.,* **61:**229–236, 1977.

20.   Anon. Nutrition and athletic performance, *Dairy Council Dig.,* **46:**7–10, 1975.

21.   G. P. Carlson and R. A. Mansmann, Serum electrolyte and plasma protein alterations in horses used in endurance rides, *J. Am. Vet. Med. Assoc.,* **165:**262–263, 1974.

22.   L. I. Rose et al., Serum electrolyte changes after marathon running, *J. Appl. Physiol.,* **29:**449–451, 1970.

23.   E. L. Hove and J. F. Herndon, Potassium deficiency in the rabbit as a cause of muscular dystrophy, *J. Nutrition,* **55:**363–374, 1955.

24.   I. Issekutz et al., Effect of prolonged bed rest on urinary calcium output, *J. Appl. Phys.,* **21:**1013–1020, 1966.

25.   N. Dalen and K. E. Olsson, Bone mineral content and physical activity, *Acta Orthop. Scand.,* **45:**170–174, 1974.

26.   A. L. Harvey and associates, Effect of work on the calcium and phosphorus retention of Percheron geldings, *J. Animal Sci.,* **2:**103–111, 1943.

27.   E. P. Hammel et al., Metabolic responses to exhaustive exercise in racing sled dogs fed diets containing medium, low, or zero carbohydrate, *Am. J. Clinical Nutrition,* **30:**309–418, 1977.

28.   N. D. Kehar and coworkers, Studies on protein metabolism, II: The effect of muscular work on endogenous protein catabolism, in cattle, *Indian J. Vet. Sci.,* **13:**263–266, 1943.

29.   A. L. Harvey and associates, The effects of limited feeding of oats and timothy hay during work on the nitrogen balance of draft geldings Proc. Am. Soc. Animal Prod., 1939, pp. 94–103.

30.   W. H. Forbes, The effects of hard physical work upon nutritional requirements, *Milbank Mem. Fund Quart,* **23:**89–96, 1945.

31.   Heinrich Kraut and Gunther Lehmann, Der Eiweissbedarf des Schwerarbeiters. I. Physiologisches und funktionelles Eiweissminimum, *Biochem. Z.,* **319:**228–246, 1949.

32.   M. Hinton, Long distance horse riding and the problem of dehydration and rhabdomyolosis, *Vet Ann.,* **17:**136–141, 1977.

SUPPLEMENTARY LITERATURE

Albanese, A. A., et al.: Nutritional and metabolic effects of physical exercise. *Nutrition Rept. Int.,* **3:**165–187, 1971.

Askew, E. W., et al.: Dietary carnitine and adipose tissue turnover rate in exercise trained rats, *J. Nutrition,* **107:**132–142, 1977.

Benzi, G., et al.: Mitochondrial enzymatic adaptation of skeletal muscle to endurance training, *J. Appl. Physiol.,* **38:**565–569, 1975.

Hollosez, J. O., and F. W. Booth: Biochemical adaptations to endurance exercise in muscle. *Ann. Rev. Physiol.,* **38:**372–392, 1976.

Keller, W. D., and H. A. Kraut: Work and nutrition, *World Rev. Nutrition Diet.,* **3:**69–81, 1962.

Kluger, J. et al.: Energy balance and lactic acid production in the exercising rabbit, *Am. J. Physiol.,* **223:**1451–1454, 1972.

Lewis, S. and B. Gutin: Nutrition and endurance, *Am. J. Clinical Nutrition,* **26:**1011–1014, 1973.

Lindholm, A. and K. Piehl: Fibre composition, enzyme activity, and concentrations of metabolites and electrolytes in muscles of Standardbred horses, *Acta Vet. Scand.,* **15:**287–309, 1974.

Milne, D. W.: Blood gases, acid-base balance and electrolyte and enzyme changes in exercising horses, *J. S. Afr. Vet. Assoc.,* **48**(4):345–354, 1974.

Murray, D. M. et al.: Energy cost of exercise in pigs, *Can. J. Animal Sci.,* **55:**201–205, 1975.

Refsum, H. E. et al.: Serum electrolyte fluid and acid-base balance after prolonged heavy exercise at low environmental temperature, *Scand. J. Clinical Invest.,* **32:**117–123, 1973.

Slade, L. M. et al.: Nutritional adaptations of horses for endurance performance, Proc. Fourth Equine Nutrition Physiol. Symp., Calif. State Polytech. U., Pomona, 1975, pp. 114–128.

Tucker, V. A.: Energetic cost of locomotion in animals, *Comp. Biochem. Physiol.,* **34:**841–846, 1970.

VanBeamont, W. et al.: Changes in total plasma content of electrolytes and proteins with maximal exercise, *J. Appl. Physiol.,* **34:**102–106, 1973.

Glossary

The following glossary contains definitions of terms that have not been defined in the text.

Achromotricia   Loss of pigment in hair.
Actin   Muscle fiber fraction that complexes with myosin to bring about muscle contraction.
Ad libitum   At pleasure or free choice of food.
Adiabatic   No gain or loss of heat.
Aerobic   Requiring oxygen.
Alopecia   Loss of hair.
Anabolism   Synthesis of compounds.
Anaerobic   Oxygen not present.
Aneuryism   Dilatation of the wall of an artery.
Anorexia   Lack or loss of appetite for food.
Arachnoid   Membrane covering the brain and spinal cord.
Bloat   A disorder of ruminants caused by excess gas in the rumen.
Canula   A tube for insertion into a body cavity.
Carnivore   Animal that eats primarily animal material.
Catabolism   Degradation of compounds.
Chemotroph   That which consumes a chemical.

Collagen Fraction of the fibers of connective tissue, cartilage of bone, forms gelatin when boiled.

Comfort zone The temperature range at which no demand is made on temperature regulating mechanism.

Compensatory growth Faster than normal weight gains following a period of underfeeding.

Congenital Existing at birth.

Coprophagy The ingestion of feces.

Critical temperature (lower) That temperature below which heat production increases in response to a fall in environmental temperature.

Diarrhea Abnormal frequency and liquidity of fecal discharge.

Diverticulitis Inflammation of a pouch of the intestine.

Dystrophy Degeneration.

Elastin The base of elastic tissue.

Endogenous Generated from within. For example, endogenous fecal calcium denotes the calcium in the feces that is from the body and not from the food.

Enteritis Inflammation of the intestine.

Entrophy Degree of randomness.

Estivation (summer hibernation) A dormant state in which some desert animals pass the hottest time of summer.

Exostosis Outgrowth of bone.

Exudative diathesis Condition in which the walls of the blood vessels are changed such that certain blood constituents pass through into adjacent tissue or spaces.

Feed additive A material added to animal rations, but which may not supply essential nutrients.

Fistula An opening, often surgically prepared for nutritional studies. For example, rumen fistula or esophageal fistula.

Goitrogens Substances which interfere with thyroxine production in the thyroid gland.

Hematopoietic An agent that promotes the formation of blood cells.

Hemorrhage Characterized by bleeding or loss of blood.

Herbivore An animal that eats primarily plant material.

Hibernation A dormant state in which some animals pass the winter, in which body temperature drops to slightly above freezing and metabolic activity falls to a very low level.

Homeotherm Warm-blooded animals in which the body temperature is relatively constant.

Homeostasis Maintenance of uniformity and stability.

Hormone A chemical substance secreted by an endocrine gland which has a specific effect on the activities of other organs.

Hydrolyze To split a compound with the introduction of water.

Hydrophobic Water hating.

Hydrophylic Water loving.

Hydroxyapatite An inorganic fraction of bone consisting of $3Ca_3(PO_4)_2Ca(OH)_2$ plus other elements such as fluoride.

Hyperlipemia An excess of fat in the blood.

Hyperostosis Hypertrophy of bone tissue.

Hypocalcemia Decreased level of calcium in the blood.

Hypokalemia Decreased level of potassium in the blood.

Hyponatremia Decreased level of sodium in the blood.

Immiscible Not mixable.

In situ Within the body.

Ketosis A condition characterized by an abnormally elevated concentration of ketone (acetone) bodies in body fluids and tissues—acetonemia.

Kyphosis Convex curvature of the spinal column.

Legume A plant that has the ability to work symbiotically with bacteria to fix nitrogen from the air.

Lordosis Concave curvature of the spinal column.

Lymph The fluid contents of the central lacteal and thoracic duct.

Metabolic body size An expression relating energy metabolism to body weight which has a relationship to body surface. It is usually expressed as a power of body weight such as $W^{0.75}$ or some other exponent.

Milieu Mixture.

Miscible Mixable.

Moiety Component, part.

Mono One.

Mucosa Mucous membrane, mucosal surface of the intestine on the absorptive side (see serosa).

Myoglobin Muscle hemoglobin.

Myosin Muscle fiber fraction that complexes with actin to bring about muscle contraction.

Nonpolar Immiscible with water.

Oligo A few.

Omnivore Literally means eats everything. Animals that eat both plant and animal materials.

Opisthotonus A condition in which the head tilts back, which may be caused by a deficiency of thiamin. Also called star-gazing.

Osteoblasts Bone forming cells.

Osteoclasts Bone resorbing cells.

Osteomalacia A decrease in the ratio of minerals to organic matrix in bone.

Osteoporosis Decreased bone density.

Otoliths Bony structures within the membranous labyrinth of the ear.

Penultimate Almost last.

Plasma Fluid portion of the blood, similar to the serum except it contains fibrogen and other clotting factors.

Polar Miscible with water.

Poly Many.

Polymers Multiple units of single compounds.

Polyneuritis Multiple neuritis (degeneration of nerves).

Porphyrin Porphin-ring derivative from complexes such as hematin and cytochrome.

Porphyrinuria Porphyrin in the urine.

Postabsorptive state No undigested or unabsorbed food remaining in the digestive tract.

Prolactin The lactation-stimulating hormone of the anterior lobe of the pituitary gland.

Prothrombin Plasma fraction that is converted to thrombin during the clotting of blood.

Rachitic Adjective used to denote the condition of rickets.

Radiant heat   Heat transmitted by radiation (as from the sun) in contrast to that transmitted by conduction or convection.

Radiogram   Radiograph or picture produced by x-rays.

Reduction   Chemically, the removal of oxygen from, or the addition of hydrogen to, a substance.

Rennin   Enzyme that coagulates milk.

Retarded growth   Slower than normal rate of growth.

Rumen culture   A liquid or dried preparation consisting of microbes from the rumen.

Scoliosis   Lateral curvature of the spinal column.

Serosa   Membrane lining the pericardial, pleural, and peritoneal cavities.

Serum   Fluid portion of blood similar to plasma except fibrinogen and other clotting factors have been removed.

Specific dynamic effect, SDE, SDA   The increase in heat production from consumption of food, usually referred to as heat increment.

Steatitis   Inflammation of the adipose tissue.

Tetany   Spasmodic muscle contractions, convulsions.

Thermogenesis   Chemical production of heat in the body.

Thermoneutrality   The temperature range or comfort zone in which the body's temperature regulating mechanisms are not required to function to either cool or warm the body.

Thrombin   Plasma fraction that converts fibrinogen to fibrin in the clotting of blood.

Ultimobranchial bodies   Thyroid primordia derivatives.

Appendix

TABLE I. Daily Nutrient Requirements of Dairy Cattle

Body weight, kg	Daily gain, g	Feed DM, kg	Protein, g	TDN, kg	DE	ME	NE$_m$	NE$_g$	Ca, g	P, g	Vitamins, I.U. A, 1000	D
					Mcal							
Growing dairy heifers												
25	300	0.45	111	0.54	2.38	2.14	0.85	0.53	6	4	1.1	165
50	500	0.76	180	0.91	4.01	3.61	1.40	0.90	9	6	2.1	330
75	700	2.10	318	1.72	7.67	6.71	1.96	1.37	15	8	3.2	495
100	700	2.80	402	2.10	9.26	8.09	2.43	1.47	18	9	4.2	660
200	700	5.20	620	3.45	15.20	13.01	4.10	1.96	21	14	8.5	1320
300	700	7.20	771	4.56	20.11	17.07	5.55	2.38	24	18	12.7	1980
400	700	8.60	864	5.45	24.03	20.40	6.89	2.66	25	20	17.0	2640
500	600	9.50	903	5.96	26.28	22.26	8.14	2.52	27	21	21.2	3300
600	200	9.58	879	5.37	23.68	19.60	9.33	0.90	25	18	25.4	3960
Growing dairy bulls												
100	800	2.80	427	2.18	9.63	8.47	2.43	1.68	19	10	4.2	660
200	1000	5.20	702	3.68	16.23	14.05	4.10	2.50	23	16	8.5	1320
400	1000	9.00	947	6.06	26.72	22.93	7.41	3.50	29	23	17.0	2640
600	700	10.80	988	7.06	31.13	26.58	10.27	3.10	29	23	25.4	3960
800	300	12.00	1040	7.13	31.44	26.35	12.74	1.35	30	23	33.9	5280

Growing veal calves

35	500	0.67	173	0.80	3.52	3.17	0.98	0.90	7	4	1.5	231
55	900	1.20	292	1.45	6.38	5.74	1.55	1.73	11	7	2.3	363
75	1050	1.48	334	1.78	7.83	7.05	1.96	2.10	15	9	3.2	495
100	1100	1.69	357	2.03	8.94	8.05	2.43	2.31	17	10	4.2	660
150	1300	2.22	428	2.66	11.75	10.58	3.30	2.99	20	12	6.4	990

Maintenance of mature bulls

600	—	8.95	766	5.01	22.09	18.29	10.74	—	23	17	25	—
800	—	11.10	942	6.17	27.20	22.52	13.32	—	29	21	34	—
1000	—	13.12	1093	7.35	32.41	26.83	15.75	—	34	25	42	—
1200	—	15.05	1244	8.43	37.17	30.77	18.05	—	39	29	51	—
1400	—	16.88	1386	9.45	41.66	34.49	20.27	—	43	33	59	—

Source: N.R.C. Nutrient Requirements of Dairy Cattle, 1978.

TABLE I (continued). Daily Nutrient Requirements of Dairy Cattle

Body weight, kg	Daily feed, kg	Protein, g	TDN, kg	DE	ME	NE$_l$	Ca, g	P, g	Vitamin A, 1000 I.U.
					Mcal				
Maintenance of mature cows									
400	5.5	373	3.15	13.86	11.90	7.16	15	13	30
500	6.5	432	3.72	16.39	14.06	8.46	18	15	38
600	7.5	489	4.27	18.79	16.12	9.70	21	17	46
700	8.5	542	4.79	21.09	18.10	10.89	24	19	53
800	9.5	592	5.29	23.32	20.01	12.03	27	21	61
Maintenance and pregnancy (last 2 months of gestation)									
400	7.2	702	4.10	17.98	15.47	9.30	26	18	30
500	8.6	821	4.84	21.25	18.29	11.00	31	22	38
600	10.0	931	5.55	24.37	20.97	12.61	37	26	46
700	11.3	1035	6.23	27.35	23.54	14.15	42	30	53
800	12.6	1136	6.89	30.24	26.02	15.64	47	34	61
Fat, %		*Nutrients per kilogram of milk*							
3.0	—	77	0.282	1.24	1.07	0.64	2.5	1.7	—
4.0	—	87	0.326	1.44	1.24	0.74	2.7	1.8	—
5.0	—	98	0.365	1.61	1.39	0.83	2.9	1.9	—
6.0	—	108	0.410	1.81	1.56	0.93	3.1	2.0	—
Body weight change during lactation									
Weight loss		−320	−2.17	−9.55	−8.25	−4.92	—	—	—
Weight gain		500	2.26	9.96	8.55	5.12	—	—	—

TABLE II. Part 1: Nutrient Requirements for Growing-Finishing Steer Calves and Yearlings (nutrient concentration in diet dry matter)

Weight, kg	Daily Gain, kg	Daily Feed, kg	Roughage, %	Total protein, %	TDN, %	ME	NE$_m$	NE$_g$	Ca, %	P, %
							Mcal/kg			
100	0.0	2.1	100	8.7	55	2.0	–	–	0.18	0.18
	0.7	2.7	50–60	14.8	70	2.5	1.60	1.00	0.70	0.48
	1.1	2.7	<15	18.2	86	3.1	2.07	1.37	1.04	0.70
150	0.0	2.8	100	8.7	55	2.0	1.17	–	0.18	0.18
	0.7	3.9	50–60	12.6	70	2.5	1.60	1.00	0.46	0.36
	1.1	3.7	<15	15.6	86	3.1	2.07	1.37	0.76	0.54
200	0.0	3.5	100	8.5	55	2.0	1.17	–	0.18	0.18
	0.7	5.7	70–80	10.8	64	2.3	1.40	0.78	0.32	0.28
	1.1	4.6	<15	13.6	86	3.1	2.07	1.37	0.59	0.43
250	0.0	4.1	100	8.5	55	2.0	1.17	–	0.18	0.18
	0.9	6.2	45–50	11.1	72	2.6	1.64	1.02	0.35	0.31
	1.3	6.0	<15	12.7	86	3.1	2.07	1.37	0.50	0.38
300	0.0	4.7	100	8.6	55	2.0	1.17	–	0.18	0.18
	0.9	8.1	55–65	10.0	70	2.5	1.56	0.95	0.27	0.23
	1.3	7.1	<15	11.7	83	3.0	1.98	1.31	0.41	0.32
350	0.0	5.3	100	8.5	55	2.0	1.17	–	0.18	0.18
	0.9	8.0	45–55	10.0	72	2.6	1.64	1.02	0.25	0.22
	1.3	8.0	<15	10.8	83	3.0	1.98	1.31	0.32	0.28
400	0.0	5.9	100	8.5	55	2.0	1.17	–	0.18	0.18
	1.0	9.4	45–55	9.4	72	2.6	1.64	1.02	0.22	0.21
	1.3	8.6	<15	10.4	86	3.1	2.07	1.37	0.29	0.26
450	0.0	6.4	100	8.5	55	2.0	1.17	–	0.18	0.18
	1.0	10.3	45–55	9.3	72	2.6	1.64	1.02	0.19	0.19
	1.3	9.3	<15	10.4	86	3.1	2.07	1.31	0.26	0.25

TABLE II (continued). Part 2: Nutrient Requirements for Growing-Finishing Heifer Calves and Yearlings (nutrient concentration in diet dry matter)

Weight, kg	Daily Gain, kg	Feed, kg	Roughage, %	Total protein, %	TDN, %	ME	NE$_m$	NE$_g$	Ca, %	P, %
						Mcal/kg				
100	0	2.1	100	8.7	55	2.0	1.17	—	0.18	0.18
	0.5	3.0	70–80	12.4	61	2.2	1.32	0.70	0.47	0.37
	0.7	2.9	50–60	14.4	69	2.5	1.56	0.95	0.66	0.48
	1.1	3.0	<15	17.8	86	3.1	2.07	1.37	0.97	0.63
150	0	2.8	100	8.7	55	2.0	1.17	—	0.18	0.18
	0.5	4.1	70–80	11.0	61	2.2	1.32	0.70	0.34	0.29
	0.7	4.0	50–60	12.4	69	2.5	1.56	0.95	0.45	0.35
	1.1	4.0	<15	15.0	86	3.1	2.07	1.37	0.70	0.50
200	0	3.5	100	8.5	55	2.0	1.17	—	0.18	0.18
	0.3	5.4	100	9.1	55	2.0	1.17	0.50	0.18	0.18
	0.7	6.0	70–80	10.2	64	2.3	1.40	0.87	0.30	0.27
	1.1	5.0	<15	12.8	86	3.1	2.07	1.37	0.50	0.38
250	0	4.1	100	8.5	55	2.0	1.17	—	0.18	0.18
	0.3	6.4	100	8.9	55	2.0	1.17	0.50	0.18	0.18
	0.7	5.8	55–65	10.5	72	2.6	1.64	1.02	0.29	0.26
	1.1	6.5	20–25	11.4	80	2.9	1.89	1.25	0.38	0.31
300	0	4.7	100	8.6	55	2.0	1.17	—	0.18	0.18
	0.3	7.4	100	8.5	55	2.0	1.17	0.50	0.18	0.18
	0.7	6.6	55–65	10.1	72	2.6	1.64	1.02	0.24	0.23
	1.1	7.5	20–25	10.4	80	2.9	1.89	1.25	0.31	0.27
350	0	5.3	100	8.5	55	2.0	1.17	—	0.18	0.18
	0.3	8.2	100	8.5	55	2.0	1.17	0.50	0.18	0.18
	0.7	7.9	55–65	9.2	69	2.5	1.56	0.95	0.19	0.19
	1.1	8.3	20–25	9.9	80	2.9	1.89	1.25	0.24	0.23
400	0	5.9	100	8.5	55	2.0	1.17	—	0.18	0.18
	0.3	9.1	100	8.5	55	2.0	1.17	0.50	0.18	0.18
	0.5	8.5	70–80	8.8	64	2.3	1.40	0.78	0.18	0.18
	0.9	8.4	20–25	9.4	77	2.8	1.81	1.18	0.20	0.20
450	0	6.4	100	8.5	55	2.0	1.17	—	0.18	0.18
	0.2	8.7	100	8.5	55	2.0	1.17	0.50	0.18	0.18
	0.8	9.1	35–45	9.0	75	2.7	1.72	1.10	0.18	0.18

TABLE II (continued). Part 3: Nutrient Requirements for Beef Cattle Breeding Herd (nutrient concentration in diet dry matter)

Weight, kg	Daily Gain, kg	Daily Feed, kg	Roughage, %	Total protein, %	TDN, %	ME	NE_m	NE_g	Ca, %	P, %
							Mcal/kg			
Pregnant yearling heifers, last third of pregnancy										
325	0.4	6.6	100	8.8	52	1.9	1.09	0.38	0.23	0.23
	0.8	9.4	85–100	9.0	58	2.1	1.24	0.60	0.23	0.21
350	0.4	6.9	100	8.8	52	1.9	1.09	0.38	0.22	0.22
	0.8	10.0	85–100	8.8	58	2.1	1.24	0.60	0.22	0.21
375	0.4	7.2	100	8.7	52	1.9	1.09	0.38	0.21	0.21
	0.8	11.0	85–100	8.7	55	2.0	1.17	0.50	0.20	0.20
400	0.4	7.5	100	8.7	52	1.9	1.09	0.38	0.21	0.21
	0.8	11.6	85–100	8.7	55	2.0	1.17	0.50	0.19	0.19
425	0.4	7.8	100	8.8	52	1.9	1.09	0.38	0.20	0.20
	0.8	12.1	85–100	8.7	55	2.0	1.17	0.50	0.18	0.18
Dry pregnant mature cows, middle third of pregnancy										
400	—	6.1	100	5.9	52	1.9	1.09	—	0.18	0.18
500	—	7.2	100	5.9	52	1.9	1.09	—	0.18	0.18
600	—	8.3	100	5.9	52	1.9	1.09	—	0.18	0.18
Dry pregnant mature cows, last third of pregnancy										
400	0.4	7.5	100	5.9	52	1.9	1.09	—	0.18	0.18
500	0.4	8.6	100	5.9	52	1.9	1.09	—	0.18	0.18
600	0.4	9.7	100	5.9	52	1.9	1.09	—	0.18	0.18
Cows nursing calves, average milking ability, first 3–4 months postpartum										
400	—	8.8	100	9.2	52	1.9	1.09	—	0.28	0.28
500	—	9.8	100	9.2	52	1.9	1.09	—	0.28	0.28
600	—	11.0	100	9.2	52	1.9	1.09	—	0.25	0.25
Cows nursing calves, superior milking ability, first 3–4 months postpartum										
400	—	10.8	100	10.9	55	2.0	1.17	—	0.42	0.38
500	—	11.8	100	10.9	55	2.0	1.17	—	0.39	0.36
600	—	12.9	100	10.9	55	2.0	1.17	—	0.36	0.34
Bulls, growth and maintenance (moderate activity)										
300	1.00	8.8	70–75	10.2	64	2.3	1.40	0.78	0.31	0.26
400	0.90	11.0	70–75	9.4	64	2.3	1.40	0.78	0.21	0.21
600	0.50	12.0	80–85	8.8	61	2.2	1.32	0.70	0.18	0.18
800	0.00	10.5	100	8.5	55	2.0	1.17	—	0.18	0.18
1000	0.00	12.4	100	8.5	55	2.0	1.17	—	0.18	0.18

Source: Nutrient Requirements of Beef Cattle, National Academy of Sciences, Washington, D.C., 1976.

TABLE III. Part 1: Nutrient Requirements of Growing-Finishing Swine Fed Ad Libitum (percent or amount per kilogram of diet)

Live weight, kg		1–5	5–10	10–20	20–35	35–60	60–100
Expected daily gain, g		200	300	500	600	700	800
Expected efficiency, g gain/ kg feed		800	600	500	400	350	270
Expected efficiency, feed/gain		1.25	1.67	2.00	2.50	2.86	3.75

Nutrients	Units						
Energy							
Digestible energy	kcal	3700	3500	3300	3300	3300	3300
Metabolizable energy	kcal	3600	3400	3200	3200	3200	3200

Requirements							
Crude protein	%	27	20	18	16	14	13
Lysine	%	1.28	0.95	0.79	0.70	0.61	0.57
Arginine	%	0.33	0.25	0.23	0.20	0.18	0.16
Histidine	%	0.31	0.23	0.20	0.18	0.16	0.15
Isoleucine	%	0.85	0.63	0.56	0.50	0.44	0.41
Leucine	%	1.01	0.75	0.68	0.60	0.52	0.48
Methionine	%	0.76	0.56	0.51	0.45	0.40	0.30
Phenylalanine	%	1.18	0.88	0.79	0.70	0.61	0.57
Threonine	%	0.76	0.56	0.51	0.45	0.39	0.37
Tryptophan	%	0.20	0.15	0.13	0.12	0.11	0.10
Valine	%	0.85	0.63	0.56	0.50	0.44	0.41
Calcium	%	0.90	0.80	0.65	0.60	0.55	0.50
Phosphorus	%	0.70	0.60	0.55	0.50	0.45	0.40
Sodium	%	0.10	0.10	0.10	0.10	0.10	0.10
Chlorine	%	0.13	0.13	0.13	0.13	0.13	0.13
Potassium	%	0.30	0.26	0.26	0.23	0.20	0.17
Magnesium	%	0.04	0.04	0.04	0.04	0.04	0.04
Iron	mg	150	140	80	60	50	40
Zinc	mg	100	100	80	60	50	50
Manganese	mg	4.0	4.0	3.0	2.0	2.0	2.0
Copper	mg	6.0	6.0	5.0	4.0	3.0	3.0
Iodine	mg	0.14	0.14	0.14	0.14	0.14	0.14
Selenium	mg	0.15	0.15	0.15	0.15	0.15	0.10
Vitamin A or	I.U.	2200	2200	1750	1300	1300	1300
β-Carotene	mg	11	11	9.0	7.0	7.0	7.0
Vitamin D	I.U.	220	220	200	200	150	125
Vitamin E	I.U.	11	11	11	11	11	11
Vitamin K	mg	2.0	2.0	2.0	2.0	2.0	2.0
Riboflavin	mg	3.0	3.0	3.0	2.6	2.2	2.2
Niacin	mg	22	22	18	14	12	10
Pantothenic acid	mg	13	13	11	11	11	11
Vitamin B_{12}	ug	22	22	15	11	11	11
Choline	mg	1100	1100	900	700	550	400
Thiamin	mg	1.3	1.3	1.1	1.1	1.1	1.1
Vitamin B_6	mg	1.5	1.5	1.5	1.1	1.1	1.1
Biotin	mg	0.10	0.10	0.10	0.10	0.10	0.10
Folic acid	mg	0.60	0.60	0.60	0.60	0.60	0.60

Source: N.R.C. Nutrient Requirements of Swine.

TABLE III. Part 2: Nutrient Requirements of Breeding Swine (percent or amount per kilogram of diet)

		Bred gilts and sows, Young and adult boars	Lactating gilts and sows
Energy and protein	*Units*		
Digestible energy	kcal	3300	3300
Metabolizable energy	kcal	3200	3200
Crude protein	%	12	13
Nutrients		Requirements	
Indispensable amino acids			
Arginine	%	0	0.40
Histidine	%	0.15	0.25
Isoleucine	%	0.37	0.39
Leucine	%	0.42	0.70
Lysine	%	0.43	0.58
Methionine	%	0.23	0.36
Phenylalanine	%	0.52	0.85
Threonine	%	0.34	0.43
Tryptophan	%	0.09	0.12
Valine	%	0.46	0.55
Mineral elements			
Calcium	%	0.75	0.75
Phosphorus	%	0.50	0.50
Sodium	%	0.15	0.20
Chlorine	%	0.25	0.30
Potassium	%	0.20	0.20
Magnesium	%	0.04	0.04
Iron	mg	80	80
Zinc	mg	50	50
Manganese	mg	10	10
Copper	mg	5	5
Iodine	mg	0.14	0.14
Selenium	mg	0.15	0.15
Vitamins			
Vitamin A	I.U.	4000	2000
or β-carotene	mg	20	10
Vitamin D	I.U.	200	200
Vitamin E	I.U.	10	10
Vitamin K	mg	2	2
Riboflavin	mg	3	3
Niacin	mg	10	10
Pantothenic acid	mg	12	12
Vitamin B_{12}	μg	15	15
Choline	mg	1250	1250
Thiamin	mg	1	1
Vitamin B_6	mg	1	1
Biotin	mg	0.1	0.1
Folic acid	mg	0.6	0.6

Source: N.R.C. Nutrient Requirements of Swine, 1979.

TABLE IV. Part 1: Nutrient Concentrations in Diets for Horses and Ponies (dry-matter basis)

	Live weight, kg	Daily gain, kg	Daily feed, kg	DE, Mcal/kg	Concentrate[a] % in diet	Crude protein, %	Ca, %	P, %	Vitamin A, I.U./kg
Maintenance (mature)	500	0	7.45	2.2	0–10	8.5	0.30	0.20	1600
Gestation, last 90 days	—	0.55	7.35	2.5	25–35	11.0	0.50	0.35	3400
Lactating, first 3 months	—	0	9.75	2.8	45–55	14.0	0.50	0.35	2800
Lactating, 3 months to weaning	—	0	9.35	2.6	30–40	12.0	0.45	0.30	2450
Creep feed	—	—	—	3.5	100	18.0	0.85	0.60	—
Foal (3 months)	155	1.2	4.2	3.25	75–80	18.0	0.85	0.60	2000
Weanling (6 months)	230	0.8	5.0	3.1	65–70	16.0	0.70	0.50	2000
Yearling (12 months)	325	0.55	6.0	2.8	45–55	13.5	0.55	0.40	2000
Long yearling (18 months)	400	0.35	6.0	2.6	30–40	11.0	0.45	0.35	2000
2 year (light training)	450	0.15	6.6	2.6	30–40	10.0	0.45	0.35	2000
Mature working horses									
Light work	—	—	—	2.5	25–35	8.5	0.30	0.20	1600
Moderate work	—	—	—	2.9	50–60	8.5	0.30	0.20	1600
Intense work	—	—	—	3.1	65–70	8.5	0.30	0.20	1600

[a] Concentrate containing 3.6 Mcal/kg, hay containing 2.2 Mcal/kg.
Source: N.R.C. Nutrient Requirements of Horses, 1978.

TABLE IV. Part 2: Dietary Minerals and Vitamins for Horses

	Adequate levels		
	Maintenance of mature horses	Growth	Toxic levels[a]
Calcium, %	0.30	0.7	
Phosphorus, %	0.20	0.5	
Sodium, %	0.35	0.35	
Potassium, %	0.4	0.5	
Magnesium, %	0.09	0.1	
Sulfur, %	0.15	0.15	
Iron, mg/kg	40	50	
Zinc, mg/kg	40	40	9,000
Manganese, mg/kg	40	40	*
Copper, mg/kg	9	9	*
Iodine, mg/kg	0.1	0.1	4.8
Cobalt, mg/kg	0.1	0.1	
Selenium, mg/kg	0.1	0.1	5.0
Fluorine, mg/kg	—	—	50+
Lead, mg/kg	—	—	80
Vitamin A	1,600	2,000	*
Vitamin D, I.U./kg	275	275	150,000
Vitamin E, mg/kg	15	15	
Thiamin, mg/kg	3	3	
Riboflavin, mg/kg	2.2	2.2	
Pantothenic acid, mg/kg	15	15	

[a]Nutrients known to be toxic to other species but without adequate information on the horse are indicated by an asterisk.

TABLE V. Nutrient Content of Diets for Sheep (nutrient concentration in diet dry matter)

Body weight, kg	Daily feed per animal, kg	Energy TDN, %	DE Mcal/kg	ME Mcal/kg	Protein, %	Ca, %	P, %	Vitamin A, I.U./kg
Maintenance of ewes								
50	1.0	55	2.4	2.0	8.9	0.30	0.28	1275
60	1.1	55	2.4	2.0	8.9	0.28	0.26	1391
70	1.2	55	2.4	2.0	8.9	0.27	0.25	1488
80	1.3	55	2.4	2.0	8.9	0.25	0.24	1569
Nonlactating and first 15 weeks of gestation								
50	1.1	55	2.4	2.0	9.0	0.27	0.25	1159
60	1.3	55	2.4	2.0	9.0	0.24	0.22	1177
70	1.4	55	2.4	2.0	9.0	0.23	0.21	1275
80	1.5	55	2.4	2.0	9.0	0.22	0.21	1360
Last 6 weeks of gestation or last 8 weeks of lactation suckling singles								
50	1.7	58	2.6	2.1	9.3	0.24	0.23	2500
60	1.9	58	2.6	2.1	9.3	0.23	0.22	2684
70	2.1	58	2.6	2.1	9.3	0.21	0.20	2833
80	2.2	58	2.6	2.1	9.3	0.21	0.20	3091
First 8 weeks of lactation suckling singles or last 8 weeks of lactation suckling twins								
50	2.1	65	2.9	2.4	10.4	0.52	0.37	2024
60	2.3	65	2.9	2.4	10.4	0.50	0.36	2217
70	2.5	65	2.9	2.4	10.4	0.48	0.34	2380
80	2.6	65	2.9	2.4	10.4	0.48	0.34	2615
First 8 weeks of lactation suckling twins								
50	2.4	65	2.9	2.4	11.5	0.52	0.37	1771
60	2.6	65	2.9	2.4	11.5	0.50	0.36	1962
70	2.8	65	2.9	2.4	11.5	0.48	0.34	2125
80	3.0	65	2.9	2.4	11.5	0.48	0.34	2267
Replacement lambs and yearlings								
30	1.3	62	2.7	2.2	10.0	0.45	0.25	981
40	1.4	60	2.6	2.1	9.5	0.44	0.24	1214
50	1.5	55	2.4	2.0	8.9	0.42	0.23	1417
60	1.5	55	2.4	2.0	8.9	0.43	0.24	1700
Replacement ram, lambs, and yearlings								
40	1.8	65	2.9	2.4	10.2	0.35	0.19	944
80	2.8	55	2.4	2.0	8.9	0.28	0.16	1214
120	2.6	55	2.4	2.0	8.9	0.33	0.18	1962

TABLE V. (Continued)

Body weight, kg	Daily feed per animal, kg	TDN, %	DE	ME	Protein, %	Ca, %	P, %	Vitamin A, I.U./kg
			Mcal/kg					
Finishing lambs								
30	1.3	64	2.8	2.3	11.0	0.37	0.23	588
40	1.6	70	3.1	2.5	11.0	0.31	0.19	638
50	1.8	70	3.1	2.5	11.0	0.28	0.17	708
Early weaned lambs								
10	0.6	73	3.2	2.6	16.0	0.40	0.27	1417
30	1.4	73	3.2	2.6	14.0	0.36	0.24	1821

Source: Nutrient Requirements of Sheep, National Academy of Sciences, Washington, D.C., 1975.

TABLE VI. Nutrient Requirements of Poultry (per kilogram feed)

Nutrient	Chickens				Turkeys		
	0–8 wk	8–18 wk	Laying	Breeding	0–8 wk	>8 wk	Breeding
Energy, Mcal	2.9	2.9	2.85	2.85	2.9	3.2	2.9
Protein, %	18	15–12	15	15	26	22–16	14
Vitamin A, I.U.	1500	1500	400	400	4000	4000	4000
Vitamin D, I.U.	200	200	500	500	900	900	900
Vitamin E, I.U.	10	5	5	10	12	10	25
Vitamin K_1, mg	0.5	0.5	0.5	0.5	1	0.8	1
Thiamin, mg	1.8	1.3	0.8	0.8	2	2	2
Riboflavin, mg	3.6	1.8	2.2	3.8	3.6	3.0	4.0
Pantothenic acid, mg	10	10	2.2	10	11	9	16
Niacin, mg	27	11	10	10	70	50	30
Pyridoxine, mg	3	3	3	4.5	4.5	3.5	4
Biotin, mg	0.15	0.10	0.10	0.15	0.2	0.1	0.15
Choline, mg	1300	500	500	500	1900	1100	1000
Folacin, mg	0.55	0.25	0.25	0.35	1.0	0.8	1.0
Vitamin B_{12}, μg	9	3	3	3	3	3	3
Linoleic acid, %	1.0	0.8	1.0	1.0	1.0	0.8	1.0
Calcium, %	0.9	0.6	3.25	2.75	1.2	0.8	2.25
Phosphorus, %	0.7	0.4	0.5	0.5	0.8	0.7	0.7
Potassium, %	0.2	0.16	0.1	0.1	0.4	0.4	0.4
Sodium, %	0.15	0.15	0.15	0.15	0.15	0.15	0.15
Chlorine, mg	800	800	800	800	800	800	800
Copper, mg	4	3	3	4	6	4	6
Iodine, mg	0.35	0.35	0.3	0.3	0.4	0.4	0.4
Iron, mg	80	40	50	80	60	40	60
Magnesium, mg	600	400	500	500	500	500	500
Manganese, mg	55	25	25	33	55	25	35
Selenium, mg	0.1	0.1	0.1	0.1	0.2	0.2	0.2
Zinc, mg	40	35	50	65	75	40	65

Source: Nutrient Requirements of Poultry, National Academy of Sciences, Washington, D.C., 1977.

TABLE VII. Nutrient Requirements of Rabbits (percentage or amount per kilogram of dry diet)

Nutrients	Growth	Maintenance	Gestation	Lactation
Digestible energy, kcal	2500	2100	2500	2500
TDN, %	65	55	58	70
Crude fiber, %	10–12	14	10–12	10–12
Fat, %	2	2	2	2
Crude protein, %	16	12	15	17
Calcium, %	0.4	—	0.45	0.75
Phosphorus, %	0.22	—	0.37	0.5
Magnesium, mg	300–400	300–400	300–400	300–400
Potassium, %	0.6	0.6	0.6	0.6
Sodium, %	0.2	0.2	0.2	0.2
Chlorine, %	0.3	0.3	0.3	0.3
Copper, mg	3	3	3	3
Iodine, mg	0.2	0.2	0.2	0.2
Manganese, mg	8.5	2.5	2.5	2.5
Vitamin A, I.U.	580	—	>1160	—
Vitamin E, mg	40	—	40	40
Vitamin K, mg	—	—	0.2	—
Niacin, mg	180	—	—	—
Pyridoxine, mg	39	—	—	—
Choline, g	1.2	—	—	—
Lysine, %	0.65	—	—	—
Methionine + cystine, %	0.6	—	—	—
Arginine, %	0.6	—	—	—
Histidine, %	0.3	—	—	—
Leucine, %	1.1	—	—	—
Isoleucine, %	0.6	—	—	—
Phenylalanine + tyrosine, %	1.1	—	—	—
Threonine, %	0.6	—	—	—
Tryptophan, %	0.2	—	—	—
Valine, %	0.7	—	—	—

Source: Nutrient Requirements of Rabbits, National Academy of Sciences, Washington, D.C., 1977.

TABLE VIII. Nutrient Requirements of the Rat (percentage or amount per kilogram of dry diet)

Nutrient	Maintenance	Growth	Gestation	Lactation
Energy				
GE, kcal/kg	4444	4444	4444	4444
ME, kcal/kg	4000	4000	4000	4000
Fat, %	5.6	5.5	5.6	5.6
Crude net protein, %	4.4	13.3	13.3	13.3
Calcium, %	—	0.56	0.67	0.67
Chlorine, %	—	0.06	0.028	0.020
Copper, mg/kg	—	5.6	?	?
Iodine, mg/kg	—	0.17	0.17	0.17
Iron, mg/kg	—	38.9	?	?
Magnesium, %	—	0.04	0.056	0.056
Manganese, mg/kg	—	55.6	56	37
Phosphorus, %	—	0.44	0.56	0.56
Potassium, %	—	0.20	0.16	0.56
Selenium, mg/kg	—	0.04	?	?
Sodium, %	—	0.06	0.06	0.06
Zinc, mg/kg	—	13.3	?	?
Vitamin A, mg/retinol/kg	—	0.67	4	4
Vitamin D, I.U./kg	—	1111	—	—
Vitamin E, mg/kg	—	39	33	22
Vitamin K, mg/kg	—	0.06	—	—
Choline chloride, mg/kg	—	833	<1111	<1111
Niacin, mg/kg	—	16.7	?	?
Calcium pantothenate, mg/kg	—	8.9	8.9	11
Riboflavin, mg/kg	—	2.8	4.4	4.4
Thiamin HCl, mg/kg	—	1.39	2.8	4.4
Vitamin B_6, mg/kg	—	7.8	0.67	0.44
Vitamin B_{12}, mg/kg	—	0.0056	0.0056	0.0056

Source: Nutrient Requirements of Laboratory Animals, National Academy of Sciences, Washington, D.C., 1972.

TABLE IX. Food and Nutrition Board, N.R.C. Recommended Daily Dietary Allowances, 1974[a]

	Age, yr	Weight, kg	Energy, Mcal	Pro-tein, g	Vitamin A, I.U.	E, I.U.	C, mg	B$_6$, mg	B$_{12}$, µg	Fola-cin, µg	Nia-cin, mg	Ribo-flavin, mg	Thia-min, mg	Cal-cium, mg	Phos-phorus, mg	Io-dine, µg	Iron, mg	Magne-sium, mg	Zinc, mg
Infants	0–0.5	6	0.117[b]	2.2[b]	1400	4	35	0.3	0.3	50	5	0.4	0.3	360	240	35	10	60	3
	0.5–1	9	0.108[b]	2.0[b]	2000	5	35	0.4	0.3	50	8	0.6	0.5	540	400	45	15	70	5
Children	1–3	13	1.3	23	2000	7	40	0.6	1.0	100	9	0.8	0.7	800	800	60	15	150	10
	4–6	20	1.8	30	2500	9	40	0.9	1.5	200	12	1.1	0.9	800	800	80	10	200	10
	7–10	30	2.4	36	3300	10	40	1.2	2.0	300	16	1.2	1.2	800	800	110	10	250	10
Males	11–14	44	2.8	44	5000	12	45	1.6	3.0	400	18	1.5	1.4	1200	1200	130	18	350	15
	15–18	61	3.0	54	5000	15	45	2.0	3.0	400	20	1.8	1.5	1200	1200	150	18	400	15
	19–22	67	3.0	54	5000	15	45	2.0	3.0	400	20	1.8	1.5	800	800	140	10	350	15
	23–50	70	2.7	56	5000	15	45	2.0	3.0	400	18	1.6	1.4	800	800	130	10	350	15
	51+	70	2.4	56	5000	15	45	2.0	3.0	400	16	1.5	1.2	800	800	110	10	350	15
Females	11–14	44	2.4	44	4000	12	45	1.6	3.0	400	16	1.3	1.2	1200	1200	115	18	300	15
	15–18	54	2.1	48	4000	12	45	2.0	3.0	400	14	1.4	1.1	1200	1200	115	18	300	15
	19–22	58	2.1	46	4000	12	45	2.0	3.0	400	14	1.4	1.1	800	800	100	18	300	15
	23–50	58	2.0	46	4000	12	45	2.0	3.0	400	13	1.2	1.0	800	800	100	18	300	15
	51+	58	1.8	46	4000	12	45	2.0	3.0	400	12	1.1	1.0	800	800	80	10	300	15
Pregnant			+.3	+30	5000	15	60	2.5	4.0	800	+2	+0.3	+0.3	1200	1200	125	18	450	20
Lactating			+.5	+20	6000	15	80	2.5	4.0	600	+4	+0.5	+0.3	1200	1200	150	18	450	25

[a] The allowances are intended to provide for individual variations among most normal persons as they live in the United States under usual environmental stresses. Diets should be based on a variety of common foods in order to provide other nutrients for which human requirements have been less well defined.

[b] Multiply these factors by body weight in kilograms to obtain the kilocalorie and protein (g) allowances. Kilojoules (kJ) = 4.2 × Mcal.

TABLE X. Table of Equivalents

I. British weights and measures with metric equivalents

Length

1 inch	=	2.54 centimeters
1 yard	=	0.9144 meter
1 mile	=	1.609 kilometers

Area

1 square foot	=	0.0929 square meter
1 square yard	=	0.8361 square meter
1 acre	=	0.4047 hectare
1 square mile	=	259.0 hectares

Capacity or volume

1 cubic inch	=	16.387 cubic centimeters
1 cubic foot	=	0.0283 cubic meter
1 fluid ounce (U.S.)	=	29.573 milliliters
1 liquid quart (U.S.)	=	0.9463 liter

Weight

1 ounce (avdp)	=	28.50 grams
1 pound (avdp)	=	453.592 grams
	=	0.4536 kilogram
1 ton (short)	=	0.907 ton (metric)
1 ton (long)	=	1.016 ton (metric)
	=	1016.05 kilograms

II. Metric weights and measures with British equivalents

Length

1 meter	=	39.37 inches
1 kilometer	=	0.6214 mile

Area

1 square centimeter	=	0.155 square inch
1 square meter	=	1.196 square yards
	=	10.764 square feet
1 hectare (10,000 m^2)	=	2.471 acres
1 square kilometer	=	247.1 acres

Capacity or volume

1 cubic centimeter	=	0.061 cubic inch
1 cubic meter	=	35.315 cubic feet
1 liter	=	33.81 fluid ounces (U.S.)
	=	1.057 quarts (U.S.)

Weight

1 gram	=	0.03527 ounce (avdp)
1 kilogram	=	2.205 pounds (avdp)
	=	0.984 ton (long)
1 metric ton (1000 kg)	=	1.102 tons (short)
	=	2204.6 pounds (avdp)

Source: Condensed from table in *J. Animal Sci.,* **25:**270–271, 1966.

TABLE XI. Composition of Selected Feeds

Feed	Dry matter, %	Protein, %	TDN, %	DE	ME	NE_m Mcal	NE_g	NE_l	Crude fiber, %	Ca, %	P, %
Alfalfa, dehydrated	93.1	16.3	61	2.69	2.20	1.31	0.69	1.39	28.4	1.32	0.24
Alfalfa, fresh	27.2	19.3	61	2.68	2.23	1.32	0.71	1.37	27.4	1.72	0.31
Alfalfa hay, mid bloom	89.2	17.1	56	2.46	2.01	1.17	0.48	1.25	30.9	1.35	0.22
Barley, grain	89.0	11–13	83	3.65	3.24	2.13	1.40	1.91	7.0	0.27	0.40
Beet pulp, dried	91.0	10.0	72	3.17	2.72	1.60	1.03	1.90	20.5	0.75	0.31
Bluegrass	30.0	17.0	72	3.17	2.72	1.36	0.76	1.64	26.4	0.39	0.38
Brewers grains, dried	92.0	28.1	66	2.90	2.45	1.42	0.83	1.50	16.3	0.29	0.54
Brewers grains, wet	23.8	28.1	67	2.95	2.49	1.46	0.89	1.52	16.3	0.29	0.54
Bromegrass	32.5	20.3	68	3.99	2.54	1.36	0.76	1.54	23.9	0.59	0.32
Bromegrass hay, late	89.7	10.0	54	2.38	1.93	1.10	0.35	1.20	35.6	0.48	0.22
Citrus pulp, dried	90.0	7.3	77	3.40	2.86	1.97	1.32	1.76	14.4	2.18	0.13
Citrus pulp, silage	19.5	7.1	83	3.65	3.24	2.15	1.42	1.91	15.9	2.04	0.15
Clover, ladino	20.6	23.0	70	3.08	2.63	1.54	0.97	1.59	19.2	1.38	0.40
Clover red, hay	27.7	14.9	64	2.82	2.36	1.39	0.80	1.45	29.6	1.01	0.27
Coconut meal	93.0	21.9	81	3.56	3.15	1.88	1.25	1.86	12.9	0.23	0.66
Corn cobs	90.4	2.5	47	2.07	1.64	1.06	0.25	1.03	35.5	0.12	0.04
Corn ears	87.0	9.3	85	3.74	3.34	2.23	1.39	1.96	9.3	0.05	0.61
Corn grain	89.0	10.0	88	3.87	3.49	2.28	1.48	2.03	2.2	0.02	0.35
Corn silage	35.0	8.1	70	3.08	2.63	1.56	0.99	1.59	24.4	0.27	0.20
Corn gluten feed	90.0	28.1	82	3.61	3.19	1.93	1.29	1.89	8.9	0.51	0.86
Cottonseed hulls	90.3	4.3	43	1.90	1.48	1.03	0.19	0.93	47.5	0.16	0.10
Cottonseed, whole	92.7	24.9	98	4.31	4.00	2.01	1.20	2.28	18.2	0.51	0.73
Cottonseed meal	94.0	43.6	78	3.43	3.00	1.81	1.20	1.79	12.8	0.17	1.28
Distillers grains, dried	92.0	29.5	84	3.70	3.29	1.99	1.33	1.94	13.0	0.10	0.40
Fish meal, menhaden	92.0	66.6	74	3.26	2.81	1.66	1.08	1.69	0.0	5.97	3.05
Hominy feed	90.6	11.8	92	4.05	3.69	2.45	1.55	2.13	5.5	0.06	0.58

(continued)

585

TABLE XI. Composition of Selected Feeds (continued)

Feed	Dry matter, %	Protein, %	TDN, %	DE	ME	NE$_m$ Mcal	NE$_g$	NE$_l$	Crude fiber, %	Ca, %	P, %
				Dry basis							
Lespedeza, fresh	25.0	16.4	55	2.42	1.97	1.29	0.66	1.23	32.0	1.35	0.21
Lespedeza, hay	93.0	15.7	50	2.20	1.76	1.22	0.55	1.10	30.7	1.19	0.26
Linseed meal	91.0	38.8	81	3.56	3.15	1.90	1.27	1.86	9.9	0.48	0.98
Meat and bone meal	94.0	58.3	72	3.17	2.72	1.61	1.03	1.64	0.0	9.01	4.44
Milk, fresh	12.0	27.3	130	5.42	4.88	4.59	2.01	—	0.0	0.94	0.78
Milk, dried skimmed	94	35.2	86	3.78	3.39	2.07	1.37	—	0.0	1.36	1.11
Millet, pearl, fresh	20.7	15.7	62	2.73	2.27	1.33	0.73	1.40	30.7	1.19	0.26
Molasses, beet	77.0	8.7	75	3.30	2.86	2.04	1.36	1.72	0.0	0.21	0.04
Molasses, citrus	65.0	10.9	77	3.40	2.96	1.97	1.32	1.76	0.0	2.01	0.25
Molasses, sugarcane	75.0	4.3	77	3.39	2.96	1.91	1.20	1.76	0.0	1.19	0.11
Napiergrass, pre bloom	14.9	11.0	63	2.51	2.06	1.36	0.76	1.42	31.5	0.60	0.41
Napiergrass, late bloom	23.0	7.8	52	2.51	2.06	1.10	0.35	1.15	39.0	0.35	0.30
Oats, grain	89.0	13.4	76	3.34	2.91	1.73	1.14	1.74	12.1	0.10	0.39
Oats, Pacific Coast	91.2	9.9	77	3.39	1.96	1.76	1.16	1.76	12.1	0.10	0.36
Oats, hay	88.2	9.2	61	2.68	2.23	1.31	0.70	1.37	31.0	0.26	0.24
Oats, silage	31.7	8.4	59	2.60	2.14	1.27	0.64	1.32	32.2	0.31	0.28
Oats, straw	90.1	4.4	52	2.29	1.85	1.11	0.35	1.15	41.0	0.33	0.10
Orchardgrass	23.8	18.4	67	2.95	2.49	1.41	0.82	1.52	23.6	0.54	0.50
Orchardgrass hay	88.3	8.1	57	2.51	2.06	1.22	0.55	1.27	34.3	0.30	0.20
Peanut meal	92.0	49.8	83	3.65	3.24	1.96	1.31	1.91	12.0	0.18	0.62

Rice bran	91.0	14.8	66	2.90	2.45	1.43	0.85	1.50	12.1	0.07	2.00
Rice hulls	92.4	3.1	11	0.48	0.34	0.52	0.0	0.15	44.5	0.09	0.08
Rye, grain	88.2	12.6	80	3.52	3.10	1.86	1.24	1.84	2.4	0.10	0.33
Rye, pasture	15.9	27.2	69	3.04	2.58	1.52	0.95	1.57	17.4	0.67	0.51
Ryegrass	24.3	19.5	66	2.90	2.45	1.44	0.86	1.50	20.0	0.65	0.40
Safflower meal	91.0	21.7	57	2.51	2.06	1.22	0.56	1.27	34.1	0.37	0.78
Sorghum, grain	88.0	7.9	81	3.56	3.14	1.89	1.26	1.86	2.2	0.02	0.28
Sorghum, grain	88.0	13.0	79	3.48	3.05	1.79	1.19	1.81	2.0	0.02	0.28
Sudangrass	17.6	16.8	70	3.08	2.63	1.56	1.00	1.59	30.9	0.65	0.46
Sudangrass hay	88.9	12.7	59	2.60	2.14	1.26	0.63	1.32	28.9	0.56	0.31
Sudangrass silage	23.3	10.2	59	2.60	2.14	1.27	0.64	1.32	34.4	0.64	0.23
Soybeans	90.0	42.1	94	4.14	3.79	2.41	1.53	2.18	5.6	0.28	0.66
Soybean hay	87.6	19.0	60	2.64	2.18	1.11	0.36	1.35	43.0	1.47	0.39
Soybean hulls	91.3	13.7	78	3.43	3.00	1.79	1.19	1.79	38.9	0.59	0.17
Soybean meal	90.0	48.7	85	3.74	3.34	2.06	1.37	1.96	6.7	0.30	0.75
Soybean straw	87.6	5.5	38	1.87	1.69	0.85	0.0	0.81	44.1	1.59	0.06
Sugarcane bagasse	91.5	1.3	28	1.24	1.02	0.71	0.0	0.57	50.9	—	—
Sunflower seed meal	93.0	44.1	70	3.08	2.63	1.54	0.97	1.59	14.0	0.46	1.12
Timothy hay, early	87.7	8.7	62	2.73	2.27	1.26	0.62	1.40	32.2	0.60	0.26
Timothy hay, late	88.0	8.3	55	2.42	1.79	1.17	0.48	1.23	32.4	0.38	0.18
Trefoil, birdsfoot hay	91.2	15.6	61	2.68	2.23	1.31	0.69	1.37	29.6	1.75	0.22
Wheat, grain	86.5	14.3	88	3.87	3.49	2.15	1.42	2.03	3.4	0.06	0.41
Wheat, bran	89.0	18.2	70	3.08	2.63	1.53	0.96	1.59	11.1	0.14	1.43
Wheat, mill run	90.0	17.0	74	3.26	2.81	1.89	1.26	1.69	8.9	0.10	1.31
Wheat, straw	90.1	3.6	46	2.02	1.60	1.03	0.19	1.01	41.5	0.17	0.08
Yeast, brewers	93.0	47.9	78	3.43	3.00	1.77	1.17	1.79	3.2	0.14	1.54

Source: N.R.C. Nutrient Requirements of Dairy Cattle. National Academy of Sciences, Washington, D.C., 1978.

Index